Modeling and Control of Modern Electrical Energy Systems

IEEE Press
445 Hoes Lane
Piscataway, NJ 08854

Modeling and Control of Modern Electrical Energy Systems

Masoud Karimi-Ghartemani
Mississippi State University
Mississippi State, MS, USA

IEEE Press Series on Power and Energy Systems
Ganesh Kumar Venayagamoorthy, Series Editor

IEEE PRESS
WILEY

Published by John Wiley & Sons, Inc., Hoboken, New Jersey.
Published simultaneously in Canada.

Library of Congress Cataloging-in-Publication Data applied for:

Hardback ISBN: 9781119883418

Cover Design: Wiley
Cover Images: Cover image provided by Masoud Karimi-Ghartemani

Set in 9.5/12.5pt STIXTwoText by Straive, Chennai, India

To my teachers, colleagues, and students; my caring parents, siblings, and friends; and my beloved wife and cheerful son.

Contents

Author Biography

Masoud Karimi-Ghartemani received his BSc and MSc degrees from Isfahan University of Technology, Isfahan, Iran, in 1993 and 1995, respectively, and his Ph.D. degree from University of Toronto, Toronto, ON, Canada, in 2004, all in Electrical and Computer Engineering. From 2005 to 2008, he was a Faculty Member with the Department of Electrical Engineering, Sharif University of Technology, Tehran, Iran. From 2008 to 2011, he was a member of the ePOWER Laboratory at Queen's University, Kingston, ON, Canada. Since 2012, he has been with the Department of Electrical and Computer Engineering, Mississippi State University (MSU), MS, USA, where he is currently a full Professor. His general research areas are electrical power and control systems, and his particular research interests include integration of distributed and renewable energy systems. He works on modeling, control, synchronization, and operation of distributed energy systems at high penetration level. He also teaches relevant courses at undergraduate and graduate levels and advises students in their research projects. Dr. Karimi has over 150 publications reporting his and his team's research results. He is the recipient of the 2020 Faculty Research Award at MSU's Bagley College of Engineering. Dr. Karimi is an Associate Editor for the *IEEE Transactions on Industrial Electronics*, *IEEE Transactions on Sustainable Energy*, and *IEEE PES Letters*.

Preface

Integration of distributed and renewable energy resource (DER) systems into the electric power system (EPS) is arguably the most critical and challenging engineering problem of the twenty-first century. The existing EPS with unidirectional power flow from the central (fossil fuel, nuclear, and hydro) generation through the extensive transmission structure to the distribution networks is not a sustainable approach. DER integration is the solution. This is, however, not easy to achieve given the scope of the problem and the infrastructure and investment made into the EPS over the past 100 years. DER integration relaxes the hierarchical structure and unidirectional power flow of the legacy EPS. However, the legacy EPS can still remain highly relevant as "the enabler" for full integration of various DER technologies and for achieving smart grid functionalities. The future EPS must enable full integration and flexible and friendly co-operation of various DER technologies to become highly resilient, robust, reliable, efficient, and sustainable.

It is now widely accepted that power electronics is the key technology for DER integration. The raw power from renewable sources is hardly suitable for immediate usage. A power electronic converter (PEC) can harvest that power and process it into a usable form. The fast and high level controllability of PEC technology can be deployed to achieve various smart and intelligent integration functionalities. Every PEC has a controller that acts as the brain for the DER and regulates its local and global functionalities by making right control decisions.

The challenge with the knowledge pertaining to the selection and design of DER control systems is rooted in the fact that its understanding and modeling demand a knowledge encompassing multiple engineering fields including the generation aspects, the power conversion processes, and the grid side aspects including the local and global DER responses in connection with other components of the EPS. This fact impedes the transfer and application of the knowledge from control theory domain to the DER integration practice. Here is where the present textbook steps in and bridges the gap.

This book presents the issues pertaining to the DER integration and formulates them from a control system perspective. It then proceeds to discuss controllers that can address them. Various practically significant aspects such as power extraction control of a renewable resource, size selection of its short-term storage component, interactions of a DER with the rest of the EPS, weak grid conditions, grid support, inertia response of a DER, power quality aspects pertaining to a DER problem, and ability of the DER to ride through the grid voltage faults and frequency swings, are formulated in terms of transparent control models in this book. Much attempts have been made to accomplish this task without requiring the reader to have too much background knowledge of various fields.

The text presented here summarizes the experiences of the author gained from working with different DER structures over the past two decades. The emphasis is, however, on solar photovoltaic (PV) and battery energy storage systems. It is written in a simple language in order to make it accessible to a wide audience. Specifically, complicated mathematical proofs are avoided. The book is primarily intended for graduate students, application engineers, and researchers who work in related areas. It may also be considered as the textbook for an advanced senior level undergraduate level. A basic course in control systems, one in power system and one in power electronics can highly help the student move throughout the course. However, those background materials have also been reviewed in the text. More advanced materials are marked with an asterisk to be skipped in the first reading. All examples come with professionally prepared simulation files. Exercises and problems are provided at the end of each chapter.

Masoud Karimi-Ghartemani
Mississippi State University
Summer 2021

Acknowledgments

I gratefully acknowledge the efforts of my teachers during many years of my schooling and beyond those. The list includes but is not limited to Dr. V. Tahani, Dr. M. Mahzoon, Dr. M. Binaye-Motlagh, Dr. H. Zangeneh, Dr. H. Alavi, Dr. S. Gazor, Dr. A. Bakhshai, Dr. P. Jain, Dr. R. Iravani, and Dr. B.T. Ooi.

I acknowledge great colleagues whom I enjoyed working with during joint research works on integration of distributed and renewable energy systems specially and foremost Dr. S.A. Khajehoddin and Dr. H. Karimi among others. I acknowledge the contribution of many graduate students including Dr. R. Sharma (who also kindly helped improve a lot of drawings in the book), Dr. P. Piya, Dr. S. Silwal, Dr. T. Qunais, Dr. M. Ebrahimi, Mr. A. Zakerian, and others.

I am grateful for the support provided at MSU particularly by our department heads, Dr. Nicolas H. Younan and Dr. Smaee U. Khan, the ECE office staff, the dean Dr. Jason Keith, and the associate dean Dr. Kari Babski-Reeves.

Last but not the least, I am highly indebted to my parents for their care, love, and affection during all my life. I am grateful for having several wonderful siblings who keep to make my personal life pleasant with their love, generosity, and sacrifice. I am grateful to my beloved wife and my cheerful son who allowed me to spend a lot of our shared family time to work on completing this book.

Masoud Karimi-Ghartemani

Acronyms

ac	Alternating current
AF	Active power filter
ANF	Adaptive notch filter
APF	All-pass filter
BES	Battery energy storage
BIBO	Bounded-input bounded-output
BPF	Band-pass filter
BSF	Band-stop filter
CB	Circuit breaker
CHP	Combined heat power
D	Derivative
dB	Decibel
dc	Direct current
DER	Distributed energy resource
DG	Diesel generator
DM	Delay margin
DOE	Department of Energy
DR	Demand response
DVR	Dynamic voltage restorer
EIA	Energy Information Administration
ePLL	Enhanced phase-locked loop
EPS	Electric power system
EV	Electric vehicle
FACTS	Flexible AC transmission system
FF	Feed forward
FVT	Final value theorem
GaN	Gallium nitride
GC	Grid-connected
GM	Gain margin
GTO	Gate-turn-off thyristor
HVDC	High-voltage DC
Hz	Hertz
I	Integrating
IEEE	Institute for Electrical and Electronic Engineers
iff	If and only if

IGBT	Insulated gate bipolar transistor
IGCT	Integrated gate-commutated thyristor
IPM	Intelligent power module
kg	Kilograms
kWh	Kilowatts hour
LPF	Low-pass filter
LQ	Linear quadratic
LQR	Linear quadratic regulator
LQT	Linear quadratic tracker
LTI	Linear time invariant
MIMO	Multi-input multi-output
MOSFET	Metal-oxide-semiconductor field-effect transistor
MPPT	Maximum power point tracking
ND	Negative-definite
NF	Notch filter
NREL	National Renewable Energy Laboratory
NSD	Negative semi-definite
OLHP	Open left half of complex plane
P	Proportional
PD	Proportional-derivative, also, positive-definite
PEC	Power electronic converter
PF	Power factor
PI	Proportional-integrating
PID	Proportional-integrating-derivative
PIR	Proportional-integrating-resonant
PLL	Phase-locked loop
PM	Phase margin
PR	Proportional-resonant
PSD	Positive semi-definite
PV	Photovoltaic
PWM	Pulse width modulation
QL	Quasi-linear
R	Resonant
s	Second (unit of time)
SA	Standalone
SCR	Silicon-controlled rectifier
SG	Synchronous generator
SiC	Silicon carbide
SISO	Single-input single-output
SOGI	Second-order generalized integrator
SRF	Synchronous reference frame
STATCOM	Static compensator
SW	Switch
TF	Transfer function
THD	Total harmonic distortion
UPFC	Unified power flow controller
UPQC	Unified power quality conditioner

UPS	Un-interruptible power supply
US	United States of America
VCO	Voltage-controlled oscillator
VSC	Voltage source converter
VSG	Virtual synchronous generator
VSM	Virtual synchronous machine
1ϕ, 3ϕ	Single phase, three phase
3ePLL	Three phase ePLL
μG	Microgrid

Symbols

$u(t)$	Control signal
$m(t)$	Modulation signal
$c(t)$	Carrier signal
ϕ	Angle (rad)
ω	Frequency (rad/s)
s	Laplace domain variable
s_d	Unity magnitude synchronizing signal
s_q	Unity magnitude quadrature signal
t	Time variable
z	Z-domain variable
x_d, x_q	dq representation of signal x
x_α, x_β	$\alpha\beta$ representation of signal x
ξ	Damping ratio of a second-order linear filter
ω_n	Natural frequency of a second-order linear filter
$\circ$	Degree
>>	Matlab command
$\dot{x}$	$\frac{d}{dt}x$ (time derivative)
J	Cost function, also, moment of inertia
$\frac{\partial}{\partial\theta}J$	Partial derivative
$\otimes$	Cross-product: $x \otimes y = x_1y_2 - x_2y_1$
$\cdot$	Dot-product: $x \cdot y = x_1y_1 + x_2y_2$
I_2	2×2 identity matrix
$v_g(t)$	Grid voltage
V_g	Magnitude of grid voltage
$v_c(t)$	dc capacitor voltage
V_c	Reference (rated) value of dc capacitor voltage
$i(t)$	Converter current
L	Inductance
C	Capacitance
R	Resistance
P	Real power
Q	Reactive power
$p(t)$	Instantaneous real power
$q(t)$	Instantaneous reactive power

Introduction

I.1 Electric Power System and the Need for Change

Electric power system (EPS) is the infrastructure to generate, transmit, and distribute electricity. The existing EPS is the largest and most complex man-made system. The bulk electricity is generated in large power plants in the form of alternating current (ac) using *synchronous generators* (or alternators). The ac voltage is then boosted using *transformers* and transmitted via long transmission lines. After transmission, the voltage levels are decreased and the electricity is used for various industrial, commercial, and residential applications. Electrification has been the greatest engineering achievement of the twentieth century [1].

I.1.1 Review of Operational Principles of EPS

The existing EPS has a hierarchical structure with central power plants generating the electricity and sending to users through the transmission system. The power flow is unidirectional from Generation to Distribution via Transmission.

- *Generation*
  - A Power Plant is an industrial plant to generate bulk electric power.
  - Dominant sources of bulk electric energy are fossil fuels (coal, natural gas, petroleum), nuclear reaction, and stored water. Corresponding turbine types are steam (for fossil fuels and nuclear) and hydro turbines.
  - Popular generator is the three-phase synchronous generator (ac, 50 or 60 Hz).
  - Unit Transformer steps up the voltage at the power plant.
- *Transmission*
  - High-voltage overhead lines (ac voltage is boosted using transformers).
  - Meshed network to increase reliability.
  - Voltage is stepped down at transmission substations.
- *Distribution*
  - Voltage is further lowered at distribution substations.
  - Medium to low-voltage transmission lines; radial feeder topology; pole transformers (three-phase and single-phase low voltage end users).
  - Industrial, commercial, and residential users.
  - Residential applications: cooling; heating; lighting; refrigerating; washing; drying; entertainment; cooking; etc.

The EPS is highly interconnected: many generation and transmission systems are connected together to form a large pool of energy resulting in a highly reliable system. Maintenance of such a large interconnected ac system in terms of synchronized, stable operation, and protection of all components while preventing cascading failures is the everyday challenge of electric utility companies.

I.1.2 Problems with the Existing EPS

According to US Energy Information Administration (EIA) [2], about 4.1×10^{12} kWh of electricity was generated in United States in the year 2016.[1] This caused a total CO_2 emission of 1.821×10^{12} kg.[2] Every kWh of electricity caused about 0.44 kg of CO_2 emission. An average house in United States consumes about 900 kWh of electricity per month which corresponds to release of about 13 kg of CO_2 per day, about 400 kg per month and about 4800 kg per year. This is a major problem with the existing EPS among several others summarized as follows.[3]

1. Environmental impacts (CO_2 emission, impacts on nature, green-house effect)
2. Unsustainable (ever increasing energy demand versus limited resources; dependence on oil market)
3. Low generation efficiency (typical efficiency of coal, petroleum, and nuclear power plants being around 30% and that of natural gas plants around 40%)
4. Transmission losses (amount to 5–10% of the total transmitted power)
5. Maintenance (synchronization, stability, protection, and cascading failures in a large interconnected ac system in addition to the cost of infrastructure).

I.2 A Potential Solution: Renewable Integration

Renewable energy[4] (from sources such as sun, wind, moving or stored water, etc.) can be converted to electricity. Only a small portion of the total energy in the solar rays reached to the earth is sufficient to supply our total energy demand [4, 5].

Dispersed or distributed generation systems are small generators interfaced with the distribution (low-voltage) or sub-transmission (medium-voltage) lines.

I.2.1 Examples of Distributed Generation Systems

1. *Renewable*: Solar photovoltaic (dc), solar thermal, wind, tidal, micro hydro[5]
2. *Nonrenewable*: Micro-turbines, diesel generators (DGs)[6]

---

1 About 34% from natural gas, 30% from coal, 20% nuclear, 15% renewable (hydro, wind, biomass, solar, geothermal), and 1% others.

2 About 70% of which came from coal and 30% from natural gas power plants. The CO_2 emission caused by the electricity generation was about 35% of the total energy-related emission.

3 The numbers for 2018 are as follows [3]. Natural-gas: 1.468×10^{12} kWh of production, 0.614×10^{12} kg of CO_2 emission. Coal: 1.146×10^{12} kWh of production, 1.156×10^{12} kg of CO_2 emission.

4 Renewable energy is an energy that is collected from a renewable source, which is naturally replenished on a human timescale.

5 Micro hydro generators convert the kinetic energy of the natural flow of water to electricity. They exist in the power range of 5–100 kW. Smaller installations are called pico hydro.

6 Microturbines are small (25–500 kW) generators that use a gas-fired combustion engine with an electric generator. They produce high-frequency ac electricity (in the frequency range of 1–2 kHz). Diesel generators (or diesel gensets) are found in wide power ranges and various types of fuels. Diesel generating sets are used in places without connection to a power grid, or as emergency backup power-supply (when grid is unavailable), as well as for more complex applications such as peak load shaving, grid support, and export to the power grid.

Geothermal generators tap to the earth heat at locations that are susceptible for it.
Biomass generators burn biological materials from nature to generate steam.
Fuel Cell technology uses chemical reactions to generate electricity (dc).

I.2.2 Additional Benefits of Deploying Distributed Generation

1. Reduce transmission loss.
2. Reduce/defer transmission system expansions.
3. Recovering the heat loss (combined heat power, CHP, systems).
4. Participation of consumers in market (producing consumers: prosumers!).
5. Autonomous (or islanded) operation of a section of distribution system: increased reliability and resilience.
6. Offering ancillary services to the grid.
7. Keeping the oil and gas prices more stable and lower for longer time.
8. Lower dependence of power industry on oil and gas industry changes/uncertainties.

I.2.3 Technical Challenges with this Solution

Technical challenges arise when high level of distributed generation is integrated.

1. Bidirectional power flow can substantially change the voltage profile. This will cause malfunctioning of voltage regulating devices. The transformer-based voltage regulators, capacitor banks, and protection equipment need to be properly upgraded and/or modified.
2. The system's responses to faults change. This will require readjustment (of settings) and rearrangement (of locations) of relays and other protection devices.
3. Grid stability problems due to uncertain and variable nature of renewable sources.
4. Grid-scale battery energy storage (BES) systems may be required to address variable and uncertain nature of the renewable resources because the conventional generators have limited ramp up/down rates and cannot respond to those variations. The BES technology is not yet fully an economical and environmental-friendly technology at large.
5. Islanding prevention (when the local EPS is unavailable, the distributed generators should not energize it).
6. Coordination and control of high number of distributed generators.

There are also the regulatory and policy-related challenges which are not discussed here. For example, the regulation of possible ancillary services that the distributed generators can provide to the grid, given the wide range and variety of services that can be possible, is an ongoing challenge.

I.3 Microgrid

A microgrid (μG) is a cluster of distributed energy resources (DERs) and loads that is connected to the grid at a single location [6]. The μG can operate in grid-connected (GC) and in islanded mode. It may also operate in isolated mode without a grid connection. From our technical discussions, islanded and isolated conditions are often similar. Thus, the term standalone (SA) is used to describe this mode.

DER includes distributed generator, distributed storage, and distributed load [7].[7] The μG concept may be the key concept to address the aforementioned challenges.

I.3.1 Properties and Advantages of μG

1. When connected to the grid, the μG performs as a controllable entity with controlled interaction with the rest of the grid. This interaction may be characterized as follows.
   - Real and reactive power exchange
   - Harmonic filtering
   - Fault ride through and grid support (and ride through grid frequency swings)
   - Grid stability support and improvement, frequency response
   - Power quality aspects: harmonics, unbalance, flickers
2. In SA mode, the μG supplies power to its own loads. Major aspects are as follows.
   - Voltage quality and stability
   - Power management and power sharing
3. The μG makes seamless transition from GC to islanded and vice versa.
4. The μG concept can help realizing the smart grid functionalities such as demand response (DR) management.

Low rotating inertia of distributed generators is a possible concern. This will reduce the total inertia of the EPS and makes it susceptible to larger frequency swings. Low over-current capability and low over-load limit of power electronic switches are other issues that must be respected. For instance, a conventional induction motor requires high level of current to start.

In order to smooth down the variable generation of renewable sources, certain amount of nonrenewable distributed generation (such as diesel and gas turbine generators) and distributed storage resources should be included in a μG. Distributed storage technologies include battery, ultra-capacitor, flywheel, pumped hydro, stored hydrogen, etc. Electronically controlled loads (e.g. active rectifiers and motor drives) may be considered among DERs as they can actively participate in the μG performance control.

I.4 Distributed Energy Resource (DER)

DERs are either directly coupled or use a power electronic converter (PEC) as the interface with the grid. A PEC converts the form of power (e.g. dc to ac) and controls the flow of power. Use of a PEC is an efficient and convenient way of converting and controlling the power extracted from the renewable sources [8, 9].

The existing grid is an ac grid. However, with the proliferation of PV and BES technologies and given the fact that many residential applications need dc power, a paradigm of dc distribution system or a hybrid grid (comprising both ac and dc distribution) has been taken into serious consideration lately [10].

Figure I.1 shows the general diagram of a DER which is interfaced using a PEC. The circuit breaker (CB) disconnects the local grid from the main grid when faults occur in the local grid. The μG should not energize the local grid when CB is open. During the short-term transitory faults, the μG must remain connected and support the local area grid. This is called the fault ride through capability of the μG.

---

7 Loads that can have an active role in the system either through using a power electronic converter and/or a demand response strategy.

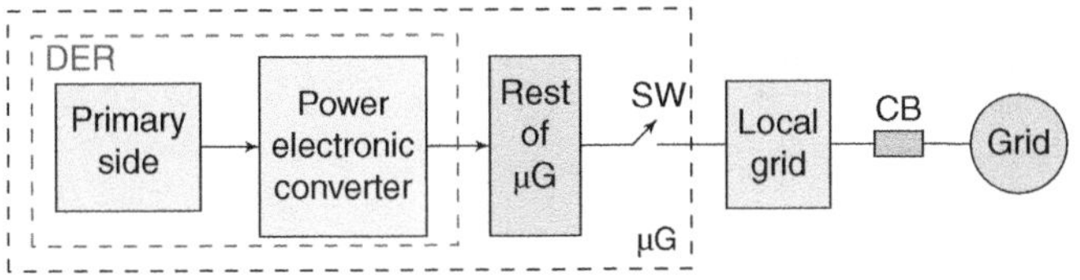

Figure I.1 General structure of a DER in a μG interfaced with the local grid.

I.4.1 Primary Side

The Primary Side of the DER may generally be one of the following cases.

- *Renewable*: Sun, wind, water, etc., e.g. PV panels, or a wind turbine and its generator, or a micro hydro turbine and its generator.
- *Nonrenewable*: Diesel, gas, etc., e.g. a micro-turbine and its generator.
- *Storage*: Battery.
- *Load*: Motor load, rectifier load, etc.

I.4.2 Power Electronic Converter (PEC)

Responsibilities of the PEC may be listed as follows.

- *Shaping the power*: Converting one form to another, e.g. dc to ac.
- Controlling the amount and flow of the power.
- Maintain the power quality at the both points of ac and ac interconnections.
- Provide ancillary services to the grid, for example, reactive power support, frequency response, and power quality improvement.
- Reducing penetration of disturbances from one side to other.
- Ride through the transient grid faults and support the grid.

The PEC is an interface between the source side (primary source) of the system to the grid side. It is responsible for making this interface efficient and strong. It must be able to minimize the adverse impacts of source side disturbances on the grid side and also to minimize the adverse impacts of grid side disturbances on the source side. Source side disturbances are those such as fast and unexpected changes in the input power due to intermittent nature of the renewable source.[8] Grid side disturbances are those such as grid voltage faults, distortions, and its frequency swings. In a more effective scenario, the DER provides grid-supporting and grid-stabilizing functions such as reactive power support and inertial response to reduce the grid transients.

I.4.3 Some Common PEC Topologies

- *Single-stage converter*: Used in some PV, battery, and FC applications (Figure I.2).
  Inverter[9] (single-phase or three-phase) is often a *voltage source converter* (VSC) where its dc side is a voltage. Therefore, a capacitor is used to support the dc side voltage. *Output Filter* is responsible for smoothing the switching ripples and noises. In order to minimize losses, it should avoid (or minimize) using dissipative elements (such as resistors). Inductors and capacitors are used.

---

8 The PV power can experience large and swift changes when the solar irradiation changes quickly [11].

9 We use the terms "inverter" and "converter" interchangeably. However, to be precise, inverter is used when the power flows from dc to ac. When the power flows from ac to dc, it is specifically called "rectifier." Converter is a general term and can include both directions of power flow.

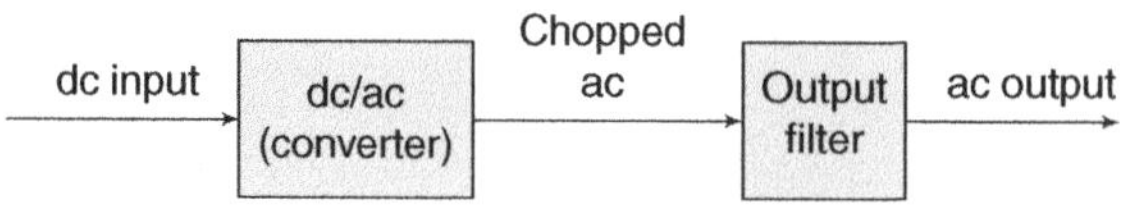

Figure I.2 Single-stage power electronic converter.

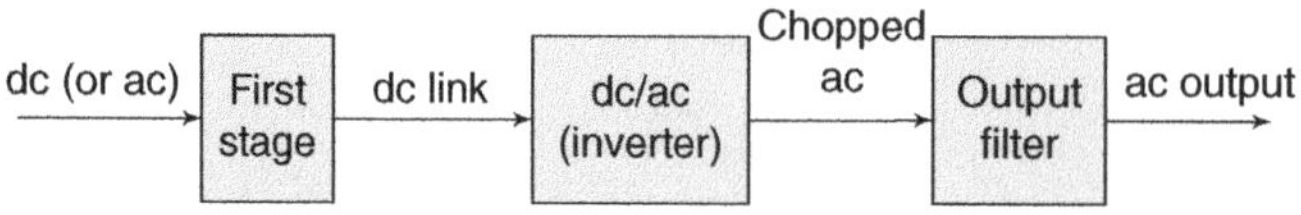

Figure I.3 Double-stage power electronic converter.

- *Double-stage converter*: Widely used in various applications (Figure I.3).
  First Stage Converter is a dc/dc converter or simply a dc converter in PV applications. In this case, its job is normally to boost or buck the voltage and also to perform maximum power point tracking (MPPT). In Wind and Micro-turbine applications, it is an ac/dc converter (or rectifier) to convert an ac variable to dc. The dc converter allows the possibility of high-frequency galvanic isolation between the source side and the grid side as well.
  Two-stage topology breaks down the control objectives so they can be addressed more efficiently by the two converters. It, however, requires more hardware which means higher cost and also lower total efficiency. The control objectives are basically the same as discussed above with the addition of controlling the dc link variable. *dc Link* is a capacitor (in a VSC).
- Multi-stage converters are also available. But the most popular ones are one- and two-stage converters. In some two-stage topologies, the first-stage converter already provides a rectified ac current and the inverter, operating at the switching frequency equal to the mains frequency, simply unfolds it. This approach has been used in some microinverter PV applications [12].

Some DERs use a low-frequency transformer (after the output is generated) to adjust the generated voltage level to the grid level. Low-frequency transformers are bulky and not very efficient and the advanced converters avoid them by integrating their function inside the converter.

I.5 Objectives and Scope of this Book

This book's overall objective is to bridge the gap between the power and the control aspects of a DER application. The power domain includes the primary source; the converter including its possible multiple stages, its dc link and its filters; the μG; and the grid. The control part consists of the entire control system on the DER that controls the interaction of the DER with the rest of the system.

The book approaches its objective by deriving simple yet efficient models that describe the power and the control. The emphasis is placed on deriving models that lend themselves well to analysis and design. After deriving the models, the control objectives and specifications are comprehensively and clearly formulated. Finally, optimal and robust controllers are designed to address them.

The book reviews the fundamentals of power electronics including the introduction of the main power electronic elements, introduction of major converters mainly the VSC, and derivation of switching and control models for such converters. In addition to that, the control theory principles in both classical Laplace domain and in modern state space time domain are reviewed and some advanced optimal control design methods are explained which will form the basis for subsequent DER control designs.

The main body of the text is devoted to system analysis and control design of DERs. In order to respect the learning curve of the readers, the book starts off with simple cases where the control models and control objectives are not much demanding. The reader is walked through the analysis and design stages gradually while more and more control objectives are pulled in and addressed. Toward the end of the book, the reader will have a deep understanding of the full requirements pertaining to a desirable interaction of a DER with the grid. Advanced topics such as fault ride-through, grid support, weak grid conditions, grid forming versus grid following controllers, and inertial response are mathematically formulated and addressed.

Exercises

I.1 Why most of the power plants are built in remote areas?

I.2 What are the main reasons for building large interconnected power systems?

I.3 Find out the typical voltage levels at the terminals of synchronous generators (before stepping up using transformers), standard transmission voltage levels, sub-transmission voltage levels, and the common distribution voltage levels.

I.4 Explain why it is required to increase the voltage level in order to transmit the power over a long distance.

I.5 Synchronous machine is widely used for generating electricity at bulk level. Explain principles of operation of this machine and its advantages as a generator.

I.6 Induction (asynchronous) machine is widely used in industrial motor applications. Explain principles of operation of this machine and its advantages as a motor.

I.7 When a μG is to transition from standalone (SA) to grid-connected (GC) mode, what is the main condition before the connection switch can be safely closed?

I.8 Discuss power quality aspects of a DER in GC operation.

I.9 Discuss power quality aspects of a DER in SA operation.

I.10 An electronically interfaced load (also called an active load) uses a converter to connect to the grid. The converter can adjust the actual voltage across the load terminals. It can also modify the PF of the load. Discuss how such a load can participate in grid functions (such as voltage and frequency support).

I.11 A hybrid dc/ac system has both of these two electricity lines available. A simple example is shown in Figure I.4 where two DERs share a dc bus. An inverter interfaces the dc bus with the ac bus. Multiple dc and ac components (loads, generators, etc.) may be interfaced to this system.

(a) Assume that Resource 1 is a solar PV, and Resource 2 is a battery storage. Discuss a sound operating scenario for this system. For example, the maximum energy is harvested from

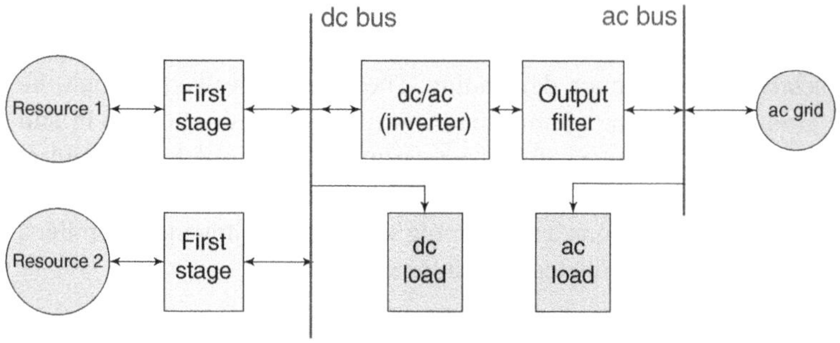

Figure I.4 A hybrid dc/ac energy system topology.

the PV and the battery is used to balance the generation versus consumption. Discuss two cases where the ac grid is or is not present.

(b) Assume that it is desired to have a 6 kW solar PV system. How many solar panels are required? How much area is needed? (Hint: Assume that the used solar panel technology generates 300 W and its area dimensions are 1 m into 2 m.)

(c) Assume that the average dc load is 0.5 kW and the average ac load is 1.5 kW over 24 hours. Also, assume that you have 10 hours of sunlight per day at your location. Determine the size of battery (in terms of kW and kWh) such that the combination of PV and battery can sustain the loads in the absence of grid.

References

1 Greatest engineering achievements of the 20th century. http://www.greatachievements.org/. Accessed: 2018-17-01.

2 U.S. Energy Information Administration. http://www.eia.gov. Accessed: 2017-06-08.

3 U.S. Energy Information Administration. U.S. energy-related carbon dioxide emissions, 2018. https://www.eia.gov/environment/emissions/carbon/pdf/2018_co2analysis.pdf. Accessed: 2020-06-01.

4 National Renewable Energy Laboratory. Solar energy basics: "that's because more energy from the sun falls on the earth in one hour than is used by everyone in the world in one year". https://www.nrel.gov/workingwithus/re-solar.html. Accessed: 2017-06-08.

5 National Renewable Energy Laboratory. U.S. solar radiation resource maps: atlas of the solar radiation data manual for flat-plate and concentrating collectors. http://rredc.nrel.gov/solar/old_data/nsrdb/1961-1990/redbook/atlas/. Accessed: 2017-06-08.

6 Nikos Hatziargyriou. *Microgrids: architectures and control*. John Wiley & Sons, 2014.

7 IEEE standard for interconnection and interoperability of distributed energy resources with associated electric power systems interfaces. IEEE *Std 1547-2018 (Revision of* IEEE *Std 1547-2003)*, pages 1–138, 2018.

8 Frede Blaabjerg, Zhe Chen, and Soeren Baekhoej Kjaer. Power electronics as efficient interface in dispersed power generation systems. IEEE *Transactions on Power Electronics*, 19(5):1184–1194, 2004.

9 Benjamin Kroposki, Christopher Pink, Richard DeBlasio, Holly Thomas, Miguel Simo es, and Pankaj K Sen. Benefits of power electronic interfaces for distributed energy systems. *IEEE Transactions on Energy Conversion*, 25(3):901–908, 2010.

10 Rob Cuzner Does DC distribution make sense? IEEE *Electrification Magazine*, 4(2), 2016.
11 Roshan Sharma and Masoud Karimi-Ghartemani. Addressing abrupt PV disturbances, and mitigating net load profile's ramp and peak demands, using distributed storage devices. *Energies*, 13(5):1024, 2020.
12 Haibing Hu, Souhib Harb, Nasser Kutkut, Issa Batarseh, and Z John Shen. A review of power decoupling techniques for microinverters with three different decoupling capacitor locations in PV systems. *IEEE Transactions on Power Electronics*, 28(6):2711–2726, 2012.

Part I

Power Electronic Conversion

Power electronics is the enabling technology for interfacing distributed energy resources (DERs) to host networks. Part I of this text presents a review of fundamentals related to power electronics. Chapter 1 reviews the concept of power electronic conversion, basic power electronic switches, various types of converters, and finally discusses applications of power electronics. Chapter 2 is devoted to principles of operation and modeling aspects of the standard converters such as buck, boost, buck–boost, and particularly the voltage source converter (VSC) which is presently the most popular type of converter used for DER applications. Single-phase (half-bridge and full-bridge) and three-phase VSC topologies are covered in detail.

Modeling and Control of Modern Electrical Energy Systems, First Edition. Masoud Karimi-Ghartemani.

1

Power Electronics

This chapter reviews fundamentals pertaining to power electronics. The concept of power electronic conversion is first introduced which is followed by a short discussion on various types of power electronic switches and different types of power electronic converters (PECs). The chapter concludes by reviewing diverse applications of power electronics in power industry.

1.1 Power Electronics Based Conversion

A *PEC* (also called a power processor) uses power electronic switches (such as power diodes and power transistors) and passive circuit components (such as inductors and capacitors) to convert one form of electrical power to another form (e.g. dc to ac, ac to dc) or change its level (dc to dc) or change its frequency, phase, magnitude (ac to ac) or change the number of phases (single-phase to three-phase). In addition to this change, the converter can control the flow of power. The flow of power may be unidirectional (only from one side to other) or bidirectional (between both sides). The PEC achieves its objective by fast turning of the switches at a rate called the *switching frequency*.

Figure 1.1 shows a block diagram for a converter with its control system. The converter works as an interface between side A and side B. The controller is a feedback system that generates the control inputs based on the system measurements and references (also called set-points or desired values). The measurements generally are taken from both sides and may also be taken from the converter as well. The control inputs directly or indirectly determine the switching instants of the power electronic switches used inside the converter.

As *an example*, assume that side A is a set of solar panels and side B is a three-phase ac grid. The PEC needs to convert the original dc power of the photovoltaic (PV) to a three-phase ac power compatible with the given grid. The flow of power is unidirectional from the PV side to the grid side. Then, the PEC can be a one-stage dc/ac three-phase converter (or inverter). The amount of power extracted from the PV panels is commonly controlled through controlling the PV panels' voltage. This means that the control variable on the side A is the voltage. The ac voltage produced by the inverter must be synchronized with the ac grid and the ac current produced by the inverter must comply with the grid code quality requirements. This implies that the control variables on the side B of the inverter generally include both the voltage and current at the point of connection to the grid. The PV side disturbances are sudden and uncertain changes of power due to change in the solar irradiation. The ac grid side disturbances are those causing sudden voltage changes (such as grid faults, and sudden load changes) and voltage distortions, unbalance, and frequency variations.

Modeling and Control of Modern Electrical Energy Systems, First Edition. Masoud Karimi-Ghartemani.

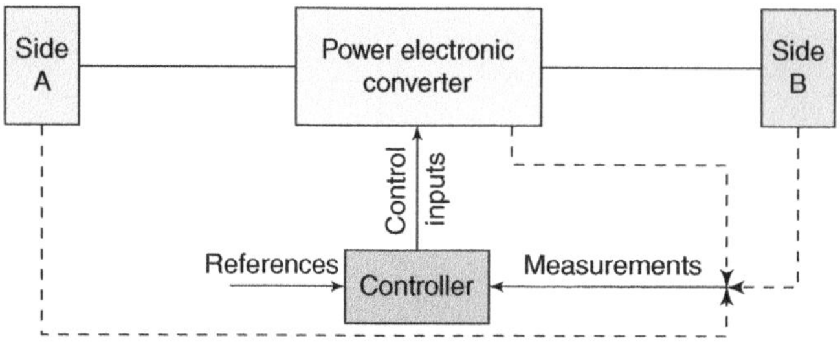

Figure 1.1 A PEC with its control as the interface between side A and side B.

A *power electronic switch* is a semiconductor device that can permit and/or interrupt the flow of current through it by the application of a gating (or activating or switching) signal as opposed to a mechanical action. Power diode is the only power electronic switch that works without a gating signal.

1.1.1 Advantages of Power Electronics

- Fast dynamics (no mechanical part)
- Good controllability
- Small footprint

1.2 Power Electronic Switches

Some common power electronic switches are briefly reviewed here.

The *diode* allows unidirectional flow of current without a gating signal. A flow of current from anode to cathode turns the diode on and a reverse voltage turns it off (Figure 1.2).

The *thyristor or silicon-controlled rectifier* (*SCR*) can only be turned on using its gate signals. It turns off when the current flow reverses. The *gate-turn-off* (*GTO*) *thyristor* can also be turned off but a large current should be drawn from its gate. The SCR and GTO have been used for high-power, high voltage dc (HVDC) applications. Their switching frequency is limited to about 1 kHz (Figure 1.3).

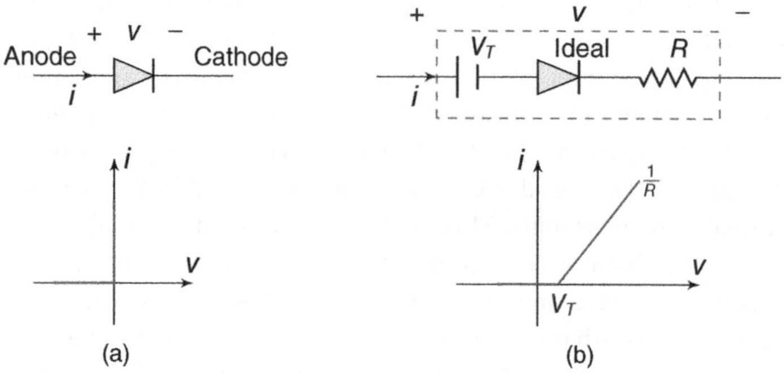

Figure 1.2 (a) Circuit symbol for diode and its ideal characteristics. (b) A practical diode model and its characteristics.

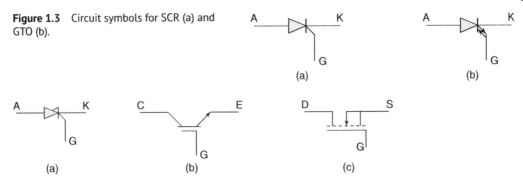

Figure 1.3 Circuit symbols for SCR (a) and GTO (b).

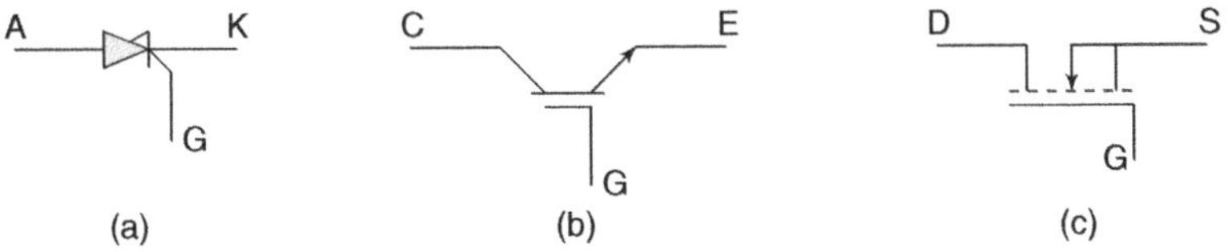

Figure 1.4 Circuit symbols for IGCT (a), IGBT (b), and MOSFET (c).

The *integrated gate-commutated thyristor (IGCT)* is basically a GTO without the limitation of large negative gate current at turn-off. Other improvements compared to GTO are faster switching frequency and lower on-state voltage drop. The *insulated gate bipolar transistor (IGBT)* is a fully controllable switch with fast switching (a few tens of kHz) and high efficiency. It can be used in medium to high power applications.

The *power MOSFET (metal oxide semiconductor field-effect transistor)* can have much higher switching frequencies with good efficiency. However, it has limited current and voltage ratings and cannot be used for high power applications.

An *intelligent power module (IPM)*, or simply a power module, includes one or more power electronic switches together with their gate-drive circuitry. Combined with the built-in protection and diagnostics, the IPMs significantly simplify converter design and implementations (Figure 1.4).

The *voltage rating* of a switch is the maximum voltage that it can stand in its off-state. Its *current rating* is the maximum current that it can allow in its on-state. Power electronic switches cannot tolerate much over-current and will be quickly damaged. This signifies one of their main limitations compared to the conventional rotating machines. Other parameters pertaining to a power electronic switch include its maximum temperature, switching frequency, on-state (or conducting) resistance, dead time, and its static voltage current characteristics.

The common semiconductor material for power electronics is presently silicon. The *wide-bandgap* semiconductor materials such as *silicon carbide* (SiC) and *gallium nitride* (GaN) are expected to be used widely in future and they are already in the market. This technology allows higher operational temperatures, faster switching rates, larger breakdown voltages, and lower power losses, leading to more compact and more efficient converters.

Switch Classification

- *Unidirectional*: Conducts/interrupts current in one direction.
  - *Unipolar*: Has a small reverse breakdown voltage.
  - *Bipolar*: Can stand larger reverse voltage.

  Anti-parallel connection of a unidirectional/unipolar switch and a diode (as shown in Figure 1.5) results in a switch that has *reverse-conducting* property. It is called a *switch cell* and most IGBT, IGCT, and MOSFET technologies have this diode built-in them. Switch cell is the common switch used in VSCs.
- *Bidirectional*: Conducts/interrupts current in both directions. Single-device technology is not yet available for bidirectional switches and they are built by anti-parallel connection of two unidirectional switches.

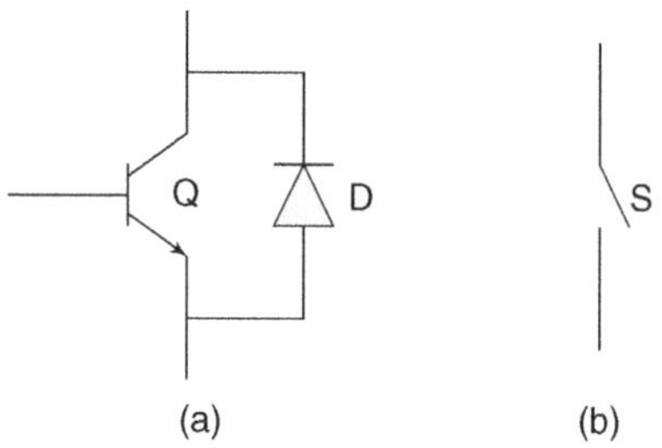

Figure 1.5 Switch cell, (a) internal structure, (b) simple circuit symbol.

1.3 Types of Power Electronic Converters

1. *dc Converter*, dc/dc (or dc-to-dc) converter, changes the level of a dc voltage. A dc/dc converter can be a *boost converter* (where its output voltage is higher than its input voltage), a *buck converter* (where the output voltage is lower than the input voltage), or a *buck-boost converter* (where the output voltage can be both higher or lower than the input voltage).
   A dc converter can normally regulate (or control) the voltage (or current) on one of its sides, e.g. unregulated input voltage but regulated output voltage. A dc converter may be *direct* or with an intermediate ac link. The latter case is used for the sake of providing a galvanic isolation using a high frequency transformer at the ac link, hence the *isolated* and *non-isolated* dc converters.
2. *Rectifier*, ac/dc (ac-to-dc) converter, converts an ac voltage to a dc voltage. An active rectifier can control its output dc voltage. On the input side, a rectifier is normally desired to draw a current from the ac supply that is sinusoidal (not distorted) and in-phase with the supply voltage, i.e. unity power factor (PF), to reduce network losses and yield a solid performance.
3. *Inverter*, dc/ac (dc-to-ac) converter, converts a dc voltage to an ac voltage. An inverter normally controls its output ac voltage or current (in terms of its magnitude, angle, and frequency) and it may control its dc side voltage as well.
4. *ac Converter*, ac/ac (ac-to-ac) converter, converts an ac voltage to another ac voltage with a change in some of the characteristics such as magnitude, phase, frequency, and number of phases. An ac converter may be *direct* or with an intermediate dc link. The latter case is used for the sake of providing a dc isolation (buffer) between the two ac sides.

1.4 Applications of Power Electronics in Power Engineering

1.4.1 Power Quality Applications

1. *Active Power Filter (AF)* provides harmonic currents to improve the quality of current drawn from the supply by a nonlinear load, Figure 1.6. The AF is normally an inverter that can generate a controlled distorted current, $i_f(t)$, that compensates fully or partially for the harmonics of the load current, $i_l(t)$, reducing the amount of distortions on the supply current, $i_s(t)$.

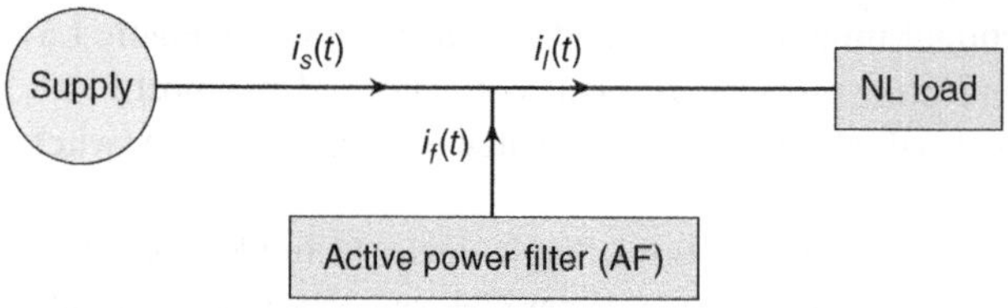

Figure 1.6 Active power filter (AF) application.

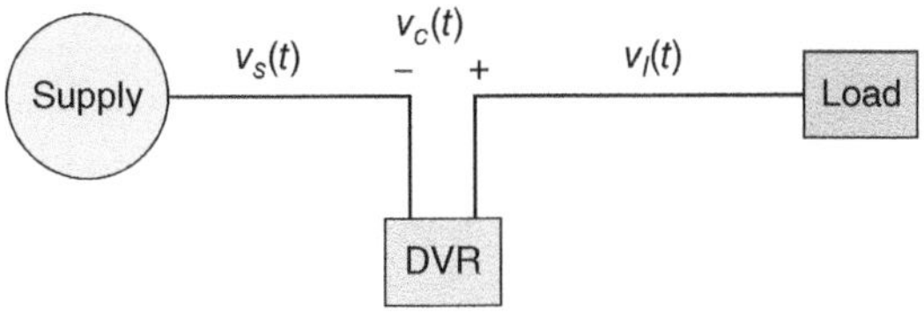

Figure 1.7 Dynamic voltage restorer (DVR) application.

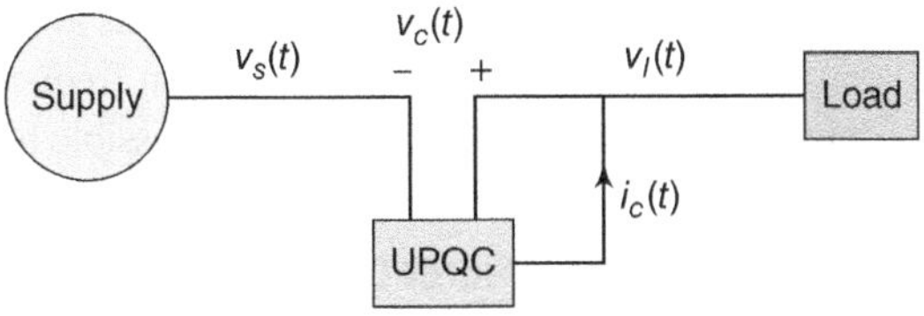

Figure 1.8 Unified power quality conditioner (UPQC) application.

2. *Static Compensator (STATCOM)* compensates for the load reactive power (or power factor). Its structure and principles of operation are much similar to the AF. It is called static compensator as opposed to the conventional rotating compensators (based on synchronous machines or synchronous condensers).
3. *Dynamic Voltage Restorer (DVR)* regulates the supply voltage for sensitive loads against supply voltage fluctuations. In Figure 1.7, $v_c(t)$ is generated by the converter and compensates for the supply voltage variations. Its power electronic structure is again an inverter which is now connected in series (as opposed to the shunt connection of the AF and STATCOM) with the supply. It generates a compensating voltage and is normally coupled through a transformer.
4. *Unified Power Quality Conditioner (UPQC)* conceptually combines the operations of AF, STATCOM, and DVR. It has a shunt and a series connection. Can provide high quality voltage for the sensitive loads against grid voltage fluctuations and short-time outages. The shunt branch prevents harmonic currents from flowing into passive filters (Figure 1.8).

1.4.2 Power System Applications

1. *Flexible ac transmission system (FACTS)* devices are power electronic based converters to improve the power transfer capability of the transmission system. A FACTS device is the unified power flow controller (UPFC) installed within the transmission system to boost the transmission capacity, Figure 1.9.
   The sending end voltage is V_S, the receiving end is V_R, and the UPFC (and line) voltage is V_C. The transmitted real and reactive powers are

$$P = \frac{V_R V_S}{X} \sin\delta, \quad Q = \frac{V_R}{X}(V_S \cos\delta - V_R), \tag{1.1}$$

   where X is the reactance of high-voltage transmission line. By controlling V_C (within the limitations of system), both power components can be controlled through V_S and δ.

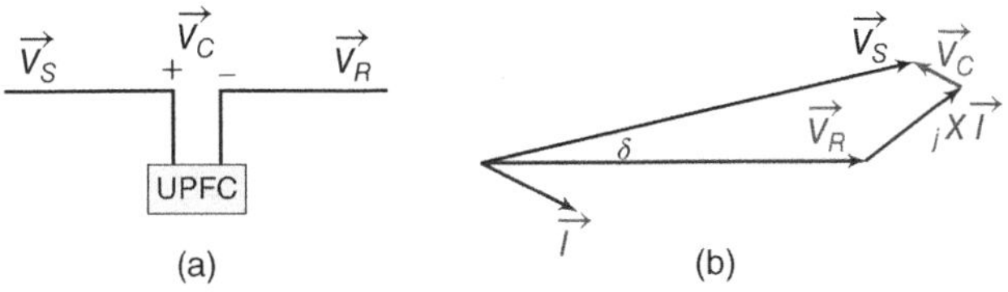

Figure 1.9 Unified power flow controller (UPFC): (a) block diagram and (b) phasor diagram.

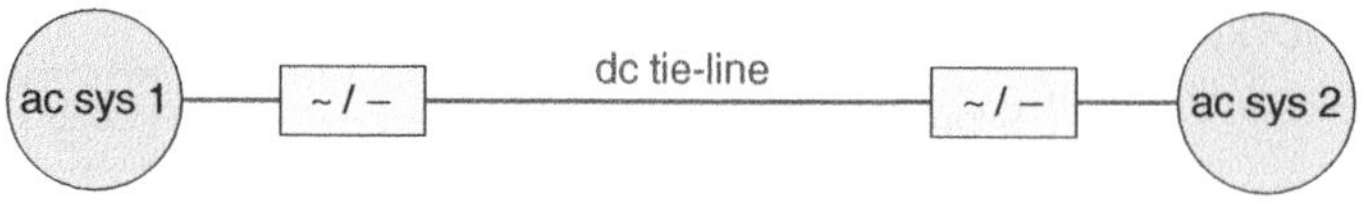

Figure 1.10 High-voltage dc (HVDC) transmission.

2. *HVDC* transmission systems are becoming more popular thanks to high-voltage semiconductor devices and due to their advantages such as lower electrical loss over very long distances, interconnection of two asynchronous or even frequency-different ac systems, and easier control of power in the tie-line [1]. This is made possible by the help of two ac/dc converters, Figure 1.10.

1.4.3 Rectifiers and Motor Drive Applications

1. *Rectifiers* are used to power dc loads. They supply a regulated dc output voltage. On the ac side, they are supposed to operate at unity (or near unity) power factor with no (or very small level of) current distortions.
2. *dc motor drives* are used to control the operation of dc motors. This is basically a controlled rectifier with flexible control over the dc output. They are used for multiple tasks such as controlling the starting transient (current inrush), speed control and position control.
3. *ac motor drives* are used to control the operation of ac motors (e.g. an induction or a synchronous motor). The emerging electric vehicle (EV) application falls in this category as well. This basically comprises an ac converter with dc link (or two back-to-back VSCs). This drive offers flexible control over the frequency and magnitude of its output voltage. They are used for multiple tasks such as controlling the starting transient (current inrush), speed or position control.

1.4.4 Backup Supply and Distributed Generation Applications

1. *Uninterruptible power supply (UPS)* systems come in various topologies. At the core, they have a battery storage system and an inverter to supply power in the event that the grid is unavailable.
2. *Distributed energy storage*, mainly in the form of battery, systems use an inverter to connect to the ac line. Alternatively, they may use dc converters to connect to a dc bus. These are converters with bidirectional power flow. They are increasingly researched for smoothing the variability of renewable energy systems. EV is a major application for the batteries.
3. *Micro-turbines* need a conversion system to convert high frequency variable ac to the smooth low frequency compatible with the existing grid. The conversion system normally comprises a rectification and an inverter stage.
4. *Wind-turbines* need a conversion system to convert low frequency variable ac to the smooth ac compatible with the existing grid. The conversion system normally comprises a rectification and an inverter stage.
5. *Solar PV* systems need to convert variable dc to the smooth ac compatible with the existing grid. The conversion system comprises either a single-stage inverter or a combination of some dc/dc and an inverter stage.

1.5 Summary and Conclusion

Power electronics technology offers an efficient and convenient platform for converting a form of power to another and control its flow in addition to addressing power quality aspects (on both sides) and achieving additional services. It is thus much needed for various power engineering applications including the DER integration. This chapter reviewed the principles of a controlled power conversion, power electronic switches, converter types, and several applications of power electronics. Further details of particular standard converters such as buck, boost, and voltage-source converter are discussed in Chapter 2.

Exercises

1.1 Consider Figure 1.1. Identify what side A, side B, and PEC are in the following applications.
(a) Powering a dc load from a single-phase ac supply.
(b) Powering a single-phase ac load from a dc supply.
(c) Charging a battery from a PV system.
(d) Connecting a PV system to a single-phase ac line.

1.2 In the applications of Exercise 1.1, identify the variable of interest to be controlled by the PEC.

1.3 In Figure 1.1, one of the jobs of a PEC is to reduce the adverse impacts of disturbances happening on one side on the other side. In the applications of Exercise 1.1, discuss the possible disturbances that may exist on each side of the system.

1.4 In Figure 1.1, assume that side A is a combined PV and battery which are both connecting to a shared dc bus through their own dc converters. This dc link is interfaced with a three-phase ac grid, i.e. side B, through a three-phase bidirectional inverter. Some local dc load is connected to the dc bus.
(a) Discuss possible services that the combined PV-battery system can provide to the ac grid.
(b) Discuss possible synergetic operating scenarios between the PV and battery, in the absence of the ac grid.
(c) Discuss possible synergetic operating scenarios between the PV and battery in the presence of the ac grid.

1.5 Consider Figure 1.6. The AF actively supplies the harmonic currents that the load requires in order to filter them from penetrating the supply. Conventionally, passive LC filters have been used to do this job and they are still used in conjunction with active filters. The passive filter is based on the fact that a series LC connection tuned at a given frequency has a zero impedance at that frequency and absorbs the current component at this frequency. Compare the active and passive filters from the cost, performance, bulkiness, and flexibility points of view.

1.6 STATCOM actively supplies the reactive currents that a load requires in order to filter them from penetrating the supply. Conventionally, the capacitor banks and synchronous condensers are used to do this job and they are still used in conjunction with STATCOMs. Compare the three power factor correction approaches from the cost, performance, bulkiness, and flexibility points of view.

1.7 Discuss advantages of having a dc tie-line between two ac systems as shown in Figure 1.10.

Problems

1.1 Prove Eq. (1.1). You may use the equation $\vec{S} = \vec{V}\vec{I}^*$ where $\vec{S} = P + jQ$ is the complex power and $\vec{V}$ and $\vec{I}$ are the voltage and current phasors. The asterisk symbol is for the complex conjugate operation. Then, substitute the current from impedance equation.

1.2 How would Eq. (1.1) be modified if the line impedance is $R + jX$ rather than jX?

Reference

1 Technical advantages of HVDC systems. http://new.abb.com/systems/hvdc/why-hvdc/technical-advantages. Accessed: 2018-17-01.

2

Standard Power Electronic Converters

This chapter is devoted to the understanding of standard power electronic converters. Topologies such as standard buck, boost, buck-boost, and the voltage source converter (VSC) are reviewed. The VSC study is divided into single-phase and three-phase VSCs. The single-phase VSC is also divided into half-bridge and full-bridge topologies. For each converter, the topology is introduced, the pulse-width modulation (PWM) technique to generate the gating signals is discussed, and finally an average control model is derived. This chapter lays the foundation for modeling of distributed energy resource (DER) systems for the purpose of analysis and control design.

2.1 Standard Buck Converter

The standard buck converter is shown in Figure 2.1 which comprises a unidirectional switch Q_1 and a diode D_2. It is the interface between side A and side B and the power flow direction is obviously from side A to side B, i.e. the input side to the output side, respectively. The inductance value is L, and R denotes the totality of cable's resistance, the inductor's parasitic resistance, and the conducting resistance of switch/diode. No actual resistance is deliberately added to the power circuits as it will cause large power losses.

When the switch Q_1 is commanded on, the voltage $v(t)$ will be equal to v_A. Then, if $v_A > v_B$, the current $i(t)$ will flow in the direction shown and will rise. When the switch Q_1 is turned off, the current will turn the diode D_2 on and will continue to flow through it and $v(t)$ will be zero. In order for the current to decline and stabilize, v_B must be positive. Thus, $0 < v_B < v_A$, meaning that the converter bucks the voltage in the direction of power flow.

2.1.1 Analysis of Operation

For a power electronic switch, the definition of the *switching function* $s(t)$ is

$$s(t) = \begin{cases} 1 & \text{switch is commanded ON (may or may not conduct)} \\ 0 & \text{switch is commanded OFF.} \end{cases}$$

The popular PWM technique is used to determine the switching instants as shown in Figure 2.2. In Figure 2.2, $m(t)$ is called the *modulating signal* and is a continuous signal with values between 0 and 1. The signal $c(t)$ is called the *carrier signal* and is a triangular waveform with frequency f_{sw} (called the switching frequency) and peak values at 0 and 1 as shown in Figure 2.3. In the PWM technique, basically, the modulating signal $m(t)$ is compared with the carrier signal $c(t)$ and if $m(t) > c(t)$ then S_1 is commanded ON.

Modeling and Control of Modern Electrical Energy Systems, First Edition. Masoud Karimi-Ghartemani.

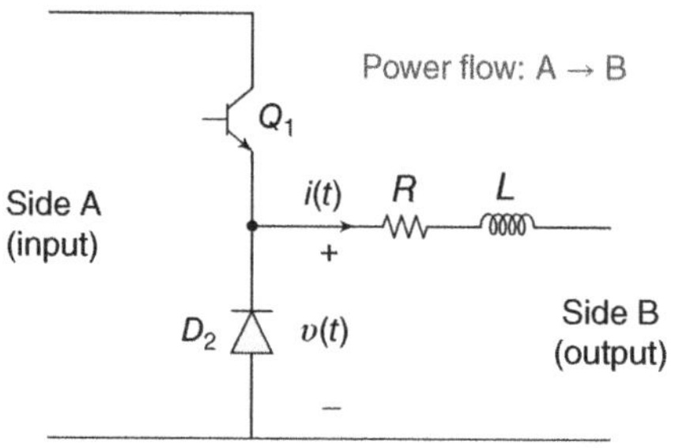

Figure 2.1 Standard buck converter interfacing the input side A with the output side B.

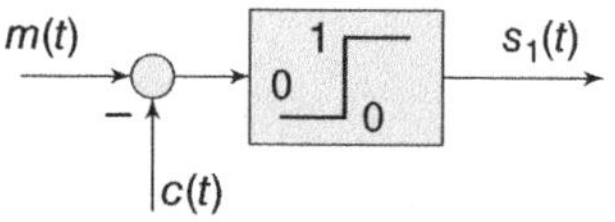

Figure 2.2 PWM for buck converter.

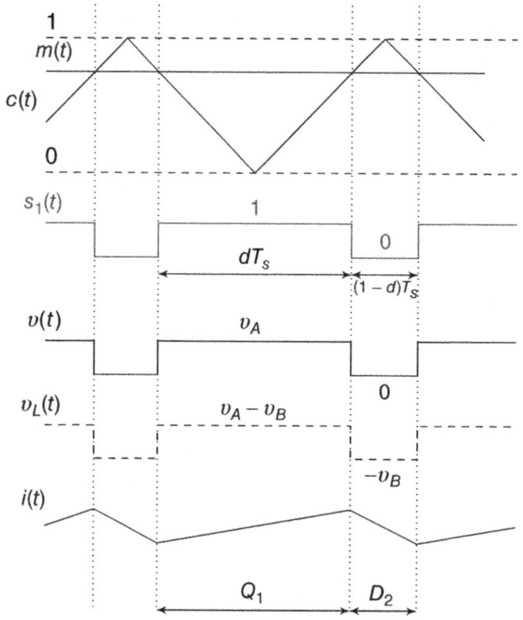

Figure 2.3 PWM operation and illustration of waveforms for buck converter.

Figure 2.3 shows a sample of signals waveform pertaining to this circuit. When $s_1 = 1$, switch Q_1 conducts and provides the path for current. Therefore, the voltage $v(t)$ is v_A. As soon as s_1 falls to zero, the diode D_2 turns on and provides a path for the current. During this period, the voltage $v(t)$ is zero.

Notice that switching frequency is high. Therefore, the voltages $v_A(t)$, $v_B(t)$ and the modulating signal $m(t)$ are considered constant during one switching cycle. This entails that the current waveform is almost linearly growing or falling due to the fact that the voltage across the inductor is $v_L = v - v_B$ and the inductor current–voltage equation is $v_L = L\frac{di}{dt}$. The voltage across L when Q_1 is ON is equal to $v_A - v_B$ and this voltage must be positive to allow a flow of current, i.e. $v_A > v_B$. When Q_1 is commanded OFF, the voltage across L is $-v_B$ and this must be negative, i.e. $0 < v_B$, to reduce the current.

Finally, the voltage $v(t)$ can be written as

$$v(t) = v_A(t)s_1(t), \tag{2.1}$$

and the circuit equation describing the relationship between current and voltage is

$$L\frac{d}{dt}i(t) + Ri(t) = v(t) - v_B(t). \tag{2.2}$$

2.1.2 Switching Model

Based on (2.1) and (2.2), an equivalent circuit may be derived for the buck converter as shown in Figure 2.4. In this circuit, the switch and diode are modeled by controlled current sources. Note that $s_{1_n}(t)$ is the logical NOT of $s_1(t)$ defined as

$$s_{1_n}(t) = 1 - s_1(t) = \begin{cases} 1 & \text{when } s_1(t) = 0 \\ 0 & \text{when } s_1(t) = 1. \end{cases}$$

This circuit may be used to simulate the buck converter at the switching level.

2.1.3 Average (or Control) Model

In the PWM approach for controlling a converter, the modulation signal $m(t)$ is considered as the control input. The output variable is chosen depending on the application. Three scenarions for the output variable, including the output current $i(t)$, the output voltage v_B, and the input voltage v_A are discussed in this section and the control block diagram is obtained for each one.

The voltage $v(t)$ is a high-frequency chopped waveform with two discrete values v_A and 0. Consider one switching cycle of this waveform and expand it periodically such that the Fourier series can be applied to it as

$$v(t) = \bar{v} + \sum_{n=1}^{\infty}[a_n \cos(n\omega_s t) + b_n \sin(n\omega_s t)] = \bar{v} + \tilde{v}(t),$$

where $\omega_s = 2\pi f_{sw}, f_{sw} = \frac{1}{T_s}$ is the switching frequency in Hz, and

$$\bar{v} = \frac{1}{T_s}\int_0^{T_s} v(t)dt, \quad a_n = \frac{2}{T_s}\int_0^{T_s} v(t)\cos(n\omega_s t)dt, \quad b_n = \frac{2}{T_s}\int_0^{T_s} v(t)\sin(n\omega_s t)dt.$$

Notice that $\bar{v}$ is the average (or dc) of $v(t)$ over a switching cycle and $\tilde{v}(t)$ is its high frequency term with zero average. The current waveform can also be written in a similar fashion: $i(t) = \bar{i} + \tilde{i}(t)$.

The voltages $v_A(t)$ and $v_B(t)$ are assumed smooth enough such that their changes over one switching cycle is negligible. Therefore, $\tilde{v}_A(t) = \tilde{v}_B(t) = 0$. This discussion concludes that (2.2) may be written as two equations

$$\begin{aligned} L\frac{d}{dt}\bar{i}(t) + R\bar{i}(t) &= \bar{v}(t) - v_B(t) \\ L\frac{d}{dt}\tilde{i}(t) + R\tilde{i}(t) &= \tilde{v}(t) \end{aligned} \tag{2.3}$$

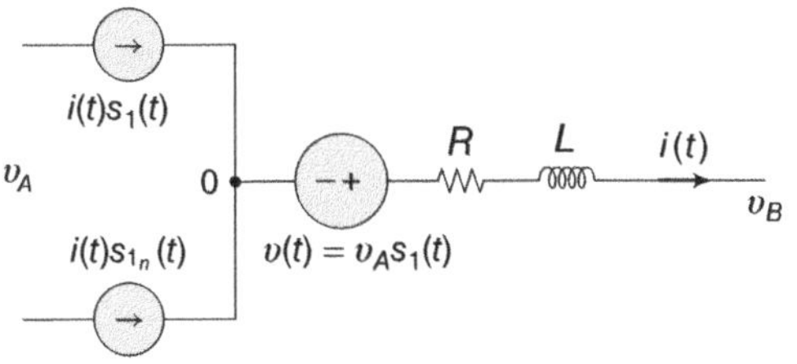

Figure 2.4 Equivalent circuit of buck converter at switching level.

assuming a linear approximation of the circuit that implies validity of superposition principle and given that the switching frequency is much higher than the supply frequency. Notice that, mathematically speaking, we have used a moving average variable over time using the definition

$$\overline{x}(t) = \frac{1}{T_s}\int_{t-T_s}^{t} x(\tau)d\tau.$$

With this definition, the average of the switching function $s_1(t)$ is

$$\overline{s}_1 = \frac{1}{T_s}\int_{0}^{T_s} s_1(t)dt = \frac{1}{T_s}dT_s = d.$$

Therefore,

$$\overline{v} = v_A\overline{s}_1 = v_A d.$$

Now, an inspection of Figure 2.3 shows that when m varies from 0 to 1, d also varies from 0 to 1. Moreover, the relationship between them is obviously linear. Therefore, they are actually equal, i.e.

$$m = d.$$

Substituting this into the previous equation results in

$$\overline{v} = v_A m. \tag{2.4}$$

Equation (2.4) describes the average model of the buck converter from the modulating signal $m(t)$ to the voltage $\overline{v}(t)$. The buck converter acts as a simple gain v_A.

2.1.3.1 Current Control Model

The control output is the converter current $\overline{i}(t)$. Combining the first equation in (2.3) and (2.4), the control transfer function model shown in Figure 2.5 is obtained. In this case, v_A and v_B are treated as external variables and are not controlled by the converter. This model is particularly suited for the case where side B represents a stiff voltage, e.g. a strong dc grid or a battery.

2.1.3.2 Output Voltage Control Model

The control output is the voltage v_B. A local load and/or a capacitance (or a combination of both) with the total impedance of $Z(s)$ is connected across the side B terminals. It may further be connected to the rest of a network to which the current i_B flows as shown in Figure 2.6a. The control model is readily obtained and is shown in Figure 2.6b.

For primary control system designs of the buck converter, the current i_B may be treated as a disturbance or it may be taken into analysis if further information about its dynamics is available. The voltage control model is suitable for designing the control to operate in standalone mode, i.e. not connected to a stiff voltage on side B. However, it may also be adjusted and used for grid-connected applications using the virtual impedance approach.

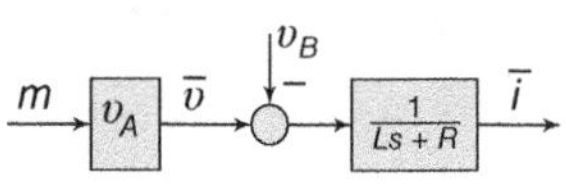

Figure 2.5 Average or control model of the buck converter.

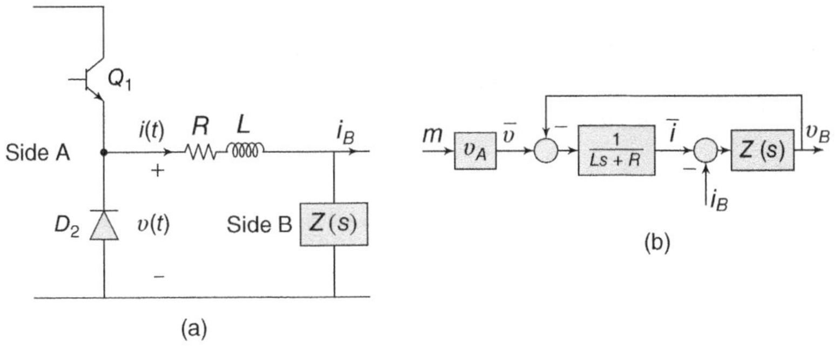

Figure 2.6 Output voltage control of buck converter: (a) circuit and (b) control model.

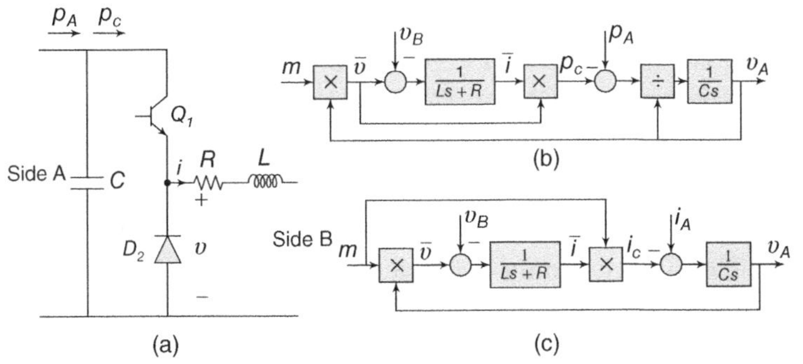

Figure 2.7 Input voltage control of buck converter: (a) circuit, (b) control model, and (c) alternative control model.

2.1.3.3 Input Voltage Control Model

The control output is the voltage v_A. This is what is done in PV applications where the PV power is determined by its voltage. A capacitor C is connected across the input terminals for voltage support, as shown in Figure 2.7a. The power balance equation is $p_A - p_c = Cv_A\dot{v}_A$. Ignoring the power loss across the switches Q_1 and D_2, p_c may be approximated by $\overline{p}_c = \overline{v}\overline{i}$. Thus, the control model is derived and is shown in Figure 2.7b. An alternative model can be derived by noting that $\overline{i}_c = m\overline{i}$ as shown in Figure 2.7c. The two models shown in Figure 2.7b,c are nonlinear due to the presence of multiplication units. Another nonlinearity, not explicitly shown in Figure 2.7, is the nonlinear relationship between p_A (or i_A) and v_A resulted from a PV panel characteristics.

2.1.4 Steady-State Analysis

In the steady-state condition, the signals settle to their steady values. Even if they are switching, e.g. $v(t)$, or have up/down switching ramps, e.g. $i(t)$, their average values settle to a constant number. This means that when the duty cycle d settles to the constant value of D, Eqs. (2.1) and (2.2) at the average level yield to

$$\overline{v} = Dv_A, \quad R\overline{i} = Dv_A - v_B, \quad \Rightarrow \quad v_B \approx Dv_A, \tag{2.5}$$

where the approximation is based on ignoring the voltage across R, i.e. $R\overline{i}$.

Example 2.1 *Buck Converter*
Consider a buck converter where on the input side A is a voltage source of $v_A = 100$ V and on the output side is a voltage source of $v_B = 40$ V. The power rating of the converter is 10 kW. The switching frequency is $f_{sw} = 4$ kHz. The inductance is $L = 400$ μH and the total parasitic resistance is estimated at $R = 10$ mΩ.

1. Find the duty cycle d at no load condition.
   At no load, no current flows in the inductor, and thus, $\bar{v} = v_B = 40$ V. This means that $d = \frac{\bar{v}}{v_A} = \frac{40}{100} = 0.4$.
2. Compute the duty cycle d at full load operation.
   At full load, the converter supplies 10 kW at 40 V which means the current of $\bar{i} = \frac{10\ 000}{40} = 250$ A. Thus, $\bar{v} = v_B + R\bar{i} = 40 + 0.01 \times 250 = 42.5$ V. This means that $d = \frac{\bar{v}}{v_A} = \frac{42.5}{100} = 0.425$.
3. How much is the peak-to-peak current ripple and its percentage with respect to rated current at full load.
   As calculated, the rated current and duty cycle are $\bar{i} = 250$ A and 0.425. The switching cycle is $T_{sw} = \frac{1}{f_{sw}} = \frac{1}{4000} = 0.25$ ms. The switch is ON for $dT_{sw} = 0.425 \times 0.25 = 0.10625$ ms and is OFF for $(1-d)T_{sw} = (1 - 0.425) \times 0.25 = 0.14375$ ms. Thus, according to the inductor equation $v = L\frac{di}{dt}$, the current ripple may be estimated during the rising stage as $\Delta i = \frac{v\Delta t}{L} = \frac{(100-40)\times 0.10625\times 0.001}{0.0004} = 15.9$ A and during the falling portion of the current as $\Delta i = \frac{v\Delta t}{L} = \frac{40\times 0.14275\times 0.001}{0.0004} = 14.4$ A. The difference between the two is due to the fact that we assume RL to be a pure L for this calculation. At any rate, the number 15 A will be a good estimate for the current ripple. The percentage current ripple is $\frac{15}{250} \times 100 = 6\%$ which is reasonable.
4. Assume that the converter is operating at full power with the calculated d and the input and output voltages are at the desired values of 100 and 40 V, respectively. The input voltage drops 2%, i.e. to 98 V, while the output voltage is still fixed at 40 V. Find the amount of exported power and the percentage of change in it.
   When the input voltage drops to 98 V, the converter internal voltage will drop to $\bar{v} = 0.425 \times 98 = 41.65$ V. This means that the current will be $\bar{i} = \frac{\bar{v}-v_B}{R} = \frac{41.65-40}{0.01} = 165$ A. The delivered power is $P_B = \bar{i}v_B = 165 \times 40 = 6.6$ kW down from the original value of 10 kW! The percentage power drop is 34%!
5. Assume that the converter is operating at full power with the calculated d and the input and output voltages are at the desired values of 100 and 40 V, respectively. The output voltage drops 2%, i.e. to 39.2 V, while the input voltage is still fixed at 100 V. Find the amount of exported power and the percentage of change in it.
   When the output voltage drops to 39.2 V, while the converter internal voltage is unchanged and equal to $\bar{v} = 42.5$ V, the current will change to $\bar{i} = \frac{\bar{v}-v_B}{R} = \frac{42.5-39.2}{0.01} = 330$ A. The delivered power is $P_B = \bar{i}v_B = 330 \times 39.2 = 12.9$ kW up from the original value of 10 kW! The percentage power rise is 29%!
6. Assume that the total parasitic resistance is actually 11 mΩ. How would this change the exported power if the calculation of d has been based on the value of 10 mΩ?
   In this case, the converter current will actually be $\bar{i} = \frac{\bar{v}-v_B}{R} = \frac{42.5-40}{0.011} = 227$ A. The delivered power is $P_B = \bar{i}v_B = 227 \times 40 = 9.1$ kW down from the desired value of 10 kW! The percentage power drop is 9%!

7. What is the settling time of the response when the converter starts from 0 or when a new change is applied in m.
   The time-constant of the transfer function is $\tau = \frac{L}{R} = \frac{0.0004}{0.01} = 40$ ms. The settling time is about four times the time-constant, i.e. 200 ms.
8. What would you conclude from this study?
   For power export application, it appears that the operation of the converter is much sensitive to parameter uncertainties or changes.

2.1.5 Sensitivity Analysis

The numerical example above showed the sensitivity of the converter to system uncertainties. In order to explain this more, a formal sensitivity analysis is presented here. Assume that y is an output variable and p is a parameter of the converter system. The sensitivity of y to p is defined as the normalized ratio of the change in y that is caused by a change in p, mathematically expressed by

$$S_p^y = \frac{\frac{\Delta y}{y}}{\frac{\Delta p}{p}} = \frac{p}{y}\frac{\Delta y}{\Delta p} = \frac{p}{y}\frac{\partial y}{\partial p}. \tag{2.6}$$

Notice that the last identity, i.e. replacing Δ with ∂, is mathematically justifiable assuming that the parameter changes are small.

For the buck converter discussed in the example, assume that the steady-state average power delivered to side B is considered as the output, i.e.

$$y = P_B = \bar{i}v_B = \frac{v_A v_B m - v_B^2}{R},$$

where $m = d$ is the modulation signal in a standard PWM approach. Thus, we can compute the sensitivity of this output to various parameters as follows.

2.1.5.1 Sensitivity to R

$$S_R^y = \frac{R}{y}\frac{\partial y}{\partial R} = \frac{R}{\frac{v_A v_B m - v_B^2}{R}} \frac{-\left(v_A v_B m - v_B^2\right)}{R^2} = -1$$

This indicates that a small unit uncertainty (or change) in R causes one small unit of change in y in the reverse direction. For example, one percent increase in R causes 1 percent decrease in power. This is consistent with the above example.

2.1.5.2 Sensitivity to v_B

$$S_{v_B}^y = \frac{v_B}{y}\frac{\partial y}{\partial v_B} = \frac{v_B}{\frac{v_A v_B m - v_B^2}{R}} \frac{v_A m - 2v_B}{R} = \frac{v_A m - 2v_B}{v_A m - v_B}$$

For the values of $v_A = 100$, $m = 0.425$, and $v_B = 40$, the sensitivity will be -15. This means that, for example, one percent decrease in v_B will cause 15% increase in the power! This is consistent with the above example.

2.1.5.3 Sensitivity to v_A

$$S^y_{v_A} = \frac{v_A}{y}\frac{\partial y}{\partial v_A} = \frac{v_A}{\frac{v_A v_B m - v_B^2}{R}}\frac{v_B m}{R} = \frac{v_A m}{v_A m - v_B}$$

For the values of $v_A = 100$, $m = 0.425$, and $v_B = 40$, the sensitivity will be 17. This means that, for example, one percents decrease in v_A will cause 17% decrease in the power! This is consistent with the above example.

2.1.6 Virtual Resistance Feedback

The high sensitivity of the converter power (or current) to uncertainties and changes can be explained by noticing that the current, in steady-state, is equal to $\bar{i} = \frac{v_A m - v_B}{R}$. The numerator is the voltage across R and is small, due to R being small. Thus, we are dividing two small numbers and any small change in any one of them can cause a big change in this division. The small voltage across R, i.e. $v_A m - v_B$, also appears in the sensitivity functions $S^y_{v_A}$ and $S^y_{v_A}$ and causes them to become large.

The virtual resistance approach, as explained below, virtually increases the resistance using a feedback of the current, as shown in Figure 2.8. This will substantially improve the robustness of current (or power) to uncertainties. In Figure 2.8, the current $\bar{i}$ is expressed as

$$\bar{i} = \frac{\frac{v_A m'}{Ls+R}}{1+\frac{v_A k}{Ls+R}} - \frac{\frac{v_B}{Ls+R}}{1+\frac{v_A k}{Ls+R}} = \frac{v_A m' - v_B}{Ls+R+kv_A} = \frac{v_A m' - v_B}{Ls+R+R_v}, \tag{2.7}$$

where $R_v = kv_A$ is called the virtual resistance. Comparing this with the original equation of $\bar{i} = \frac{v_A m - v_B}{Ls+R}$, we realize that this feedback is virtually increasing R to $R + R_v = R + kv_A$.

For this feedback-compensated system, we can readily verify that

$$S^y_R = -\frac{R}{R+R_v}, \quad S^y_{v_B} = \frac{v_A m' - 2v_B}{v_A m' - v_B}, \quad S^y_{v_A} = \frac{v_A m'}{v_A m' - v_B}.$$

The denominator term $v_A m' - v_B = v_A m - v_B + R_v \bar{i} = (R + R_v)\bar{i}$ is the voltage across $R + R_v$. This voltage is now large and under control, meaning that the sensitivity will decrease.

The feedback loop strategy of Figure 2.8 may be thought of adding a virtual resistance or equivalently of pushing the pole of the system from the original point $-\frac{R}{L}$ to $-\frac{R+R_v}{L}$. As the pole moves further away from the imaginary axis to the left side, the system's transient responses become faster as well. Obviously, R_v must not be chosen excessively large so as not compromise other aspects of the system, such as its noise immunity.

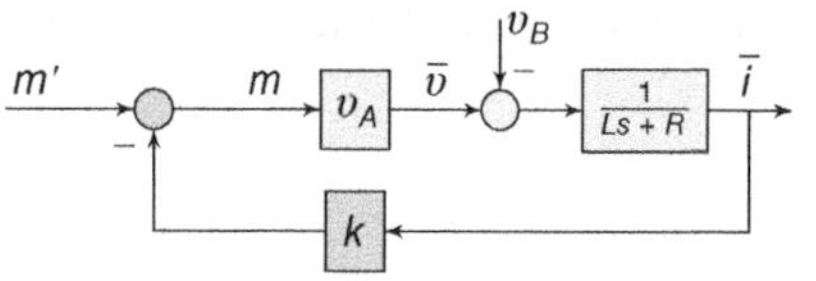

Figure 2.8 Feedback of current to virtually increase R.

Example 2.2 ***Buck Converter with Feedback***

Consider the buck converter example studied in Example 2.1 which was shown to be sensitive to uncertainties. To reduce this sensitivity, we uses a virtual resistance feedback strategy shown in Figure 2.8.

1. Find the value of feedback gain k such that $R_{\upsilon} = 20R = 20 \times 0.01 = 0.2$.
   $k = \frac{R_{\upsilon}}{\upsilon_A} = \frac{0.2}{100} = 0.002$
2. Find the modified duty cycle (or modified modulation index) m' for the rated operation.[1]
   $m' = m + k\bar{i} = 0.425 + 0.002 \times 250 = 0.925$
3. Assume that the converter is operating at full power with the calculated m' and the input and output voltages are at the desired values of 100 and 40 V, respectively. The input voltage drops 2%, i.e. to 98 V, while the output voltage is still fixed at 40 V. Find the amount of exported power and the percentage of change in it.
   The converter average current is $\bar{i} = \frac{\upsilon_A m' - \upsilon_B}{Ls+R+R_{\upsilon}}$ in Laplace domain where m' and υ_B are considered as inputs. According to the final value theorem (FVT),
   $$\bar{i}(\infty) = \lim_{s\to 0} s\frac{\upsilon_A \frac{m'}{s} - \frac{\upsilon_B}{s}}{Ls + R + R_{\upsilon}} = \frac{\upsilon_A m' - \upsilon_B}{R + R_{\upsilon}}.$$
   When the input voltage drops to 98 V, the current will be $\bar{i} = \frac{98\times0.925-40}{0.01+0.002\times98} = 246$ A. The delivered power is $P_B = \bar{i}\upsilon_B = 246 \times 40 = 9.83$ kW down from the original value of 10 kW. The percentage power drop is only 1.7%.
4. Assume that the converter is operating at full power with the calculated m' and the input and output voltages are at the desired values of 100 and 40 V, respectively. The output voltage drops 2%, i.e. to 39.2 V, while the input voltage is still fixed at 100 V. Find the amount of exported power and the percentage of change in it.
   When υ_B drops to 39.2 V, the current changes to $\bar{i} = \frac{100\times0.925-39.2}{0.01+0.2} = 253.8$ A. The delivered power is $P_B = \bar{i}\upsilon_B = 253.8 \times 39.2 = 9.95$ kW down from the original value of 10 kW! The percentage power drop is only 0.5%!
5. Assume that the total parasitic resistance is actually 11 mΩ. How would this change the exported power if the calculation of m' has been based on the value of 10 mΩ?
   In this case, the converter current will actually be $\bar{i} = \frac{\upsilon_A m' - \upsilon_B}{R+R_{\upsilon}} = \frac{100\times0.925-40}{0.011+0.2} = 248.8$ A. The delivered power is $P_B = \bar{i}\upsilon_B = 248.8 \times 40 = 9.95$ kW down from the desired value of 10 kW. The percentage power drop is only 0.5%!
6. What will be the settling time of the response in the feedback-compensated system?
   The time-constant of the feedback compensated transfer function is $\tau = \frac{L}{R+R_{\upsilon}} = \frac{0.0004}{0.21} = 1.9$ ms. The settling time is about four times the time-constant, i.e. 7.6 ms.
7. What would you conclude from this study?
   The virtual resistance approach using a feedback of current effectively improves the system robustness to uncertainties and changes, as well as the speed of its transient response.

Figure 2.9 shows the simulation results of the buck converter. The switching level and average model outputs are shown for two cases of no-feedback and with-feedback. The converter parameters and the control gain k are selected according to the numerical examples discussed above: $\upsilon_A = 100$, $\upsilon_B = 40$, $L = 0.4$ mH, $R = 10$ mΩ, $f_{sw} = 4$ kHz, and $k = 0.002$. Two disturbances happen in the system: υ_B decreases to 39.2 V at $t = 0.25$ seconds, υ_B is restored to 40 V and υ_A is reduced to

1 Notice that m' does not have to be between 0 and 1.

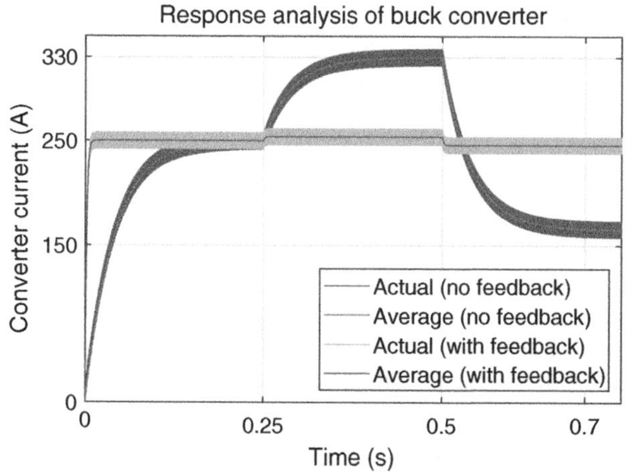

Figure 2.9 Responses of buck converter to disturbances in v_A and v_B in two cases of without and with feedback of virtual resistance.

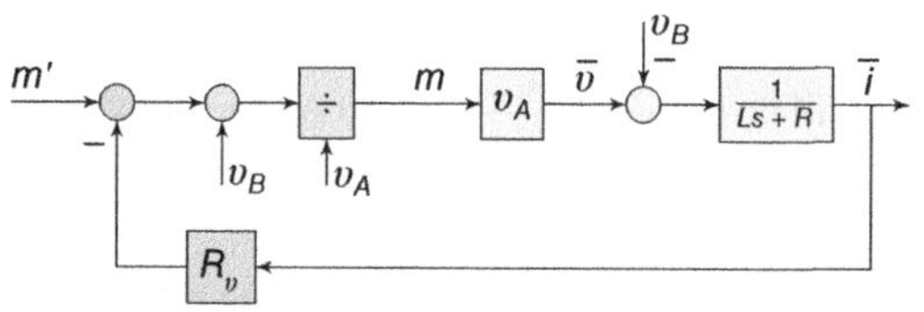

Figure 2.10 Virtual resistance feedback combined with input feedback linearization.

98 V at $t = 0.5$ seconds. The simulation results shown in Figure 2.9 are in perfect agreement with the calculations presented above.

2.1.7 Input Feedback Linearization

The virtual resistance technique significantly improves the transient response and the performance robustness of the converter. It is also possible to completely cancel the impact of input and output voltages, v_A and v_B, on the system responses by measuring them in real-time and compensating them when calculating the modulation index. This approach is shown in Figure 2.10 and may be called input feedback linearization. The division block cancels the impact of v_A and the summation block calcels the impact of v_B. The systems transfer function can be written as

$$\frac{\bar{i}}{m'} = \frac{1}{Ls + R + R_v}. \tag{2.8}$$

Note that in Figure 2.10, $m' = mv_A - v_B + R_v\bar{i}$. Therefore, for example, for the numerical values of the previous example, it will be equal to $m' = 0.425 \times 100 - 40 + 0.2 \times 250 = 52.5$.

2.2 Standard Boost Converter

The standard boost converter is shown in Figure 2.11 which comprises a unidirectional switch Q_2 and a diode D_1. It interfaces the input side A with the output side B. The inductance value is L, and R denotes the totality of the cable's resistance, the inductor's parasitic resistance and the conducting resistance of switche/diode.

Figure 2.11 Standard boost converter.

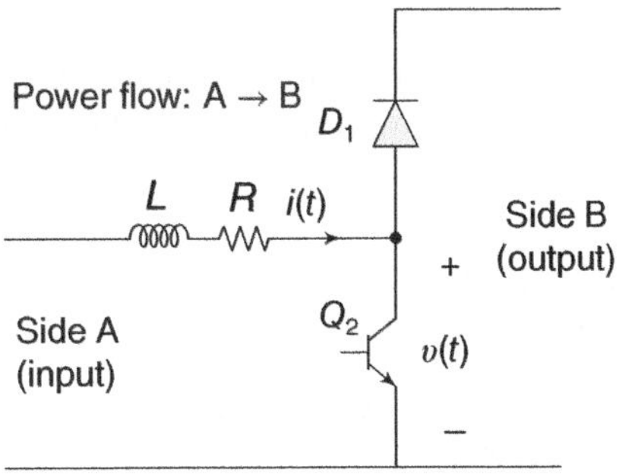

When the switch Q_2 is commanded on, the voltage $v(t)$ will be equal to zero. Then, the current $i(t)$ will flow in the direction shown and will rise, assuming $v_A > 0$. When the switch Q_2 is turned off, the current will turn the diode D_1 on and will continue to flow through it. During this time, $v(t)$ is equal to v_B. In order for the current to decline and stabilize, the voltage v_B must be larger than v_A. Thus, $0 < v_A < v_B$, meaning that it will boost the voltage in the direction of power flow.

2.2.1 Analysis of Operation

Assume that the PWM technique is used to determine the switching instants as shown in Figure 2.2 where $m(t)$ is the modulating signal and has continuous values between 0 and 1, and $c(t)$ is the triangular carrier signal at the switching frequency f_{sw} and peak values at 0 and 1 as shown in Figure 2.12. When $s_2 = 1$, switch Q_2 conducts and provides the path for current. Therefore, the voltage $v(t)$ is zero. As soon as s_2 falls to zero, the diode D_1 turns on and provides a path for the current. During this period, the voltage $v(t)$ is equal to v_B.

Since the switching frequency is high, the voltages $v_A(t)$, $v_B(t)$ and the modulating signal $m(t)$ are considered constant during one switching cycle. This entails that the current waveform is almost linearly rising and falling within a switching cycle due to the fact that the voltage across the inductor

Figure 2.12 PWM operation and illustration of waveforms for boost converter.

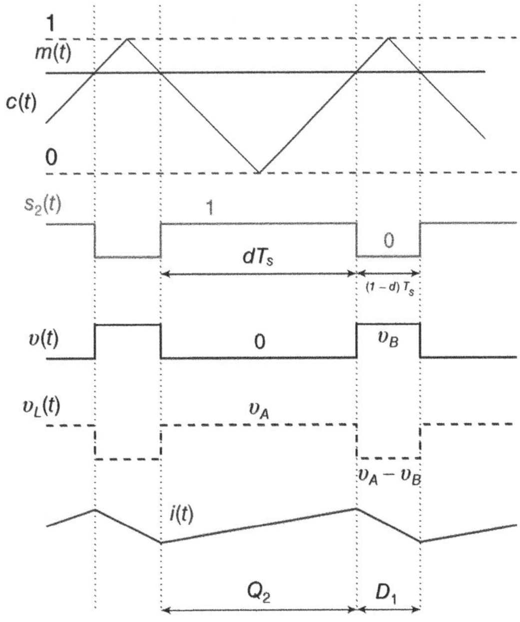

is $v_L = v_A - v$ which switches between v_A and $v_A - v_B$. The inductor current voltage equation is $v_L = L\frac{di}{dt}$ which results in the current constantly ramping up and down within a switching cycle.

The above discussion also results in that the voltage $v(t)$ can be written as

$$v(t) = v_B(t)(1 - s_2(t)) = v_B(t)s_{2_n}(t), \tag{2.9}$$

where $s_{2_n}(t) = 1 - s_2(t)$ is the logical NOT of $s_2(t)$. The circuit equation describing the current and voltage is

$$L\frac{d}{dt}i(t) + Ri(t) = v_A(t) - v(t). \tag{2.10}$$

Equations (2.9) and (2.10) summarize the operation of the boost converter.

2.2.2 Steady-State Analysis

When the circuit reaches its steady state, and the duty cycle d settles to the constant value of D, Eqs. (2.9) and (2.10) at the average level yield to

$$\bar{v} = v_B(1 - D),\ R\bar{i} = v_A - \bar{v},\ \Rightarrow\ v_B \approx \frac{1}{1 - D}v_A, \tag{2.11}$$

where the approximation is based on ignoring the voltage across R, i.e. $R\bar{i}$.

2.2.3 Switching Model

Based on (2.9) and (2.10), an equivalent circuit may be derived for the boost converter as shown in Figure 2.13. The switch Q_2 and diode D_1 are modeled by controlled current sources.

2.2.4 Average (or Control) Model

The voltage $v(t)$ is a high-frequency chopped waveform with two discrete values v_B and 0. Similar to the averaging approach taken for the buck converter, (2.10) may be written as two equations

$$\begin{aligned} L\frac{d}{dt}\bar{i}(t) + R\bar{i}(t) &= v_A(t) - \bar{v}(t) \\ L\frac{d}{dt}\tilde{i}(t) + R\tilde{i}(t) &= -\tilde{v}(t). \end{aligned} \tag{2.12}$$

The average of the switching function $s(t)$ is

$$\bar{s}_2 = \frac{1}{T_s}\int_0^{T_s} s_2(t)dt = \frac{1}{T_s}dT_s = d.$$

Therefore,

$$\bar{v} = v_A\bar{s}_{2_n} = (1 - d)v_B.$$

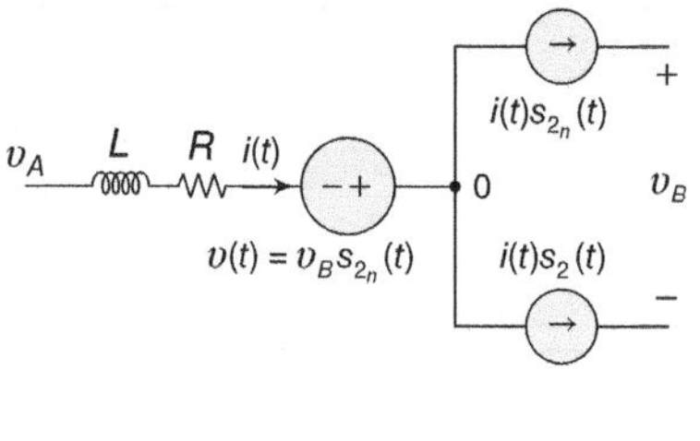

Figure 2.13 Equivalent circuit of boost converter at switching level.

Now, an inspection of Figure 2.3 shows that when m varies from 0 to 1, d also varies from 0 to 1. Moreover, the relationship between them is obviously linear. Therefore, they must be equal, i.e.,

$$m = d.$$

Substituting this into the previous equation results in

$$\overline{v} = (1 - m)v_B. \tag{2.13}$$

Equation (2.4) describes the average model of the boost converter from the modulating signal $m(t)$ to the voltage $v(t)$.

2.2.4.1 Current Control Model

Here the control output is the converter current $\overline{i}(t)$. In this case, v_A and v_B are treated as external variables and are not controlled by the converter. The control model is as derived above and shown in Figure 2.14. This model is particularly suited for the case where the input and output voltages are relatively stable, e.g. a battery connected to a dc grid (dc supply).

2.2.4.2 Input Voltage Control

The control output is the voltage v_A. This is what is done in PV applications where the PV power is determined by its voltage. A capacitor C is connected across the input terminals for voltage support, as shown in Figure 2.15a. The current balance equation is $i_A - i = C\dot{v}_A$. Thus, the control model is derived and is shown in Figure 2.15b. In the PV application, the input current i_A is a nonlinear function of the voltage, $i_A = f(v_A)$ corresponding to the PV characteristics.

2.2.4.3 Output Voltage Control

The control output is the voltage v_B. A local load and/or a capacitance (or a combination of both) with the total impedance of $Z(s)$ is connected across the output terminals. It may further be connected to the rest of a network to which the current i_B flows as shown in Figure 2.16a. The current flowing into the impedance $Z(s)$ is equal to $(1 - m)\overline{i} - i_B$. Thus, the control model is readily obtained and is shown in Figure 2.16b.

For primary control system designs of the buck converter, the current i_B may be treated as a disturbance or it may be taken into analysis if further information about its dynamics is available.

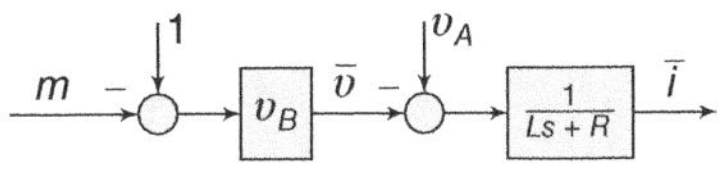

Figure 2.14 Average or control model of the boost converter.

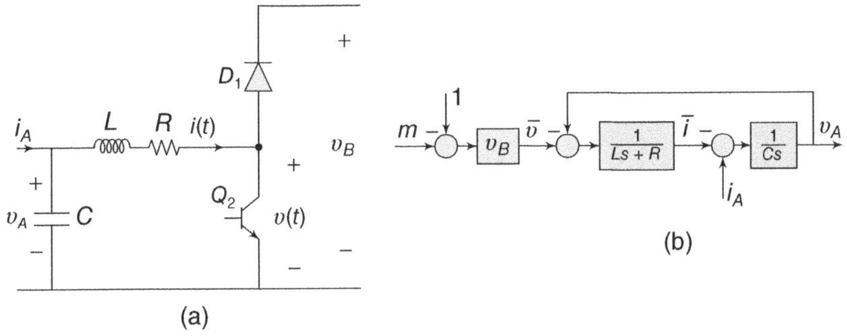

Figure 2.15 Input voltage control of boost converter: (a) circuit and (b) control model.

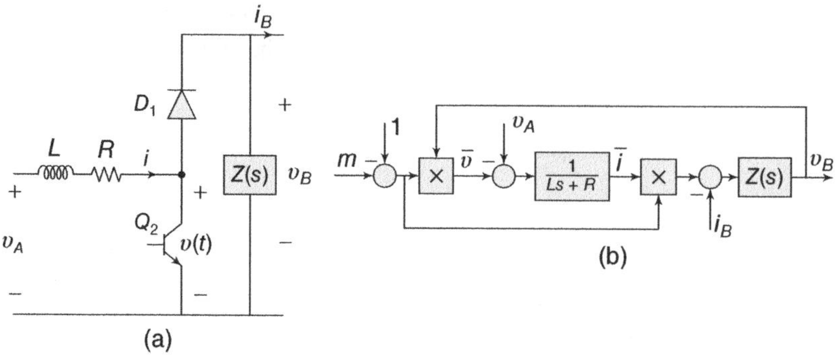

Figure 2.16 Output voltage control of boost converter: (a) circuit and (b) control model.

The voltage control model is suitable for designing the control to operate in standalone mode, i.e. not connected to a stiff voltage on side B. However, it may also be adjusted and used for grid-connected applications using the virtual impedance approach. Notice that this model is nonlinear due to the presence of two multiplications shown in Figure 2.16b.

2.3 Standard Inverting Buck-Boost Converter*

The standard inverting buck-boost converter is shown in Figure 2.17 which comprises a unidirectional switch Q_1 and a diode D_2. It is the interface between side A and side B and the power flow direction is obviously from side A to side B, i.e. the input side to the output side, respectively. The inductance value is L, and R denotes the totality of cable's resistance, the inductor's parasitic resistance and the conducting resistance of switch/diode.

When the switch Q_1 is commanded on, the voltage $v(t)$ will be equal to v_A. Then, the current $i(t)$ will flow in the direction shown and will rise. When the switch Q_1 is turned off, the current will turn the diode D_2 on and will continue to flow through it and $v(t)$ will be equal to v_B. In order for the current to decline and stabilize, v_B must be negative. Thus, $v_A > 0$ and $v_B < 0$ meaning that the converter is inverting the voltage in the direction of power flow.

2.3.1 Analysis of Operation

Figure 2.18 shows a sample of signals waveform pertaining to this circuit. When $s_1 = 1$, switch Q_1 conducts and provides the path for current. Therefore, the voltage $v(t)$ is equal to v_A. As soon as s_1 falls to zero, the diode D_2 turns on and provides a path for the current. During this period, the voltage $v(t)$ is equal to v_B.

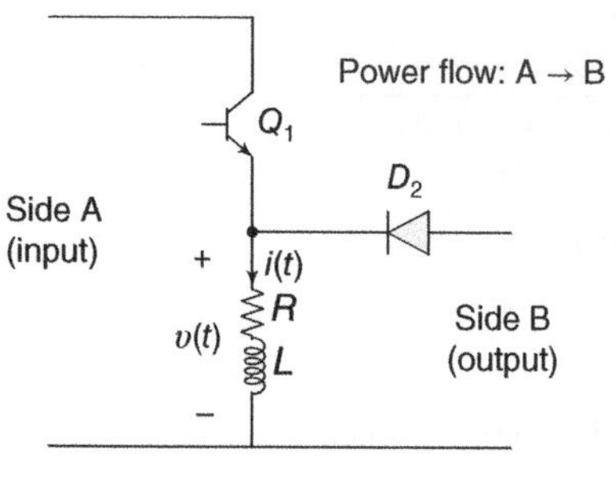

Figure 2.17 Standard inverting buck-boost converter.

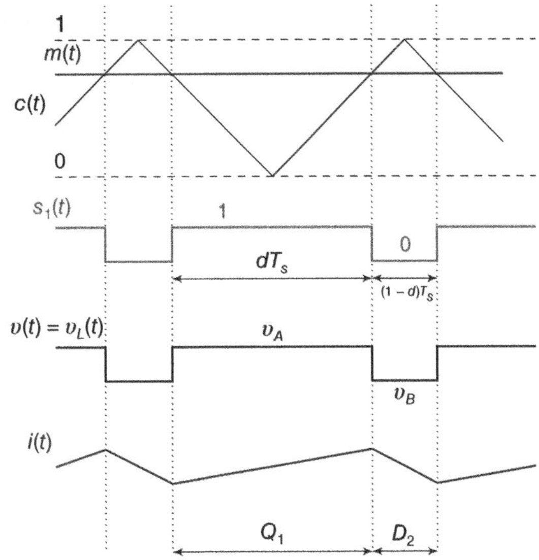

Figure 2.18 PWM operation and illustration of waveforms for buck-boost converter.

We assume that the voltages $v_A(t)$, $v_B(t)$ and the modulating signal $m(t)$ are considered constant during one short switching cycle. This entails that the current waveform is almost linearly growing or falling due to the fact that the voltage across the inductor is $v_L = v$ and the inductor current–voltage equation is $v_L = L\frac{di}{dt}$. The voltage across L when Q_1 is ON is equal to v_A and this voltage must be positive to allow a flow of current, i.e. $v_A > 0$. When Q_1 is commanded OFF, the voltage across L is v_B and this must be negative, i.e. $v_B < 0$, to reduce the current.

Finally, the voltage $v(t)$ can be written as

$$v(t) = v_A(t)s_1(t) + v_B(t)[1 - s_1(t)], \tag{2.14}$$

and the circuit equation describing the relationship between current and voltage is

$$L\frac{d}{dt}i(t) + Ri(t) = v(t). \tag{2.15}$$

2.3.2 Steady-State Analysis

When the circuit reaches its steady state, and the duty cycle d settles to the constant value of D, Eqs. (2.14) and (2.15) at the average level yield to

$$\bar{v} = v_A D + v_B(1 - D), \; R\bar{i} = \bar{v}, \; \Rightarrow \; v_B \approx -\frac{D}{1 - D}v_A, \tag{2.16}$$

where the approximation is based on ignoring the voltage across R, i.e. $R\bar{i}$. This confirms that while v_B is negative, its magnitude can be smaller and larger than v_A depending on whether D is below 0.5 or above 0.5, respectively.

2.3.3 Switching Model

Based on (2.14) and (2.15), an equivalent circuit may be derived for the buck converter as shown in Figure 2.19. In this circuit, the switch and diode are modeled by controlled current sources. Note that $s_{1_n}(t)$ is the logical NOT of $s_1(t)$ defined as $s_{1_n}(t) = 1 - s_1(t)$. This circuit may be used to simulate the buck converter at the switching level.

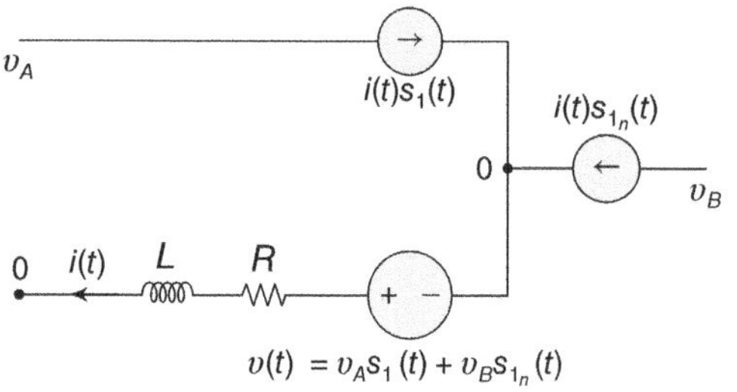

Figure 2.19 Equivalent circuit of buck-boost converter at switching level.

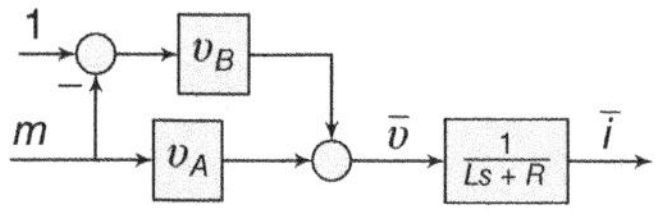

Figure 2.20 Average or control model of the buck boost converter.

2.3.4 Average (or Control) Model

In the PWM approach for controlling a converter, the modulation signal $m(t)$ is considered as the control input. The output variable is chosen depending on the application. Three scenarios for the output variable, including the output current $i(t)$, the output voltage v_B, and the input voltage v_A are discussed in this section and the control block diagram is obtained for each one.

Taking the average from (2.14) and (2.15) results in

$$L\frac{d}{dt}\bar{i}(t) + R\bar{i}(t) = \bar{v} = v_A\bar{s}_1 + v_B(1-\bar{s}_1) = v_A d + v_B(1-d) = v_A m + v_B(1-m). \tag{2.17}$$

2.3.4.1 Current Control Model

The control output is the converter current $\bar{i}(t)$. Based on (2.17), the control transfer function model shown in Figure 2.20 is obtained. In this case, v_A and v_B are treated as external variables and are not controlled by the converter. This model is particularly suited for the case where side B represents a stiff voltage, e.g. a strong dc grid or a battery.

2.4 Standard Four-Switch Buck-Boost Converter*

The standard four-switch buck-boost converter is shown in Figure 2.21 which comprises two unidirectional switch Q_1 and Q_2 and two diodes D_1 and D_2. It is the interface between side A and side B and the power flow direction is obviously from side A to side B, i.e. the input side to the output side, respectively. The inductance value is L, and R denotes the totality of cable's resistance, the inductor's parasitic resistance and the conducting resistance of switch/diode.

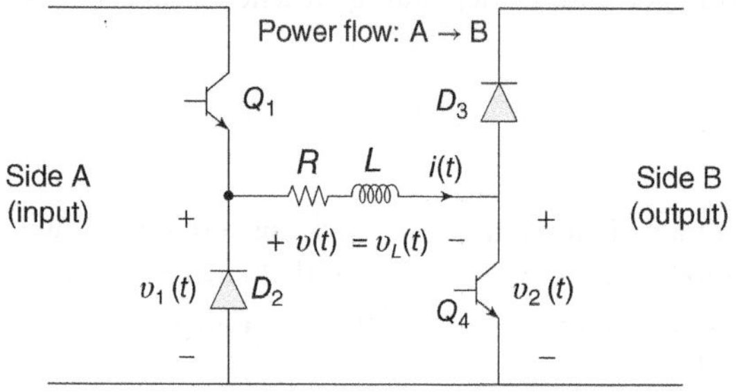

Figure 2.21 Standard four-switch buck-boost converter.

When the switches Q_1 and Q_4 are commanded on, the voltage $v_1(t)$ will be equal to v_A and the voltage $v_2(t)$ is equal to zero. Then, the current $i(t)$ will flow in the direction shown through Q_1 and Q_4 and will rise because the voltage across the inductor is $v_L = v_1 - v_2 = v_A$. When the switches Q_1 and Q_4 are turned off, the current will turn the diodes D_2 and D_3 on and will continue to flow through them and $v_1(t)$ will be equal to zero and $v_2(t)$ will be equal to v_B. The voltage across the inductor will be $v_L = v_1 - v_2 = -v_B$. In order for the current to decline and stabilize, v_B must be positive. Thus, $v_A > 0$ and $v_B > 0$ meaning that the converter is non-inverting in the direction of power flow.

2.4.1 Analysis of Operation

Figure 2.22 shows a sample of signals waveform pertaining to this circuit. When $s_1 = s_4 = 1$, switches Q_1 and Q_4 conduct and provide the path for current. Therefore, the inductor voltage $v_L(t) = v(t) = v_1(t) - v_2(t)$ is equal to v_A. As soon as $s_1 = s_4$ falls to zero, the diodes D_2 and D_3 turn on and provide a path for the current. During this period, the voltage $v(t)$ is equal to $-v_B$.

We assume that the voltages $v_A(t)$, $v_B(t)$ and the modulating signal $m(t)$ are considered constant during one short switching cycle. This entails that the current waveform is almost linearly growing or falling due to the fact that the voltage across the inductor is $v_L = v$ and the inductor current–voltage equation is $v_L = L\frac{di}{dt}$. The voltage across L when Q_1 and Q_4 are ON is equal to v_A and this voltage must be positive to allow a flow of current, i.e. $v_A > 0$. When Q_1 and Q_4 are commanded OFF, the voltage across L is $-v_B$ and this must be negative, i.e. $v_B > 0$, to reduce the current.

Finally, the voltage $v(t)$ can be written as

$$v(t) = v_1(t) - v_2(t) = v_A(t)s_1(t) - v_B(t)[1 - s_1(t)], \tag{2.18}$$

and the circuit equation describing the relationship between current and voltage is

$$L\frac{d}{dt}i(t) + Ri(t) = v(t). \tag{2.19}$$

Figure 2.22 PWM operation and illustration of waveforms for four-switch buck-boost converter.

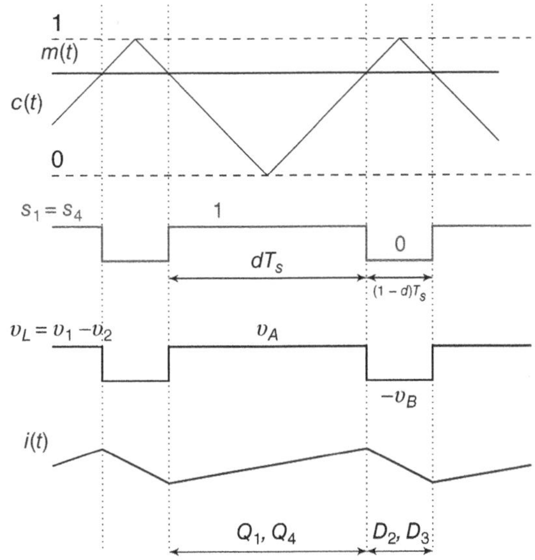

2.4.2 Steady-State Analysis

When the circuit reaches its steady state, and the duty cycle d settles to the constant value of D, Eqs. (2.18) and (2.19) at the average level yield to

$$\bar{v} = v_A D - v_B(1 - D), \; R\bar{i} = \bar{v}, \; \Rightarrow \; v_B \approx \frac{D}{1 - D} v_A, \tag{2.20}$$

where the approximation is based on ignoring the voltage across R, i.e. $R\bar{i}$. This confirms that while v_B is positive, its magnitude can be smaller and larger than v_A depending on whether D is below 0.5 or above 0.5, respectively.

2.4.3 Switching Model

Based on (2.18) and (2.19), an equivalent circuit may be derived for the buck converter as shown in Figure 2.23. In this circuit, the switches and diodes are modeled by controlled current sources. Note that $s_{1_n}(t)$ is the logical NOT of $s_1(t)$ defined as $s_{1_n}(t) = 1 - s_1(t)$, and $s_1(t) = s_4(t)$. This circuit may be used to simulate the buck converter at the switching level.

2.4.4 Average (or Control) Model

In the PWM approach for controlling a converter, the modulation signal $m(t)$ is considered as the control input. The output variable is chosen depending on the application. Three scenarios for the output variable, including the output current $i(t)$, the output voltage v_B, and the input voltage v_A are discussed in this section and the control block diagram is obtained for each one.

Taking the average from (2.18) and (2.19) results in

$$L\frac{d}{dt}\bar{i}(t) + R\bar{i}(t) = \bar{v} = v_A\bar{s}_1 - v_B(1 - \bar{s}_1) = v_A d - v_B(1 - d) = v_A m - v_B(1 - m). \tag{2.21}$$

2.4.4.1 Current Control Model

The control output is the converter current $\bar{i}(t)$. Based on (2.21), the control transfer function model shown in Figure 2.24 is obtained. In this case, v_A and v_B are treated as external variables and are not controlled by the converter. This model is particularly suited for the case where side B represents a stiff voltage, e.g. a strong dc grid or a battery.

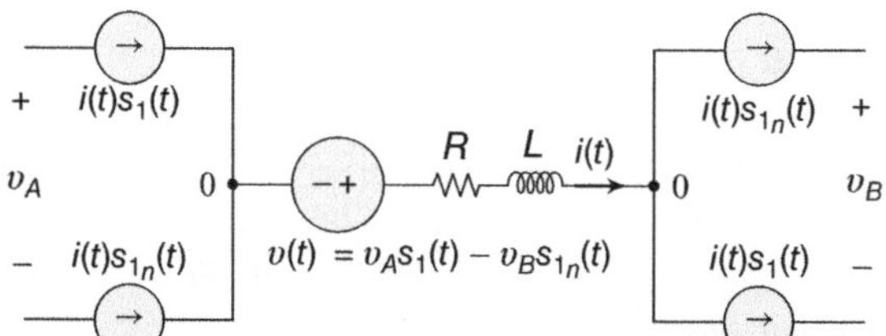

Figure 2.23 Equivalent circuit of four-switch buck-boost converter at switching level.

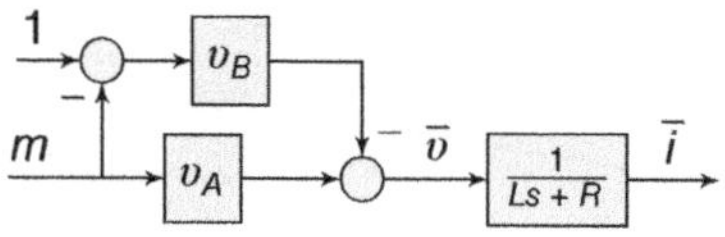

Figure 2.24 Average or control model of the four-switch buck boost converter.

2.5 Standard Bidirectional Converter

The standard buck and boost (and buck-boost) converters are unidirectional meaning that the power flow is from one side to another. It is possible to upgrade their switches, as shown in Figure 2.25, to allow bidirectional power flow. Each switch comprises a transistor and a reverse conducting diode, i.e. is a switch cell. The switching functions must be complementary, i.e. $s_2(t) = 1 - s_1(t) = s_{1_n}(t)$, to prevent a short circuit. For the power flow from side A to side B, Q_1 and D_2 conduct, and for the power flow from side B to side A, Q_2 and D_1 conduct. In either case, the condition $v_A > v_B > 0$ must obviously be satisfied to ensure that the current remains stable, as shown in Figure 2.26.

For the PWM operation depicted in Figure 2.26, it is clear that $v(t) = s_1(t)v_A$. Therefore, the equivalent circuit at the switching level and also the control (average) model of the bidirectional converter of Figure 2.25 is the same as the buck converter, i.e. Figures 2.4 and 2.5, respectively, with the difference that here $i(t)$ can be both positive and negative while in the buck converter $i(t)$ is only positive.

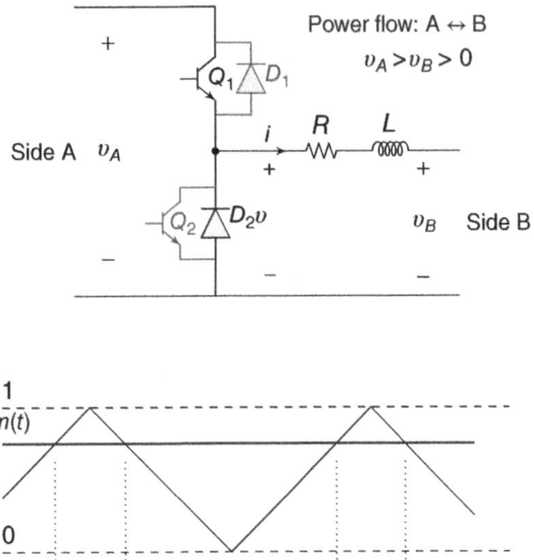

Figure 2.25 Standard buck (or boost) converter with bidirectional power flow.

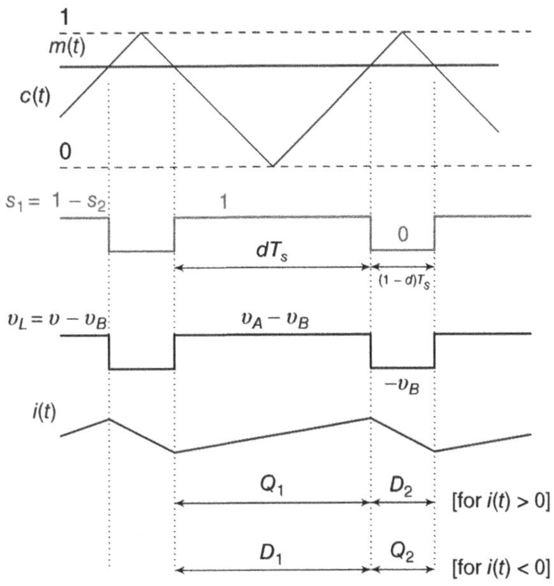

Figure 2.26 PWM operation and illustration of waveforms for bidirectional buck or boost converter, Figure 2.25.

2.6 Single-Phase Half-Bridge VSC

Single-phase half-bridge voltage source converter (HB-VSC) comprises two switch cells S_1 and S_2 as shown in Figure 2.27. The side A voltage $v_A > 0$ is broken down to two half voltages and their mid-point is taken as the negative output terminal. Inductor L is the interfacing filter and v_B is the side B voltage. The resistor R models the compound effect of the cable's resistance, the inductor's parasitic resistance and the conducting resistance of switches. The HB-VSC of Figure 2.27 allows the voltage $v(t)$ to switch bipolar, that is $\pm\frac{v_A}{2}$. Therefore, $v_B(t)$ can be negative. As a result, this converter can be a dc/ac converter (or an inverter) as well.

2.6.1 Analysis of Operation

Obviously, $s_1(t)$ and $s_2(t)$ cannot be 1 at the same time because this means a short circuit. The common PWM technique adopts $s_1(t) + s_2(t) = 1$ as shown in Figure 2.28. In Figure 2.28, $m(t)$ is called the *modulating signal* and is a continuous signal with values between −1 and 1. The signal $c(t)$ is called the *carrier signal* and is a triangular waveform with frequency f_{sw} (called the switching frequency) and peak values at −1 and 1 as shown in Figure 2.29. In the PWM technique, basically, the modulating signal $m(t)$ is compared with the carrier signal $c(t)$ and if $m(t) > c(t)$ then S_1 is commanded ON, otherwise S_2 is commanded ON.

Figure 2.29 shows a sample of signals waveform pertaining to this circuit. In Figure 2.29, the switching function s_2 is complement of s_1. This diagram is shown for the case where the current i is positive, i.e. flowing from side A to side B. In this case, when $s_1 = 1$, switch S_1 conducts (through its transistor Q_1) and provides the path for current and S_2 is not conducting. Therefore, the voltage $v(t)$ is $0.5v_A$. As soon as s_1 falls to zero, S_1 turns off and the reverse diode of S_2 (called D_2) turns on and provides a path for the current. During this period, the voltage $v(t)$ is $-0.5v_A$.

Notice that switching frequency is high. Therefore, the voltages v_A, v_B and the modulating signal $m(t)$ are considered constant during one switching cycle. This entails that the current waveform is almost linearly rising or falling due to the fact that the voltage across the inductor is $v_L = v - v_s$ and the inductor current-voltage equation is $v_L = L\frac{di}{dt}$. The voltage across L when S_1 is ON is equal to $0.5v_A - v_B$ and this voltage must be positive to allow a rising of current. This requires that

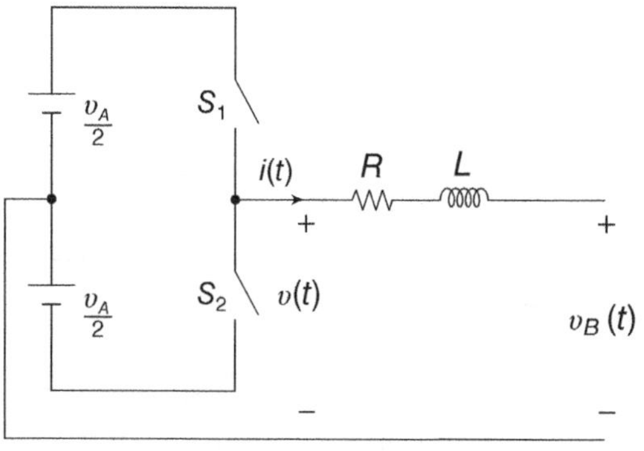

Figure 2.27 Half-bridge voltage source converter.

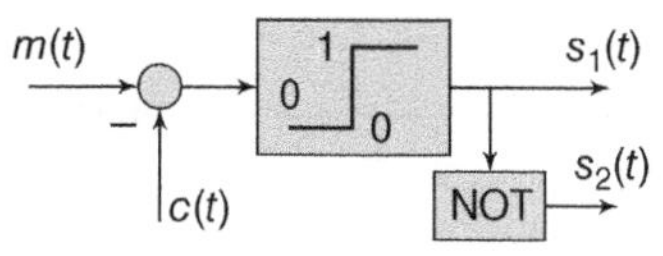

Figure 2.28 Pulse width modulation for HB-VSC.

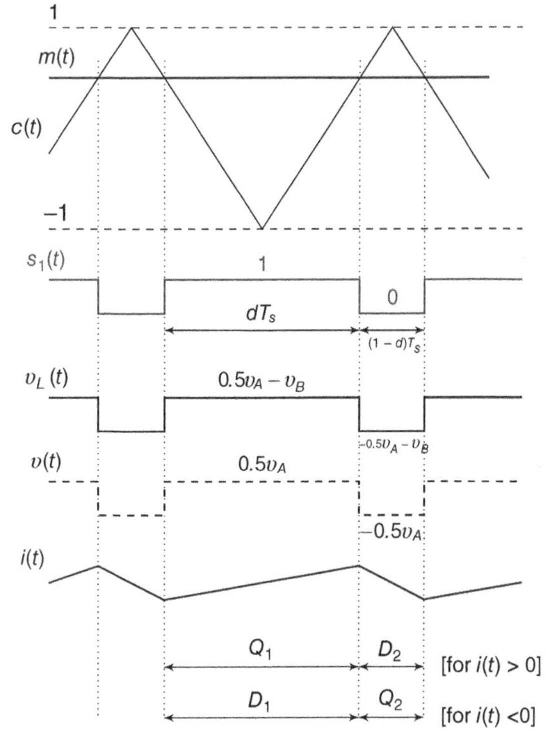

Figure 2.29 PWM operation and illustration of waveforms for the HB-VSC.

$0.5v_A > v_B$. When S_1 is commanded OFF, the voltage across L is $-0.5v_A - v_B$ and this must be negative which requires $-0.5v_A < v_B$. Thus, if $-0.5v_A < v_B < 0.5v_A$, $\forall t$, it satisfies both conditions.

When the current $i(t)$ is negative, the situation is slightly different: when $s_2 = 1$, switch S_2 is conducting (through its transistor Q_2) and provides the path for the current. As soon as s_2 falls to zero, the reverse diode of S_1 (called D_1) turns on and provides a path for the current.

The above discussion concludes that, regardless of direction of current flow, the VSC voltage $v(t)$ can be written as

$$v(t) = \frac{V_{dc}}{2}s_1(t) - \frac{V_{dc}}{2}s_2(t) = \frac{V_{dc}}{2}[s_1(t) - s_2(t)] = \frac{V_{dc}}{2}[2s_1(t) - 1], \tag{2.22}$$

and the circuit equation describing the current is

$$v(t) = Ri(t) + L\frac{d}{dt}i(t) + v_B(t). \tag{2.23}$$

2.6.2 Switching Model

Based on the two equations (2.22) and (2.23), an equivalent circuit may be derived for the HB-VSC as shown in Figure 2.30. In this circuit, the switches are modeled by current sources controlled by the switching functions. Notice that $i(t)$ can be both positive or negative.

2.6.3 Average (or Control) Model

The circuit equations describing the HB-VSC is given by (2.22) and (2.23). The voltage $v(t)$ is a high-frequency chopped waveform with two discrete values $\pm\frac{v_A}{2}$. Using the moving average

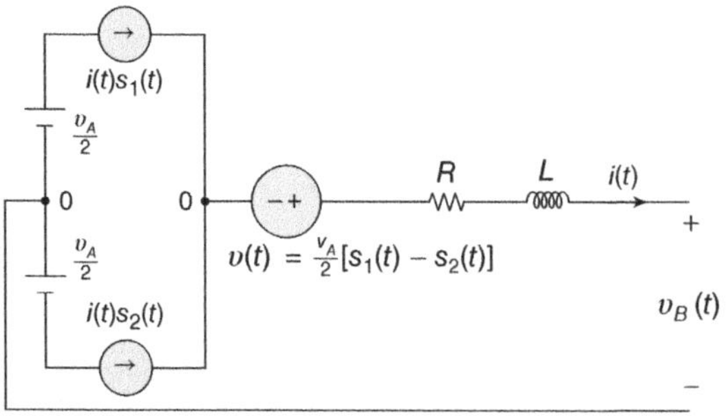

Figure 2.30 Equivalent circuit of HB-VSC at switching level.

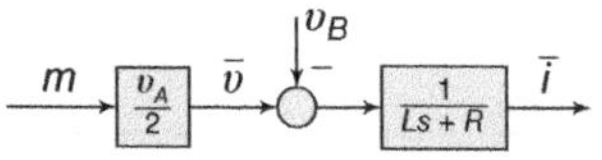

Figure 2.31 Average or control model of the HB-VSC.

variable over time by the definition $\overline{x}(t) = \frac{1}{T_s}\int_{t-T_s}^{t} x(\tau)d\tau$, it is observed that the averages of switching functions are $\overline{s}_1 = \frac{1}{T_s}\int_0^{T_s} s_1(t)dt = \frac{1}{T_s}dT_s = d$ and $\overline{s}_2 = 1 - d$. Therefore,

$$\overline{v} = \frac{v_A}{2}(\overline{s}_1 - \overline{s}_2) = (2d - 1)\frac{v_A}{2}.$$

Now, an inspection of Figure 2.29 shows that when m varies from −1 to 1, d varies from 0 to 1. Moreover, the relationship between them is obviously linear. Therefore, they are related through

$$m = 2d - 1.$$

Substituting this into the previous equation results in

$$\overline{v} = \frac{v_A}{2}m. \tag{2.24}$$

Equation (2.24) describes the average model of the HB-VSC from the modulating signal $m(t)$ to the voltage $v(t)$. The HB-VSC acts as a simple gain $\frac{v_A}{2}$. Finally, the HB-VSC circuit of Figure 2.27 can be represented by the transfer function model shown in Figure 2.31. This is called the average model or the control model and is used for control analysis, design and simulations.

Example 2.3 ***Single-Phase VSC***

Consider a single-phase HB-VSC with the following numerical values:
$v_A = 1200$ V, $v_B = 400$ V, $L = 700$ μH, $R = 0.005$ Ω, $f_{sw} = 1600$ Hz.

1. The VSC is initially planned to export a power of 640 kW. Find the values for the modulating signal m, the duty-cycle d, the (average) value of the transferred current, and the peak-to-peak current ripples.
   Average current is $\overline{i} = \frac{P}{v_B} = \frac{640000}{400} = 1600$ A.
   Average converter voltage is $\overline{v} = v_B + R\overline{i} = 400 + 0.005 \times 1600 = 408$ V.
   This voltage is related to the dc side voltage by the converter modulation equation $\overline{v} = \frac{v_A}{2}m$ which means $m = \frac{2\overline{v}}{v_A} = \frac{2\times408}{1200} = 0.68$.
   The duty cycle is calculated from $d = \frac{m+1}{2} = \frac{1.68}{2} = 0.84$.
   The current ripples may be calculated by calculating the voltage across the inductor L during its rising (or falling) stage. When S_1 is ON, $\overline{v}$ is 600 V and 600 − 400 = 200 V is across the RL branch. This voltage is almost entirely across the inductor because the voltage across the resistor constitutes only a small portion of it.[2] This voltage is across the inductor for the whole duration

2 Notice that the reactance (or impedance) of the inductor at the switching frequency is $L\omega_s = 0.0007 \times 2\pi \times 1600 = 7$ Ω which is much larger than $R = 0.005$ Ω.

of the duty cycle: $dT_s = 0.84 \times 1600^{-1} = 0.000\,525$ seconds. Now, notice that for an inductor we have $v = L\frac{di}{dt}$ or $\Delta i = \frac{v\Delta t}{L}$. Therefore, we can say that the current ripple is approximately equal to $i_{\text{ripple}} = \frac{200\times0.000525}{700\times10^{-6}} = 150$ A.
This was calculated during the rising stage of the current. It may be calculated during the falling stage too: $\Delta t = (1-d)T_s = 0.16 \times 1600^{-1} = 0.0001$, $v = -600 - 400 = -1000$, $i_{\text{ripple}} = \frac{1000\times0.0001}{700\times10^{-6}} = 142.9$ A. Thus the value of about 150 A is a good approximation of the peak-to-peak current ripples. Note that the difference is due to the fact that we neglected the impact of R in the series connection of RL, for this calculation.

2. If the converter starts at $t = 0$ with this m, how long does it take to reach its steady-state operation?
   The time-constant is $\frac{L}{R} = \frac{0.7}{5} = 0.14$ seconds and the settling-time is about $4 \times 0.14 = 0.64$ seconds.
3. Assume that there is an unknown error of 1 mΩ in the value of R. In other words, the actual R is 6 mΩ. How much will be the exported power at the same operating condition of part 1?
   In this case, the current will be $\bar{i} = \frac{408-400}{0.006} = 1333.33$ A and the power will be $P = \bar{i}v_B = 1333.33 \times 400 = 533.3$ kW.
4. Assume that there is a disturbance (or uncertainty) in the system that causes v_B to change to 404 V (1% change). How much will be the exported power at the same operating condition of part 1?
   In this case, the current will be $\bar{i} = \frac{408-404}{0.005} = 800$ A and the power will be $P = \bar{i}v_B = 800 \times 404 = 323.2$ kW.
5. Assume that there is a disturbance (or uncertainty) in the system that causes v_A to change to 1212 V (1% increase). How much will be the exported power at the same operating condition of part 1?
   In this case, the converter voltage is $\bar{v} = 0.68 \times 606 = 412.08$ V and the current will be $\bar{i} = \frac{412.08-400}{0.005} = 2416$ A and the power will be $P = \bar{i}v_B = 2416 \times 400 = 966.4$ kW.
6. What would you conclude?!
   The system is very sensitive to uncertainties and disturbances. This is due to the fact that R is small and the pole of the transfer function is relatively close to the imaginary axis: $-\frac{R}{L} = -\frac{0.005}{0.0007} \approx -7$. In a system such as this, a feedback control is required to constantly observe the system conditions and adjust d accordingly to ensure that the desired level of power is being transferred at all conditions.

A simulation result is shown in Figure 2.32. In this study, first the 1% step jump to v_B is applied at $t = 0.7$ seconds. Then, this jump is cancelled and the 1% step jump to v_A is applied at $t = 1.4$ seconds.

2.6.4 Sensitivity Analysis and Role of Feedback

The numerical example above showed the sensitivity of the converter to system uncertainties. This is the same phenomenon that happened in the standard buck converter and was studied and resolved using feedback in Section 2.1. Assume that the steady-state average power delivered to side B is considered as the output:

$$y = P_B = \bar{i}v_B = \frac{0.5v_A v_B m - v_B^2}{R}.$$

Thus, we can compute the sensitivity of this output to various parameters as follows.

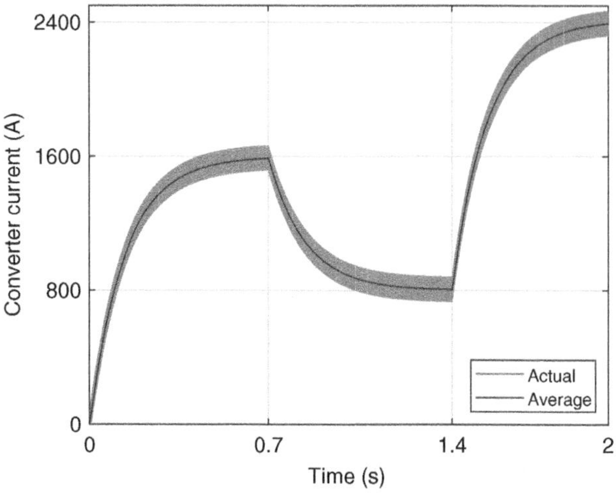

Figure 2.32 Responses of the HB-VSC to 1% change in v_B at $t = 0.07$ seconds and v_A at $t = 1.4$ seconds.

2.6.4.1 Sensitivity to *R*

$$S_R^y = \frac{R}{y}\frac{\partial y}{\partial R} = \frac{R}{\frac{0.5v_Av_Bm - v_B^2}{R}}\frac{-(0.5v_Av_Bm - v_B^2)}{R^2} = -1$$

$$S_{v_B}^y = \frac{v_B}{y}\frac{\partial y}{\partial v_B} = \frac{v_B}{\frac{0.5v_Av_Bm - v_B^2}{R}}\frac{0.5v_Am - 2v_B}{R} = \frac{0.5v_Am - 2v_B}{0.5v_Am - v_B}$$

For the values of $v_A = 1200$, $m = 0.68$, and $v_B = 400$, the sensitivity will be -49.

$$S_{v_A}^y = \frac{v_A}{y}\frac{\partial y}{\partial v_A} = \frac{v_A}{\frac{0.5v_Av_Bm - v_B^2}{R}}\frac{0.5v_Bm}{R} = \frac{0.5v_Am}{0.5v_Am - v_B}$$

For the values of $v_A = 1200$, $m = 0.68$, and $v_B = 400$, the sensitivity will be 51.

A feedback of the current, as shown in Figure 2.33, can substantially improve this situation. In Figure 2.33, the current $\bar{i}$ is expressed as

$$\bar{i} = \frac{0.5v_Am' - v_B}{Ls + R + 0.5kv_A} = \frac{0.5v_Am' - v_B}{Ls + R + R_v}. \tag{2.25}$$

Comparing this with the original equation of $\bar{i} = \frac{0.5v_Am - v_B}{Ls+R}$, we realize that this feedback is virtually increasing R to $R + R_v = R + 0.5kv_A$. Now, the denominator of the sensitivity functions with respect to v_A and v_B will change to $0.5v_Am' - v_B = 0.5v_Am - v_B + R_v\bar{i} = (R + R_v)\bar{i}$ which is increased from $R\bar{i}$. This means that the sensitivity will decrease.

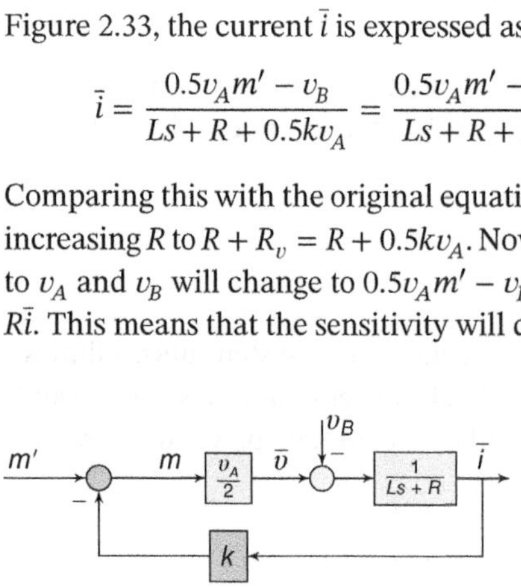

Figure 2.33 Feedback of current to virtually increase *R*.

The feedback loop strategy of Figure 2.33 pushes the pole of the system from the original point $-\frac{R}{L}$ to $-\frac{R+0.5kv_A}{L}$. As the pole moves further away from the imaginary axis to the left side, the system response becomes faster as well. Obviously, k must not be chosen excessively large so as not compromise other aspects of the system, such as its noise immunity.

Example 2.4 ***Single-Phase VSC with Feedback***

For Example 2.3, assume that a virtual resistance in terms of a current feedback as shown in Figure 2.33 is used to improve its robustness and transient responses. The value of k is chosen such that $R_v = 0.5kv_A$ is equal to $18R = 18 \times 0.005 = 0.09$. This means $k = \frac{18\times0.005}{600} = 0.00015$. Here we repeat the items in the example for this feedback-compensated system.

1. The VSC is initially planned to export a power of 640 kW. Find the values for the modulating signal m, the duty-cycle d, the modified modulating signal m', the (average) value of the transferred current, and the peak-to-peak current ripples.
   Average current is $\bar{i} = \frac{P}{v_B} = \frac{640000}{400} = 1600$ A.
   Average converter voltage is $\bar{v} = v_B + R\bar{i} = 400 + 0.005 \times 1600 = 408$ V.
   From $\bar{v} = \frac{v_A}{2}m$ we get $m = \frac{2\bar{v}}{v_A} = \frac{2\times408}{1200} = 0.68$.
   The duty cycle is calculated from $d = \frac{m+1}{2} = \frac{1.68}{2} = 0.84$.
   The current ripple is approximately equal to $i_{\text{ripple}} = \frac{200\times0.000525}{700\times10^{-6}} = 150$ A.
   The modified modulation index is $m' = m + k\bar{i} = 0.68 + 0.00015 \times 1600 = 0.92$.
2. If the converter starts at $t = 0$ with this m', how long does it take to reach its steady-state operation?
   The time-constant is $\frac{L}{R+R_v} = \frac{0.0007}{0.005\times19} = 0.0074$ seconds and the settling-time is about $4 \times 0.14 = 0.0295$ seconds which means 19 times faster than before this compensation.
3. Assume that there is an unknown error of 1 mΩ in the value of R. In other words, the actual R is 6 mΩ. How much will be the exported power at the same operating condition of part 1?
   The current in the steady state will be $\bar{i} = \frac{0.5v_Am'-v_B}{R+R_v} = \frac{600\times0.92-400}{0.006+0.09} = 1583.3$ A and the power will be $P = \bar{i}v_B = 1583.3 \times 400 = 633.3$ kW.
4. Assume that there is a disturbance (or uncertainty) in the system that causes v_B to change to 404 V. How much will be the exported power at the same operating condition of part 1?
   In this case, the current will be $\bar{i} = \frac{0.5v_Am'-v_B}{R+R_v} = \frac{600\times0.92-404}{0.005+0.09} = 1558$ A and the power will be $P = \bar{i}v_B = 1558 \times 404 = 629$ kW.
5. Assume that there is a disturbance (or uncertainty) in the system that causes v_A to change to 1212 V. How much will be the exported power at the same operating condition of part 1?
   In this case, the current will be $\bar{i} = \frac{0.5v_Am'-v_B}{R+R_v} = \frac{606\times0.92-400}{0.005+0.09} = 1658$ A and the power will be $P = \bar{i}v_B = 1658 \times 400 = 663$ kW.
6. What would you conclude?!
   The virtual resistance feedback has substantially improved the performance robustness to uncertainties as well as the transient response of the converter.
   A simulation result of the feedback compensated HB-VSC using virtual resistance concept is shown in Figure 2.34. In this study, first the 1% step jump to v_B is applied at $t = 0.7$ seconds. Then, this jump is cancelled and the 1% step jump to v_A is applied at $t = 1.4$ seconds. Compared with Figure 2.32, significant improvement in the transient response as well as robustness to parameter uncertainties/changes is observed.

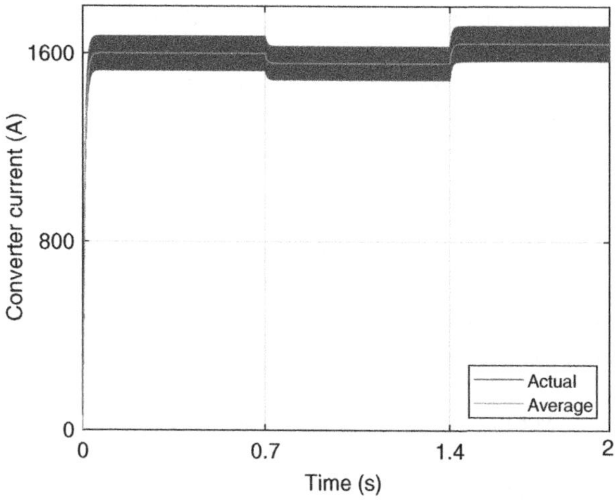

Figure 2.34 Responses of the feedback-compensated (using virtual resistance concept) HB-VSC to 1% change in v_B at $t = 0.07$ seconds and v_A at $t = 1.4$ seconds.

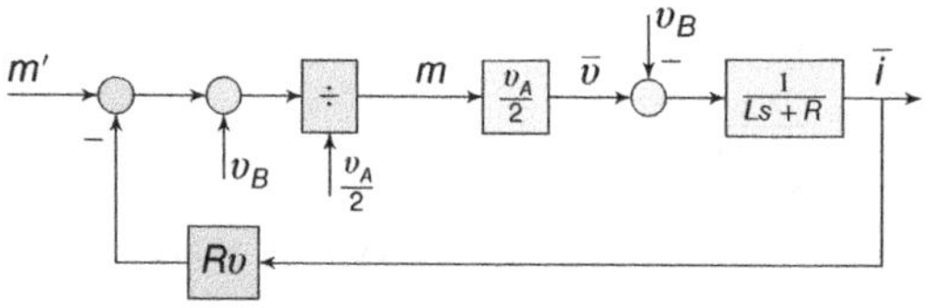

Figure 2.35 Virtual resistance feedback combined with input feedback linearization.

The virtual resistance technique significantly improves the transient response and the performance robustness of the converter. It is also possible to completely cancel the impact of input and output voltages, v_A and v_B, on the system responses by measuring them in real-time and compensating them when calculating the modulation index. This approach is shown in Figure 2.35 and may be called input feedback linearization. The division block cancels the impact of v_A and the summation block cancels the impact of v_B. The systems transfer function can be written as

$$\frac{\bar{i}}{m'} = \frac{1}{Ls + R + R_v}. \tag{2.26}$$

Note that in Figure 2.35, $m' = m\frac{v_A}{2} - v_B + R_v\bar{i}$. Therefore, for example, for the numerical values of the previous example, it will be equal to $m' = 0.68 \times 600 - 400 + 0.09 \times 1600 = 152$.

To fully obtain a robust control system, another outer feedback loop on the current is used, as shown in Figure 2.36. Here the compensator $C(s)$ must be properly structured and designed to achieve desired control specifications. For example, in order to completely remove the steady-state

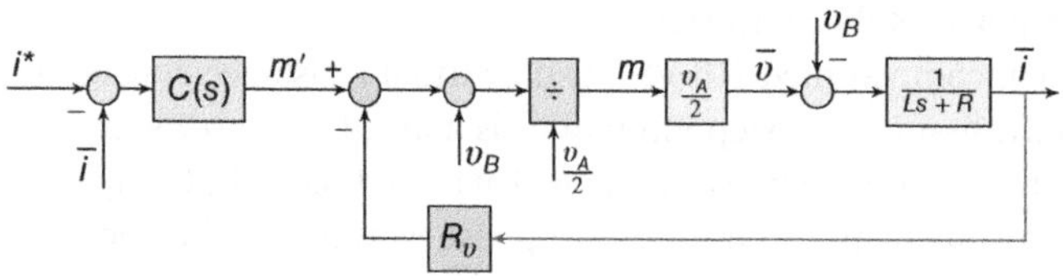

Figure 2.36 Full feedback current control structure.

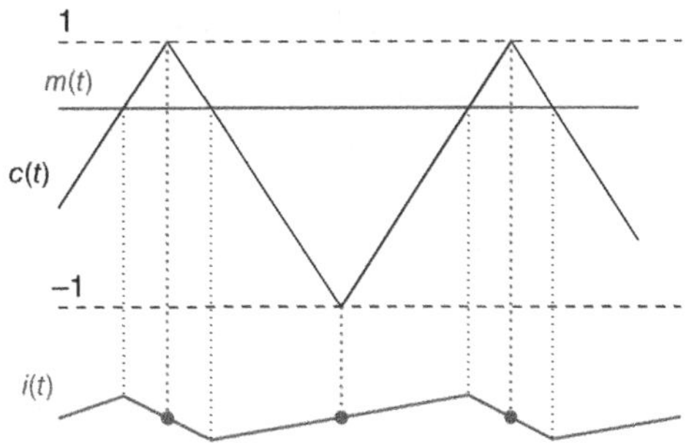

Figure 2.37 Synchronized sampling of current to provide a smooth waveform for feedback.

errors to step commands in i^*, the compensator $C(s)$ must have an integrating function. Optimal design of such compensators as well as R_v will be discussed in Chapters 4 and 5 of the book.

2.6.5 Synchronized Sampling

The converter current waveform has ripples due to discontinuous voltage $v(t)$. Therefore, directly feeding it to the feedback loop can add to the system noise and compromise the quality of responses. To mitigate this issue, and at any rate because digital control approach is widely used, the synchronized sampling technique may be used. As shown in Figure 2.37, the current waveform is synchronously sampled at the peaks of the triangular carrier waveform. This way, the samples represent the average of current and the switching ripple is filtered. As observed from Figure 2.37, the sampling frequency will be twice the switching frequency. Notice that conventional filtering techniques are not effective in this situation because they may cause too much delay to the current samples.

2.7 Full-Bridge VSC

Single-phase full-bridge voltage source converter (FB-VSC) comprises four switch cells S_1 to S_4 as shown in Figure 2.38. It interfaces the dc side (or side A) with voltage v_A to side B shown by v_B (which can be dc, ac, source, or load). The resistor R models the cable's resistance, the inductor's parasitic resistance, and the conducting resistance of switches and it is very small.

2.7.1 Bipolar PWM Operation

The bipolar PWM technique adopts $s_1(t) = s_4(t)$, $s_2(t) = s_3(t)$ and $s_1(t) + s_2(t) = 1$ as shown in Figure 2.39. In other words, the switches 1 and 4 are turned on and off simultaneously; and the

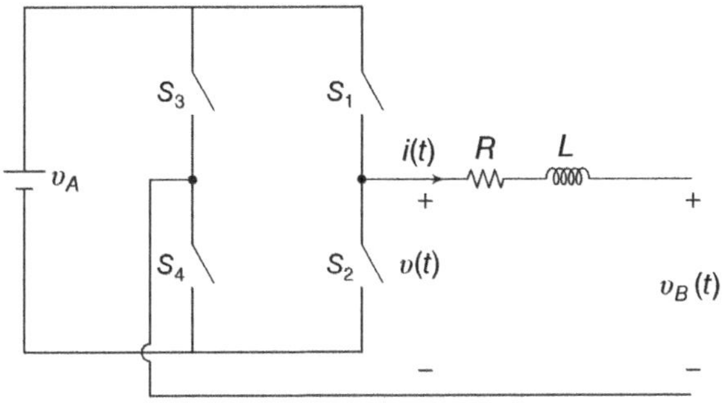

Figure 2.38 Full-bridge voltage source converter.

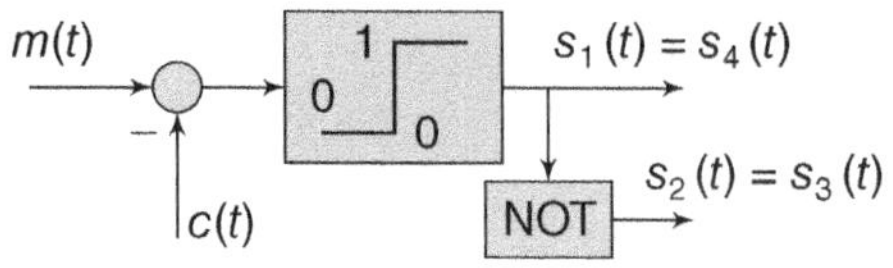

Figure 2.39 Bipolar PWM for FB-VSC.

switches 2 and 3 as well. Moreover, when the switches 1 and 4 are ON, the switches 2 and 3 are OFF, and vice versa.

When the current $i(t)$ is positive, when $s_1 = s_4 = 1$, switches S_1 and S_4 conduct (through their transistors Q_1 and Q_4) and provide the path for the current. Therefore, the voltage $v(t)$ is v_A. As soon as $s_1 = s_4$ falls to zero, S_1 and S_4 turn off and the reverse diodes of S_2 and S_3 (i.e. D_2 and D_3) turn on and provide a path for the current. During this period, the voltage $v(t)$ is $-v_A$. When the current $i(t)$ is negative, when $s_2 = s_3 = 1$, switches S_2 and S_3 conduct (through their transistors) and provide the path for the current while S_1 and S_4 are off. As soon as $s_2 = s_3$ falls to zero, S_2 and S_3 turn off and the reverse diodes of S_1 and S_4 turn on and provide a path for the current. Thus, the same pattern of Figure 2.29 is valid with the difference that $0.5v_A$ is now increased to v_A.

The above discussion concludes that the bipolar-PWM FB-VSC has the voltage $v(t)$ as

$$v(t) = v_A s_1(t) - v_A s_2(t) = v_A[s_1(t) - s_2(t)] = v_A[2s_1(t) - 1]. \tag{2.27}$$

Therefore,

$$\bar{v} = v_A(\bar{s}_1 - \bar{s}_2) = (2d - 1)v_A = mv_A. \tag{2.28}$$

Equation (2.28) describes the average model of the FB-VSC from the modulating signal $m(t)$ to the voltage $v(t)$. The FB-VSC acts as a simple gain v_A. Finally, the FB-VSC circuit of Figure 2.38 can be represented by the transfer function model shown in Figure 2.40. This is used for control analysis, design and control system simulations.

2.7.2 Unipolar PWM Operation

The unipolar PWM technique uses *zero states* where are those switching states that result in a zero output voltage. When S_1 and S_3 (or S_2 and S_4) are simultaneously ON, the output voltage $v(t)$ is zero. The PWM technique to generate the switching functions for this method is shown in Figure 2.41. Notice that in this method, similar to bipolar, the switches on one leg are turned on and off opposite to each other to avoid a short circuit. The following analysis shows that the output voltage $v(t)$ switches between 0 and V_{dc} (or between 0 and $-V_{dc}$) and that is why it is called unipolar.

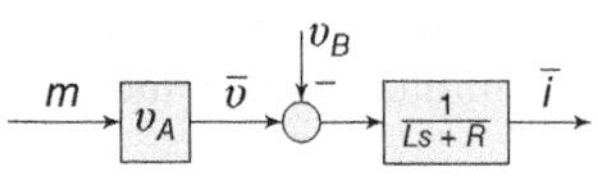

Figure 2.40 Average or control model of the FB-VSC.

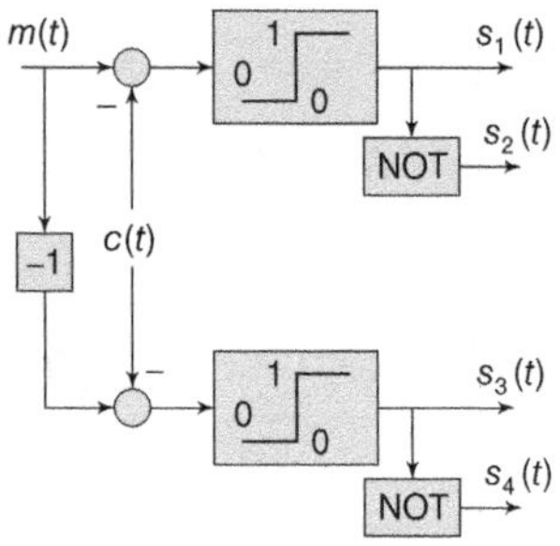

Figure 2.41 Unipolar PWM for FB-VSC.

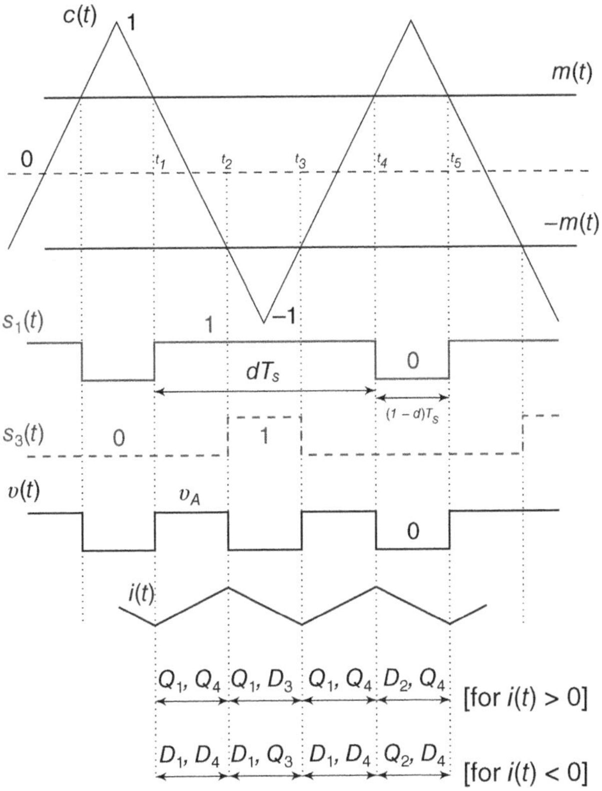

Figure 2.42 Unipolar PWM and illustration of waveforms of FB-VSC for $m(t) > 0$.

Switching functions and some of the signals of the unipolar PWM for FB-VSC are shown in Figure 2.42. The functions $s_1(t)$ and $s_3(t)$ are directly obtained by comparing $m(t)$ and $-m(t)$ with $c(t)$, respectively. Then, $s_2(t)$ and $s_4(t)$ are simply NOT of $s_1(t)$ and $s_3(t)$, respectively. Four operating conditions are observed in one full switching cycle (t_1, t_5) from Figure 2.42: (i) When $m > c$ but $-m < c$, i.e. $t \in (t_1, t_2)$, then $s_1 = 1$ and $s_4 = 1$, switches S_1 and S_4 conduct and provide the path for the current. Therefore, the voltage $v(t)$ is v_A. (ii) At t_2, as soon as $-m > c$, s_3 rises to one, S_4 turns off and D_3 turns on to provide a path for the current. During this period, the voltage $v(t)$ is zero. (iii) At t_3, as soon as $-m < c$, s_3 falls to zero, we go back to condition (i). (iv) At t_4, as soon as $m < c$, s_1 falls to zero, D_2 turns on to provide a path for the current. During this period, the voltage $v(t)$ is zero. The zero states occur during (t_2, t_3), where $m > c$ and $-m > c$, and during (t_4, t_5), where $m < c$ and $-m < c$. This whole discussion is with the assumption that $i(t)$ is positive (flowing from side A to side B). A similar discussion may be done when the current $i(t)$ is negative.

Remark 2.1 ***doubled effective switching frequency*** It is observed from examining the signals waveform of unipolar PWM of Figure 2.42 and comparing with the bipolar PWM of Figure 2.29 that in every switching cycle of unipolar PWM, the output voltage $v(t)$ experiences twice as many switchings than the bipolar PWM. Therefore, from the voltage and current waveform perspectives, the switching frequency is effectively doubled in the unipolar PWM. This is a desirable effect which allows selection of a lower value for the inductance L for a given level of current ripples.

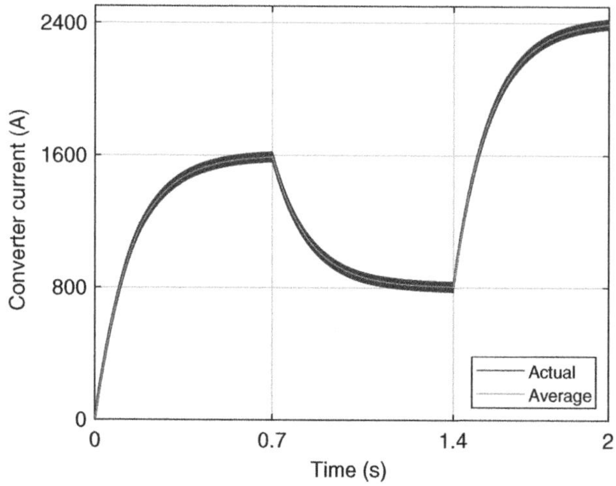

Figure 2.43 Responses of the (unipolar PWM) FB-VSC to 1% change in v_B at $t = 0.07$ seconds and v_A at $t = 1.4$ seconds.

Remark 2.2 ***core losses*** The ac voltage across the inductor is halved in unipolar PWM comparing with the bipolar PWM. This means that the core loss in the inductor will become smaller as it depends on the magnitude of voltage to the power of $\alpha > 1$ despite the fact that the ac frequency is doubled. The number of switchings have not practically increased which implies the overall similar switching losses of the two PWMs.

The above discussion concludes that the FB-VSC voltage $v(t)$ can be written as

$$v(t) = v_A s_1(t) - v_A s_3(t) = v_A[s_1(t) - s_3(t)]. \tag{2.29}$$

Now, the average of these two switching functions is $\bar{s}_1 = d$ and $\bar{s}_3 = 1 - d$. We also notice that when m varies from 0 to 1, d varies from 0.5 to 1. Since m and d are linearly related, we can write $d = \frac{m}{2} + \frac{1}{2}$. Therefore,

$$\bar{v} = v_A(\bar{s}_1 - \bar{s}_3) = (2d - 1)v_A = mv_A. \tag{2.30}$$

Equation (2.30) describes the average model of the FB-VSC from the modulating signal $m(t)$ to the voltage $v(t)$. This model is valid for both unipolar and bipolar PWMs. The FB-VSC acts as a simple gain v_A in both cases. Finally, the same transfer function model shown in Figure 2.40 is valid for unipolar PWM converter.

Example 2.5 ***Full-Bridge VSC***
A simulation result of the FB-VSC with unipolar PWM is shown in Figure 2.43 where the same simulation scenario of Figure 2.32 is considered. It is observed that due to doubling of effective switching frequency, the current ripples are reduced.

2.8 Three-Phase VSC

The three-phase voltage source converter (3Ph-VSC) comprises six switch cells S_1 to S_6 as shown in Figure 2.44 (three single-phase HB-VSCs connected in parallel). It interfaces the dc side with voltage V_{dc} to the the ac side shown by $v_s(t)$ (which can be a three-phase ac source, load, or a

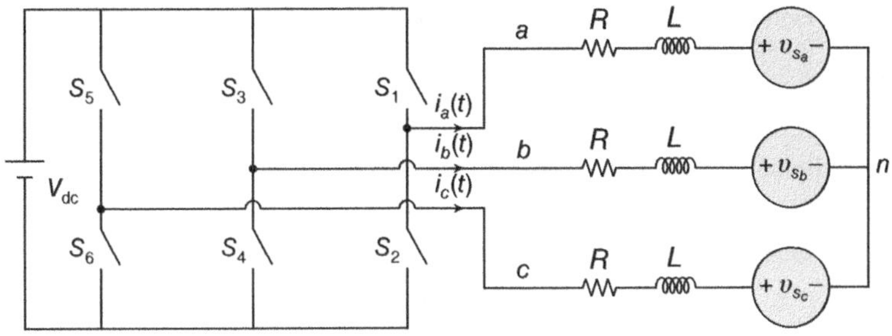

Figure 2.44 Three-phase voltage source converter.

general three-phase ac network). It is a bidirectional converter. The interface filter is shown as a simple L filter but other variations are also possible. The resistor R models the cable's resistance, the inductor's parasitic resistance and the conducting resistance of switches and it is very small.

2.8.1 Modeling in Stationary Domain

Assume that v_a, v_b, and v_c are the converter output voltages with respect to a reference point. In practice, often the midpoint of the dc supply is taken as the reference point. Therefore, the system is described by

$$\begin{aligned} L\frac{d}{dt}i_a(t) &= v_a(t) - Ri_a(t) - v_{S_a}(t) - v_n(t) \\ L\frac{d}{dt}i_b(t) &= v_b(t) - Ri_b(t) - v_{S_b}(t) - v_n(t) \\ L\frac{d}{dt}i_c(t) &= v_c(t) - Ri_c(t) - v_{S_c}(t) - v_n(t). \end{aligned} \tag{2.31}$$

Since the system is three-wire, the sum of all currents is zero at every time: $i_a(t) + i_b(t) + i_c(t) = 0$. Also, if we assume that the voltage v_s is balanced, it implies $v_{S_a}(t) + v_{S_b}(t) + v_{S_c}(t) = 0$. These two facts together with (2.31) results in

$$v_n(t) = \frac{v_a(t) + v_b(t) + v_c(t)}{3}.$$

If the converter generates a balanced voltage (at the average level, i.e. $\overline{v}_a(t) + \overline{v}_b(t) + \overline{v}_c(t) = 0$) then the voltage v_n will be zero (at the average level, $\overline{v}_n = 0$). Therefore, at the average level, (2.31) reduces to the equations pertaining to three separated single-phase VSCs.

The bipolar (sinusoidal) PWM technique is shown in Figure 2.45 for the leg-a of the converter where m_a is the modulating signal and $c(t)$ is the carrier signal. With this PWM, the 3Ph-VSC is simply equivalent to three single-phase HB-VSCs as shown in Figure 2.46.

Figure 2.45 Bipolar PWM for each leg of three-phase VSC.

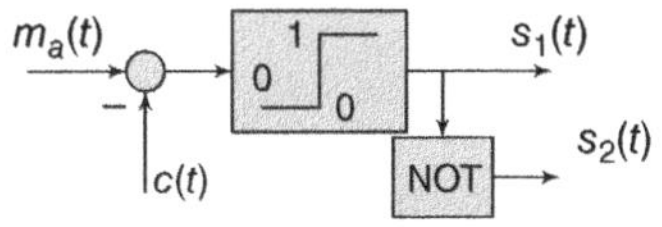

Figure 2.46 Per-phase average or control model of the 3Ph-VSC (phase-a is shown).

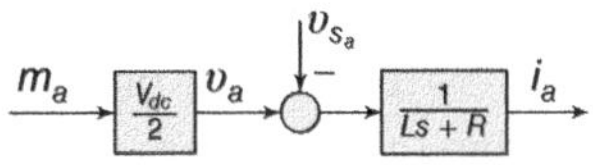

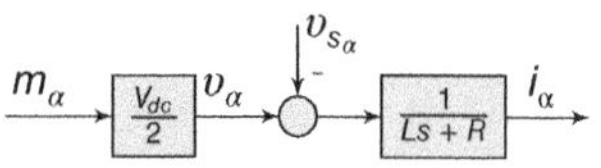

Figure 2.47 Per-phase average or control model of the 3Ph-VSC in $\alpha\beta$ (or stationary) domain.

The model can further be reduced to two single-phase models using

$$x_{\alpha\beta} = \frac{2}{3}\left(x_a + x_b e^{j\frac{2\pi}{3}} + x_c e^{-j\frac{2\pi}{3}}\right) \tag{2.32}$$

called the *space phasor*. For a three-phase balanced waveform $x_a = X\cos(\omega t + \delta)$, $x_b = X\cos\left(\omega t + \delta - \frac{2\pi}{3}\right)$, $x_c = X\cos\left(\omega t + \delta + \frac{2\pi}{3}\right)$, the space phasor is

$$x_{\alpha\beta} = x_\alpha + jx_\beta = X\cos(\omega t + \delta) + jX\sin(\omega t + \delta) = Xe^{j(\omega t+\delta)},$$

which is a rotating vector in complex plane. Since this is an algebraic linear transformation, (2.31) implies that

$$L\frac{d}{dt}i_{\alpha\beta}(t) = v_{\alpha\beta}(t) - Ri_{\alpha\beta}(t) - v_{s_{\alpha\beta}}(t). \tag{2.33}$$

The equivalent circuit can then be derived in $\alpha\beta$ domain as shown in Figure 2.47 for α signal. The circuit for β signal is also similar and decoupled from the α circuit. So, here the system is modeled by two decoupled single-input single-output (SISO) systems.

2.8.2 Modeling in Rotating Synchronous Frame

The rotating synchronous frame model can be achieved by defining the rotating synchronous reference frame variables as

$$x_{dq} = e^{-j\theta}x_{\alpha\beta}. \tag{2.34}$$

Obviously, if $\theta = \int \omega dt = \omega t$, where ω is the instantaneous frequency of the signal x, then x_{dq} will be a dc signal for a balanced positively-sequenced x. In particular, if θ is exactly equal to the angle of x, then x_{dq} will be a real dc signal, i.e. $x_q = 0$. Differentiating (2.34) yields to

$$\dot{x}_{dq} = -j\omega x_{dq} + e^{-j\theta}\dot{x}_{\alpha\beta}. \tag{2.35}$$

Multiplying into L and substituting from (7.43) results in

$$\begin{aligned} L\frac{d}{dt}i_{dq} &= -j\omega Li_{dq} + e^{-j\theta}L\frac{d}{dt}i_{\alpha\beta} \\ &= -j\omega Li_{dq} + e^{-j\theta}[v_{\alpha\beta}(t) - Ri_{\alpha\beta}(t) - v_{s\alpha\beta}(t)] \\ &= -j\omega Li_{dq} + v_{dq} - Ri_{dq} - v_{sdq}, \end{aligned} \tag{2.36}$$

which simplifies to the following two real equations

$$\begin{aligned} L\frac{d}{dt}i_d &= -Ri_d + L\omega i_q + v_d - v_{s_d} \\ L\frac{d}{dt}i_q &= -Ri_q - L\omega i_d + v_q - v_{s_q}. \end{aligned} \tag{2.37}$$

These two equations summarize the *dq*-frame or rotating synchronous frame control model of the three-phase VSC which is shown in Figure 2.48. Notice that the *dq*-frame model represents a two-input two-output model similar to $\alpha\beta$ frame. However the *dq*-frame model is coupled due to those $L\omega$ terms while the $\alpha\beta$-frame model is decoupled. The *dq*-frame model is popular due to the fact that the signals are of dc nature. A phase-locked loop (PLL) may be necessary to measure θ.

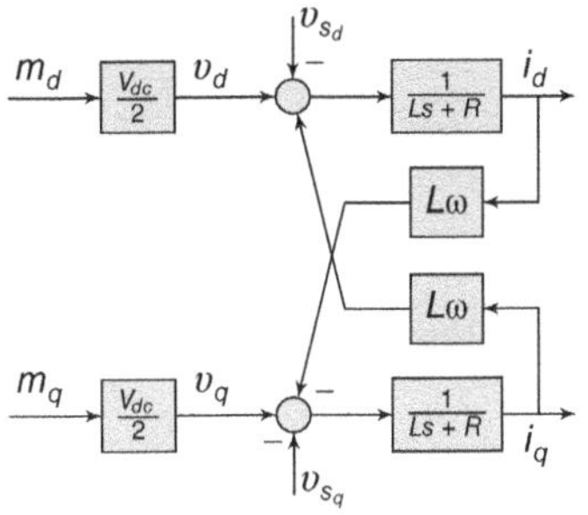

Figure 2.48 Average or control model of 3Ph-VSC in *dq* (or synchronous) domain.

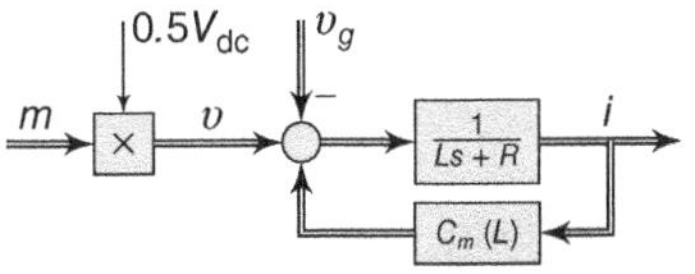

Figure 2.49 Control block diagram of VSC: valid for both stationary and rotating frame.

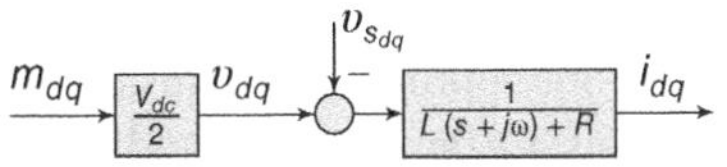

Figure 2.50 Compact presentation of the average or control model of 3Ph-VSC in *dq* (or synchronous) domain using complex transfer function.

Both the stationary frame model of Figure 2.47 and the rotating frame model of Figure 2.48 can be combined ad shown in one format, as depicted in Figure 2.49. This diagram is valid for both $\alpha\beta$ and dq frames with the note that the coupling matrix C_m is zero in $\alpha\beta$ and is equal to $C_m(L) = \begin{bmatrix} 0 & L\omega \\ -L\omega & 0 \end{bmatrix}$ in dq domain.

2.8.3 Compact Modeling Using Complex Transfer Functions*

It is possible to present the two channel equations of (2.37) and Figure 2.48 in a compact for using transfer functions with complex coefficients. This approach is useful for some control system analysis and design purposes. Taking Laplace transform from (2.36) leads to

$$I_{dq}(s) = \frac{V_{dq}(s) - V_{S_{dq}}(s)}{L(s + j\omega) + R}. \tag{2.38}$$

The corresponding block diagram is shown in Figure 2.50. In this model, every variable x_{dq} is defined as $x_{dq} = x_d + jx_q$. Therefore, the two control channels d and q are combined into one channel using a transfer function $\frac{1}{L(s+j\omega)+R}$ which has a complex coefficient, i.e. $j\omega L$. Notice that this diagram indeed is equivalent to two coupled loops of transfer functions with real coefficients, i.e. the model shown in Figure 2.48 or Figure 2.49.

2.9 Modeling of Converter Delays

In deriving all the modeling approaches discussed in this chapter, the averaging over one switching cycle is used. Due to the discrete nature of the PWM technique, it has a limited bandwidth and it introduces a delay which may be roughly approximated as about half a switching cycle. The digital implementation of the controller on a digital signal processor will also cause one sampling delay. So, it is customary to consider the presence of an overall delay of 1.5 switching cycle in the

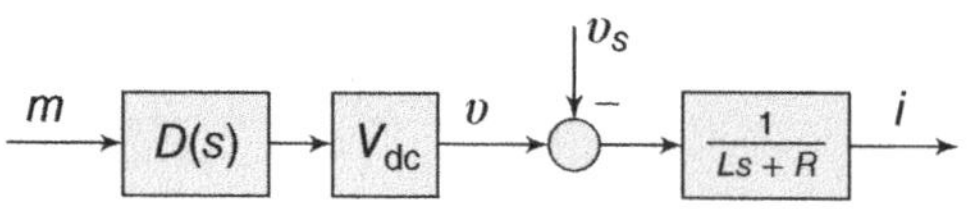

Figure 2.51 Full-bridge VSC model with the loop delays.

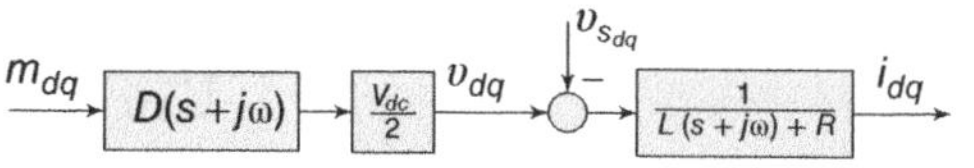

Figure 2.52 Compact presentation of 3Ph-VSC in *dq* (or synchronous) domain using complex transfer function with the model of loop delays.

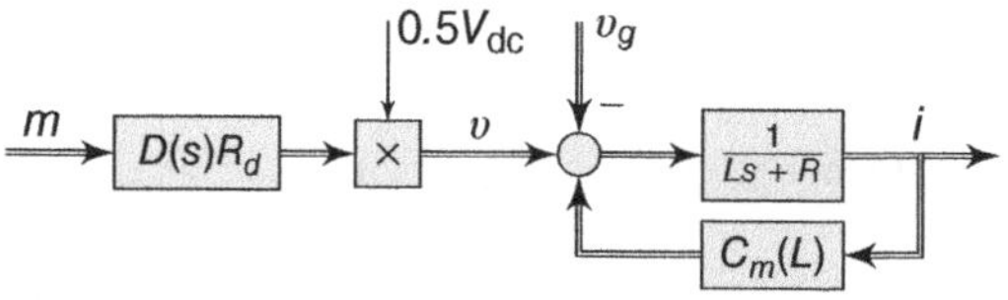

Figure 2.53 Control block diagram of 3Ph-VSC with the delay modeling (real multivariable transfer functions).

control loop. If the switching frequency is high enough, compared with the desired bandwidth of the control loop, this delay will not cause a problem even if it is not considered during the design stage. However, for small switching frequencies, and/or when the control system bandwidth is large, one approach is to consider this delay during the design stage to ensure that sufficient stability margins are obtained to handle the delay.

Figure 2.51 shows how the control loop delays may be modeled in a full-bridge VSC (or a buck converter). Here, $D(s)$ is $e^{-T_d s}$ where T_d is the amount of delay. This function may be approximated with $\frac{1}{T_d s+1}$ or $\frac{0.5T_d s-1}{0.5T_d s+1}$ or some higher order Pade approximation of the delay transfer function.

Figure 2.52 shows the same concept for the three-phase VSC in rotating frame [1, 2]. This is the compact representation using complex transfer functions.

For a non-compact presentation, i.e. with matrix transfer functions, the matrix transfer function $D(s)R_d$, where R_d is the rotation matrix with angle ωT_d, models the delay [2], as shown in Figure 2.53.

2.10 Summary and Conclusion

This chapter reviewed the basic standard power electronic converters including the buck, boost, buck-boost, and the VSC concept. Single-phase and three-phase VSCs are studied and their mathematical models are derived. The average models are used in Chapters 5–7 to develop mathematical models for DER control systems and to carry our analysis and design of such systems.

Exercises

2.1 Redraw Figure 2.27 and show that if the switch cell S_1 is replaced with a unidirectional switch (without reverse conduction) and switch cell S_2 is replaced by a diode, (and lower dc supply is merged into the upper one), the HB-VSC converter reduces to the popular buck converter.

2.2 Derive (2.25) from the feedback loop structure of Figure 2.33.

2.3 Using the definition of sensitivity function, (2.6), prove that S_L^y is zero in the steady-state where $y = P_B$ is the power delivered by a VSC to side B.

2.4 Compute the sensitivity functions S_R^y, $S_{v_A}^y$, and $S_{v_B}^y$ for $y = P_B$ in the feedback-compensated system of Figure 2.33. Compare them with those of the non-compensated system, Figure 2.31.

2.5 Explain advantages of using a unipolar PWM as opposed to a bipolar PWM in a single-phase full-bridge VSC.

2.6 Explain what the zero states are in a single-phase full-bridge VSC.

2.7 Compare the models of a three-phase VSC in stationary and synchronous domains. Note that the dq variables have a dc nature while the $\alpha\beta$ variables are ac.

Problems

2.1 The steady-state approximate equations (2.5), (2.11), (2.16), and (2.20) are derived in this chapter based on averaging the algebraic and differential equations and also putting the time-derivative terms to zero in the steady state. Prove that each one of these equations can also be derived based on the following reasoning: the average of inductor voltage over one full switching cycle must be zero.

2.2 Consider the single-phase pulse-width modulation half-bridge voltage-source converter (HB-VSC) with dc-side voltage of V_{dc}. It is connected to another dc supply of V_s through an inductor L. The inherent resistors of the inductor, of the line, and of the conducting-state of switches are aggregated and denoted by R. Consider the following numerical values:
$V_{dc} = 1200$ V, $V_s = 400$ V, $L = 700$ μH, $R = 0.005$ Ω, $f_{sw} = 1600$ Hz.

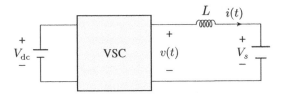

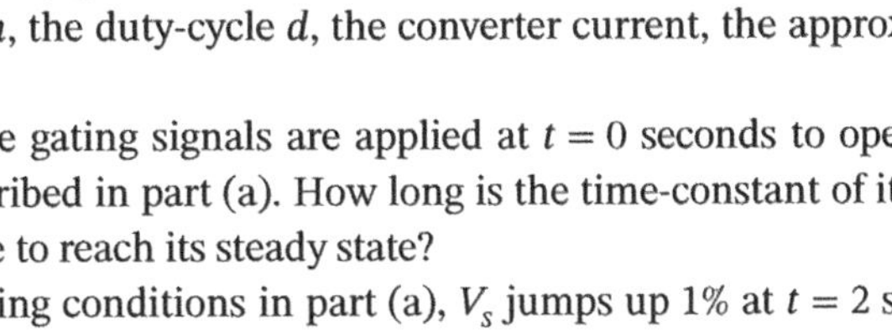

(a) The VSC is initially planned to export a power of 640 kW. Find the values for the modulating signal m, the duty-cycle d, the converter current, the approximate peak-to-peak current ripple.
(b) Assume that the gating signals are applied at $t = 0$ seconds to operate the VSC at the conditions described in part (a). How long is the time-constant of its response and how long does it take to reach its steady state?
(c) While at operating conditions in part (a), V_s jumps up 1% at $t = 2$ seconds. What is the new steady state values of the power and current? Discuss!
(d) While at operating conditions in part (b), V_{dc} jumps up 1% at $t = 3$ seconds. What is the new steady state values of the power and current?
(e) Build a simulation file that comprises two different models: average (or control) model using transfer functions; real model using electric circuit components (switches, diodes, etc.). Confirm your answers to (a)–(d) by showing the responses of both systems.
(f) Repeat items (a)–(d) if a full-bridge VSC with $V_{dc} = 600$ V is used (other parameters unchanged). Consider two cases of bipolar and unipolar PWM.

(g) Investigate the sensitivity of the VSC operation to other circuit components such as L and R.

(h) Use a current feedback structure as shown in Figure 2.33. Pick a number for R_v for example $R_v = 10R = 0.05$ and then find k. Repeat all the items above for this new system and explain how the system responses improve.

2.3 *(Clark Transformation)* For the space phasor definition (2.32), prove the following facts.

(a) It can be represented in the matrix form as

$$\begin{bmatrix} x_\alpha \\ x_\beta \end{bmatrix} = \frac{2}{3}\begin{bmatrix} 1 & -\frac{1}{2} & -\frac{1}{2} \\ 0 & \frac{\sqrt{3}}{2} & -\frac{\sqrt{3}}{2} \end{bmatrix}\begin{bmatrix} x_a \\ x_b \\ x_c \end{bmatrix}. \tag{2.39}$$

(b) If we further define the zero component as $x_o = \frac{1}{3}(x_a + x_b + x_c)$, and also for simplicity of notations, define

$$x_{abc} = \begin{bmatrix} x_a & x_b & x_c \end{bmatrix}^T, \quad x_{\alpha\beta o} = \begin{bmatrix} x_\alpha & x_\beta & x_o \end{bmatrix}^T,$$

then the abc to $\alpha\beta o$ transformation is defined as $x_{\alpha\beta o} = T_{abc}^{\alpha\beta o} x_{abc}$ where $T_{abc}^{\alpha\beta o}$ is a 3×3 matrix. Prove that

$$T_{abc}^{\alpha\beta o} = \frac{2}{3}\begin{bmatrix} 1 & -\frac{1}{2} & -\frac{1}{2} \\ 0 & \frac{\sqrt{3}}{2} & -\frac{\sqrt{3}}{2} \\ \frac{1}{2} & \frac{1}{2} & \frac{1}{2} \end{bmatrix}. \tag{2.40}$$

(c) The $\alpha\beta o$ to abc transformation is defined by $x_{abc} = T_{\alpha\beta o}^{abc} x_{\alpha\beta o}$. Show that

$$T_{\alpha\beta o}^{abc} = \begin{bmatrix} 1 & 0 & 1 \\ -\frac{1}{2} & \frac{\sqrt{3}}{2} & 1 \\ -\frac{1}{2} & -\frac{\sqrt{3}}{2} & 1 \end{bmatrix}. \tag{2.41}$$

(d) If the instantaneous power (in the natural abc domain) is $p(t) = v_a i_a + v_b i_b + v_c i_c$, it is equal to $p(t) = \frac{3}{2}(v_\alpha i_\alpha + v_\beta i_\beta + 2v_o i_o)$ which will be equal to $p(t) = \frac{3}{2}(v_\alpha i_\alpha + v_\beta i_\beta)$ in a three-wire system (i.e. a three-phase system without neutral connection) where no zero-sequence current flows, i.e. $i_o = \frac{1}{3}(i_a + i_b + i_c) = 0$.

(e) If we want to have a *power-invariant transformation* (that is $p(t) = v_\alpha i_\alpha + v_\beta i_\beta$), it is obvious that we must change $\frac{2}{3}$ (in the definition of space phasor) to $\sqrt{\frac{2}{3}}$. Show how the above matrices change.

2.4 *(Park Transformation)* For the synchronous reference frame vector defined by (2.34), prove the following facts.

(a) It can be represented in the matrix form as

$$\begin{bmatrix} x_d \\ x_q \end{bmatrix} = \begin{bmatrix} \cos\theta & \sin\theta \\ -\sin\theta & \cos\theta \end{bmatrix}\begin{bmatrix} x_\alpha \\ x_\beta \end{bmatrix}. \tag{2.42}$$

(b) The *abc* to *dqo* transformation is defined as $x_{dqo} = T_{abc}^{dqo} x_{abc}$ where

$$T_{abc}^{dqo} = \frac{2}{3}\begin{bmatrix} \cos\theta & \cos(\theta - \frac{2\pi}{3}) & \cos(\theta + \frac{2\pi}{3}) \\ -\sin\theta & -\sin(\theta - \frac{2\pi}{3}) & -\sin(\theta + \frac{2\pi}{3}) \\ \frac{1}{2} & \frac{1}{2} & \frac{1}{2} \end{bmatrix}. \tag{2.43}$$

(c) The *dqo* to *abc* transformation is defined as $x_{abc} = T_{dqo}^{abc} x_{dqo}$ where

$$T_{dqo}^{abc} = \begin{bmatrix} \cos\theta & -\sin\theta & 1 \\ \cos\left(\theta - \frac{2\pi}{3}\right) & -\sin\left(\theta - \frac{2\pi}{3}\right) & 1 \\ \cos\left(\theta + \frac{2\pi}{3}\right) & -\sin\left(\theta + \frac{2\pi}{3}\right) & 1 \end{bmatrix}. \tag{2.44}$$

(d) If $x_a = X\cos(\omega t)$, $x_b = X\cos(\omega t - \frac{2\pi}{3})$, and $x_c = X\cos(\omega t + \frac{2\pi}{3})$, then

$$x_d = X\cos(\omega t - \theta), \quad x_q = X\sin(\omega t - \theta). \tag{2.45}$$

(e) If the instantaneous power (in natural abc domain) is $p(t) = v_a i_a + v_b i_b + v_c i_c$, it is equal to $p(t) = \frac{3}{2}(v_d i_d + v_q i_q + 2v_o i_o)$ and will become equal to $p(t) = \frac{3}{2}(v_d i_d + v_q i_q)$ in a three-wire system where no zero-sequence current flows, i.e. $i_o = \frac{1}{3}(i_a + i_b + i_c) = 0$.

(f) If we want to have a *power-invariant transformation* [that is $p(t) = v_d i_d + v_q i_q$], it is obvious that we must change $\frac{2}{3}$ (in the definition of space phasor) to $\sqrt{\frac{2}{3}}$. Show how the above matrices change.

(g) Equation (2.45) shows that if the angle of transformation, i.e. θ, is chosen to be equal to the angle of the signal, i.e. $\theta = \omega t$, then $x_d = X$ and $x_q = 0$. In some literature, e.g. literature on motor control, people would prefer to have $x_d = 0$ and $x_q = X$. What should be the angle of transformation to yield this?

2.5 Prove the equivalence of Figures 2.52 and 2.53 where $D(s) = e^{-T_d s}$ and

$$R_d = \begin{bmatrix} \cos(\omega T_d) & -\sin(\omega T_d) \\ \sin(\omega T_d) & \cos(\omega T_d) \end{bmatrix}.$$

References

1 Javier Samanes, Andoni Urtasun, Ernesto L Barrios, David Lumbreras, Jesús López, Eugenio Gubia, and Pablo Sanchis. Control design and stability analysis of power converters: the MIMO generalized bode criterion. *IEEE Journal of Emerging and Selected Topics in Power Electronics*, 8(2):1880–1893, 2019.

2 Francisco D Freijedo, Ana Vidal, Alejandro G Yepes, Josep M Guerrero, Óscar López, Jano Malvar, and Jesús Doval-Gandoy. Tuning of synchronous-frame PI current controllers in grid-connected converters operating at a low sampling rate by MIMO root locus. *IEEE Transactions on Industrial Electronics*, 62(8):5006–5017, 2015.

Part II

Feedback Control Systems

Power electronic converters require robust and strong feedback control systems to enable them to operate as desirable interfaces for distributed energy resource (DER) applications. Such control systems are generally designed based on either the conventional frequency-domain approach or the state-space models or a combination of both. Part II of this text presents a review of fundamentals related to feedback control systems and theories. Chapter 3 reviews the more classic approach that is based on Laplace transform and transfer function concepts. Chapter 4 is devoted to the approach based on state-space concept. The optimal control approach has been given a particular attention. Both chapters also contain some common and useful Matlab commands for analysis and design of linear control systems. The optimal control approach is used as the preferred method throughout this textbook.

Modeling and Control of Modern Electrical Energy Systems, First Edition. Masoud Karimi-Ghartemani.

3

Frequency-Domain (Transfer Function) Approach

This chapter presents a review of the classical feedback control approach that is based on the concepts of Laplace transform and transfer functions (TFs). The fundamental concepts pertaining to this approach including the definitions of stability, poles and zeros, and transfer functions are explained. The main analysis and design tools such as root-locus method and Bode diagrams are reviewed. The stability margins in this domain such as the phase margin, gain margin, and delay margin are explained. The popular PID controller is introduced and its properties are explained.

3.1 Key Concepts

A *system* (plant or process) is a combination of components that are connected and work together through cause–effect relationships.

An *output* in this system is a variable that is of interest to us in the sense that we want it to perform or behave in a particular desired way.

An *input* or control input is a variable that impacts the output variable and it can be manipulated to change (or control) the output. This should not be confused with the set-point (or command, or reference).

For example, when we want to keep the temperature of a room close to a desired value using a heater with thermostat; the entire room, with the heater is the system. The actual room temperature is the output. The heat energy supplied by the heater is the input. And the desired temperature set on the thermostat panel is the reference, or set-point, or the desired output.

3.1.1 Transfer Function

Transfer function block diagram representation of a linear time invariant (LTI) system is shown in Figure 3.1 where $G(s)$ is the *transfer function* (TF) and is normally in the form of

$$G(s) = \frac{b_m s^m + b_{m-1} s^{m-1} + \cdots + b_1 s + b_0}{s^n + a_{n-1} s^{n-1} + \cdots + a_1 s + a_0}, \tag{3.1}$$

in which s is the Laplace transform variable, a complex variable $s = \sigma + j\omega$.[1]

When $m \leq n$, the TF is called *proper*. If $m < n$, it is called *strictly proper*.

The signals $u(t)$ and $y(t)$ are the control input and the output of interest, respectively.

MATLAB: ≫ `G = tf([bm, bm-1, ..., b1, b0], [1, an-1, ..., aa, a0])`

1 The one-sided (or unilateral) Laplace transform defined as $F(s) = \mathcal{L}[f(t)] = \int_0^\infty f(t)e^{-st}dt$ is used in control systems.

Modeling and Control of Modern Electrical Energy Systems, First Edition. Masoud Karimi-Ghartemani.

3.1.1.1 Differential Equation

The differential equation associated with (3.1) is

$$\frac{d^n y}{dt^n} + a_{n-1}\frac{d^{n-1}y}{dt^{n-1}} + \cdots + a_1\frac{dy}{dt} + a_0 y = b_m\frac{d^m u}{dt^m} + b_{m-1}\frac{d^{m-1}u}{dt^{m-1}} + \cdots$$
$$+ b_1\frac{du}{dt} + b_0 u.$$

Figure 3.1 Transfer function block diagram of an LTI system.

Important property of Laplace transform: $\mathcal{L}[\dot{x}(t)] = sX(s) - x(0)$ for any arbitrary function $x(t)$. The transfer function approach ignores the initial conditions and is able to express the input–output relationship of an LTI system in the convenient form of $Y(s) = G(s)U(s)$.

3.1.1.2 Definition of Zeros and Poles of a TF or an LTI System

If $G(s) = \frac{p(s)}{q(s)}$, roots of $p(s)$ are zeros and roots of $q(s)$ are poles of $G(s)$. For the above TF, according to the fundamental theorem of Algebra, there are exactly m zeros and n poles on the complex plane C (all zeros and poles may not necessarily be distinct).

MATLAB: ≫ `p = pole(G),  z = zero(G)`

3.1.1.3 Partial Fraction Expansion (PFE)

If p_i for $i = 1, \ldots, n$ are distinct poles of the TF of (3.1), it can be written as

$$G(s) = \frac{r_1}{s - p_1} + \frac{r_2}{s - p_2} + \cdots + \frac{r_n}{s - p_n}.$$

The coefficients r_i's are called the residues. The response of the system to the input $u(t)$ is given by

$$y(t) = \mathcal{L}^{-1}[G(s)U(s)].$$

For the impulse input, $u(t) = \delta(t)$ whose Laplace transform is 1, the impulse response is

$$g(t) = r_1 e^{p_1 t} + r_2 e^{p_2 t} + \cdots + r_n e^{p_n t}.$$

Impulse response, and even in general the system response to any input is characterized by p_i's, whence the name *poles*! The functions $e^{p_i t}$ which form the basis for the system responses are called the system *modes*.

MATLAB: ≫ `y = impulse(G),  y = step(G),  y = lsim(G, u)`

≫ $\texttt{b} = [\texttt{b}_\texttt{m}, \texttt{b}_{\texttt{m-1}}, \ldots, \texttt{b}_1, \texttt{b}_0]$, $\texttt{a} = [\texttt{a}_\texttt{n}, \texttt{a}_{\texttt{n-1}}, \ldots, \texttt{a}_\texttt{a}, \texttt{a}_0]$, `r = residue(b, a)`

3.1.2 Stability

A system is bounded-input bounded-output (BIBO) stable if its output is bounded for all bounded inputs. Mathematically, it can be expressed as

$$|u(t)| < \infty \quad \forall t \geq 0 \Rightarrow |y(t)| < \infty \quad \forall t \geq 0.$$

BIBO stability theorem: An LTI system is BIBO stable if and only if (iff) all its poles reside on the open left half of complex plane (OLHP). (Open left half excludes the imaginary axis.) Mathematically, it can be expressed as

$$\text{BIBO Stability} \Leftrightarrow \text{Real}\{p_i\} < 0 \quad i = 1, \ldots, n.$$

3.1.3 Disturbance

Disturbance is a force that impacts the system's responses. It is present in almost all practical systems. It often has some unpredictable (un-deterministic or stochastic) aspect to it; for example its time of occurrence, or its magnitude, or some other attributes are unknown. Therefore, it can often be formulated as a deterministic function of an unknown variable. For instance, a step function where its time of inception or its magnitude is unknown. Or a sinusoidal function with unknown magnitude or angle or frequency.

Common ways of incorporating (or modeling) the disturbance in the TF block diagram are shown in Figure 3.2. In all these three structures, disturbance is modeled as an additive external input.

In structure (I): $Y(s) = G(s)U(s) + G(s)D(s)$
In structure (II): $Y(s) = G(s)U(s) + D(s)$
In structure (III): $Y(s) = G_2(s)[G_1(s)U(s) + D(s)]$

In structure (I), if the disturbance is measurable (or can be estimated from measurements), its reverse can be fed to input to cancel it directly. This is called a feed-forward approach to cancel the disturbance as shown in Figure 3.3.

$\hat{D}(s)$: measured or estimated disturbance

$$Y(s) = G(s)D(s) + G(s)U(s) = G(s)D(s) - G(s)\hat{D}(s) + G(s)V(s) \approx G(s)V(s)$$

Thus, the design of control can be done for $Y(s) = G(s)V(s)$ and then the actual control will be $U(s) = V(s) - \hat{D}(s)$.

In structure (II), the disturbance may not be canceled completely because $G(s)$ is often not invertible in an accurate and/or physically realizable way. However, a similar structure can be used and a gain factor of $G(0)^{-1}$ is incorporated in the feed-forward path, i.e. $G(0)^{-1}\hat{D}(s)$. This will mitigate the dc (or low frequency) disturbances. For structure (III), the feed-forward gain should be $G_1(0)^{-1}$.

3.1.4 Uncertainty

The transfer function model uses the "nominal" values of system parameters to perform the design. Those parameters are often known to certain accuracy. They may also be subject to change due to aging or other system conditions or modifications. Therefore, the constituting elements of a TF are subject to uncertainty. This type of uncertainty is called the structural and unmodeled uncertainty.

Other type of uncertainty is due to the fact that the TF model can only represent the system to certain level. There are three phenomena that are not included in this way of modeling such as (i) high frequency dynamics, (ii) nonlinear dynamics, and (iii) time-varying dynamics. These types of uncertainty are unstructured uncertainties.

If high-frequency dynamics are to be included in the TF, the order of the TF keeps increasing and it may no longer be efficiently useful for design and control purposes. A simpler model is

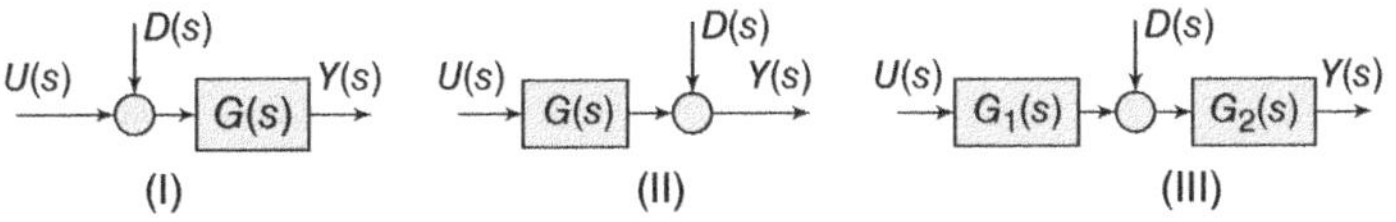

Figure 3.2 Disturbance modeling.

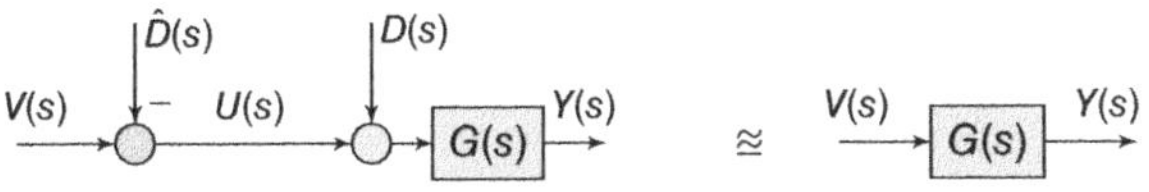

Figure 3.3 Feedforward cancelation of disturbance.

preferred as long as it can offer an efficient basis for control analysis and design purposes. This needs a trade-off between the model simplicity and the level of uncertainties.

3.1.5 Statement of Control Problem

Assume that the desired trajectory for the output variable is given by $y_d(t)$. Find the control input $u(t)$ such that the actual output of the system, $y(t)$, remains close to $y_d(t)$ for all times t.

The function $y_d(t)$ is called the reference or the command signal. The output is to follow the command, thus the term *Command Following*.

The above control objective should be achieved despite the presence of disturbance. In other words, the control input should reject the disturbances as much as possible and prevent them from disturbing the output. Thus, the term *Disturbance Rejection*.

The above control objective should also be achieved despite the system uncertainties and changes. In other words, the control input should ensure that the system response remains robustly close to the desired value despite the system uncertainties and changes. Thus, the term *Performance Robustness*.

Finally, the control input $u(t)$ should be practical. This means that it should be realizable using a practical actuator and the limitations imposed by the actuator and the system itself.

Example 3.1 ***dc Motor Speed Control***

As an example to demonstrate the control concepts, the dc motor is briefly introduced and modeled. The dc motor is supplied by a dc power and produces mechanical rotation. It is used to perform some heavy and tough industrial works. It is also used as an actuator for various control applications where it is called a *servomotor*.

Operation of a dc motor is based on interactions between electrical and magnetic fields. The magnetic field is generated by either permanent magnets or field windings located on the stator. The armature (or power) winding is located on the rotor. The access to the armature winding is made possible through graphite brushes that push on the conducting slip rings (attached to the shaft of the rotor). This setting also rectifies the ac current (or the ac torque) of the rotor winding.

The torque generated by a dc motor is expressed by $\tau = k_1\phi i_a$ where k_1 is a constant, ϕ is the magnetic flux of the machine, and i_a is the current in the armature winding. The induced voltage in the armature winding (also called back electromotive force – emf) is $e_a = k_2\phi\omega_m$ where k_2 is a constant and ω_m is the mechanical speed of the rotor mass in rad/s. These are the two main equations of the dc motor.

The equivalent circuit of a dc motor is shown in Figure 3.4 where the index a refers to armature circuit and f to the field circuit. Assume that the moment of inertia of the rotor mass and the load together is J and its friction coefficient is b. Thus, the Newton's equation is

$$\tau = J\frac{d}{dt}\omega_m + b\omega_m + \tau', \tag{3.2}$$

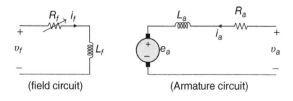

Figure 3.4 Equivalent circuit of a dc motor.

where τ' represents an external torque not included in (J, b) pair. It can represent unexpected forces on the shaft of motor and be considered as the disturbance in this system. The electrical equation describing the armature circuit is

$$v_a = R_a i_a + L_a \frac{d}{dt} i_a + e_a. \tag{3.3}$$

Assume that the field circuit voltage v_f is constant. This means that the magnetic flux ϕ is constant. (A permanent magnet-based machine will also have a constant magnetic flux.) Therefore,

$$e_a = k_2 \phi \omega_m = k_e \omega_m, \quad \tau = k_1 \phi i_a = k_\tau i_a. \tag{3.4}$$

Taking Laplace transform of (3.2) and (3.3) and substituting from (3.4) result in

$$\Omega_m(s) = \frac{T(s) - T'(s)}{Js + b} = \frac{k_\tau I_a(s) - T'}{Js + b}, \quad I_a(s) = \frac{V_a(s) - E_a(s)}{L_a s + R_a} = \frac{V_a(s) - k_e \Omega_m(s)}{L_a s + R_a}, \tag{3.5}$$

where $\Omega_m(s)$, $T(s)$, and $T'(s)$ are Laplace transform of $\omega_m(t)$, $\tau(t)$, and $\tau'(t)$, respectively. Equation set (3.5) can be represented by the block diagram of Figure 3.5. This diagram shows the armature voltage as the input (or control input) and the speed as the output. It is called the speed control block diagram of the dc motor.

The two equations (3.5) may be combined, by omitting $I_a(s)$, and written as

$$\begin{aligned}
\Omega_m(s) &= \frac{k_\tau I_a(s) - T'}{Js + b} = \frac{k_\tau \dfrac{V_a(s) - k_e \Omega_m(s)}{L_a s + R_a} - T'}{Js + b} \\
&= \frac{k_\tau V_a(s)}{(L_a s + R_a)(Js + b)} - \frac{k_\tau k_e \Omega_m(s)}{(L_a s + R_a)(Js + b)} - \frac{T'(s)}{Js + b} \\
&\Rightarrow \frac{(L_a s + R_a)(Js + b) + k_e k_\tau}{(L_a s + R_a)(Js + b)} \Omega_m(s) = \frac{k_\tau V_a(s)}{(L_a s + R_a)(Js + b)} - \frac{T'(s)}{Js + b} \\
&\Rightarrow \Omega_m(s) = \frac{L_a s + R_a}{(L_a s + R_a)(Js + b) + k_e k_\tau} \left[\frac{k_\tau V_a(s)}{L_a s + R_a} - T'(s) \right].
\end{aligned} \tag{3.6}$$

This latter equation is in the form of $Y(s) = G_2(s)[G_1(s)U(s) + D(s)]$ and the block diagram of (III) in Figure 3.2 (repeated in Figure 3.6) may be developed where

$$G_2(s) = \frac{L_a s + R_a}{(L_a s + R_a)(Js + b) + k_e k_\tau}, \quad G_1(s) = \frac{k_\tau}{L_a s + R_a}, \quad D(s) = -T'(s). \tag{3.7}$$

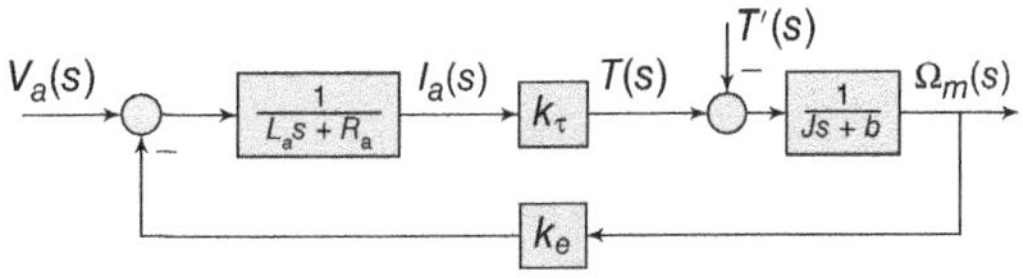

Figure 3.5 Block diagram representing dc motor speed control.

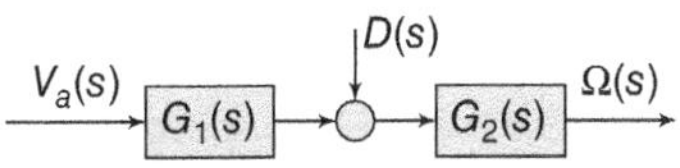

Figure 3.6 Alternative block diagram representation for the dc motor speed control with definitions (3.7).

3.2 Open-Loop Control

In open-loop control method, the system measurements are not used. The controller is blind to the variations of the output. It only uses the information on the desired value of the output to process the control input.

The block diagram of an open-loop control is shown in Figure 3.7, where $C(s)$ is the controller's transfer function. The total transfer function is $T(s) = \frac{Y(s)}{Y_d(s)} = G(s)C(s)$.

Advantages of open-loop control:

- Simple and economic structure which avoids using sensors.
- No stability concerns if G and C are stable.

Disadvantages of open-loop control:

- Open-loop control is generally sensitive to the presence of disturbance. If disturbance is strong (and it is not measured and canceled by the controller), it can cause strong impact on the output.
- If the system parameters are uncertain and/or change over time, the controller may not be able to satisfy the error requirements.
- Open-loop control cannot change the basic dynamics of the system. Poles of $G(s)$ are retained in $T(s)$ and not affected by the controller unless they are cancelled by the zeros of $C(s)$. Such possible cancelations can be problematic if they are uncertain and not sufficiently damped. Particularly, the open-loop control cannot be used to control a system that is initially unstable such as the inverted pendulum system.

Generally speaking, the open-loop control method is only suitable for systems where the levels of disturbance and uncertainty are sufficiently small and the system is initially well-behaved!

Example 3.2 ***Open-Loop Speed Control of dc Motor***
Open-loop control of dc motor is shown in Figure 3.8. A more complete version of the diagram showing how the armature voltage is processed is shown in Figure 3.9. Here, the *Drive* or *Actuator* supplies a variable dc voltage for the armature winding. Depending on the power rating of the motor and the form of primary supply, it can have various structures: a simple manual voltage divider; a

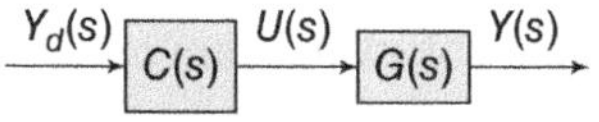

Figure 3.7 Open-loop control.

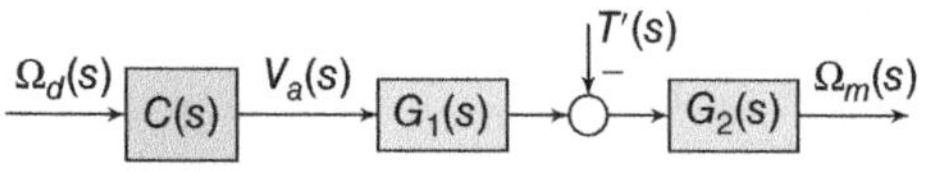

Figure 3.8 Open-loop speed control of dc motor.

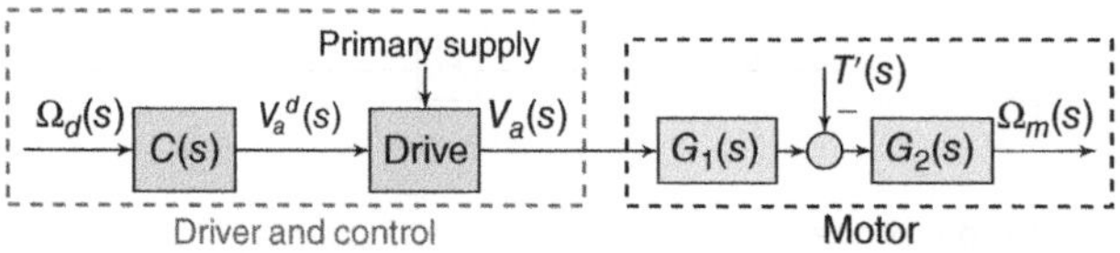

Figure 3.9 Open-loop speed control of dc motor (driver system is shown).

dc/dc or an ac/dc converter.

$$G_2(s) = \frac{L_a s + R_a}{(L_a s + R_a)(Js + b) + k_e k_\tau}, \; G_1(s) = \frac{k_\tau}{L_a s + R_a}.$$

A proportional controller $C(s) = K$ can be used and adjusted to ensure that the set-point is reached in the steady state. The total TF is

$$\frac{\Omega_m(s)}{\Omega_d(s)} = KG_1(s)G_2(s) = KG(s) = K\frac{k_\tau}{(L_a s + R_a)(Js + b) + k_e k_\tau}.$$

The dc (or steady-state) gain of this TF is

$$KG_1(0)G_2(0) = \frac{Kk_\tau}{R_a b + k_e k_\tau}.$$

To make this gain unity, K should be

$$K = \frac{R_a b + k_e k_\tau}{k_\tau}.$$

This gain is heavily dependent on the system parameters. The speed will easily shift from the desired value if there is an uncertainty or change in these parameters. Moreover, the disturbance is not addressed by the controller. When it happens, it will change the speed unexpectedly. Specifically, if a disturbance with magnitude τ' occurs, it will deviate the speed as much as $\tau' G_2(0) = \frac{\tau' R_a}{R_a b + k_e k_\tau}$. Finally, this controller has very limited impact on the system response. It does not change the poles. If the motor has some oscillatory or sluggish initial modes, this controller cannot modify them.

3.3 Closed-Loop (or Feedback) Control

In closed-loop control method, the output and possibly some other variables from the system are measured and used in a feedback loop structure to modify the control input to address the changes. Generally speaking, almost all natural and/or engineering systems work based on some sort of feedback loop. So, this is a natural way of addressing the control problem.

3.3.1 Feedback Philosophy

The typical output feedback structure is shown in Figure 3.10 where $C(s)$ is the controller's (or compensator's) transfer function and $E(s)$ is the tracking error.

Advantages of closed-loop control:

- Feedback can in principle respond to disturbances.
- Feedback can in principle respond to uncertainties and changes.
- Feedback can modify the original dynamics of $G(s)$.

Disadvantages of closed-loop control:

- Sensor adds to the cost and injects measurement noise to the loop.
- Design stage can become a challenge.

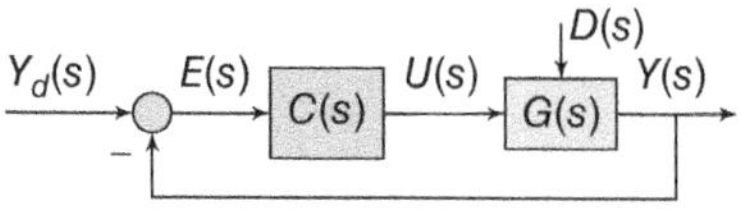

Figure 3.10 Closed-loop or feedback control.

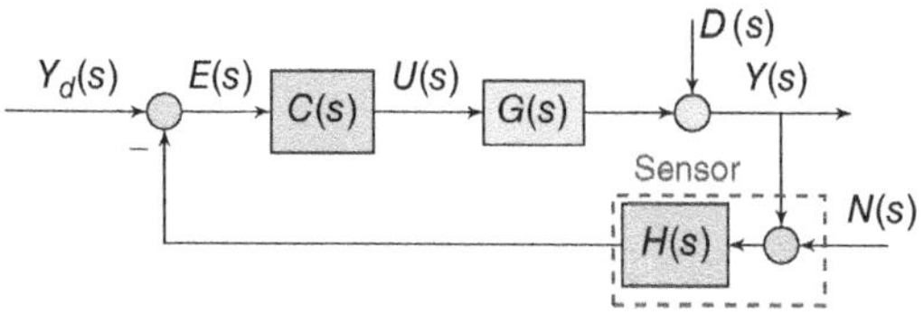

Figure 3.11 Detailed diagram of closed-loop or feedback control.

If $G(s) = \frac{p(s)}{q(s)}$, $C(s) = \frac{p_c(s)}{q_c(s)}$, then

$$Y(s) = G(s)C(s)E(s) = G(s)C(s)Y_d(s) - G(s)C(s)Y(s)$$

$$\Rightarrow \ T(s) = \frac{Y(s)}{Y_d(s)} = \frac{G(s)C(s)}{1 + G(s)C(s)} = \frac{L(s)}{1 + L(s)} = \frac{p(s)p_c(s)}{q(s)q_c(s) + p(s)p_c(s)},$$

where $L(s) = G(s)C(s)$ is the open-loop transfer function also called the open-loop gain or simply *the loop gain* or the loop transfer function.

Zeros of closed-loop TF are the roots of $p(s)p_c(s)$. This is the union of the zeros of plant and those of the controller. The poles of the closed-loop TF are the roots of $q(s)q_c(s) + p(s)p_c(s)$. This in general has no explicit relationship with the poles of the plant and those of the controller.

A more detailed block diagram is shown in Figure 3.11, where $H(s)$ represents the sensor TF. The following TFs can be identified for this system.

Loop gain: $L(s) = G(s)H(s)C(s)$, assume $H(s) = 1$ for simplicity of notations.

Closed loop TF: $T(s) = \dfrac{Y(s)}{Y_d(s)} = \dfrac{L(s)}{1 + L(s)}$

Error (or sensitivity) TF: $S(s) = \dfrac{E(s)}{Y_d(s)} = \dfrac{1}{1 + L(s)}$

Disturbance TF: $T_d(s) = \dfrac{Y(s)}{D(s)} = \dfrac{1}{1 + L(s)}$

Noise (or sensor) TF: $T_n(s) = \dfrac{Y(s)}{N(s)} = -\dfrac{L(s)}{1 + L(s)} = -T(s)$

Remark The sensitivity transfer function $S(s)$ is a measure of the robustness of $T(s)$ with respect to uncertainties in $G(s)$:

$$S(s) = \frac{\frac{\Delta T(s)}{T(s)}}{\frac{\Delta G(s)}{G(s)}} = \frac{G(s)}{T(s)}\frac{\Delta T(s)}{\Delta G(s)} = \frac{G(s)}{T(s)}\frac{\partial T(s)}{\partial G(s)} = \frac{1}{1 + L(s)}$$

If loop gain $L(s)$ is "large," $T(s)$ is close to unity: *command following* is achieved.

If loop gain $L(s)$ is "large," $S(s)$ is close to zero: *disturbance rejection* and *robustness* to structural uncertainties.

If loop gain $L(s)$ is "large", sensor TF is large: sensitivity to sensor noise!

Fortunately, measurement noise has a high-frequency spectrum while disturbances and commands have a low-frequency nature. Therefore, a loop gain that exhibits a magnitude function like Figure 3.12 can satisfy all requirements.

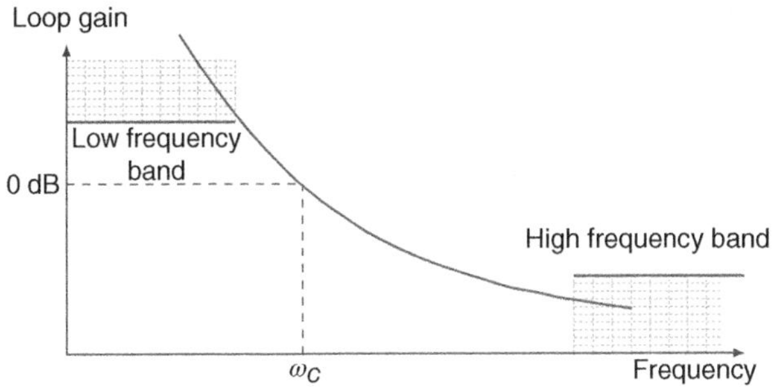

Figure 3.12 Loop gain versus frequency.

Open-loop Bode plots show $20 \log |L(j\omega)|$ (in dB) and $\angle L(j\omega)$ (in degrees) versus the frequency ω and are much useful tool for frequency-domain analysis and design.

MATLAB: ≫ `bode(L)`

3.3.2 Stability Margins

Gain Margin (GM) is the maximum gain that the loop can tolerate before it becomes unstable. *Delay Margin* (DM) is the maximum delay that the loop can tolerate before it becomes unstable. *Phase Margin* (PM) is the maximum angle that the loop can tolerate before it becomes unstable. Relationship between DM and PM is

$$\text{PM} = \text{DM} \times \omega_c,$$

where ω_c is the crossover frequency where the magnitude crosses 0 dB. Thus, the stability margins are readily derived from the loop transfer function.

MATLAB: ≫ `margin(L)`

Stability margins are indications of (i) how distant the closed-loop poles are from the imaginary axis and (ii) how sensitive they are with respect to system uncertainties and changes. Therefore, they are regarded as indices for both the transient response and for performance robustness. However, there is some ambiguity with these indices because it may be a case that poles are close to the imaginary axis but less sensitive to system uncertainties as opposed to another case where poles are more distant from the axis but also more sensitive to the uncertainties. Or a system may have a large PM but at the same time a very small DM because its bandwidth is large. Despite these facts, the stability margins can be used with some care for various analysis and design applications in frequency domain.

Example 3.3 ***Closed-Loop Speed Control of dc Motor***
A *closed-loop speed control* of dc motor is shown in Figure 3.13. The closed-loop TF is

$$T_{\text{cl}}(s) = \frac{\Omega_m(s)}{\Omega_d(s)} = \frac{C(s)G_1(s)G_2(s)}{1 + C(s)G_1(s)G_2(s)}.$$

Several cases for selection of $C(s)$ are discussed below.

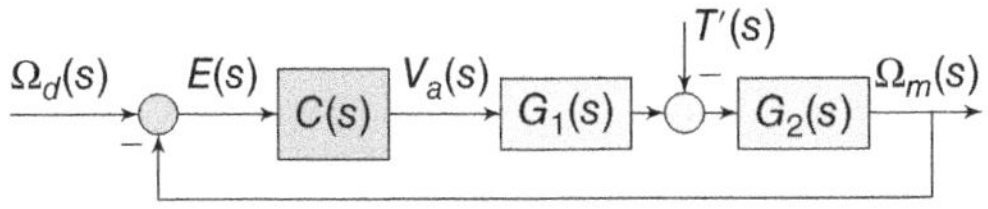

Figure 3.13 Closed-loop speed control of dc motor.

3.3.2.1 Case I: Proportional Control $C(s) = K_p$

The closed loop TF is

$$\frac{\Omega_m(s)}{\Omega_d(s)} = \frac{K_p \frac{k_\tau}{(L_a s + R_a)(Js + b) + k_e k_\tau}}{1 + K_p \frac{k_\tau}{(L_a s + R_a)(Js + b) + k_e k_\tau}} = \frac{K_p k_\tau}{(L_a s + R_a)(Js + b) + k_e k_\tau + K_p k_\tau}.$$

Notice that the dc gain of the closed-loop system is $\frac{K_p k_\tau}{R_a b + k_e k_\tau + K_p k_\tau}$ and is not unity. Therefore, to have zero steady-state error, a forward gain equal to inverse of this dc gain must be used in front of the loop. The open loop TF (or the loop gain) is

$$L(s) = \frac{\Omega_m(s)}{E(s)} = K_p \frac{k_\tau}{(L_a s + R_a)(Js + b) + k_e k_\tau}.$$

Root-locus of the closed-loop poles when K_p varies from 0 to 8 is shown in Figure 3.14. And the Bode diagram of $L(s)$ for $K_p = 8$ is shown in Figure 3.15.[2]

MATLAB: ≫ `rlocus(L)` and `rlocfind(L)`

Increasing K_p has the following impacts.

1. Increases bandwidth (look at the Bode diagram of Figure 3.15). The responses become swifter but with more oscillations. This is confirmed by the root-locus diagram of Figure 3.14 where shows decrease in the damping of the poles.
2. Decreases the steady-state error. The dc gain is equal to $T_{\text{cl}}(0) = \frac{K_p k_\tau}{R_a b + k_e k_\tau + K_p k_\tau}$ which tends to unity as K_p increases.
3. Decreases the phase margin. This, together with the increase in bandwidth, indicates a stronger decrease in the delay margin.
4. Increased bandwidth indicates increase in the system sensitivity to sensor noise, and also to un-modeled and high frequency dynamics.

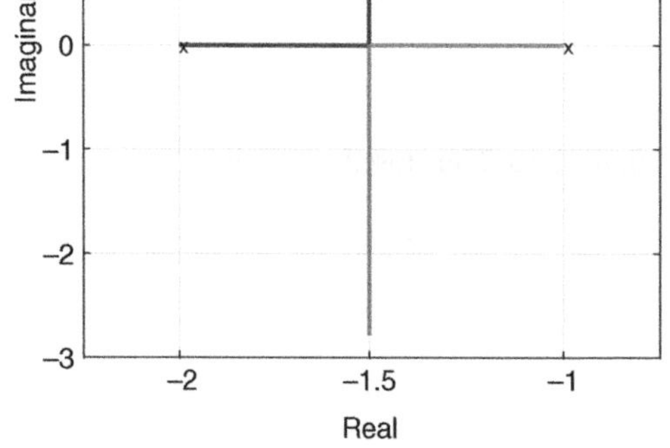

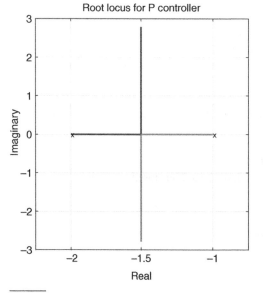

Figure 3.14 Root locus of dc motor control with P controller: K_p varies from 0 to 8.

2 For these and the subsequent several graphs, we have chosen the numerical values such that

$$\frac{k_\tau}{(L_a s + R_a)(Js + b) + k_e k_\tau} = \frac{1}{s^2 + 3s + 2}.$$

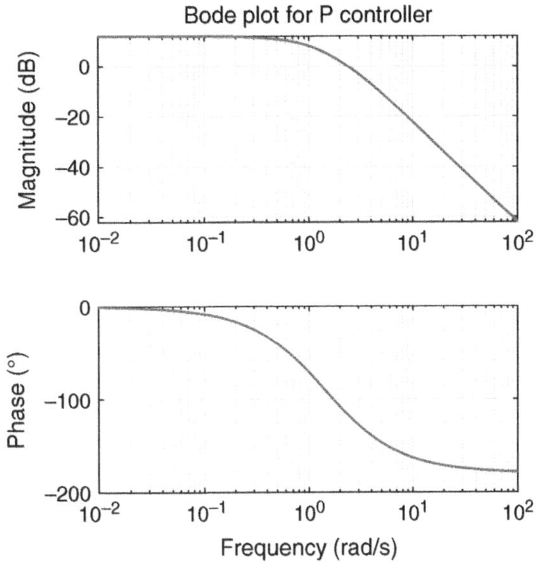

Figure 3.15 Bode plot of dc motor control at $K_p = 8$ (closed-loop poles at $-1.5 \pm j2.8$).

3.3.2.2 Case II: Proportional-Derivative Control $C(s) = K(s + z) = K_d s + K_p$

$$\frac{\Omega_m(s)}{\Omega_d(s)} = \frac{(K_d s + K_p)\frac{k_\tau}{(L_a s+R_a)(Js+b)+k_e k_\tau}}{1 + (K_d s + K_p)\frac{k_\tau}{(L_a s+R_a)(Js+b)+k_e k_\tau}} = \frac{(K_d s + K_p)k_\tau}{(L_a s + R_a)(Js + b) + k_e k_\tau + (K_d s + K_p)k_\tau}$$

Properties of proportional-derivative (PD) Controller (Figures 3.16 and 3.17):

1. Does not change the steady-state error. The dc gain is still equal to $\frac{K_p k_\tau}{R_a b+k_e k_\tau+K_p k_\tau}$
2. Moves the poles further to the left and achieve faster responses without necessarily causing oscillations.
3. Increases the phase margin. May or may not increase the delay margin since the bandwidth is also increased.

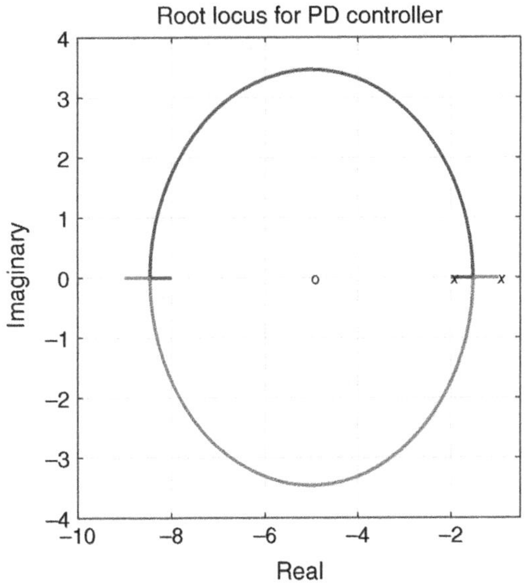

Figure 3.16 Root locus of dc motor control with PD controller: $C(s) = K(s + 5)$, K varies from 0 to 14.

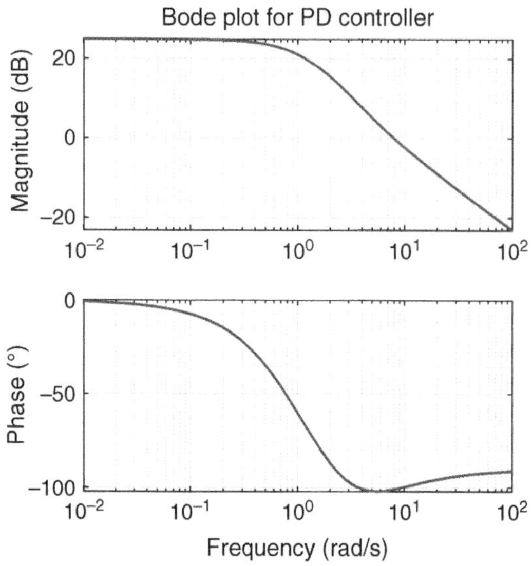

Figure 3.17 Bode plots of dc motor control with PD controller: $C(s) = 7(s + 5)$ (closed-loop poles are at $-5 \pm j3.46$.)

4. The derivative operator has a positive phase angle (phase lead effect) and it functions as if it can predict the system's response for future times. This is why it can significantly increase the response speed.
5. Increases the bandwidth and the loop gain at high frequencies making the system sensitive to sensor noise and un-modeled and high frequency dynamics.

Remark The derivative function should be implemented with care because it introduces very high gain at large frequencies. In practice, the term s is approximated by $\frac{s}{\tau s+1}$ where τ is a small number. Without changing it much at low frequencies, this modification limits the derivative gain at high frequencies.

3.3.2.3 Case III: Proportional-Integrating Control $C(s) = K\frac{s+z}{s} = K_p + \frac{K_i}{s}$

$$\frac{\Omega_m(s)}{\Omega_d(s)} = \frac{\left(K_p + \frac{K_i}{s}\right)\frac{k_\tau}{(L_a s + R_a)(Js + b) + k_e k_\tau}}{1 + \left(K_p + \frac{K_i}{s}\right)\frac{k_\tau}{(L_a s + R_a)(Js + b) + k_e k_\tau}} = \frac{(K_p s + K_i)k_\tau}{s[(L_a s + R_a)(Js + b) + k_e k_\tau] + (K_p s + K_i)k_\tau}$$

Properties of proportional-integrating (PI) Controller (Figures 3.18 and 3.19):

1. Completely removes the steady-state error! The dc gain is unity: $\frac{K_i k_\tau}{K_i k_\tau} = 1$. This is called a *robust property* because it is achieved regardless of values of the system parameters. It is the big advantage of the PI controller.
2. Cannot move the poles further to the left. Faster response is not achieved.
3. Decreases the bandwidth and the loop gain at high frequencies. Can make the loop robust to sensor noise and un-modeled and high frequency dynamics.

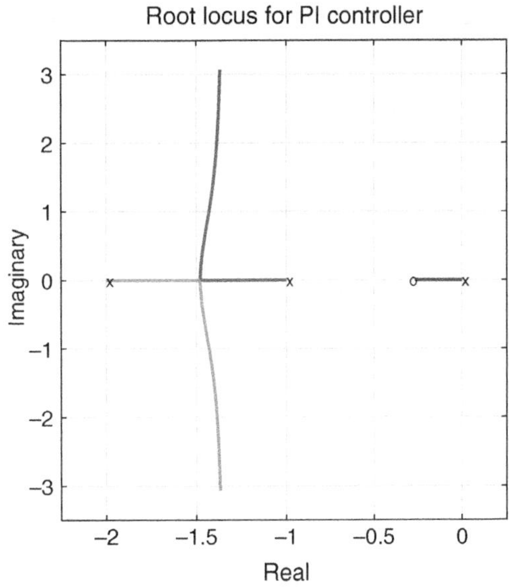

Figure 3.18 Root locus of dc motor control with PI controller: $C(s) = K\frac{s+0.3}{s}$, K varies from 0 to 10.

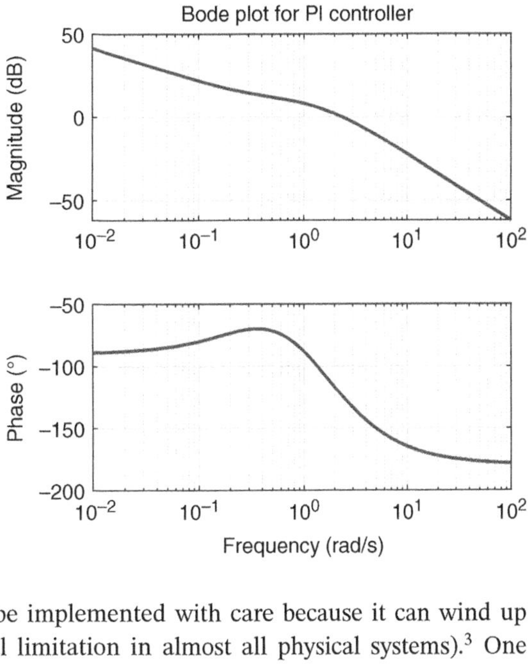

Figure 3.19 Bode plots of dc motor control with PI controller: $C(s) = 8\frac{s+0.3}{s}$ (closed-loop poles are $-1.4 \pm j2.7, -0.26$.)

Remark The integrating function should be implemented with care because it can wind up due to saturation of the actuator (a practical limitation in almost all physical systems).[3] One anti-windup way to deal with this is shown in Figure 3.24.

3 Consider a case where the tracking error is positive, big such that the actuator is saturated on the positive side. Then, the output of integrators keeps growing without having any impact on the system. Eventually, when the error becomes small and the actuator enters into linear region, it takes long time for the integrator to wind-down. This will introduce delays and oscillations in the closed-loop response.

3.3.2.4 Case IV: PID Control $C(s) = K\frac{(s+z_1)(s+z_2)}{s} = K_d s + K_p + \frac{K_i}{s}$

$$
\begin{aligned}
\frac{\Omega_m(s)}{\Omega_d(s)} &= \frac{\left(K_d s+K_p+\frac{K_i}{s}\right)\dfrac{k_\tau}{(L_a s+R_a)(Js+b)+k_e k_\tau}}{1+\left(K_d s+K_p+\frac{K_i}{s}\right)\dfrac{k_\tau}{(L_a s+R_a)(Js+b)+k_e k_\tau}} \\
&= \frac{(K_d s^2+K_p s+K_i)k_\tau}{s[(L_a s+R_a)(Js+b)+k_e k_\tau]+(K_d s^2+K_p s+K_i)k_\tau}
\end{aligned}
$$

Properties of the Proportional-Integrating-Derivative (PID) Controller (Figures 3.20 and 3.21):

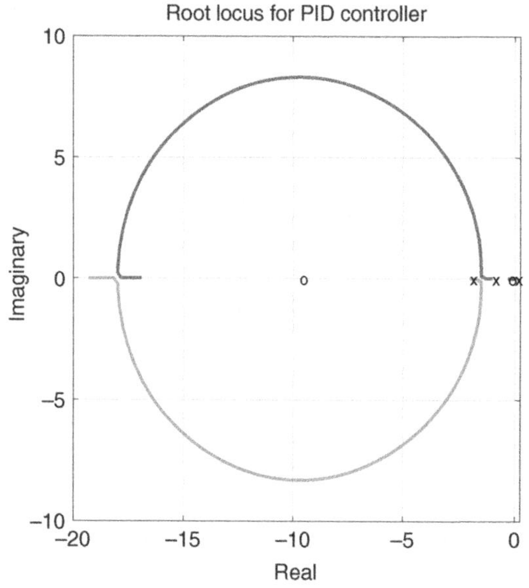

Figure 3.20 Root locus of dc motor control with PID controller: $C(s) = K\frac{s^2+10s+2.91}{s}$ (zeros at −0.3 and −9.7), K varies from 0 to 33.5.

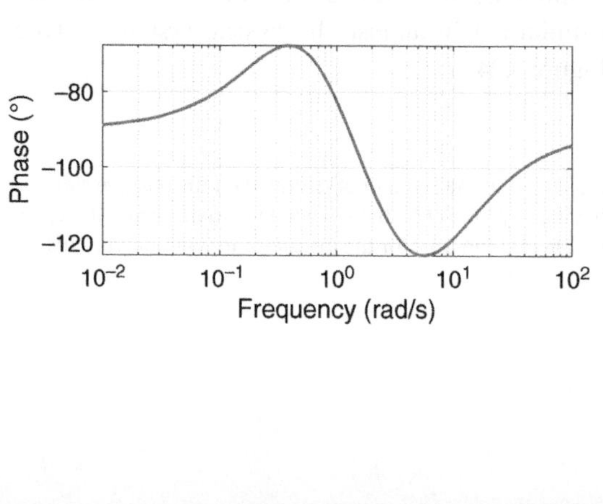

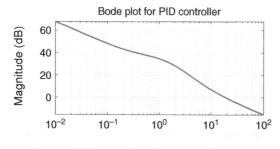

Figure 3.21 Bode plots of dc motor control with PID controller: $C(s) = 16.75\frac{s^2+10s+2.91}{s}$ (closed-loop poles are $-9.7 \pm j8.3, -0.298$.)

1. Completely removes the steady-state error (due to the presence of the I term).
2. Moves the poles to the left and achieve fast response (due to the presence of the D term).
3. The gains should be designed properly to achieve a desired trade-off between fast response, low oscillations, and low sensitivity to noise and high frequency dynamics.

Step responses of the open-loop and the closed-loop systems with different P, PD, PI, and PID controllers discussed above are shown in Figure 3.22. This confirms the points discussed above. The Bode plots of all five different cases are also drawn on a single graph and shown in Figure 3.23.

A block diagram implementation of the PID controller is shown in Figure 3.24. The D controller is modified to prevent too large a gain at high frequencies. An anti-windup mechanism is also

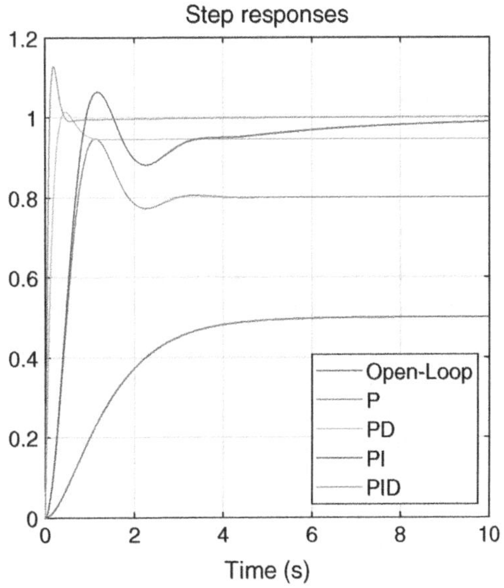

Figure 3.22 Step responses of the open-loop and closed-loop control systems.

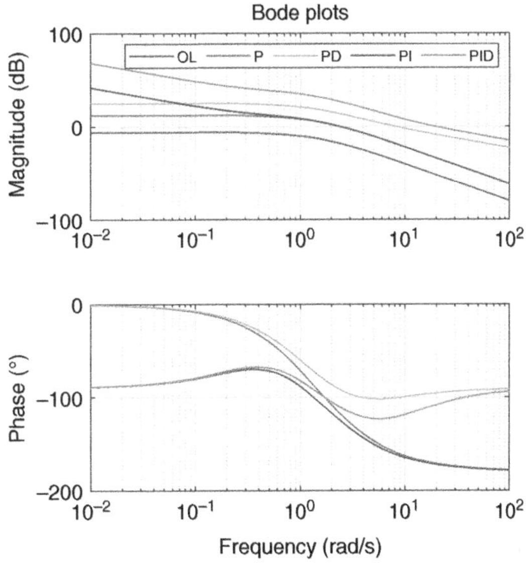

Figure 3.23 Bode plots of the open-loop and closed-loop control systems.

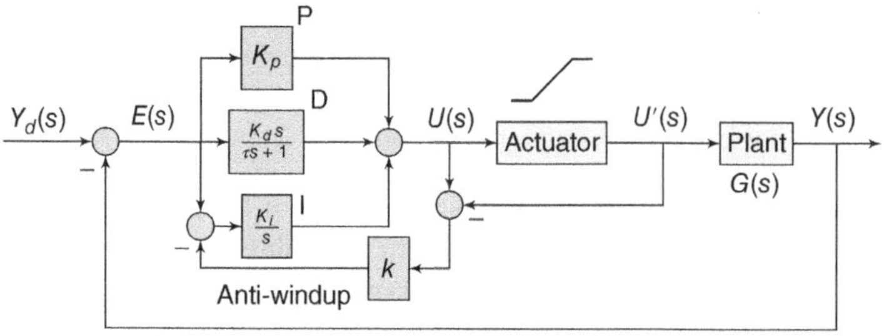

Figure 3.24 PID controller with filtered derivative and anti-windup.

employed to prevent the integrator from increasing (or decreasing) indefinitely when the actuator is saturated.

Remark 3.1 PD controller is a special case of the first-order *lead controller*. PI controller is a special case of the first-order *lag controller*. PID controller is a special case of the second-order *lead/lag controller*.

Remark 3.2 The P controller can be distributed by getting feedback from two signals E and Y as shown in Figure 3.25. While this does not impact the location of closed-loop poles, it will change the zero in the transfer function from the reference input to the output. By adjusting the value of α ($0 \leq \alpha \leq 1$), one can fine-tune the transient response of the system in response to sudden command signal changes. Larger α basically means that the sudden changes in Y_d are quickly transferred to U and are responded to by the loop faster. One particular application of this methodology is to improve the start-up transient of the system.

General Remark: Control design process involves methods to find the best values for the controller gains such that the closed-loop system has desirable responses. The desirable response characteristics are normally specified in terms of transient response characteristics (rise-time, settling-time, overshoot, etc.), and steady-state responses (steady-state error to commands, disturbances, and noise), robustness indices (phase margin, gain margin, etc.). It often happens that the designer has difficulty deciding which one of these many factors is the more important one and moreover, it is difficult to characterize and establish the trade-offs among these factors. These are things that make the design stage challenging. An engineer with lots of experience with particular systems can use his/her knowledge to accomplish a good design. State space designs and optimal control techniques developed based on state-space models address those types of challenges to a relatively good extent.

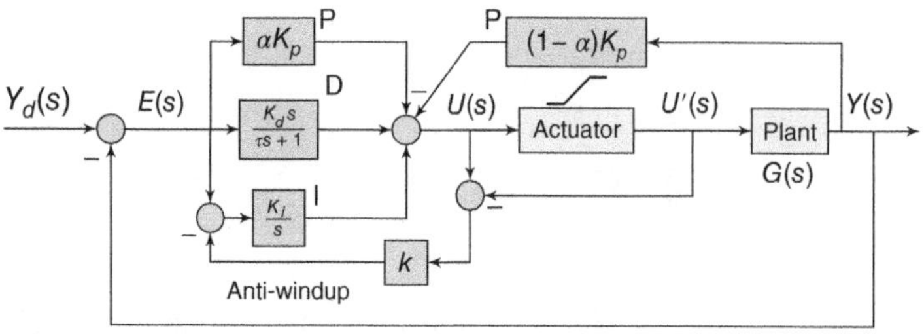

Figure 3.25 PID controller with distributed P ($0 \leq \alpha \leq 1$).

3.4 Some Feedback Loop Properties

Some feedback loop properties are studied in this section. The conditions to remove the steady-state error, and the relationship between the locations of poles of a transfer function with its form of response are discussed.

3.4.1 Removal of Steady-State Error

Consider the standard closed-loop system of Figure 3.26. The transfer functions of this system are:

$$T(s) = \frac{Y(s)}{Y_d(s)} = \frac{L(s)}{1 + L(s)}, \quad T_D(s) = \frac{Y(s)}{D(s)} = \frac{G_2(s)}{1 + L(s)},$$

where $G(s) = G_1(s)G_2(s)$ is the plant's forward TF and $L(s) = C(s)G(s)$ is the loop TF. It can easily be observed that if $C(s)$ has infinite (or very large) gain at a given frequency ω_0, then $T(s)$ tends to unity and $T_D(s)$ tends to zero at this frequency. This means that perfect command following and disturbance rejection is achieved at that frequency in the steady-state situation. In other words, there will be no (or very small) steady-state error.

- For the popular case of step commands and disturbances, infinite gain is achieved by using an integrator, i.e. $\frac{1}{s}$, in the loop.
- For a single nonzero frequency, that is an ac sinusoidal signal, infinite gain is achieved by using a second-order generalized integrator (SOGI), i.e. $\frac{1}{s^2+\omega_0^2}$ or $\frac{s}{s^2+\omega_0^2}$ or their combinations, in the loop. This is also called the *Resonant* controller.

In some practical systems,[4] $Y_d(s)$ and $D(s)$ may comprise different components (such as a dc component and some ac components) where the objective is to track one particular component of the command (e.g. its dc) and reject the other components of the command as well as all components of the disturbance. In such cases, the structure of Figure 3.26 does not work. An appropriate structure for this case is shown in Figure 3.27.

In Figure 3.27, the controller comprises two parts: $C(s)$ and $H(s)$. One works on the error signal and the other on the output signal. The function $C(s)$ has infinite (or very large) gain at the frequency of the command that is desired to be tracked. The function $H(s)$ has infinite (or very large gain) at frequency (or frequencies) of the disturbance that need to be rejected. The input to the block $H(s)$ should eventually have no frequency component at those frequencies that H is

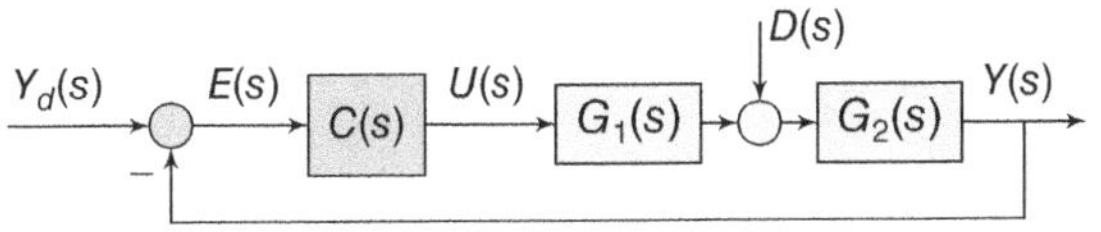

Figure 3.26 A standard closed-loop system.

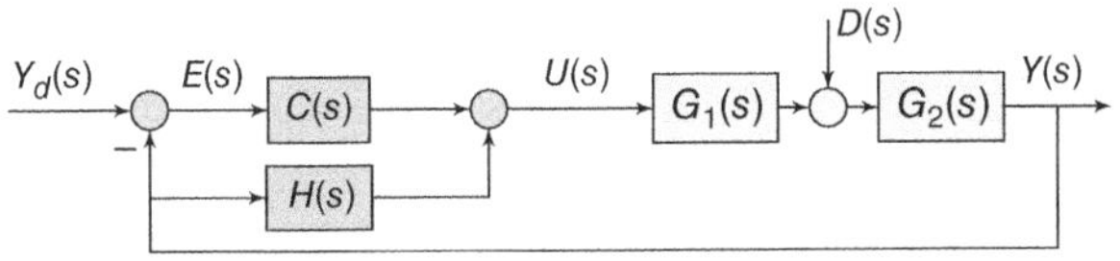

Figure 3.27 Alternative feedback structure for a closed-loop system.

4 Such as a grid-connected inverter as we will discuss later.

infinite at. Therefore, all those components are removed from the output regardless of whether they are originated by the command or disturbance. This is the conceptual way of understanding how it works. Mathematically, the transfer functions are

$$T(s) = \frac{Y(s)}{Y_d(s)} = \frac{L(s)}{1 + L(s) + H(s)G(s)}, \quad T_D(s) = \frac{Y(s)}{D(s)} = \frac{G_2(s)}{1 + L(s) + H(s)G(s)},$$

where $G(s) = G_1(s)G_2(s)$ is the plant's forward TF and $L(s) = C(s)G(s)$ is the loop TF. Both T_D and T are zeros at the frequencies where H is infinity which means those frequencies will be rejected from both the disturbance or command. For example, let s assume that we want to track the dc and reject an ac at frequency ω_o. Then, $C(s)$ has a $\frac{1}{s}$ and $H(s)$ has a term $\frac{1}{s^2+\omega_o^2}$. Then, at frequency of zero, T is unity and T_D is zero. At frequency ω_o, both T and T_D are zero.

3.4.2 Pole Location and Transient Response

A system with a first order TF of $\frac{p}{s+p}$ where $p > 0$ has a pole at $-p$. Its step response approaches unity at the time-constant of $\tau = \frac{1}{p}$ seconds as shown in Figure 3.28. The response is expressed by $y(t) = 1 - e^{-pt}$ in time domain. By $t = \tau$, the response has reached about 63% of its final value. The response settles within 98% of its final value by the time of 4τ. This is called the settling time.

Consider a system with a second order TF of $\frac{b}{s^2+as+b} = \frac{\omega_n^2}{s^2+2\zeta\omega_n s+\omega_n^2}$ where $\zeta > 0$ is called the damping ratio and $\omega_n > 0$ is called the natural frequency. Location of the poles on the complex plane is shown in Figure 3.29. Step response of the system has a form

$$y(t) = 1 - e^{\zeta\omega_n t}\sin(\omega_n\sqrt{1-\zeta^2}t) \tag{3.8}$$

that is shown in Figure 3.30.

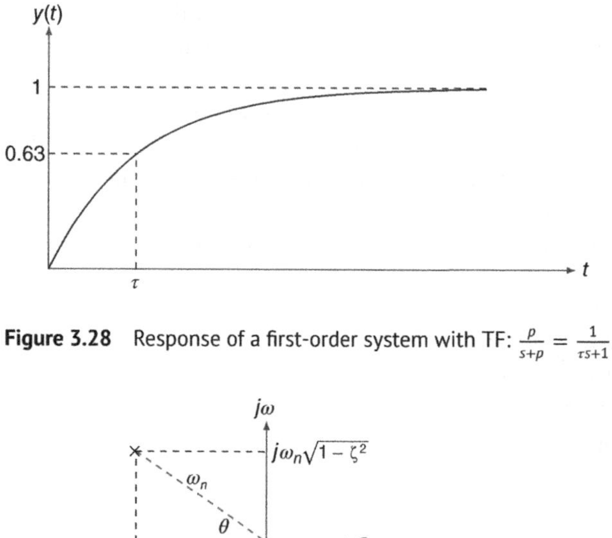

Figure 3.28 Response of a first-order system with TF: $\frac{p}{s+p} = \frac{1}{\tau s+1}$.

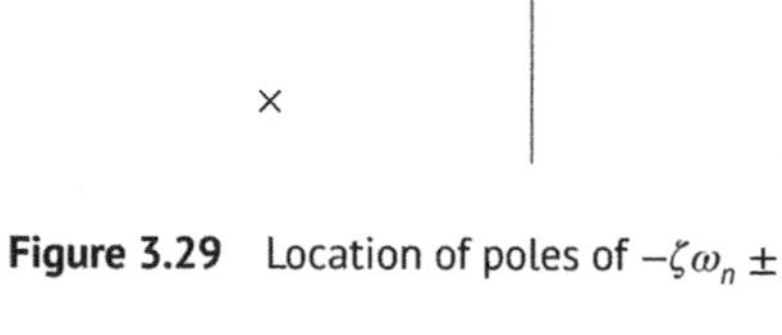

Figure 3.29 Location of poles of $-\zeta\omega_n \pm j\omega_n\sqrt{1-\zeta^2}$.

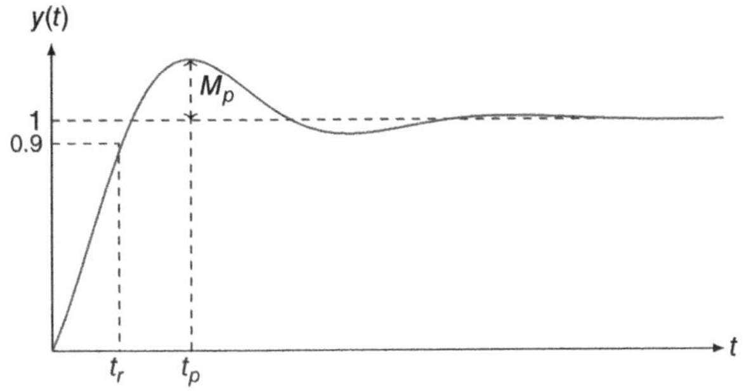

Figure 3.30 Response of a second-order system with TF: $\frac{\omega_n^2}{s^2+2\zeta\omega_n s+\omega_n^2}$.

Notice that the response for $\zeta > 1$ is not oscillatory and it is called over-damped. For $\zeta = 1$, the response is critically damped. And for $0 < \zeta < 1$, the response has exponentially damping oscillations and the system is called under-damped. Case of $\zeta \leq 0$ signifies unstable system and is not of interest. From a purely control system point of view, a value of ζ around 0.7 exhibits a very fast response without excessive oscillations and it is desirable.

It can easily be shown that the peak time and overshoot are obtained from

$$t_p = \frac{\pi}{\omega_n\sqrt{1-\zeta^2}}, \quad M_p = e^{-\frac{\pi\zeta}{\sqrt{1-\zeta^2}}}. \tag{3.9}$$

The rise-time of the system is the time that the system reaches 90% of its final value and it is a measure of swiftness of the response. An approximate equation for that is given by $t_r = \frac{1.8}{\zeta\omega_n}$ for "good" ζ's! Similarly, an approximate equation for the settling-time is given by $T_s = \frac{4}{\zeta\omega_n}$.

Notice that too small a value of ζ signifies a very swift response that takes a very long time to settle while a slightly larger ζ reduces the settling-time at the expense of rise time. (This recalls the rabbit-turtle race!) Therefore, it is important to be precise and be clear when we talk about speed and fastness of the systems and compare them.

3.5 Summary and Conclusion

This chapter reviewed the classical control system analysis and design using transfer function approach. The advantage of this approach is to avoid differential equations and convert them to algebraic analysis of a complex variable, i.e. the Laplace variable s or the Fourier variable ω. Another advantage is fewer number of sensors, compared to the state-space domain where full state feedback is used. The disadvantage is loss or compromise of clear insight and transparency. Specifically, the relations between frequency domain and time domain characteristics are not clearly defined. The existing tools such as root-locus, Bode diagrams, and Nyquist plots offer efficient means for analysis and design of linear control systems in this domain. However, it is still generally challenging to simultaneously address stability, robustness, and transient response characteristics.

Problems

3.1 For the transfer function $G(s) = \frac{1}{s^2+3s+2}$, find the poles and zeros of the system. Write down its partial fraction expansion and find its modes.

3.2 Use Matlab commands introduced in the chapter to obtain and draw the step response of the TF given in Problem 3.1.

3.3 Augment $G(s) = \frac{1}{s^2+3s+2}$ with P compensator of $C(s) = K_p$ and close the feedback loop. Draw the root locus when K_p varies from 0 to 8 and confirm the results shown in Figure 3.14. Draw the Bode diagram at $K_p = 8$ and confirm the results shown in Figure 3.15.

3.4 Augment $G(s) = \frac{1}{s^2+3s+2}$ with PD compensator of $C(s) = K(s + z)$ and close the feedback loop. Draw the root locus $z = 5$ and K varies from 0 to 14 and confirm the results shown in Figure 3.16. Draw the Bode diagram at $K = 7$ and confirm the results shown in Figure 3.17.

3.5 Augment $G(s) = \frac{1}{s^2+3s+2}$ with PI compensator of $C(s) = K\frac{s+z}{s}$ and close the feedback loop. Draw the root locus $z = 0.3$ and K varies from 0 to 10 and confirm the results shown in Figure 3.18. Draw the Bode diagram at $K = 8$ and confirm the results shown in Figure 3.19.

3.6 Augment $G(s) = \frac{1}{s^2+3s+2}$ with PID compensator of $C(s) = K\frac{(s+z_1)(s+z_2)}{s}$ and close the feedback loop. Draw the root locus $z_1 = 0.3$, $z_2 = 9.7$ and K varies from 0 to 33.5 and confirm the results shown in Figure 3.20. Draw the Bode diagram at $K = 15.9$ and confirm the results shown in Figure 3.21.

3.7 Draw the step responses of the systems you analyzed in Problems 3.3–3.6 and confirm Figure 3.22.

3.8 The MATLAB command *sisotool* is a useful tool for analysis and design of feedback controllers. In a basic form, *sisotool(G,C)* opens a panel where the root-locus, Bode diagram (with marked stability margins), and the closed-loop step response are all shown. While an initial value for $C(s)$ is selected, these tools allow you to grab and change the numbers in $C(s)$ directly from the panel. Use the Matlab sisotool command to redo Problems 3.3–3.6 and try to fine tune your controllers.

3.9 Use the Matlab command *lsim* to obtain and draw the response of $G(s) = \frac{1}{s^2+3s+2}$ to a sinusoidal input signal with magnitude 1 and frequency of 1 Hz.

3.10 Derive (3.8) by taking the inverse Laplace transform of $Y(s) = \frac{\omega_n^2}{s^2+2\zeta\omega_n s+\omega_n^2}\frac{1}{s}$.

3.11 Derive (3.9) by differentiating (3.8) with respect to time and putting it to zero.

3.12 Consider the speed control of dc motor using the armature voltage at constant field. The differential equations are summarized as

$$\text{Armature circuit equation:}\quad v_a(t) = L_a\frac{d}{dt}i_a(t) + R_a i_a(t) + e_a(t)$$
$$\text{Load rotation equation:}\quad \tau(t) = J\frac{d}{dt}\omega(t) + b\omega(t) + \tau'(t)$$

where v_a and i_a are the armature voltage and current, R_a and L_a are its resistance and inductance, e_a is the induced voltage, τ is the induced torque, ω is the rotor speed, J and b are the moment of inertia and the coefficient of friction of the mechanical load and the rotor shaft, and τ' is the external opposing torque (regarded as disturbance). The induced voltage and

torque satisfy $e_a = k_e\omega$ and $\tau = k_\tau i_a$ where k_e and k_τ are motor constants. The armature voltage v_a is the control input and ω is the output.
Consider the following set of numerical values whenever needed:

$$R_a = 1\ \Omega,\ L_a = 0.2\ \text{H},\ J = 10\ \text{kgm}^2,\ b = 0.1,\ k_e = k_\tau = 2.$$

(a) Represent this control problem in the block diagram of Figure 3.31. Specify the transfer functions and the control input, the output, and the disturbance.
(b) For the numerical values given, find the poles and zeros of the system's input–output transfer function and its dc gain.
(c) How much armature voltage is required to have a speed of $n_d = 1000$ rpm in the steady-state in the absence of disturbance torque? Calculate the value of a simple open-loop gain K_f placed in front of the system, see Figure 3.32, that converts ω_d (in rad/s) to the desired V_a. What are the values of armature current, induced voltage, and induced torque in the steady-state?
(d) Simulate the open-loop controlled system and show its response to a 1000 rpm speed command with zero initial conditions and zero disturbance. How much is the approximate rise-time of the response? Check and verify the values for the variables you calculated in part c.
(e) While the motor is running at steady state of 1000 rpm, apply a disturbance torque of $\tau' = 10$ N m in your simulation and show the traces of speed and armature current. Zoom on the graphs to clearly observe the impacts of disturbance. How much steady-state error is caused?
(f) Close the feedback loop and use a simple proportional gain inside the loop so that the closed loop poles have a damping ratio of about 0.7. Also include a forward gain (before the loop) to obtain a unity dc gain, refer to Figure 3.33. Simulate the closed-loop system and show its responses to the command of 1000 rpm and the disturbance of 10 N m (disturbance is applied when the system has reached its steady-state operation). Compare the results with the open-loop system in terms of its speed of response and the amount of steady-state error caused by the same disturbance.

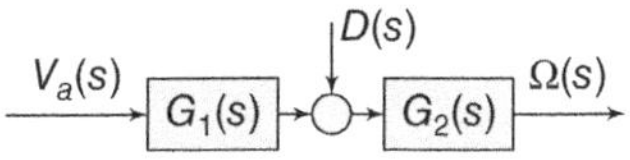

Figure 3.31 Block diagram representation for the dc motor speed control.

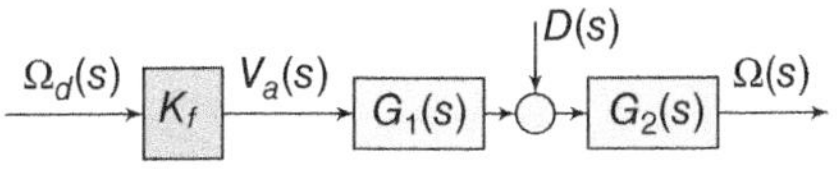

Figure 3.32 Open-loop control with just a feed-forward gain.

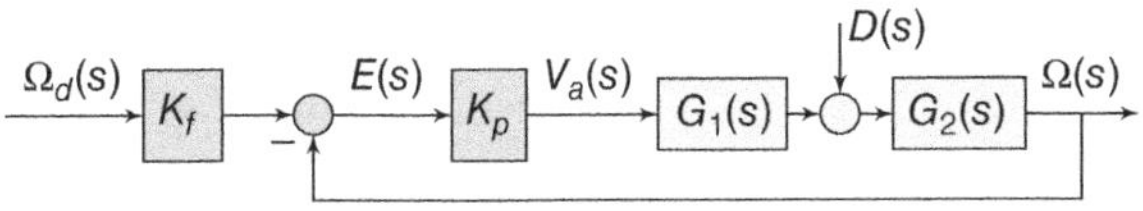

Figure 3.33 Closed-loop control using a P controller and a feed forward.

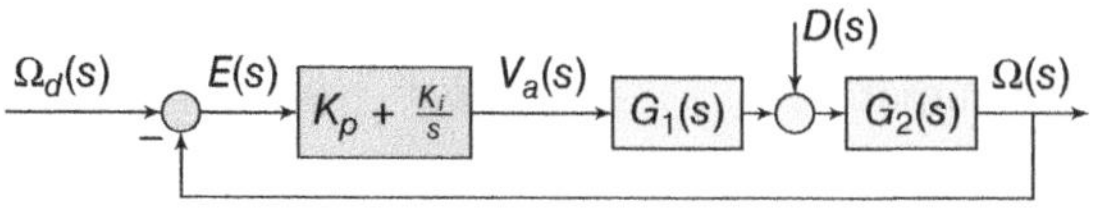

Figure 3.34 Closed-loop control using a PI controller.

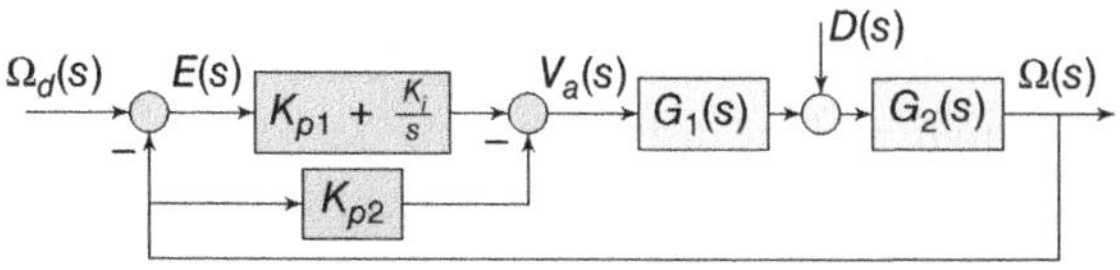

Figure 3.35 Alternative closed-loop control using a PI controller.

Repeat this part and place the poles such that they have a damping ratio of unity (critically damped). Also repeat for the damping ratio of 0.25.

(g) Close the feedback loop and use a PI controller: $C(s) = k\frac{s+z}{s}$, Figure 3.34. Choose the zero of the PI close to the origin and draw the root-locus for k varying from zero up to where the two other poles become identical. (Suggestion: $z = 0.4$ and $k = 5$). Do not include a forward gain (before the loop). Simulate the closed-loop system and show its responses to the command of 1000 rpm and the disturbance of 10 N m. Compare the results with the open-loop system in terms of its speed of response and the amount of steady-state error caused by the same disturbance.

(h) The P gain of the PI controller may be distributed between the error and the output terminals, Figure 3.35 with $K_{p_1} + K_{p_2} = K_p$. Show that by transferring part of this gain to the output path, the start-up response of the system can be modified. Repeat the simulation in part g by distributing the P gain and observing the responses. Show that you can reduce the peak value of the armature voltage (and current) by increasing the share of P on the output path. This will of course increase the starting transient time.

4

Time-Domain (State Space) Approach

This chapter reviews the basic concepts pertaining to state-space approach for the analysis and design of feedback control systems. The state-space approach has the advantage of working in time domain, and also allowing optimal design of controllers without worrying about stability, and even robustness (to certain degree). It is also well extendable to multivariable systems. Concepts such as linear state feedback, state estimation, and linear quadratic optimal control designs are discussed.

4.1 State Space Representation and Properties

State of a system at time t_0 is defined as the minimum amount of information that, together with the input $u(t)$ for all $t \geq t_0$ uniquely determine the system.

State-space representation of an LTI system is

$$\text{state space description}: \begin{cases} \dot{x}(t) = Ax(t) + Bu(t) & \leftarrow \text{state equation} \\ y(t) = Cx(t) + Du(t) & \leftarrow \text{output equation}, \end{cases} \tag{4.1}$$

where A, B, C, and D are constant matrices of dimensions $n \times n$, $n \times m$, $p \times n$, and $p \times m$, respectively. The n-dimensional vector $x(t)$ is called the *state vector*, and each element of it is a *state variable*. The variables $u(t)$ and $y(t)$ signify the control input and the output of interest. They are m and p number of them, respectively, in this formulation in general signifying a multi-input multi-output (MIMO) system. For a system with a single input and a single output (SISO), $m = p = 1$. In this discussion, all the vectors are normally column vectors unless stated otherwise.

MATLAB: ≫ `Sys = ss(A, B, C, D)`

The solution to state equations is

$$x(t) = \Phi(t)x(0) + \int_0^t \Phi(t-\tau)Bu(\tau)d\tau, \tag{4.2}$$

where $\Phi(t) = e^{At} = I + tA + \frac{1}{2}t^2A^2 + \frac{1}{6}t^3A^3 + \cdots$ is called the state transition matrix or the fundamental matrix, and I denotes the identity matrix.[1]

Remark The above equation confirms that if $x(0)$ and $u(\tau)$, $0 \leq \tau \leq t$ are known, then $x(t)$ and $y(t)$ are uniquely determined. Thus, $x(t)$ clearly satisfies the definition of "state."

1 See Problem 4.1.

Modeling and Control of Modern Electrical Energy Systems, First Edition. Masoud Karimi-Ghartemani.

4.1.1 Relationship between SS and TF

Take Laplace transform from the SS equations (4.1) results in

$$\begin{cases} sX(s) - x(0) = AX(s) + BU(s) \\ Y(s) = CX(s) + DU(s) \end{cases}$$

which leads to

$$\begin{cases} X(s) = (sI - A)^{-1}BU(s) + (sI - A)^{-1}x(0) \Rightarrow \Phi(t) = e^{At} = \mathscr{L}^{-1}\{(sI - A)^{-1}\} \\ Y(s) = [C(sI - A)^{-1}B + D]U(s) + C(sI - A)^{-1}x(0) \end{cases}$$

$$\Rightarrow G(s) = C(sI - A)^{-1}B + D. \tag{4.3}$$

Note that to obtain the TF for $Y(s) = G(s)U(s)$, the initial condition $x(0)$ is set to zero. Thus, for a given set of (A, B, C, D), the transfer function $G(s)$ is determined uniquely. Conversion from TF to SS is not, however, unique. Matlab gives a standard set of matrices (called the controllable canonical form) when it generates the matrices.

MATLAB: ≫ `[p, q] = ss2tf(Sys)`, `[A, B, C, D] = tf2ss(G)`

Eigenvalues of A are defined as the scalar (possibly complex) numbers λ that satisfy

$$Av = \lambda v$$

for a nonzero vector v. This equation may be written as

$$(\lambda I - A)v = 0.$$

The inverse matrix $(sI - A)^{-1}$ can be expressed as $\frac{1}{\Delta(s)}$ Adj$(sI - A)$ where $\Delta(s) = \det(sI - A)$ is given by

$$\Delta(s) = s^n + a_{n-1}s^{n-1} + \cdots + a_1 s + a_0 = (s - \lambda_1)(s - \lambda_2)\cdots(s - \lambda_n)$$

where λ_i for $i = 1,2, \ldots, n$ are eigenvalues of A. The matrix denoted by Adj$(sI - A)$ has polynomial entries and is called the adjoint matrix. Based on this analysis, the following facts can be inferred.

4.1.2 Facts

- Poles of $G(s)$ are included in the set of eigenvalues of A.
- The system is *bounded-input bound-state* stable if and only if all eigenvalues of A are on the open-left-half-plane (OLHP). Such a matrix A is called a Hurwitz or stable matrix. Notice that if the state variables are bounded, the output is necessarily bounded. The opposite is not necessarily true. A system may have bounded-input bounded-output (BIBO) stability (output being bounded) but some of its state variables not being bounded. In such cases, we say the system is internally unstable. A simple example is shown in Figure 4.1. The input–output TF of this system is $\frac{1}{s+1}$ and is stable. However, an internal variable (the one between the two blocks) is unstable.

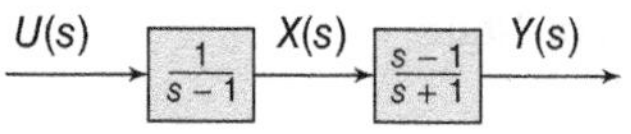

Figure 4.1 An internally unstable but input–output stable system.

Example 4.1 ***dc Motor Equations***
Consider the speed control of dc motor using the armature voltage at constant field as discussed in Chapter 3. The differential equations are summarized as

$$\begin{aligned} &\text{Armature circuit equation :} \quad v_a(t) = L_a \tfrac{d}{dt} i_a(t) + R_a i_a(t) + e_a(t) \\ &\text{Load rotation equation :} \quad \tau(t) = J \tfrac{d}{dt}\omega_m(t) + b\omega_m(t) + \tau'(t), \end{aligned} \tag{4.4}$$

where v_a and i_a are the armature voltage and current, R_a and L_a are its resistance and inductance, e_a is the induced voltage, τ is the induced torque, ω is the rotor speed, J and b are the moment of inertia and the coefficient of friction of the mechanical load and the rotor shaft, and τ' is the external opposing torque (regarded as disturbance). The induced voltage and torque satisfy $e_a = k_e\omega$ and $\tau = k_\tau i_a$ where k_e and k_τ are motor constants. The armature voltage v_a is the control input and ω_m is the output.

(a) Define the state variables as $x_1(t) = i_a(t)$ and $x_2(t) = \omega_m(t)$. Derive the state-space description of the dc motor as $\dot{x} = Ax + Bu + B_w w$, $y = Cx + Du + D_w w$ where w is the disturbance and write down the matrices A, B, B_w, C, D, and D_w.
The state-space description is

$$\dot{x}(t) = Ax(t) + Bu(t) + B_w w(t), \quad y(t) = Cx(t) + Du(t) + D_w w,$$

where the matrices and constants are easily identified by inspecting (4.4) as

$$A = \begin{bmatrix} -\frac{R_a}{L_a} & -\frac{k_e}{L_a} \\ \frac{k_\tau}{J} & -\frac{b}{J} \end{bmatrix}, \quad B = \begin{bmatrix} \frac{1}{L_a} \\ 0 \end{bmatrix}, \quad B_w = \begin{bmatrix} 0 \\ -\frac{1}{J} \end{bmatrix}, \quad C = [0 \;\; 1], D = 0, D_w = 0$$

and

$$x = \begin{bmatrix} i_a \\ \omega \end{bmatrix}, \quad u = v_a, \quad w = \tau', \quad y = \omega_m.$$

(b) Use the state space equations in part (a) to calculate the tracking and disturbance transfer functions $G(s) = \frac{Y(s)}{U(s)}$ and $G_2(s) = \frac{Y(s)}{W(s)}$.

$$G(s) = C(sI - A)^{-1}B + D = [0 \;\; 1]\begin{bmatrix} s + \frac{R_a}{L_a} & \frac{k_e}{L_a} \\ -\frac{k_\tau}{J} & s + \frac{b}{J} \end{bmatrix}^{-1} \begin{bmatrix} \frac{1}{L_a} \\ 0 \end{bmatrix}$$

$$= [0 \;\; 1]\frac{1}{\left(s + \frac{R_a}{L_a}\right)\left(s + \frac{b}{J}\right) + \frac{k_e k_\tau}{JL_a}} \begin{bmatrix} s + \frac{b}{J} & -\frac{k_e}{L_a} \\ \frac{k_\tau}{J} & s + \frac{R_a}{L_a} \end{bmatrix} \begin{bmatrix} \frac{1}{L_a} \\ 0 \end{bmatrix}$$

$$= \frac{\frac{k_\tau}{JL_a}}{\left(s + \frac{R_a}{L_a}\right)\left(s + \frac{b}{J}\right) + \frac{k_e k_\tau}{JL_a}} = \frac{k_\tau}{(L_a s + R_a)(Js + b) + k_e k_\tau}$$

$$G_2(s) = C(sI - A)^{-1}B_w + D_w = [0 \;\; 1]\begin{bmatrix} s + \frac{R_a}{L_a} & \frac{k_e}{L_a} \\ -\frac{k_\tau}{J} & s + \frac{b}{J} \end{bmatrix}^{-1} \begin{bmatrix} 0 \\ -\frac{1}{J} \end{bmatrix}$$

$$= [0 \;\; 1]\frac{1}{\left(s + \frac{R_a}{L_a}\right)\left(s + \frac{b}{J}\right) + \frac{k_e k_\tau}{JL_a}} \begin{bmatrix} s + \frac{b}{J} & -\frac{k_e}{L_a} \\ \frac{k_\tau}{J} & s + \frac{R_a}{L_a} \end{bmatrix} \begin{bmatrix} 0 \\ -\frac{1}{J} \end{bmatrix}$$

$$= -\frac{\frac{1}{J}\left(s + \frac{R_a}{L_a}\right)}{\left(s + \frac{R_a}{L_a}\right)\left(s + \frac{b}{J}\right) + \frac{k_e k_\tau}{JL_a}} = -\frac{L_a s + R_a}{(L_a s + R_a)(Js + b) + k_e k_\tau}$$

which are consistent with the derivations in Chapter 3, e.g. Eq. (3.7).

4.2 State Feedback

Assuming that the state vector is available, it can be used to construct a sort of internal feedback loop shown in Figure 4.2 and described by

$$u(t) = -Kx(t) + u'(t), \tag{4.5}$$

where K is an $m \times n$ matrix (for a single-input system, K is a row vector).

The modified system has the following state equations

$$\begin{aligned} \dot{x}(t) &= Ax(t) + Bu(t) = Ax(t) + B[-Kx(t) + u'(t)] = (A - BK)x(t) + Bu'(t) \\ y(t) &= Cx(t) + Du(t) = Cx(t) + D[-Kx(t) + u'(t)] = (C - DK)x(t) + Du'(t). \end{aligned}$$

This means that the matrix A is changed to $A - BK$. Since this matrix determines the basic stability and response properties of the system, the state feedback can significantly impact it.

4.2.1 Concept of Controllability

The system described by $\dot{x}(t) = Ax(t) + Bu(t)$ is said to be *controllable* if for any arbitrary initial and final conditions x_o and x_f and final time t_f, there exists a control input $u(t)$ for $0 \le t \le t_f$ that transfers the system from the initial condition x_o at time zero to final condition x_f at time t_f.

Theorem 4.1 *The system $\dot{x} = Ax + Bu$ is controllable if and only if the matrix $M_c = [B\ AB\ A^2B\ \cdots\ A^{n-1}B]$ is full rank. (Notice: this matrix is a square $n \times n$ matrix for a single-input system and the condition simply states that the matrix should be invertible.)*

Theorem 4.2 *The system $\dot{x} = Ax + Bu$ is controllable if and only if the matrix $[B\ \lambda I - A]$ is full rank for all λ. It is sufficient to check this condition for all λ belonging to the set of eigenvalues of A.*

Theorem 4.3 *If the system $\dot{x} = Ax + Bu$ is controllable, there exists a state feedback gain vector K that places the eigenvalues of $A - BK$ to any arbitrary set of locations in the complex plane. This state feedback gain is unique in a single-input system.*

MATLAB: ≫ `K = acker(A, B, p), K = place(A, B, p), p : desired eigenvalues`

4.2.2 Concept of Stabilizability

If the system $\dot{x} = Ax + Bu$ is not controllable, it means that there are some eigenvalues of A that cannot be moved by the state feedback. Those eigenvalues are all those that do not satisfy the rank condition in Theorem 4.2. Now, if all such uncontrollable eigenvalues (or modes) are on the OLHP, it implies that the state feedback can still entail a stable system. Otherwise, the state feedback cannot stabilize the system. So, *a stabilizable system is one whose uncontrollable modes are stable.* If a system is not stabilizable, the linear state feedback cannot stabilize it.

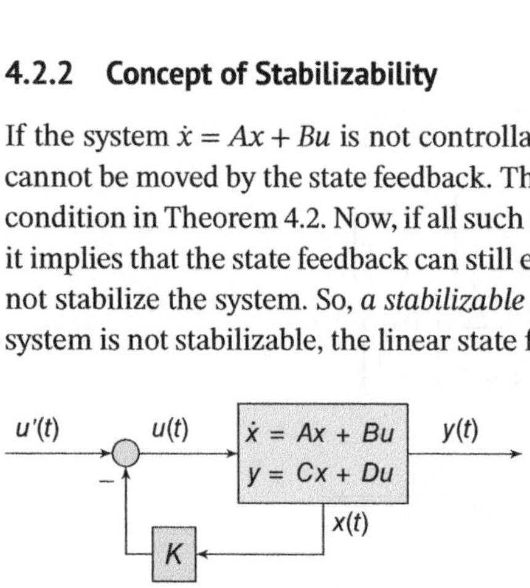

Figure 4.2 State feedback concept.

Example 4.2 ***Controllability of dc Motor Control***
Show that the state-space model of dc motor of Example 4.1 is controllable.

$$AB = \begin{bmatrix} -\frac{R_a}{L_a} & -\frac{k_e}{L_a} \\ \frac{k_\tau}{J} & -\frac{b}{J} \end{bmatrix} \begin{bmatrix} \frac{1}{L_a} \\ 0 \end{bmatrix} = \begin{bmatrix} -\frac{R_a}{L_a^2} \\ \frac{k_\tau}{JL_a} \end{bmatrix}, \quad [B \;\; AB] = \begin{bmatrix} \frac{1}{L_a} & -\frac{R_a}{L_a^2} \\ 0 & \frac{k_\tau}{JL_a} \end{bmatrix}$$

which has its determinant equal to $\frac{k_\tau}{JL_a^2}$ which is obviously nonzero. Therefore, the matrix $[A \;\; AB]$ is nonsingular meaning that the system is controllable.

4.2.3 Removing Steady-State Error

The previous discussion entails that if a system is controllable, its modes (or eigenvalues or poles) that basically determine its transient responses can be completely manipulated using a full-state linear feedback of $-Kx(t)$. As far as the steady-state error is concerned, if the desired output or the command function $y_d(t)$ has a dc nature, we can use a feedforward scalar gain k_o (for a SISO system) as shown in Figure 4.3 to ensure zero steady-state error.

The state-space equations of the system of Figure 4.3 are

$$\dot{x}(t) = (A - BK)x(t) + Bk_o y_d(t)$$
$$y(t) = (C - DK)x(t) + Dk_o y_d(t)$$

which amount to the transfer function of

$$\frac{Y(s)}{Y_d(s)} = k_o[(C - DK)(sI - A + BK)^{-1}B + D].$$

In order to have a unity gain, k_o should satisfy

$$k_o = \frac{1}{(C - DK)(-A + BK)^{-1}B + D}.$$

This way of adjusting the steady-state gain is, however, not robust to system parameters and disturbances. The output may readily shift away from the desired value due to either of these factors. An alternative structure to achieve robust steady-state error removal is to use an integrator as shown in Figure 4.4. This structure also allows more stringent control of the output variable y through proper design of controller gains.

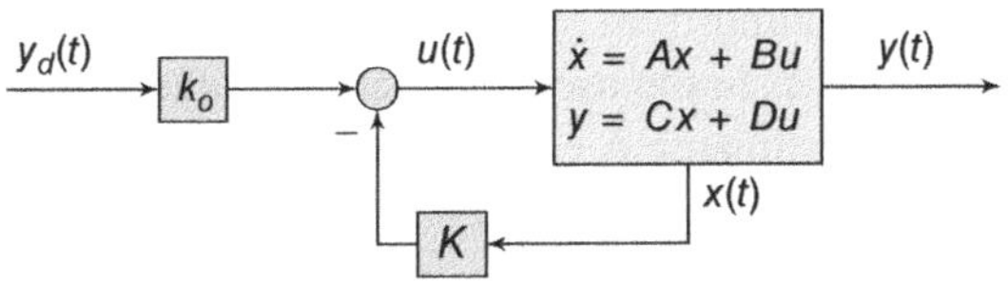

Figure 4.3 Feedforward gain to remove steady-state error.

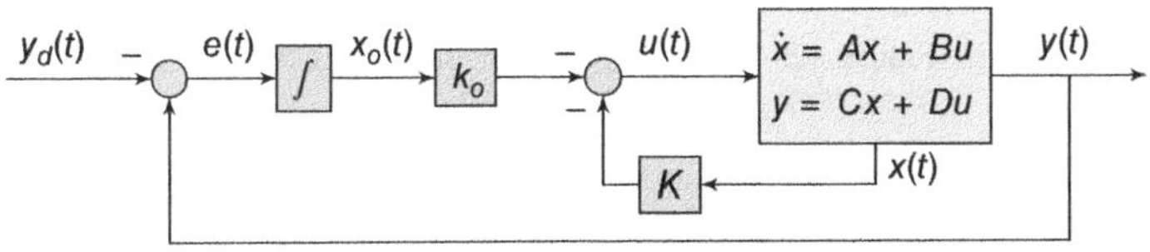

Figure 4.4 Combined state feedback and output feedback.

In system of Figure 4.4, $e(t) = y(t) - y_d(t) = Cx(t) + Du(t) - y_d(t)$ is the tracking error. The state equation for the integrator variable is $\dot{x}_o(t) = e(t) = Cx(t) + Du(t) - y_d(t)$. The whole system has $n+1$ state variables. Let us augment them and define $\bar{x}(t) = \begin{bmatrix} x_o(t) \\ x(t) \end{bmatrix}$. Then,

$$\dot{\bar{x}}(t) = \begin{bmatrix} 0 & C \\ 0 & A \end{bmatrix} \bar{x}(t) + \begin{bmatrix} D \\ B \end{bmatrix} u(t) + \begin{bmatrix} -1 \\ 0 \end{bmatrix} y_d(t) = \bar{A}\bar{x}(t) + \bar{B}u(t) + \bar{B}_d y_d(t) \tag{4.6}$$

and the expression for the control input is

$$u(t) = -k_o x_o(t) - Kx(t) = -[k_o \quad K]\bar{x}(t) = -\bar{K}\bar{x}(t).$$

Therefore, $\bar{K}$ can be readily obtained based on any selection of $n+1$ arbitrary locations for the eigenvalues of the closed-loop system, for example by running the Matlab command $Kbar = acker(Abar, Bbar, p)$. Of course the assumption is that $(\bar{A}, \bar{B})$ must be controllable. It can be proved that if (A, B) is controllable and (C, A) is observable, then $(\bar{A}, \bar{B})$ is controllable, see Problem 4.2.

4.2.4 Challenges with State Feedback Method

1. State variables should all be available for feedback. Many number of sensors may be required. Solution: state estimator or observer.
2. How to choose desired location of closed-loop poles? Solution: optimal control.

4.3 State Estimator

Objective of the state estimator (or observer) is to estimate (or observe) the state of the system using its input–output measurements. Consider a plant with the state-space description of

$$\begin{aligned} \dot{x}(t) &= Ax(t) + Bu(t) \\ y(t) &= Cx(t) + Du(t). \end{aligned}$$

The state feedback loop using a state estimator is shown in Figure 4.5 where $\hat{x}(t)$ is the estimated state vector.

The estimator should follow the plant dynamics. Therefore, it basically emulates the plant. However, in order to reduce the possible errors originating from initial conditions mismatch between the plan and the estimator, or due to uncertainties in the used parameters, or due to the plant disturbances, an error term is also needed by the estimator. A common estimator structure known as the Luenberger estimator is described by

$$\begin{aligned} \dot{\hat{x}}(t) &= A\hat{x}(t) + Bu(t) + H[y(t) - \hat{y}(t)] \\ \hat{y}(t) &= C\hat{x}(t) + Du(t). \end{aligned}$$

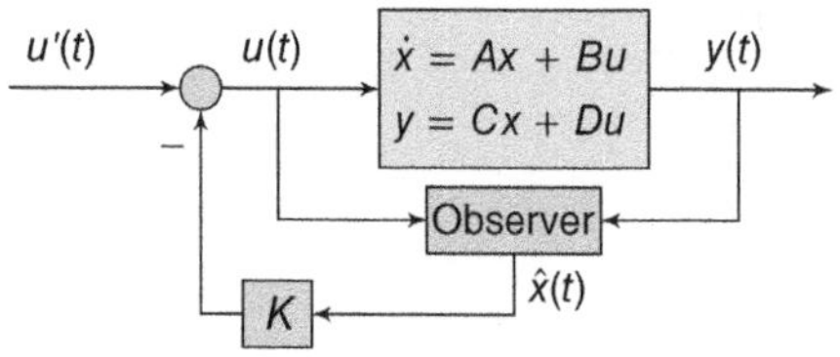

Figure 4.5 State feedback using state estimator (or observer.)

In order to understand why this works, consider the estimation error $e(t) = x(t) - \hat{x}(t)$, then

$$\begin{aligned}\dot{e}(t) &= \dot{x}(t) - \dot{\hat{x}}(t)\\ &= Ax(t) + Bu(t) - A\hat{x}(t) - Bu(t) - H[Cx(t) + Du(t) - C\hat{x}(t) - Du(t)]\\ &= (A - HC)e(t).\end{aligned}$$

Thus, if the matrix $A - HC$ is stable (Hurwitz), the estimation error goes to zero.
Note: $(A - HC)^T = A^T - C^T H^T$ where T denotes matrix transposition. We know that eigenvalues of a matrix do not change when it is transposed. Therefore, the same acker command in Matlab command can easily be adjusted to calculate H as

MATLAB: ≫ `H = (acker(A', C', q))'`

where q is the location of estimator poles and $'$ shows matrix transposition in Matlab.

4.3.1 How to Choose the Estimator's Poles?

The estimator should be faster than the control loop. In other words, the estimated state should be available before being used by the feedback. Therefore, a rule of thumb says that the estimator poles should be several times (3–10 times) faster than the controller poles. Fastness is determined by the distance from the imaginary axis. This rule still does not give more information on where to actually choose them! Furthermore, are the estimator poles separated from the controller poles. In other words, can the estimator gain H and the controller gain K be designed separately?

4.3.2 Separation Property

This property shows that the estimator and controller are theoretically separated and can be designed separately. The system of Figure 4.5 is described by

$$\begin{aligned}\dot{x}(t) &= Ax(t) + Bu(t)\\ \dot{\hat{x}}(t) &= A\hat{x}(t) + Bu(t) + H[y(t) - \hat{y}(t)]]\\ y(t) &= Cx(t) + Du(t)\\ \hat{y}(t) &= C\hat{x}(t) + Du(t)\\ u(t) &= -K\hat{x}(t) + u'(t)\end{aligned}$$

which shows a state space equation of order $2n$. If we augment the plant and the estimator states as $\begin{bmatrix} x(t) \\ \hat{x}(t) \end{bmatrix}$, we end up with

$$\begin{bmatrix} \dot{x}(t) \\ \dot{\hat{x}}(t) \end{bmatrix} = \begin{bmatrix} A & -BK \\ HC & A - BK \end{bmatrix} \begin{bmatrix} x(t) \\ \hat{x}(t) \end{bmatrix} + \begin{bmatrix} B \\ B \end{bmatrix} u'(t).$$

Alternatively, if we augment the plant and the estimator states as $\begin{bmatrix} x(t) \\ e(t) \end{bmatrix}$ where $e(t) = x(t) - \hat{x}(t)$, we end up at

$$\begin{bmatrix} \dot{x}(t) \\ \dot{e}(t) \end{bmatrix} = \begin{bmatrix} A - BK & BK \\ 0 & A - HC \end{bmatrix} \begin{bmatrix} x(t) \\ e(t) \end{bmatrix} + \begin{bmatrix} B \\ 0 \end{bmatrix} u'(t).$$

This last equation is interesting in the sense that it shows that the eigenvalues of the combined estimator/controller system is simply the union of the eigenvalues of the controller and those of the estimator.[2] Therefore, the two components seem to be somewhat separate! It means that the estimator can be designed independently from the controller.

2 Notice that eigenvalue of a block-diagonal matrix is equal to the union of eigenvalues of the submatrices on its diagonal.

In practice, uncertainties may contribute to violate this separation property. Therefore, the estimator design should be performed with care in systems with high level of uncertainties. Avoiding unnecessarily high-speed estimator can help reduce the side effects of uncertainties and unmodeled dynamics on causing undesired interactions between the estimator and the controller.

4.3.3 Conditions for Existence of Estimator Gain *H*

Based on the above derivations, it is readily concluded that the vector H uniquely exists for any set of observer eigenvalues q if the pair (A^T, C^T) is controllable. According to Theorem 4.1, this means the matrix

$$\left[C^T \; A^T C^T \; \left(A^T\right)^2 C^T \; \cdots \; \left(A^T\right)^{n-1} C^T\right]$$

should have full rank. Full-rankness of a matrix and its transpose are equivalent. Therefore, the matrix

$$M_o = \begin{bmatrix} C \\ CA \\ CA^2 \\ \vdots \\ CA^{n-1} \end{bmatrix}$$

should be full rank.

4.3.4 Concept of Observability

The system described by

$$\dot{x}(t) = Ax(t) + Bu(t)$$
$$y(t) = Cx(t) + Du(t)$$

is called *observable* if the initial condition $x(0)$ can be determined based on the output measurement $y(t)$ during a finite time.

Theorems 4.4–4.6 can be stated in parallel with the controllability theorems.

Theorem 4.4 *The system is observable if and only if the matrix M_o (given above) is full rank. (Notice: this matrix is a square $n \times n$ matrix for a single-input system and the condition simply states that the matrix should be invertible.)*

Theorem 4.5 *The system is observable if and only if the matrix $\begin{bmatrix} C \\ \lambda I - A \end{bmatrix}$ is full rank for all λ belonging to the set of eigenvalues of A.*

Theorem 4.6 *If the system is observable, there exists a observer gain vector H (unique for single-output systems) that places the eigenvalues of $A - HC$ to any arbitrary set of locations in the complex plane.*

4.3.5 Concept of Detectability

If the system is not observable, it means that there are some modes of the system that cannot be observed by the estimator. Those modes correspond to those eigenvalues of A that do not satisfy the rank condition in Theorem 4.5. Now, if all such unobservable eigenvalues (or modes) are on the OLHP, it implies that the estimator can entail a stable system. Otherwise, the observer is unstable. So, *a detectable system is one whose unobservable modes are stable.*

Conclusion is that if a system is observable, all its modes can be completely estimated using the above estimator. If it is detectable, the estimator cannot estimate all the modes but it can entail a stable loop. If the system is not detectable, this estimator cannot work and the loop will be unstable.

Remark There are observers other than Luenberger observer. Reduced-order observers are used to estimate only a subset of state variables (e.g. in case that some variables are available through measurements or are unobservable). Integral observers add an additional integral term in the estimator dynamics: $H[y(t) - \hat{y}(t)] + L \int [y(t) - \hat{y}(t)]dt$. Their design is more complicated but they can offer better performance. Kalman filter concept and linear quadratic Gaussian (LQG) theory offer methods for optimizing the estimator design. Finally, in some practical applications where the measured variables are noisy, their estimated values (together with the other estimated variables) may be used in the state feedback since, in any case, the estimator acts like a filter and can improve the noisy conditions.

Example 4.3 ***Observability of dc Motor Speed Control***
Show that the state-space model of dc motor in Example 4.1 is observable.

$$CA = [0 \;\; 1] \begin{bmatrix} -\frac{R_a}{L_a} & -\frac{k_e}{L_a} \\ \frac{k_\tau}{J} & -\frac{b}{J} \end{bmatrix} = \begin{bmatrix} \frac{k_\tau}{J} & -\frac{b}{J} \end{bmatrix}, \quad \begin{bmatrix} C \\ CA \end{bmatrix} = \begin{bmatrix} 0 & 1 \\ \frac{k_\tau}{J} & -\frac{b}{J} \end{bmatrix}$$

which has its determinant equal to $-\frac{k_\tau}{J}$ which is obviously nonzero. Therefore, the system is observable.

4.4 Optimal Control

Optimal control theory deals with formulating and finding the control input that when applied to the system makes the system perform in an optimal way. This optimal way is normally expressed in terms of minimizing a cost function or a performance measure. Here, the standard linear quadratic regulator (LQR) problem together with some new developments on it are reviewed in a brief yet effective way.

4.4.1 Linear Quadratic Regulator (LQR)

Consider the system of Figure 4.6 and assume that no external input is applied. If K is such that $A - BK$ is stable, then $x(t)$ will tend to zero as t increases. In other words, $x(t)$ will be *regulated* to zero. In other words, if the system is released from the initial condition $x(0)$, and if $A - BK$ is stable, then $x(t) \to 0$ as $t \to \infty$. The control input $u(t)$ will also obviously go to zero. There are infinite solutions for K to make $A - BK$ stable and to regulate the state to zero. Which one is the "best"? Can we perform the regulation process in some optimal fashion?

To answer this question, one way is to define a quadratic cost function as

$$J = \int_0^\infty \left[x(t)^T Q x(t) + u(t)^T R u(t)\right] dt \tag{4.7}$$

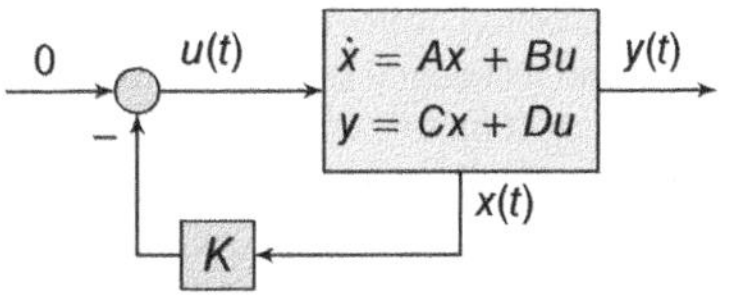

Figure 4.6 Regulation problem.

where Q is a positive semi-definite (PSD) matrix of dimension $n \times n$ and R is a positive definite (PD) matrix of dimension $m \times m$ where n is the number of state variables and m is the number of inputs.[3] A common form of PD (or PSD) matrix is a diagonal matrix with positive (or non-negative) numbers on the diagonal. Also notice that for a single-input system, $m = 1$, and R is a real positive number. Without losing generality, it can be assumed equal to unity and the cost function reduces to

$$J = \int_0^\infty \left[x(t)^T Qx(t) + u(t)^2\right] dt = \int_0^\infty \left[q_1x_1(t)^2 + q_2x_2(t)^2 + \cdots + q_nx_n(t)^2 + u(t)^2\right] dt \tag{4.8}$$

where q_i's are the diagonal elements of the matrix Q and they are all nonnegative. This cost function obviously has clear meaning. Integral of square of a variable represents the energy.[4] Minimizing this function means achieving the regulation in an optimal fashion. The coefficients q_i's simply serve as weight factors to impose different levels of significance on different variables.[5]

Is there a K that minimizes J? If so, such a K is of interest in the sense that minimizing J means *keeping state vector $x(t)$ close to zero without excessive expenditure of control $u(t)$*. Each state variable that has a larger q is given a stronger weight in the cost function.

Theorem 4.7 *Write Q as $Q = Q_1^TQ_1$. If (A, B) is controllable (even stabilizable) and (Q_1, A) is observable (even detectable), there exists a unique K that minimizes the cost function J. Moreover, the eigenvalues of $A - BK$ (closed-loop system) are on the OLHP. Furthermore, the optimal K is derived from the following matrix equation, called the algebraic Riccati equation*

$$Q + LA + A^TL - LBR^{-1}B^TL = 0, \quad K = R^{-1}B^TL.$$

Remark For a diagonal Q with diagonal elements q_1 to q_n, $Q_1 = [\sqrt{q_1}\ \sqrt{q_2}\ \cdots\ \sqrt{q_n}]$.

Remark The LQR solution defined above is an optimal solution that results in a closed-loop with "good" robustness properties such as phase margin and gain margin and also robustness to nonlinearities and unmodeled dynamics [1, 2].

MATLAB: ≫ `K = lqr(A, B, Q, R)`

---

3 A symmetric matrix M ($M = M^T$) is called PD if for all nonzero vector x in n-dimensional space, the product x^TMx is positive. In other words, M is PD if and only if $x^TMx > 0$ for all real nonzero vectors x. Similarly, the symmetric matrix M is PSD if and only if $x^TMx \geq 0$ for all real nonzero vectors x. A symmetric matrix M is negative definite (ND) or negative semi-definite (NSD) if $-M$ is PD or PSD, respectively.
Following properties can be easily proved.

- All eigenvalues of a real symmetric matrix are real.
- All eigenvalues of a PD matrix are real and positive.
- All eigenvalues of a PSD matrix are real and nonnegative.
- For any symmetric matrix M, we have $\underline{\lambda} \parallel x\parallel^2 \leq x^TMx \leq \overline{\lambda} \parallel x\parallel^2$ where $\underline{\lambda}$ is the smallest eigenvalue and $\overline{\lambda}$ is the largest eigenvalue of M. This equation implies that $\underline{\lambda} \leq \frac{x^TMx}{\|x\|^2} \leq \overline{\lambda}$. Thus the function $\frac{x^TMx}{\|x\|^2}$ has its maximum (or minimum) equal to $\overline{\lambda}$ (or $\underline{\lambda}$) and is achieved when x is equal to the eigenvector corresponding to the maximum (or minimum) eigenvalue.

4 For example, consider the voltage across a resistor or the current flowing in it. Its square is proportional to the power and its integral is proportional to the energy consumed by the resistor.
5 It is clear from this explanation why R must be PD while Q is allowed to be PSD because the optimal controller cannot ignore the energy of the control input, otherwise it may end up at a solution that is not practical, i.e. requires infinite energy.

4.4.2 Linear Quadratic Tracker (LQT)

There could be multiple approaches to form a control loop structure and also to the design of its gain using LQR approach. We put them all under linear quadratic tracker (LQT) terminology. Some LQT approaches are discussed in this section first and then a robust LQT approach is developed and discussed.

4.4.2.1 LQT Without Direct Output Feedback

Consider the state-feedback and forward gain structure shown in Figure 4.7. We already discussed that if $k_o = \frac{1}{(C-Dk)(-A+Bk)^{-1}B+D}$, the system will have no steady-state error to step commands.

In Figure 4.7, any state feedback gain that makes $A - BK$ stable will remove the steady state error if k_o is properly calculated. The tracking error $e(t) = y(t) - y_d$ will go to zero as t increases. Is it possible to find a K to perform the tracking in an optimal way? Notice that here $x(t)$ does not go to zero and the cost function (4.7) is not a well-defined function because it will diverge as time increases. However, the output $y(t)$ will eventually settle to y_d. Consequently, $x(t)$ will settle to x_d and $u(t)$ will settle to u_d. One candid cost function may be defined as

$$J = \int_0^\infty \left[qe(t)^2 + \tilde{u}(t)^2\right] dt, \tag{4.9}$$

where $e(t) = y(t) - y_d$ is the tracking error, q is a positive constant and $\tilde{u}(t) = u(t) - u_d$. Note that both $e(t)$ and $\tilde{u}(t)$ tend to zero as t increases.

Problem

Find a K that minimizes (4.9).

Solution:

Define $\tilde{x}(t) = x(t) - x_d$ then

$$\dot{\tilde{x}}(t) = \dot{x}(t) = Ax(t) + Bu(t) = A[\tilde{x}(t) + x_d] + B[\tilde{u}(t) + u_d] = A\tilde{x}(t) + B\tilde{u}(t)$$

since $Ax_d + Bu_d = 0$. Meanwhile, the tracking error can be written as

$$e(t) = y(t) - y_d = Cx(t) + Du(t) - y_d = C[\tilde{x}(t) + x_d] + D[\tilde{u}(t) + u_d] - y_d = C\tilde{x}(t) + D\tilde{u}(t)$$

since $y_d - Cx_d - Du_d = 0$. Therefore, the cost function can be rewritten as

$$J = \int_0^\infty \left[q\tilde{x}^T C^T C\tilde{x} + \tilde{u}^2 + 2\tilde{x}^T C^T D\tilde{u}\right] dt.$$

For the common case of $D = 0$, it reduces to

$$J = \int_0^\infty \left[q\tilde{x}(t)^T C^T C\tilde{x}(t) + \tilde{u}(t)^2\right] dt.$$

Moreover, notice that

$$\tilde{u}(t) = u(t) - u_d = -Kx(t) + k_o y_d - u_d = -K[\tilde{x}(t) + x_d] + k_o y_d - u_d = -K\tilde{x}(t)$$

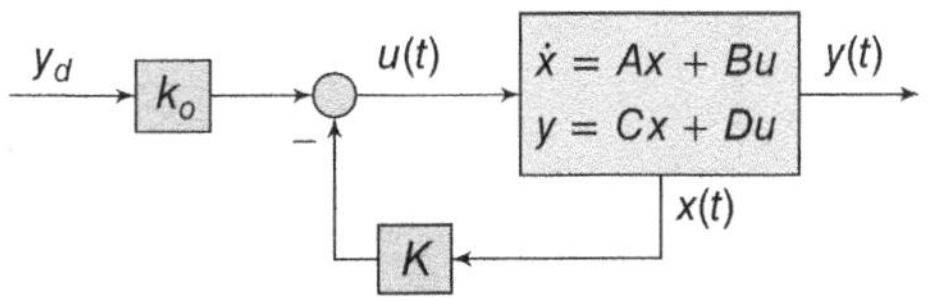

Figure 4.7 Tracking problem (feedforward gain with no output feedback.)

because $-Kx_d + k_o y_d - u_d = 0$. Therefore we are encountering a pure LQR problem and a solution to minimize this cost function can be obtained from LQR theory. In Matlab, we need to run $K = lqr(A, B, Q, 1)$ where $Q = qC^T C$. The conditions for existence/uniqueness of a stabilizing solution is that (A, B) be stabilizable and (C, A) be detectable.

As mentioned previously, the way to remove the steady state error as shown in Figure 4.7 is not robust. Presence of an uncertainty and/or a disturbance will drift the output away from its desired value because k_o is designed based on the nominal values of system parameters and the disturbance is not formulated. Moreover, this structure does not offer a direct way of manipulating and having control on the output. The following approach, LQT with robust tracking, addresses these issues.

4.4.2.2 Robust LQT with Direct Output Feedback

In order to achieve an optimal yet *robust* tracking of step commands in the presence of disturbances and uncertainties, an output feedback with an integrating block is used as shown in Figure 4.8. Notice that we have included the disturbance input $w(t)$ in the formulation in order to directly address it as well. Here, we assume that the command and the disturbance have a dc nature and are modeled by step functions.

In system of Figure 4.8, $e(t) = y - y_d(t) = Cx(t) + Du(t) - y_d$ is the tracking error. The state equation for the integrator variable is $\dot{x}_o(t) = e(t) = Cx(t) + Du(t) - y_d$. The whole system has $n + 1$ state variables. Let us augment them and define $\bar{x}(t) = \begin{bmatrix} x_o(t) \\ x(t) \end{bmatrix}$ as the state vector of the entire system. Then,

$$\begin{aligned}\dot{\bar{x}}(t) &= \begin{bmatrix} 0 & C \\ 0 & A \end{bmatrix} \bar{x}(t) + \begin{bmatrix} D \\ B \end{bmatrix} u(t) + \begin{bmatrix} -1 \\ 0 \end{bmatrix} y_d(t) + \begin{bmatrix} 0 \\ B_w \end{bmatrix} w(t) \\ &= \bar{A}\bar{x}(t) + \bar{B}u(t) + \bar{B}_d y_d(t) + \bar{B}_w w(t)\end{aligned} \tag{4.10}$$

and the expression for the control input is

$$u(t) = -k_o x_o(t) - Kx(t) = -[k_o \;\; K]\bar{x}(t) = -\bar{K}\bar{x}(t).$$

We discuss two approaches to address this problem. Will show that the first approach, while appearing to be perfectly sound, does not yield a stable solution! Then, an innovative approach is presented.

4.4.2.3 Elementary Design Approach (Unstable!)

It is shown in this section that an attempt to develop an elementary approach for optimally designing the controller gains of Figure 4.8 fails. An efficient method is then presented in Section 4.4.2.4.

Assume that for a solution $\bar{K}$, the loop is stable. Also, for a given values of y_d and w, let us assume that the state $\bar{x}(t)$ converges to $\bar{x}_d$ and the control input $u(t)$ converges to u_d. Notice that the tracking error goes to zero as long as the loop is stable. Define $\tilde{x}(t) = \bar{x}(t) - \bar{x}_d$ and $\tilde{u}(t) = u(t) - u_d$. Notice that $\bar{A}\bar{x}_d + \bar{B}u_d + \bar{B}_d y_d + \bar{B}_w w = 0$ and $u_d = -\bar{K}\bar{x}_d$. Then, by combining these facts, it can readily be concluded that

$$\dot{\tilde{x}}(t) = \bar{A}\tilde{x}(t) + \bar{B}\tilde{u}(t). \tag{4.11}$$

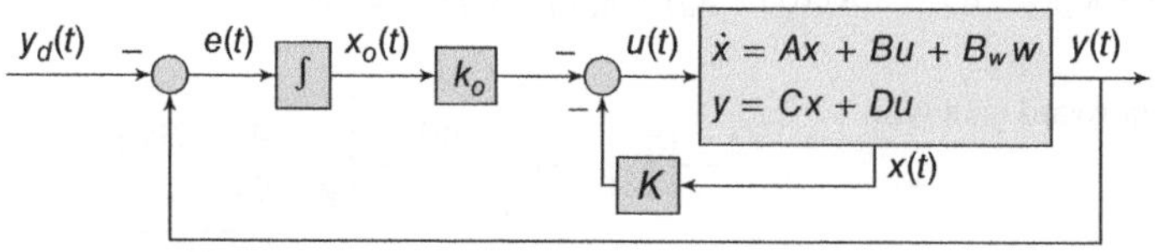

Figure 4.8 Tracking problem (robust tracking using output feedback and integrator).

Also notice that $\tilde{u}(t) = u(t) - u_d = -\overline{K}\overline{x} - \overline{K}\overline{x}_d = -\overline{K}\tilde{x}$. These two facts mean that now we are dealing with a regulation problem.

Problem

Find $\overline{K}$ that minimizes the cost function

$$J = \int_0^{\infty} \left[qe(t)^2 + \tilde{u}(t)^2\right] dt \tag{4.12}$$

for the system of (4.11).

Solution:
Notice that $e(t) = Cx(t) + Du(t) - y_d = \overline{C}\overline{x}(t) + Du(t) - y_d$ where $\overline{C} = [0 \;\; C]$. Subsequently, $e(t) = \overline{C}\tilde{x}(t) + D\tilde{u}(t)$. Let $D = 0$ for simplicity. Then, $e(t) = \overline{C}\tilde{x}(t)$. Thus, the cost function may be written as

$$J = \int_0^{\infty} \left[q\tilde{x}(t)^T\overline{C}^T\overline{C}\tilde{x}(t) + \tilde{u}(t)^2\right] dt. \tag{4.13}$$

This means that the LQR problem with matrix $Q = q\overline{C}^T\overline{C}$ should be solved.

Theorem 4.8 *The above LQR problem does not have a stabilizing solution.*

Proof: For the solution to make the loop stable, $(\bar{A}, \overline{B})$ should be stabilizable and $(\overline{C}, \bar{A})$ should be detectable. We show that the latter condition does not hold [3]. To do this, we show that there is some λ such that the matrix $\begin{bmatrix} \lambda I - \bar{A} \\ \overline{C} \end{bmatrix}$ is not full rank. This means that there exists a nonzero vector v such that

$$\begin{bmatrix} \lambda I - \bar{A} \\ \overline{C} \end{bmatrix} v = \begin{bmatrix} \lambda & -C \\ 0 & \lambda I - A \\ 0 & C \end{bmatrix} \begin{bmatrix} v_1 \\ v_2 \end{bmatrix} = 0.$$

The value $\lambda = 0$ is an eigenvalue of $\bar{A}$ (due to the presence of integrator). Choose $v_1 = 1$ and $v_2 = 0$. For this choice and for the value of $\lambda = 0$ (which is an unstable one), the above equation holds. Therefore, $(\overline{C}, \bar{A})$ is not detectable. ■

4.4.2.4 LQT Design for Step Commands and Step Disturbances

Consider Eq. (4.10) and apply the operator $\frac{d}{dt}$ to both sides of it. Note that since y_d and w are constant values, their derivative is zero. Therefore,

$$\dot{z}(t) = \bar{A}z(t) + \overline{B}v(t) \tag{4.14}$$

where $z(t) = \dot{\overline{x}}(t)$ and

$$v(t) = \dot{u}(t) = \frac{d}{dt}\left[-\overline{K}\overline{x}(t)\right] = -\overline{K}z(t). \tag{4.15}$$

Equations (4.14) and (4.15) describe a regulation problem. In other words, by applying the transformation $\frac{d}{dt}$, the tracking problem is converted to a regulation problem. More interesting than that is the fact that the first element of the transformed state vector $z(t)$ is equal to the tracking error! In order to see this, it is enough to notice that the first element of $\overline{x}(t)$ is $x_o(t)$.

Problem

Find $\overline{K}$ that minimizes the cost function

$$J = \int_0^{\infty} \left[qe(t)^2 + v(t)^2 \right] dt \tag{4.16}$$

for the system of (4.14).

***Solution*:**

Notice that both $e(t)$ and $v(t)$ tend to zero as time passes if the loop is stable. Therefore, this cost function is well-defined. Based on the above discussion, this cost function is equal to

$$J = \int_0^{\infty} \left[qz_1(t)^2 + v(t)^2 \right] dt, \tag{4.17}$$

where $z_1(t)$ is the first element of $z(t)$. This immediately concludes that the solution corresponds to an LQR problem with a matrix Q whose only nonzero element is q at the first row first column location:

$$Q = \begin{bmatrix} q & 0 & \cdots & 0 \\ 0 & 0 & \cdots & 0 \\ \vdots & \vdots & \ddots & \vdots \\ 0 & 0 & \cdots & 0 \end{bmatrix}$$

and R is unity. Is there a solution for this problem?

Theorem 4.9 *The LQR problem with matrices $\bar{A}$, $\overline{B}$, above Q and $R = 1$ has a unique and stable solution.*

Proof: For the solution to make the loop stable, $(\bar{A}, \overline{B})$ should be stabilizable and $(Q_1, \bar{A})$ should be detectable where $Q_1 = [1 \;\; 0 \;\; \cdots \;\; 0]$. The proof can be found in [[3], Theorem 3]. ■

This approach is the true way of addressing the robust LQT problem. Some developments and applications of this method for various applications are reported in [3, 4].

Example 4.4 ***LQT Controller Design for dc Motor***

For the dc motor speed control of Example 4.1, design an optimal and robust tracking controller using LQT approach discussed above. A step response with rise-time of less than 0.5 seconds and an over-shoot not exceeding 10% are desired. Consider the following set of numerical values

$$R_a = 1\ \Omega,\ L_a = 0.2\ \text{H},\ J = 10\ \text{kg m}^2,\ b = 0.1,\ k_e = k_\tau = 2.$$

The state-space description is

$$\dot{x}(t) = Ax(t) + Bu(t) + B_w w(t), \;\; y(t) = Cx(t) + Du(t) + D_w w,$$

where

$$A = \begin{bmatrix} -\dfrac{R_a}{L_a} & -\dfrac{k_e}{L_a} \\ \dfrac{k_\tau}{J} & -\dfrac{b}{J} \end{bmatrix} = \begin{bmatrix} -5 & -10 \\ 0.2 & -0.01 \end{bmatrix}, \; B = \begin{bmatrix} \frac{1}{L_a} \\ 0 \end{bmatrix} = \begin{bmatrix} 5 \\ 0 \end{bmatrix}, \; B_w = \begin{bmatrix} 0 \\ -\frac{1}{J} \end{bmatrix} = \begin{bmatrix} 0 \\ -0.1 \end{bmatrix}$$

and $C = [0 \;\; 1]$, $D = D_w = 0$. The matrices $\bar{A}$ and $\overline{B}$ are defined in (4.10) and calculated as

$$\bar{A} = \begin{bmatrix} 0 & C \\ 0 & A \end{bmatrix} = \begin{bmatrix} 0 & 0 & 1 \\ 0 & -5 & -10 \\ 0 & 0.2 & -0.01 \end{bmatrix}, \; \overline{B} = \begin{bmatrix} -D \\ B \end{bmatrix} = \begin{bmatrix} 0 \\ 5 \\ 0 \end{bmatrix}.$$

The locus of closed-loop poles (which are the eigenvalues of $\bar{A} - \bar{A}\overline{K}$) versus q are shown in Figure 4.9. The poles are originally at −0.45, −4.56 and 0. They move as shown in Figure 4.9 when q is increased up to 10^4. Step responses of the closed-loop system for three values of q are shown in Figure 4.10. The response becomes faster (without noticeable change in its overshoot) as q increases. For these three values of $q = 10^4,\ 10^5,\ 10^6$, the state feedback gains and the location of closed-loop poles are equal to $\overline{K} = [100,\ 1.13,\ 44.1]$, $-2.5 \pm j3.4, -5.6$; $\overline{K} = [316,\ 1.92,\ 94]$, $-3.6 \pm j5.5, -7.4$; and $\overline{K} = [1000,\ 3.14,\ 201]$, $-5.2 \pm j8.4, -10.4$; respectively. The 0.5 seconds rise-time and below 10% overshoot design goals are met at $q = 10^5$.

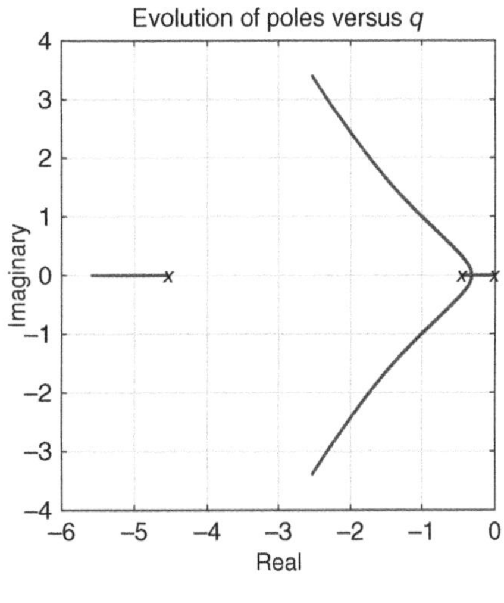

Figure 4.9 Locus of closed-loop poles of dc motor control with optimal robust control design when q increases up to 10^4.

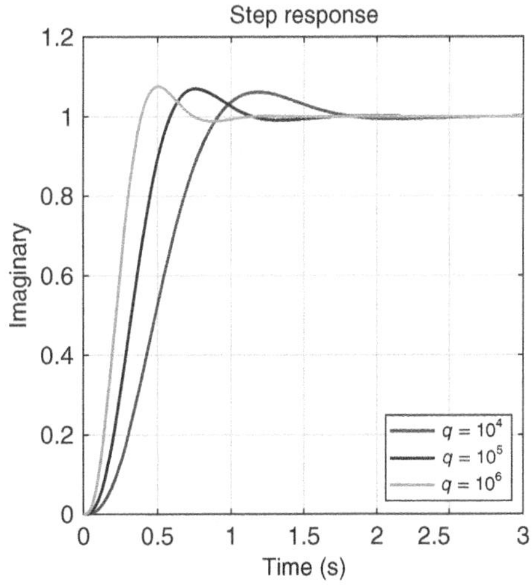

Figure 4.10 Step responses of robust optimal dc motor controller for three different values of q.

4.4.2.5 LQT Design for Sinusoidal References and Disturbances

Above formulation for LQT was for the case where an integrator was used in the outer loop, i.e. for constant commands and disturbances. It can easily be extended to general cases as is done in [3]. For the case of a sinusoid at frequency ω_o, the integrator should be replaced by a second-order generalized integrator (SOGI) at this frequency. Consider a SOGI with the input of the tracking error $e(t)$. It can be characterized by two state variables that are

$$X_{o_1}(s) = \frac{1}{s^2 + \omega_o^2} E(s), \quad X_{o_2}(s) = \frac{s}{s^2 + \omega_o^2} E(s). \tag{4.18}$$

This means $sX_{o_1}(s) = X_{o_2}$ and $sX_{o_2}(s) = -\omega_o^2 X_{o_1}(s) + E(s)$. The state-state description for this SOGI is given by

$$\dot{x}_o(t) = \begin{bmatrix} 0 & 1 \\ -\omega_o^2 & 0 \end{bmatrix} x_o(t) + \begin{bmatrix} 0 \\ 1 \end{bmatrix} e(t) = A_o x_o(t) + B_o e(t), \tag{4.19}$$

where $e(t)$ is the tracking error.[6] Here, when the state variables x_o and x are augmented, $\overline{x}$ becomes of dimension $n + 2$. In this system, $e(t) = y(t) - y_d = Cx(t) + Du(t) - y_d$ is the tracking error. Therefore, the state equation for the SOGI variables can be written as

$$\dot{x}_o(t) = A_o x_o(t) + B_o e(t) = A_o x_o(t) + B_o[Cx(t) + Du(t)] - B_o y_d.$$

Augment the state vectors and define $\overline{x}(t) = \begin{bmatrix} x_o(t) \\ x(t) \end{bmatrix}$ to yield

$$\begin{aligned} \dot{\overline{x}}(t) &= \begin{bmatrix} A_o & B_o C \\ 0 & A \end{bmatrix} \overline{x}(t) + \begin{bmatrix} B_o D \\ B \end{bmatrix} u(t) + \begin{bmatrix} -B_o \\ 0 \end{bmatrix} y_d(t) + \begin{bmatrix} 0 \\ B_w \end{bmatrix} w(t) \\ &= \bar{A}\overline{x}(t) + \overline{B}u(t) + \overline{B}_d y_d(t) + \overline{B}_w w(t) \end{aligned} \tag{4.20}$$

and the expression for the control input is

$$u(t) = -k_0 x_o(t) - Kx(t) = -[k_0 \quad K]\overline{x}(t) = -\overline{K}\overline{x}(t),$$

where k_0 has two elements. The transformation $\frac{d^2}{dt^2} + \omega_o^2$ should be applied to eliminate y_d and w and convert the equations to an LQR problem. The transformed equation is

$$\dot{z}(t) = \bar{A}z(t) + \overline{B}v(t),$$

where

$$z(t) = \left(\frac{d^2}{dt^2} + \omega_o^2\right)\overline{x}(t), \quad v(t) = \left(\frac{d^2}{dt^2} + \omega_o^2\right)u(t) = -\overline{K}z(t).$$

Thus, we are now dealing with a pure LQR problem. Moreover, it is also readily observed from (4.18) that

$$z_1(t) = e(t), \quad z_2(t) = \dot{e}(t).$$

Thus, for instance, to minimize a general cost function

$$J = \int_0^{\infty} \left[qe(t)^2 + q_1\dot{e}(t)^2 + v(t)^2\right] dt, \tag{4.21}$$

6 Notice that this is simply the controllable canonical form of second-order integrator (or resonant controller).

the matrix Q must be chosen as

$$Q = \begin{bmatrix} q & 0 & 0 & \cdots & 0 \\ 0 & q_1 & 0 & \cdots & 0 \\ 0 & 0 & 0 & \cdots & 0 \\ \vdots & \vdots & \vdots & \ddots & \vdots \\ 0 & 0 & 0 & \cdots & 0 \end{bmatrix}.$$

Further details may be found in [3]. This method will be well practiced in the context of DERs connected to ac grids in Chapters 6 and 7 of this book.

4.5 Summary and Conclusion

This chapter reviewed the state-space approach for control system analysis and design. Compared to the transfer function (or frequency domain) approach, the state-space approach operates in the time domain and works with differential equations. However, the powerful concepts such as the state, state feedback, controllability, observability, and finally optimal control theory (specially the linear quadratic theory) offer a much convenient framework for the analysis and design in this domain. One great advantage is that the optimal designs can integrate multiple aspects (such as transient responses and various robustness indices) and address them all through a well-defined cost function or performance measure. The concept of LQR and LQT were reviewed. It is true that the cost function itself is a function of some parameters to be properly selected, i.e. Q. However, selection of this matrix is way simpler than selecting the original control gains because (i) the stability of closed-loop is always guaranteed as long as Q is PSD, and (ii) by observing the location of poles while elements of Q are varied, one gets a comprehensive view of the transient responses of the closed-loop system. It must also be noted that the elements of Q must not be excessively and unnecessarily increased as this may compromise the robustness properties. Their application to various energy system applications are discussed in Chapters 5–7 of the book.

Problems

4.1 Prove that (4.2) is the solution for the state equation in (4.1).

4.2 Consider Figure 4.4 and the augmented equation (4.6). Prove that if (A, B) is controllable, then $(\bar{A}, \overline{B})$ is also controllable. (Refer to [3] for similar proofs.)

4.3 Derive (4.18) by taking Laplace transform from (4.19).

4.4 Consider the system of Figure 4.1 and apply a unit step function as the input to it. Compute its output and show that it remains bounded. Compute the internal signal $x(t)$ and show that it is not bounded.

4.5 Prove that if M is a real symmetric matrix, all its eigenvalues are real numbers.

4.6 Example 4.4 designed a state feedback optimal design for the dc motor speed control. For the same system, design a P controller to meet the desired specifications (maximum 0.5 seconds

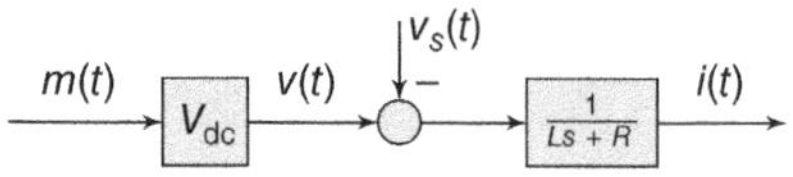

Figure 4.11 Average or control model of the FB-VSC.

rise-time and maximum 10% overshoot). Compare the two controllers from the standpoints of phase margin, delay margin, gain margin, bandwidth, disturbance response, etc.

4.7 Example 4.4 designed a state feedback optimal design for the dc motor speed control. For the same system, design a PI controller to meet the desired specifications (maximum 0.5 seconds rise-time and maximum 10% overshoot). Compare the two controllers from the standpoints of phase margin, delay margin, gain margin, bandwidth, disturbance response, etc.

4.8 Example 4.4 designed a state feedback optimal design for the dc motor speed control. For the same system, design a PD controller to meet the desired specifications (maximum 0.5 seconds rise-time and maximum 10% overshoot). Compare the two controllers from the standpoints of phase margin, delay margin, gain margin, bandwidth, disturbance response, etc.

4.9 Example 4.4 designed a state feedback optimal design for the dc motor speed control. For the same system, design a PID controller to meet the desired specifications (maximum 0.5 seconds rise-time and maximum 10% overshoot). Compare the two controllers from the standpoints of phase margin, delay margin, gain margin, bandwidth, disturbance response, etc.

4.10 Theorem 4.8 proves that a stabilizing solution does not exist for the cost function (4.13). Check this for the dc motor speed control system by trying to get a solution using Matlab *lqr* command.

4.11 Consider the single-phase full-bridge VSC control model described in Chapter 2, redrawn in Figure 4.11. If connected to an ac grid, $v_s(t)$ is the grid voltage and can be treated as a sinusoidal disturbance. Assume that $i_d(t)$ is the desired current reference to be tracked by $i(t)$. Design an LQT (using resonant controller) for this system to track the sinusoidal command and reject the sinusoidal disturbance. Consider the numerical values of $L = 5$ mH, $R = 10$ mΩ, $V_{dc} = 400$ V, $\omega_o = 120\pi$ rad/s, $V_s = 240$ V (rms).
Steps: Write down the state-space description of the resonant control, given in (4.19), and combine with those of the converter to reach the augmented equations (4.20). Follow the procedure to define the cost function (4.21). First, set $q_1 = 0$ and increase q from a small number and observe the closed-loop poles. If the poles do not end up in a desired region, fix q and increase q_1.

References

1 Brian DO Anderson and John B Moore. *Optimal control: linear quadratic methods*. Courier Corporation, 2007.
2 Feng Lin. *Robust control design: an optimal control approach*, volume 18. John Wiley & Sons, 2007.

3 M Karimi-Ghartemani, S A Khajehoddin, P Jain, and A Bakhshai. Linear quadratic output tracking and disturbance rejection. *International Journal of Control*, 84(8):1442–1449, 2011.

4 Sayed Ali Khajehoddin, Masoud Karimi-Ghartemani, and Mohammad Ebrahimi. Optimal and systematic design of current controller for grid-connected inverters. *IEEE Journal of Emerging and Selected Topics in Power Electronics*, 6(2):812–824, 2017.

Part III

Distributed Energy Resources (DERs)

Depending on the operating conditions and requirements where a distributed energy resource (DER) is desired to function, the power electronic converter, and the control system pertaining to it should be designed properly. Some converters such as a voltage source converter (VSC) can function properly within various operating conditions if the control system is capable of directing it properly. Part III of this textbook directly engages the control systems pertaining to DER integration. Chapter 5 will cover various DER integration scenarios to a dc system. Chapter 6 covers integration of single-phase ac DERs. Three-phase ac DERs are covered in Chapter 7.

Modeling and Control of Modern Electrical Energy Systems, First Edition. Masoud Karimi-Ghartemani.

5

Direct-Current (dc) DERs

In a dc distributed energy resource (DER) application, the power electronic converter interfaces two sides which are of dc nature, such as a solar photovoltaic (PV) panel, a battery, the rectified voltage of a wind turbine, a dc load, or a dc microgrid. Furthermore, vector control of a three-phase DER in synchronous reference frame (*dq* frame) is also largely similar to a dc DER application. This chapter provides an understanding of the control system objectives and challenges pertaining to such a system. Various scenarios are studied and their control objectives are taken into consideration and addressed.

Special attention is paid to the ability of converter controller to limit its current against severe disturbances such as transient short circuits. Grid support functions including dynamic (inertia) and static (droop) supports are treated. Grid-forming versus grid-following controllers are studied. Aspects with using higher order filters (such as an LCL filter) are also studied. The PV application is studied in further particular details.

We use the linear quadratic (LQ) optimal control theory for the design of controllers. This leads to a systematic way of designing modular controllers which enjoy an inherent optimality and robustness as demonstrated throughout the chapter using various studies.

5.1 Introduction

An introduction comprising the description and definition of the dc DER concept is given first. Then, a classification of control objectives in a dc DER is presented.

5.1.1 System Description

Figure 5.1 shows the schematic of the DER system. It has a "dc side" and a "grid side."[1] If the grid side is dc, we call it a dc DER. And if the grid side is ac, we call it an ac DER. This chapter is devoted to dc DERs.

On the "grid side assets," we may have a dc source (a battery), a dc load, or in general a dc network comprising loads and sources (i.e. a dc grid). On the "dc side assets," we may have a battery, a dc load, a PV array, another converter (e.g. in a wind turbine DER), or even a dc grid comprising multiple dc loads and sources.

The voltage source converter (VSC) topology is considered for the studies in this chapter to allow bidirectional power flow. The filter is in general a combination of inductor, capacitor (LC) elements.

1 The terminology of "dc side" versus "grid side" is common because grid is commonly an ac grid.

Modeling and Control of Modern Electrical Energy Systems, First Edition. Masoud Karimi-Ghartemani.

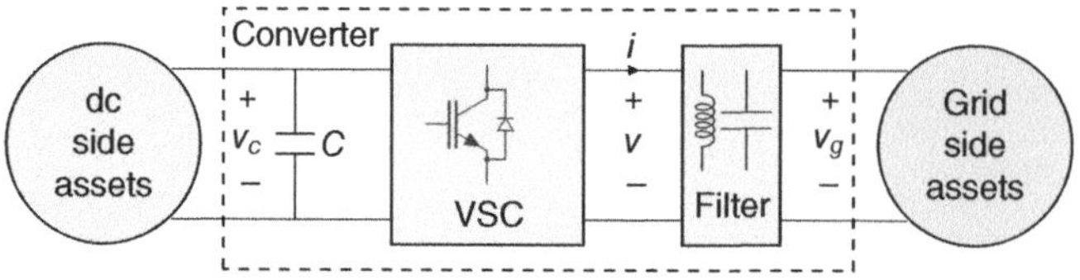

Figure 5.1 General structure of a dc DER.

The filter mitigates the switching ripples on the converter current i and provides a means of dynamically achieving the control objectives. The converter voltage before the filter is denoted by v. The grid voltage is denoted by v_g and the dc side voltage is denoted by v_c. The nominal (or rated) value for the grid voltage is denoted by V_g. The capacitor C is considered across the dc side terminals for supporting the dc voltage, for mitigating its possible noises and providing a means of dynamically reaching the control objectives.

5.1.2 General Statement of Control Objectives

In the dc DER system introduced in Figure 5.1, the controller shown in Figure 5.2 determines the turn on and off instants of the power electronic switches to achieve certain control objectives explained as follows.

1. **Track set-points:** The set-points may be provided locally or through communication from a higher level controller. The most common is a power set-point, denoted by P_{set}. The direction of power may be from either side to the other. The power control may be done through other intermediary variables. For instance, it may be translated into a current control problem. Or, in a PV system, it can be translated into dc side voltage control.
2. **Limit converter current:** The converter current i must be maintained within the acceptable range. This is to protect the switches and to prevent nuisance tripping/interruption of the converter during system transients.
3. **Provide support:** The converter shall provide dynamic and static voltage support to the grid and to the dc side. This is crucial for achieving a stable and robust grid intended to host large number of DERs.
4. **Remain robust:** The above three objectives must be achieved within uncertain, variable, or disturbed system conditions. Such conditions may include uncertainty in assumed system parameters, slow or fast variations in its operating conditions, and transient faulty conditions.

As we will demonstrate during this chapter, it is possible to define and develop an overarching control approach that is able to combine all control objectives. Before going further into details of control, let us briefly get familiar with a solar PV generation system that will help understanding the control requirements more accurately.

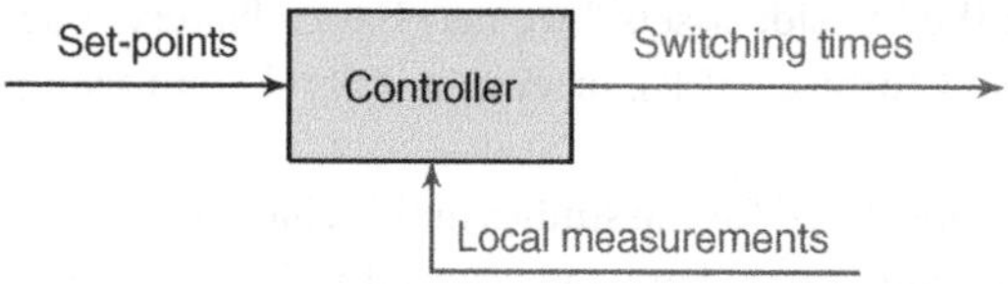

Figure 5.2 General structure of a dc DER controller.

5.2 Overview of a Solar PV Conversion System

This section gives a general overview on PV power generation and the power electronic conversion systems used for them.

5.2.1 Photovoltaic Effect and Solar Cell

PV effect is the generation of electric field in a material upon exposure to light. Modern PV materials are based on *silicon* and the solid-state devices such as *photodiodes*. In most PV applications, the radiation is sunlight and the device is called a *solar cell*, Figure 5.3a. The light is absorbed in a solar cell and causes the electrons to jump into the conduction band and become free. They diffuse and reach a rectifying p-n junction and lead to the generation of an electromotive force. Those electrons which do not gain sufficient energy to move to a higher energy level will lose their energy as heat and will remain in the same energy level. A solar cell has a typical current-versus-voltage (V–I) and a power-versus-voltage (V–P) characteristics shown in Figure 5.3b.

A solar cell consists of a thin sheet of semiconductor, very often made of suitably treated silicon. Conversion efficiency of commercial solar cells lies between 13% and 18%. The physical thickness is between 0.25 and 0.35 mm, and the surface area between 100 and 225 cm^2. At the solar irradiance[2] of 1 kW/m^2, and the temperature of 25 °C, a solar cell can produce a current between 3 and 4 A at approximately 0.5 V voltage, totalling to a maximum power of 1.5–2 W. Cells are connected in series and parallel to form solar modules (or panels), and the panels are connected to form an array with larger voltage, current, and power, as shown in Figure 5.3a. A common PV panel rates at about 200–250 W, 30–40 V, and 6–7 A.

When the terminals are open, the voltage is V_{oc} and that is the maximum voltage called the open-circuit voltage. When the terminals are short-circuited, the current is I_{sc} and this signifies

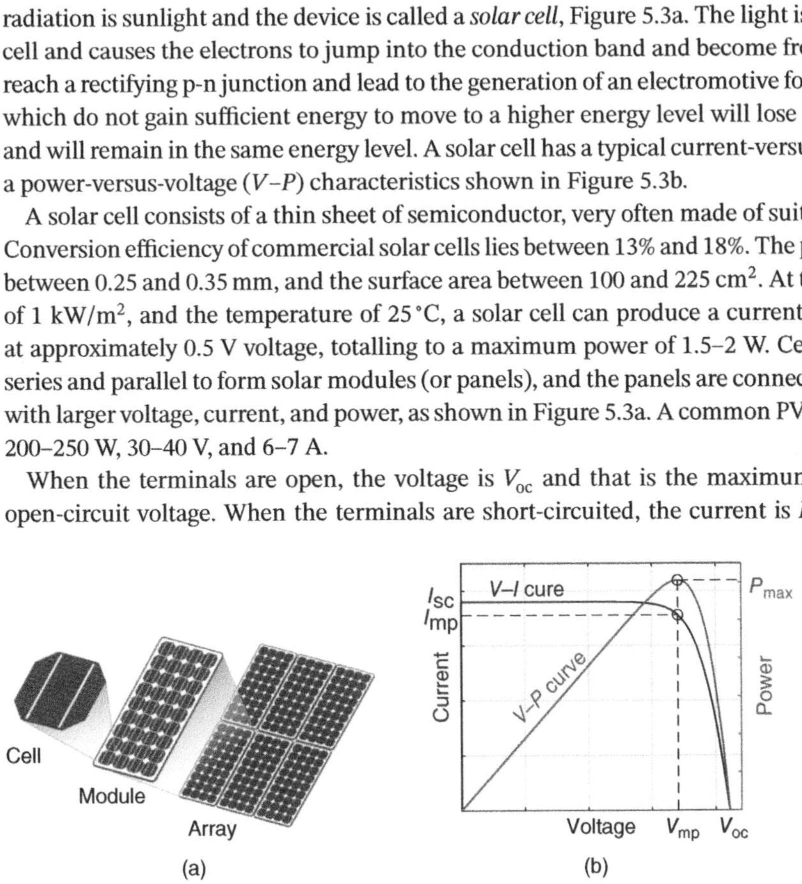

Figure 5.3 A solar PV cell, a PV module and a PV array (a). Typical V–I and V–P characteristics of a PV cell (b).

2 *Solar irradiance* is the power per unit area received from the sun in the form of electromagnetic radiation as reported in the wavelength range of the measuring instrument. Solar irradiance is often integrated over a given time period in order to report the radiant energy emitted into the surrounding environment during that time period. This integrated solar irradiance is called *solar irradiation, solar exposure, solar insolation,* or *insolation*. The sun's rays are attenuated as they pass through the atmosphere, leaving maximum normal surface irradiance at approximately 1000 W/m^2 at sea level on a clear day. When the sun is at the zenith in a cloudless sky, the global radiation on a horizontal surface at ground level is about 1120 W/m^2. Ignoring clouds, the daily average insolation for the earth is approximately 6 kWh/m^2 [1].

the maximum possible current of the module called the short-circuit current. Note that in these two extreme cases, there is no power output.

The characteristics is a function of the solar irradiance. The short-circuit current increases from zero up to the capacity of the device as the solar irradiance goes up from zero, Figure 5.4a. The open-circuit voltage is also a function of the irradiance level but it does not experience very wide variations when the irradiance varies. When the ambience temperature changes, it impacts the characteristics as also shown in Figure 5.4b. Increased temperature results in slight increase of the short-circuit current and rather larger reduction of the open-circuit voltage.

The amount of power extracted from the module is equal to the product of its voltage and current. The voltage-power characteristics of the solar module versus the irradiance and temperature are also shown in Figure 5.4 using dashed or dotted lines. The power increases from zero when the output voltage goes up. It reaches a maximum value called the maximum power point (MPP). At this point, the voltage is called the maximum power voltage V_{mp}. The power decreases to zero when the voltage is increased beyond V_{mp} and approaches V_{oc}.

The equivalent circuit of a solar cell in ideal case comprises the parallel connection of a current source with a diode, as shown in Figure 5.5. The light depending current is shown by I_L and the diode current is I_D. The cell current will then be

$$I = I_L - I_D = I_L - I_o(e^{\frac{qV}{nkT}} - 1) \approx I_L - I_o e^{\frac{qV}{nkT}},$$

where I_o is reverse saturation current of the diode, n is the diode ideality factor (1 for an ideal diode), q is the elementary charge, k is the Boltzmann's constant ($1.380\,649 \times 10^{-23}$ J/K), and T is the absolute temperature (in Kelvins). The ratio $V_T = \frac{kT}{q}$ is called the thermal voltage and is approximately equal to 0.0259 V at 25 °C.

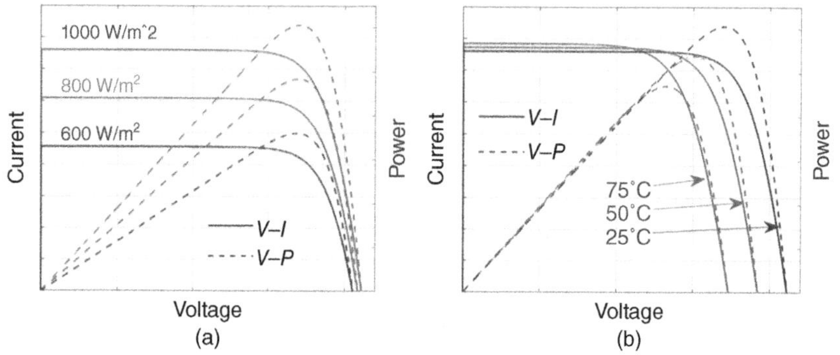

Figure 5.4 *V–I* characteristics of a solar photovoltaic module for various irradiance levels (a) and ambience temperatures (b).

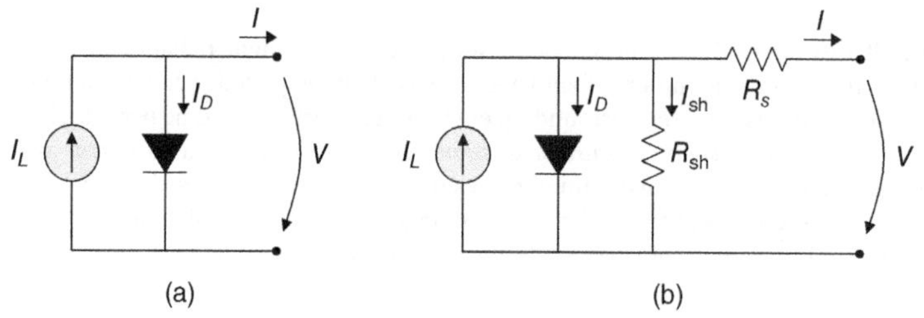

Figure 5.5 Solar PV equivalent circuit (a) ideal and (b) practical.

In practice, a series and a shunt resistors, R_s and R_{sh}, respectively, are added to the circuit, as shown in Figure 5.5b. Here the equation will be modified to

$$I = I_L - I_D - I_{sh} = I_L - I_o \left[e^{\frac{q(V+R_S I)}{nkT}} - 1 \right] - \frac{V + R_s I}{R_{sh}}.$$

5.2.2 General PV Converter Structures

The common converter structures for a PV system are shown in Figure 5.6. It is either a double-stage or a single-stage structure. The converter is supposed to achieve the following two primary, general tasks:

- *Task 1*: Control the *amount* of extracted power.
- Task 2: Control the *quality* of output power (voltage or current).

Task 1 of the converter (stated above) is often accomplished by controlling the voltage across the panels. As discussed above, this voltage directly controls the power. A capacitor is normally inserted at the output terminals of the PV module to support the voltage and filter the possible converter noises. Then, the converter controls this voltage according to the system requirements.

One typical approach for grid-connected PV systems is to extract the maximum power from the module. In other words, the module is biased at the maximum power voltage. This voltage, however, is normally unknown because it is a function of solar irradiance, temperature, partial shading, module structural uncertainties and also can change as the module ages. Therefore, an adaptive (i.e. a feedback) mechanism should be in place to find this voltage continuously. This mechanism is *maximum power point tracking* (MPPT) algorithm. We will discuss this topic later in detail.

In the double-stage structure, the first-stage converter is responsible for biasing the module at the right point to extract the right amount of power from it (Task 1). This is the control of voltage across capacitor C_1 in Figure 5.6. This stage may also be responsible for other secondary tasks such as voltage boosting and high-frequency isolation.

The second-stage converter is responsible for Task 2: shaping the output power and ensuring its quality. In an ac grid-connected situation for instance, this objective is translated to smooth

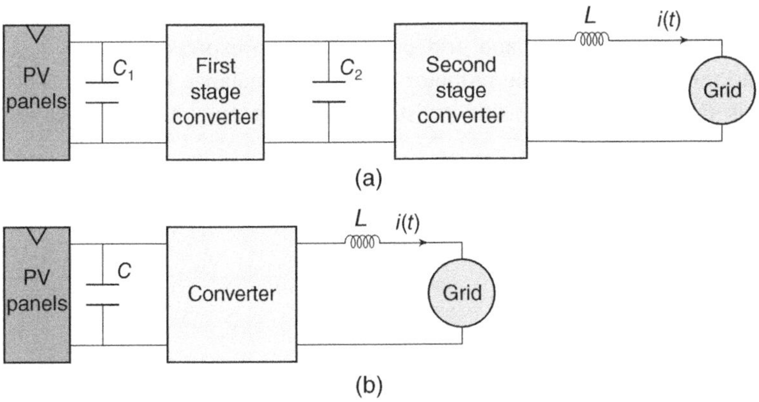

Figure 5.6 Typical structure of a solar photovoltaic conversion system: double-stage (a), single-stage (b). Grid may be dc or ac (single-phase or three-phase).

sinusoidal current generation. In islanded (not connected to grid) applications, the power quality is often translated into the quality of the output voltage of the converter.

The first-stage converter is also called the source-side converter. The second-stage converter is also called the grid-side or the load-side converter.

The capacitor C_2 serves as a link between the two stages. It serves as a short-term power storage that supports the dc voltage required by the second-stage converter. The voltage of this *dc link* capacitor should be high enough because the second-stage may not be able to boost the voltage (e.g. when the second-stage is a VSC).

Power is extracted by first-stage converter and sent to the dc link. This will increase the dc link voltage. The second-stage converter picks that power and sends to the output. If these two powers balance, the dc link voltage remains stable. If the output power is more than the input power, the dc link voltage decreases. Therefore, by regulating (or stabilizing) the dc link voltage, the second-stage converter can ensure that the right amount of power is transferred. This discussion concludes that controlling the dc link voltage is a secondary job of the second-stage converter.

The single-stage converter is normally used in applications where the voltage boosting and isolation are not required. It controls the amount of extracted power (by controlling the voltage across C through an MPPT algorithm for instance) and ensures quality of output power. Many high-power three-phase PV converters adopt this topology. At low-power, single-phase applications, a double-stage topology is more popular due to its ability to decouple the double-frequency power from the PV side (as will be discussed in details later).

The two-stage converter tends to have a lower efficiency and higher cost due to employment of more number of power electronic components. However, it extends the voltage operational range of the PV because a dc/dc converter can start working for smaller input voltage levels compared to the case where only one inverter is used. Therefore, during the low irradiation levels (such as early in the morning, late afternoon, and cloudy intervals), it can extend the low-voltage harvesting capability of the converter.

5.3 Power Control via Current Feedback Loop

To embark on the discussion of control systems for dc DERs, we begin with the most preliminary yet most important case of controlling the power exchange. The converter intends to execute the power set-point P_{set} while satisfying the other objectives stated in Section 5.1.2.

As observed from Figure 5.7, the converter and grid powers may be expressed by $p = iv_g$.[3] In this section, we assume the filter is a simple inductor with the inductance value of L. The small resistance R models the inductor's parasitic resistance and the on-state resistance of switches.

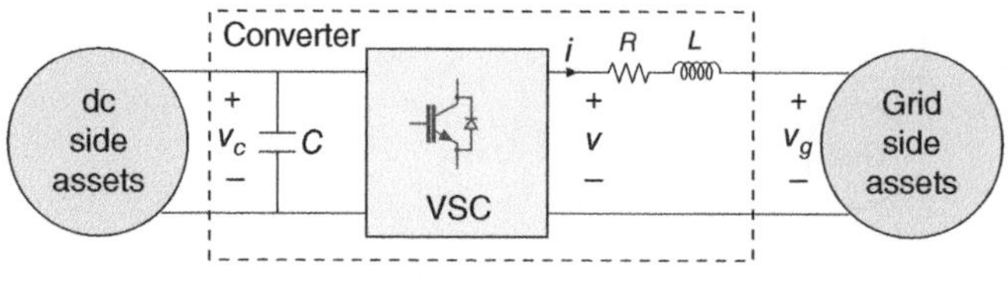

Figure 5.7 General structure of a dc interface.

3 The power across the filter is neglected.

5.3.1 Control Objectives

The control objectives can be stated more specifically as follows.

- Control the amount of the exchanged power at the set-point (or command) P_{set}.
- Maintain this objective robustly despite system uncertainties including those in L, R, v_c, v_g, etc.
- Respond to changes in power commands in a quick and smooth manner.
- The starting transient of the converter should be smooth without transient over-currents.
- Ride through grid disturbances and faults, that cause transient voltage sags and swells, without causing converter over-current.
- Limit the converter current within $[I_{min} \;\; I_{max}]$ for all times.
- Provide dynamic and static support to the grid.

5.3.2 Control Approach

The control approach discussed in this section is called the *current control approach* (or the current feedback approach) where the power is controlled via the medium of current. The block diagram of the system (without controller) is shown in Figure 5.8 where $m(t)$ is the modulation signal for a standard PWM.[4] The dc side voltage v_c is assumed to be relatively constant around the number V_c volts.[5] In case of using a half-bridge converter, v_c should be replaced with $0.5v_c$. The grid side voltage v_g is also assumed to be relatively constant around the number V_g volts.

The common control structure is shown in Figure 5.9. The converter current i is measured and used in the feedback loop. This will allow current limiting as we will see shortly. In this diagram, P_{set} shows the set-point (desired or reference) value for the exchanged power. The desired or reference value for the current is $i_{ref} = \frac{P_{set}}{V_g}$ where V_g is the known value for the dc grid voltage. According to the robust property of the integrating controller, it is able to achieve zero current steady-state error despite the system uncertainties and changes as long as the loop is stable.

5.3.2.1 Robust Tracking and Current Limiting

If v_g experiences some change, the control diagram of Figure 5.9 will experience some steady-state error (as far as the exact power tracking is concerned) because the forward block $\frac{1}{V_g}$ cannot see this change. To remedy this problem, v_g should be measured and used in this block as shown in Figure 5.10.

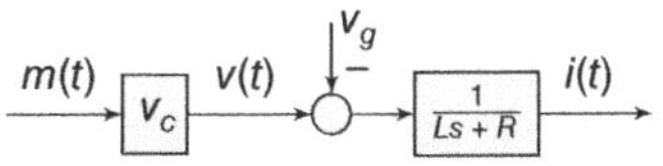

Figure 5.8 Average model block diagram of a VSC with dc side voltage v_c, the filter inductance L, and connected to a dc grid with voltage v_g.

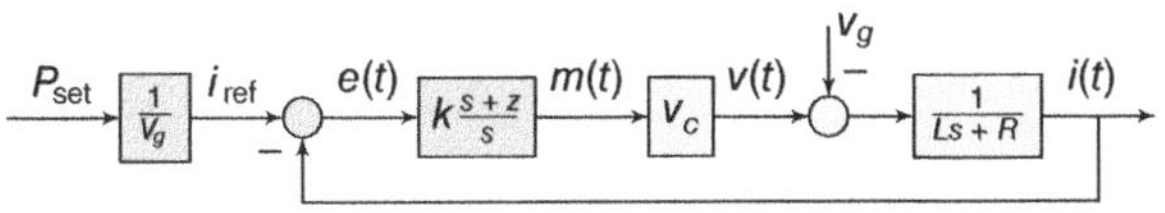

Figure 5.9 Closed-loop block diagram with a common PI controller.

4 The bar signs indicating the average functions are removed to simplify the notations.
5 If this voltage is not constant and experiences fast and wide range variations, the system will be technically nonlinear. But this nonlinearity can easily be removed by measuring the dc side voltage and dividing it in the loop.

One more issue to address is the over-current problem caused by transient grid faults. In order to understand the root cause of this problem, assume for example that the supply voltage v_g drops unexpectedly as a result of a transient fault or disturbance or any other system condition, e.g. the sensor problem. Therefore, for a given amount of P_{set}, i_{ref} will significantly go up. In an extreme case of a short circuit at v_g, i_{ref} will go to infinity. The converter will attempt to feed very large current and will cross the limits and interrupted by the protection schemes. In order to prevent this problem, a limiter unit must be added as shown in Figure 5.10. The control loop must be fast enough to limit the current, as will be discussed in Section 5.3.5.

5.3.2.2 Soft Start Control

The control diagram of Figure 5.10 may have a starting problem. In order to understand this, assume that at the start of operation of the VSC, the reference power is zero. Therefore, the points denoting e and m and v are also initially zero. Zero value of v means that a large current will start to flow to the VSC (that is the reverse direction). It will take some time for the PI controller to build up its output and correct this negative current if the controller is not fast enough to respond to it in time. If the transient current is large and exceeds the switch protection limits, the protection circuit stops the converter's function. The converter will not practically start functioning. To remedy this problem, a feed-forward term may be included in the controller as shown in Figure 5.11. A term equal to $\frac{v_g(0)}{v_c(0)}$ is added to offset the initial value of v to $v_g(0)$. In this way, the converter will exhibit a soft start.

If the converter is intended to start at a nonzero current command, say $i_{ref}(0)$, the PI output at time zero will be $ki_{ref}(0)$. This means that the feed-forward term must be corrected to $\frac{v_g(0)}{v_c(0)} - ki_{ref}(0)$. The whole idea is to ensure that the initial value of m at $t = 0$ is such that $m(0)v_c(0) = v_g(0)$. During this whole discussion, we assumed that the initial condition of the integrator in the PI compensator is set to zero. Otherwise, the feed-forward term must be adjusted accordingly.

5.3.3 Design of Feedback Gains Using TF Approach

Figure 5.12 shows the simplified block diagram of the control system pertaining to the VSC connected to a dc supply. Notice that all additional feed-forward blocks are removed as they do not impact the feedback controller design. Moreover, the dc side voltage v_c is substituted with its nominal value V_c.

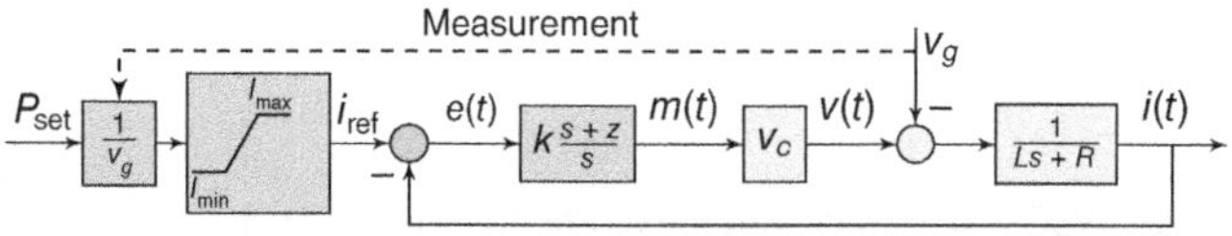

Figure 5.10 Closed-loop block diagram when the gain $\frac{1}{v_g}$ is adjusted adaptively through the measurements to ensure zero power steady-state error despite changes in v_g. The limiter limits the current reference in cases where an excessive drop in v_g may occur.

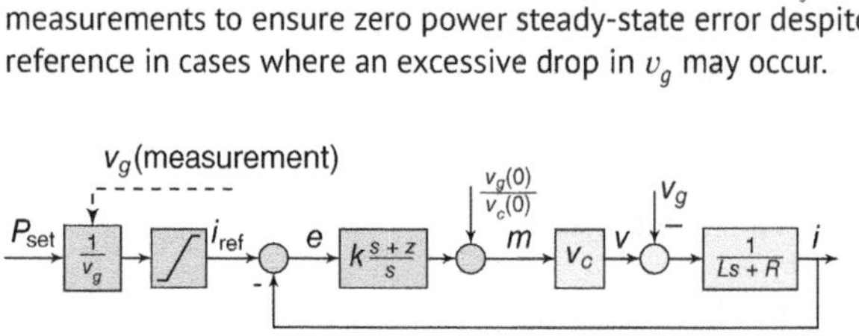

Figure 5.11 Control diagram with a feed-forward term to reduce the starting transient.

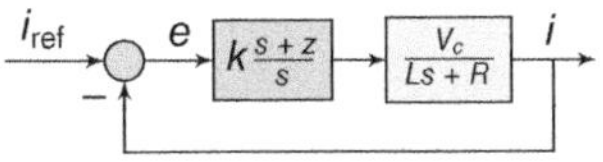

Figure 5.12 Simplified closed-loop block diagram with a PI compensator.

A simple (although not the best) way to choose the PI controller gains k and z using the transfer function (TF) approach is explained as follows. If an estimate of R is available, then a selection of $z = \frac{R}{L}$ may be made. Then, the loop is characterized by one pole that moves to the left as k increases:

$$1 + k\frac{s+z}{s}\frac{V_c}{Ls+R} = 1 + \frac{kV_c}{Ls} = 0.$$

This pole is given by $s = -\frac{kV_c}{L}$. When this pole moves to the left, the system's output response speeds up. A practical rule-of-thumb states that the time-constant of the response should be at least several switching cycles, i.e.

$$\frac{L}{kV_c} \geq \beta\frac{1}{f_{sw}} \Rightarrow k \leq \frac{Lf_{sw}}{\beta V_c}, \tag{5.1}$$

where f_{sw} is the switching frequency and β is a real value greater than 4 or 5.

Example 5.1 ***Design of PI Gains Using TF Approach***
The parameters of the converter system are given in Table 5.1. For this set of parameters, $z = R/L = 6$ and $k < \frac{Lf_{sw}}{\beta V_c} = \frac{0.005\times4000}{400\beta}$ or $k < \frac{0.05}{\beta}$. For a $\beta = 5$, $k < 0.01$. Therefore, if $k = 0.01$ is selected, the response will have a time-constant that spans over 5 switching cycles that is $5 \times \frac{1}{4000} = 1.25$ ms. Time-response of the converter with this controller is shown in Figure 5.13 where a command of 20 A is applied at time 2 ms. The response has a first-order form with a time-constant of around 1.25 ms as expected.

5.3.4 LQT Approach and Design

Figure 5.14 shows the simplified block diagram of the control system pertaining to the VSC connected to a dc supply. In this diagram, the proportional gain is relocated properly in order to make the feedback loop in the standard form of a state feedback where we can apply state feedback tools such as pole placement and optimal control design. Design of k_1 and k_2 can be done using pole placement or using the LQT technique discussed in Chapter 4. In the following, the LQT method for this system is described and applied.

Table 5.1 Converter system parameters used for the examples in this section

Parameter	Symbol	Value	Unit
dc side voltage	V_c	400	V
Grid voltage	V_g	300	V
Inductance	L	5	mH
Parasitic resistance	R	30	mΩ
Switching frequency	f_{sw}	4	kHz
Power rating	P_{rate}	6	kW

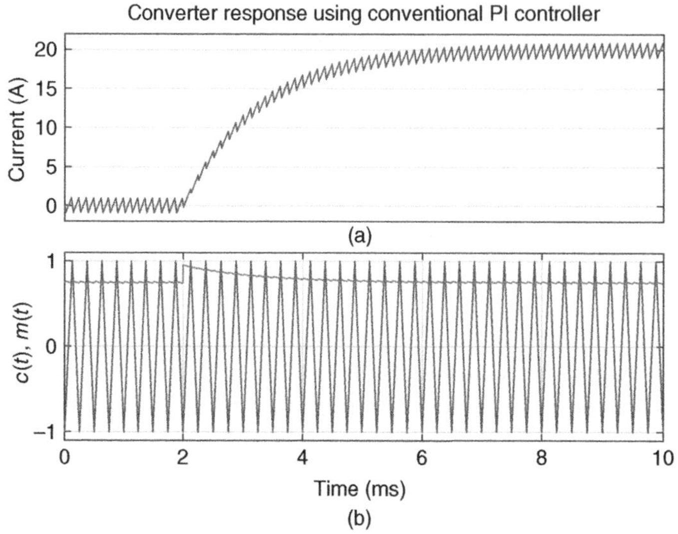

Figure 5.13 Time-response of the converter with a PI compensator designed using TF approach (Example 5.1): converter current (a), carrier and modulation signals (b).

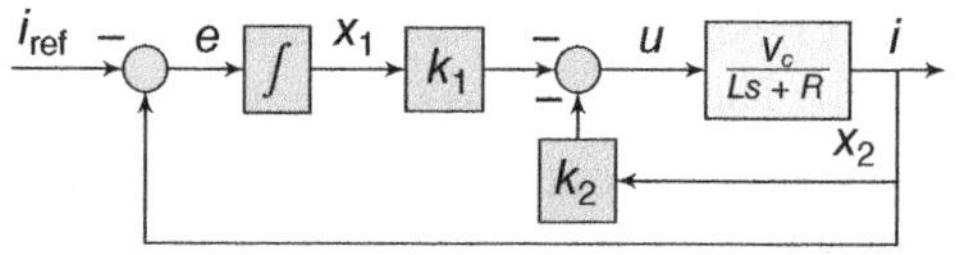

Figure 5.14 Closed-loop block diagram with LQT state feedback.

The state equations for this system are

$$\dot{x}_1(t) = e(t) = i(t) - i_{\text{ref}} = x_2(t) - i_{\text{ref}}$$
$$\dot{x}_2(t) = -\tfrac{R}{L}x_2(t) + \tfrac{V_c}{L}u(t),$$

where $x_2(t) = i(t)$ is defined and used. Apply the operator $\frac{d}{dt}$ to both sides of the two above equations to obtain

$$\dot{z}_1(t) = z_2(t)$$
$$\dot{z}_2(t) = -\tfrac{R}{L}z_2(t) + \tfrac{V_c}{L}w(t),$$

where $z_1(t) = \dot{x}_1(t) = e(t)$, $z_2(t) = \dot{x}_2(t)$ and $w(t) = \dot{u}(t)$. Notice that $\frac{d}{dt}i_{\text{ref}} = 0$. Also notice that

$$w(t) = \dot{u}(t) = -k_1\dot{x}_1(t) - k_2\dot{x}_2(t) = -k_1z_1(t) - k_2z_2(t) = -Kz(t),$$

where $K = [k_1\ k_2]$ and $z(t) = \begin{bmatrix} z_1(t) \\ z_2(t) \end{bmatrix}$. Therefore, the state space representation in terms of z variables is

$$\dot{z}(t) = Az(t) + Bw(t), \text{ where } A = \begin{bmatrix} 0 & 1 \\ 0 & -\frac{R}{L} \end{bmatrix} B = \begin{bmatrix} 0 \\ \frac{V_c}{L} \end{bmatrix}.$$

Let us minimize the cost function $J = \int_0^\infty [qe^2(t) + w^2(t)]dt$ for some $q > 0$. This cost is equal to $J = \int_0^\infty [qz_1^2(t) + w^2(t)]dt$. Therefore, the matrix Q should be chosen as

$$Q = \begin{bmatrix} q & 0 \\ 0 & 0 \end{bmatrix}$$

and the LQR problem is solved in MATLAB: ≫ `K = lqr(A, B, Q, 1)`.

Remark It is readily observed that (A, B) is controllable and (Q_1, A) is observable where $Q_1 = [1 \;\; 0]$. (Notice $Q = Q_1^T Q_1$). Therefore, this LQR problem has a unique solution and makes the closed-loop asymptotically stable.

If the matrix Q is set to

$$Q = \begin{bmatrix} q & 0 \\ 0 & q_2 \end{bmatrix},$$

the cost function changes to

$$J = \int_0^\infty [q z_1^2(t) + q_2 z_2^2(t) + w^2(t)]dt.$$

Notice that $z_2(t) = \dot{x}_2(t)$ is the derivative of the current signal. By including it into the cost function, we allow further reduction of the oscillations (fast rates) of this function. The example below illustrates these points.

Example 5.2 ***Design of Controller Gains Using LQT Approach***
Figure 5.15 show locus of closed-loop poles (eigenvalues of $A - BK$) when q varies from 10^{-12} to 10^{-8} (on the left graph) and when varies from 10^{-5} to 10^{0} (on the right graph). The poles start at 0 and $-\frac{R}{L}$ when q starts from zero. They meet at some point between the two and depart. They move to the left as q increases making the response faster and faster. Interestingly, the poles move very close to the damping ratio of about 0.7 which confirm optimality of the system according to our engineering experience. When q_2 is added, the result is that the controller introduces more damping into the closed-loop poles as shown in Figure 5.16.(For this graph, the numerical values of Table 5.1 are used.)

Example 5.3 ***Performance Comparison of PI and LQT Controllers***
The PI controller and the LQT optimal controller are designed, simulated, and their responses are studied in this example. The PI controller is designed such that its zero cancels the system's pole and its gain is adjusted for a time-constant of 5 switching cycle, that is $\beta = 5$ in (5.1). Its transfer function will be $G(s) = k\frac{s+z}{s} = 0.01\frac{s+6}{s}$. The closed-loop pole of this system will be at $\frac{kV_c}{L} = \frac{0.01\times400}{0.005} = 800$ which results in a time constant of 5 times the switching cycle, i.e. 1.25 ms. The LQT controller is designed by increasing q to 10^2 and subsequently increasing q_2 to $10^{-3.85}$. This will move its eigenvalues as shown in Figure 5.17. The final location of roots are around $-800 \pm j400$ (exact location is $-791 \pm j417$) and the controller gains are $K = [k_1 \;\; k_2] = [10 \;\; 0.0197]$.

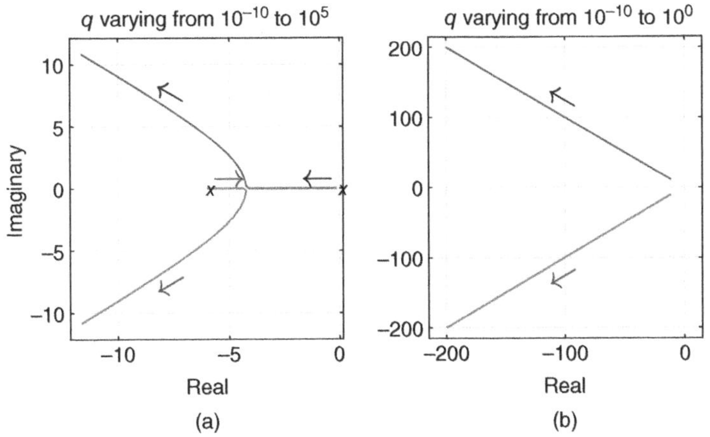

Figure 5.15 Closed-loop poles of Example 5.2 when q varies.

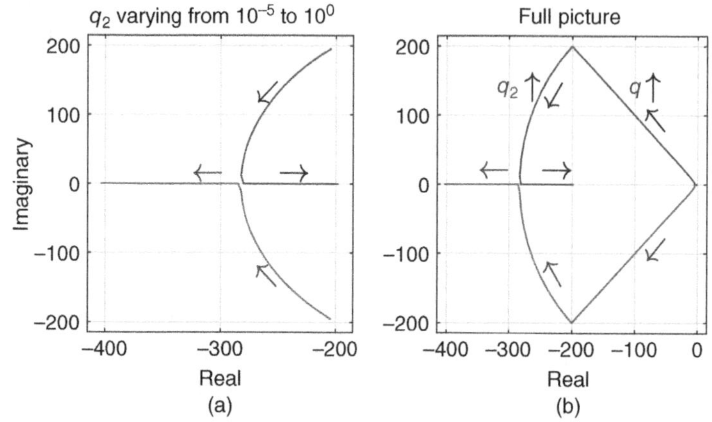

Figure 5.16 Closed-loop poles of Example 5.2: when q_2 varies (a), full picture (b).

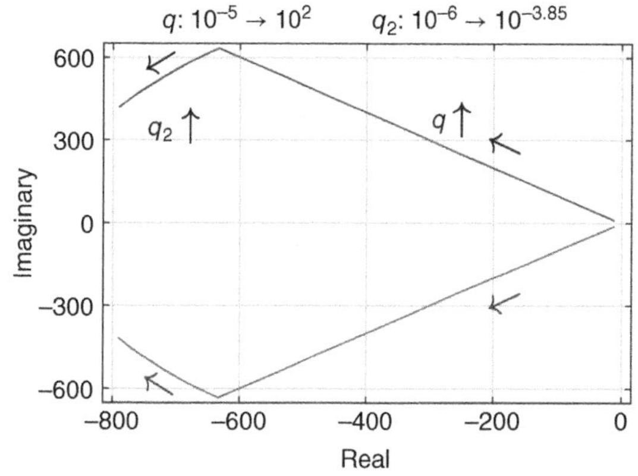

Figure 5.17 Closed-loop poles of Example 5.3: q varying up to 10^2 and q_2 varying up to $10^{-3.85}$. Final location: $-791 \pm j417$, Controller gains: $K = [k_1 \ k_2] = [10 \ \ 0.0197]$

Time responses of both control systems are shown in Figure 5.18. On the top panel, a step command of 20 A is applied at $t = 2$ ms. Both controllers respond quickly and with no steady-state error. The PI controller tends to be swifter with a time-constant of 1.25 ms as expected. On the middle panel, the responses of both systems to a sudden drop of V_c from 400 V to 380 V at $t = 0.1$ seconds are shown. The LQT controller exhibits a fast and strong response while PI controller exhibits a slow response. On the bottom panel, the responses of both systems to a sudden drop of V_g from 300 V to 280 V at $t = 0.6$ seconds are shown. The LQT controller exhibits a fast and strong response while PI controller responds sluggishly with a transient error which is much larger than that of the LQT controller.

In order to understand the sluggish response of PI controller, we notice that, for example, the transfer function from V_g to I is given by

$$\frac{I(s)}{V_g(s)} = \frac{\frac{1}{Ls+R}}{1+\frac{kV_c}{s}} = \frac{1}{L}\frac{s}{\left(s+\frac{R}{L}\right)\left(s+\frac{kV_c}{L}\right)}$$

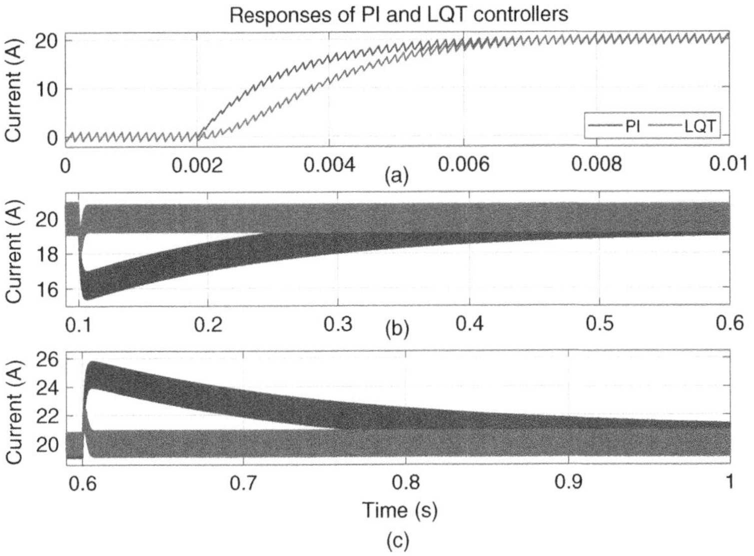

Figure 5.18 Time responses of the system of Example 5.3 with PI and LQT controllers. (top) step responses to a command of 20 A at time $t = 2$ ms (middle) responses to a v_c sudden drop from 400 V to 380 V at $t = 0.1$ seconds (bottom) responses to a sudden fall in v_g from 300 V to 280 V at $t = 0.6$ seconds

which shows a poles at $s = -\frac{R}{L} = -6$. This indicates a sluggish response at time-constant of one sixth of a second. Moreover, this pole depends on R and L and it can make the response worse if, for example, R decreases or L increases. On the other hand, the same transfer function for the LQT controller is given by

$$\frac{I(s)}{V_g(s)} = \frac{\dfrac{1}{Ls+R}}{1 - \dfrac{k_1 V_c}{s(Ls+R)} + \dfrac{k_2 V_c}{Ls+R}} = \frac{1}{L}\frac{s}{(s+p_1)(s+p_2)},$$

where p_1 and p_2 are the poles of closed-loop system, i.e. around $-800 \pm j400$. Therefore, disturbance rejection of the LQT system is strong, fast, and does not depend on the values of R and L. Similar sort of analysis may be done for other disturbances and uncertainties in this system.

This simple example clearly confirms some of the advantages of an optimal controller compared to a conventional controller. As the size of the systems grow and their models become larger and more complicated, the advantages of optimal controllers also become more pronounced.

Example 5.4 ***Version 2 and Version 3 of LQT Controller****

Example 5.3 and the simulation result shown in Figure 5.18a might cause a concern that the LQT has a slower response rising when a sudden command change is applied. This is due to the fact that the feedback gain k_2 in the standard LQT approach of Figure 5.14 does not operate on the error signal while in the PI, it does. If it is desired to speed up the rising of the LQT response to the command changes, the input of the gain k_2 can be connected to the error point, e, as shown in Figure 5.19 called version 2 LQT as apposed to the standard LQT loop of Figure 5.14. In version 3, the gain is distributed using a factor α as shown Figure 5.20 where $0 \leq \alpha \leq 1$. Version 3 reduces to the Standard one when $\alpha = 1$ and to version 2 when $\alpha = 0$. Notice that all these three versions have identical closed-loop poles.

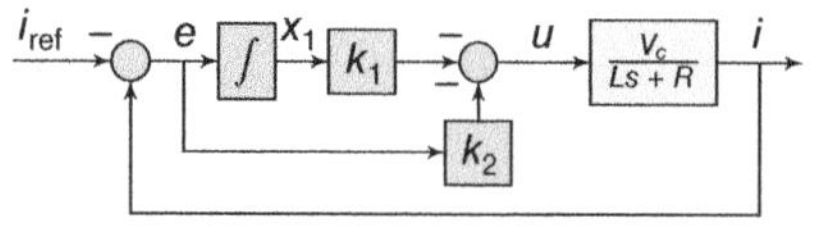

Figure 5.19 Closed-loop LQT state feedback: version 2, Example 5.4.

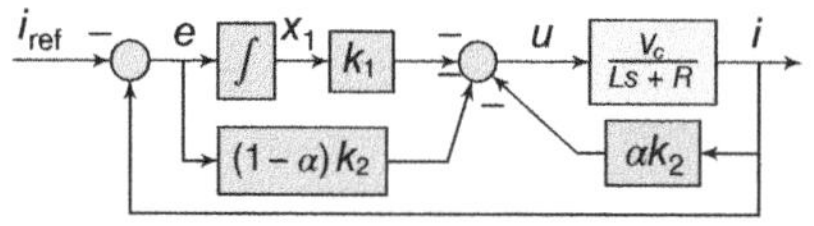

Figure 5.20 Closed-loop LQT state feedback: version 3, Example 5.4.

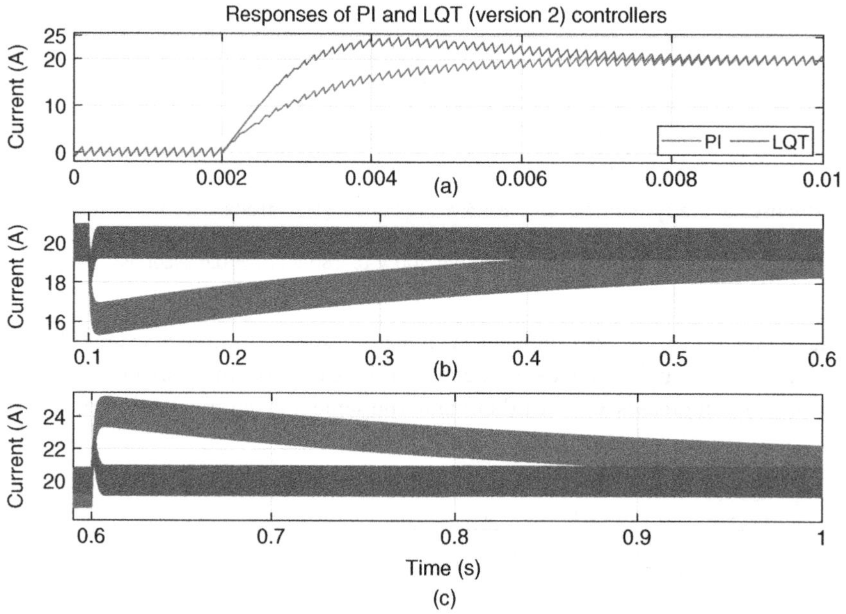

Figure 5.21 Time responses of the system of Example 5.4 with PI and (version 2) LQT controllers. (a) step responses to a command of 20 A at time $t = 2$ ms (b) responses to a v_c sudden drop from 400 V to 380 V at $t = 0.1$ seconds (c) responses to a sudden drop in v_g from 300 to 280 V at $t = 0.6$ seconds.

Responses of these versions of LQT and the PI are shown in Figures 5.21 and 5.22 for the same simulation scenario used in Example 5.3. The version 2 LQT shows a very fast rising stage with a small overshoot. The version 3 with $\alpha = 0.5$ has a transient response almost identical to that of the PI controller as shown in Figure 5.22.

Remark The version 2 and version 3 of the LQT discussed in Example 5.4 are only to explain a transient behavior. In the rest of this textbook, unless explicitly specified, the standard LQT form will be used.

5.3.5 Control Design Requirements for Current Limiting

The above study shows that when the grid voltage v_g experiences a sudden fall, the converter current experiences a transient over-current. For a 20 V voltage drop, the LQT controller showed just over 2 A while the PI approach showed just over 5 A peak over-current. If the grid voltage

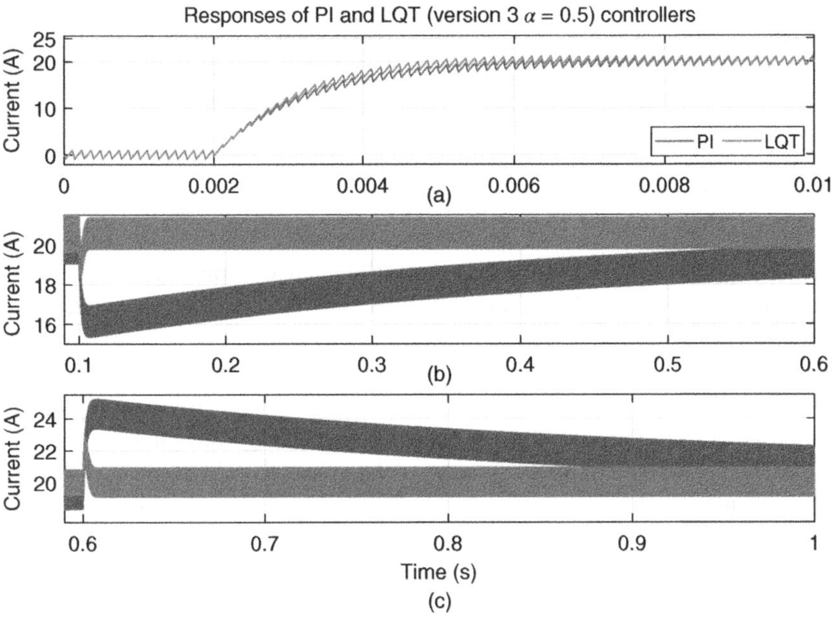

Figure 5.22 Time responses of the system of Example 5.4 with PI and (version 3) LQT controllers (same simulation scenario as Figure 5.21).

undergoes a large transient voltage dip (which may occur frequently in practical systems), the converter current can experience an extremely large transient over-current. This will cause the converter over-current protection system to shut down the converter. However, if the control system can by itself limit this transient over-current, the converter will continue its operation during such transient low-voltage faults.

To understand the requirements for the controller to be able of limiting the transient currents cause by low-voltage faults, assume for a moment that the control is not in place, the converter is operating in a normal state and feeding the current I_0 to the grid while the grid voltage suddenly drops to zero at $t = 0$. The interfacing filter impedance $Ls + R$ will experience a sudden voltage rise across it and its current starts rising at a time-constant of $\tau = \frac{L}{R}$. The initial voltage across RL before the incident was very small (practically around zero) while the voltage becomes almost around V_s when the fault occurs. Thus, the current will "theoretically" rise as an exponential function $i(t) = I_0 + I_f(1 - e^{-t/\tau})$ where $I_f = \frac{V_g}{R}$ is a huge number. For $V_g = 300$ V, and $L = 5$ mH and $R = 30$ mΩ, for instance, $I_f = \frac{300}{0.030} = 10\,000$ A and it means that it will already increase $10\,000 \times (1 - e^{-0.001\times 30/5}) = 60$ A even in short time-interval of 1 ms! The situation worsens for smaller τ's.

The converter control system can only prevent the over-current if its speed is higher than this natural current rise time in the system. A real eigenvalue (pole) of the closed loop system around -1000 corresponds to a time-constant of around 1 ms. For a pair of complex poles, assuming that they have sufficient damping, this is still roughly valid. Therefore, the more the poles of the closed-loop system are pushed away from the imaginary axis, the tighter the controller will limit the transient currents.

A faster closed-loop system will limit the current transients more tightly. However, this is at the expense of the two things: (i) lower delay margin[6], (ii) more sensitivity to noises, especially the sensor noise and switching noise. The digital implementation of the controller on DSP causes one

6 Refer to Stability Margins in Chapter 3.

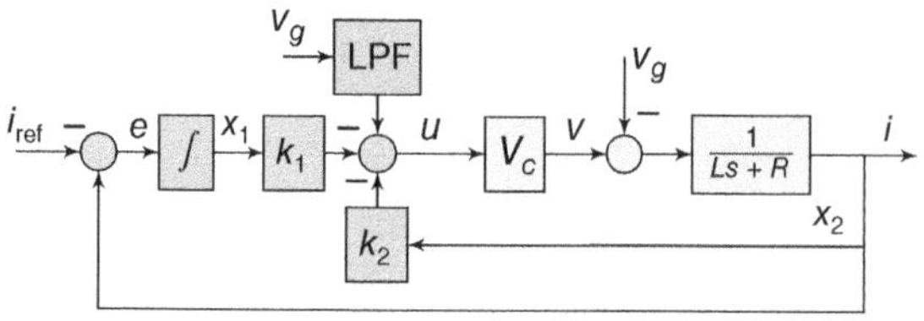

Figure 5.23 Feed-forward of grid voltage to mitigate over-current issue.

sampling time of delay. The PWM and converter switching itself introduces some delay. Therefore, a too-fast design may readily start oscillating and lead to instability in real world conditions. Moreover, its modulation signal $m(t)$ may quickly go beyond the limits $[-1\ \ 1]$ during transients and cause irregular and distorted transient responses.

Therefore, a too-fast design must be avoided. One mitigating solution is to use a filtered-version of v_g as the feed-forward term as shown in Figure 5.23. When the grid voltage experience an abrupt and large change, e.g. at short-circuit instant, the feed-forward term quickly transmits this change to the controller and enables it to respond to the change quickly. Thus, the LPF must have a relatively large bandwidth.[7] It may be chosen as a simple first-order transfer function $H_{\text{LPF}}(s) = \frac{1}{\tau s+1}$ where for instance a $\tau = 0.001$ allows to transmit the changes within a 1 ms time frame. The combination of this mitigating approach and a relatively fast (but not too fast) feedback controller allows to have current limiting without making the system sensitive to delays and noises. These details are illustrated in the following example.

Example 5.5 ***Limiting Current During Grid Low-Voltage Transients***

Consider the same converter system parameters of Table 5.1. An optimal LQT-based designed controller is used. Four different designs are considered as shown in Figure 5.24. Design 1 places the

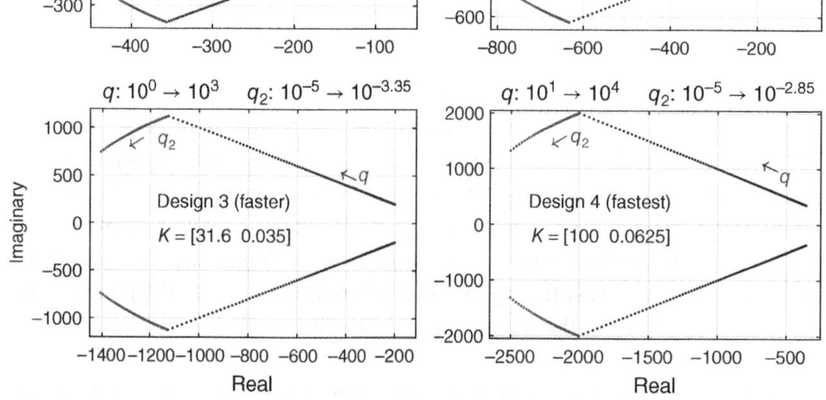

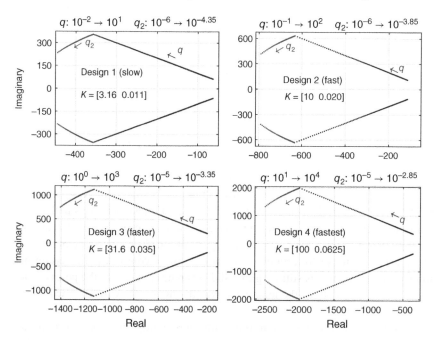

Figure 5.24 LQT-based controller at four designs which become progressively faster.

---

7 However, its bandwidth must be sufficiently small to filter noises that are present on v_g in a practical system.

poles at $-445 \pm j235$ with $K = [3.16 \;\; 0.011]$. Design 2 places the poles at $-791 \pm j417$ with $K = [10 \;\; 0.0197]$. [8] Design 3 places the poles at $-1407 \pm j742$ with $K = [31.6 \;\; 0.0350]$. Design 4 places the poles at $-2502 \pm j1320$ with $K = [100 \;\; 0.0625]$. Designs are progressively becoming faster by distancing the poles from the imaginary axis.

Responses of these four controllers to a full power command at $t = 0.05$ seconds, a V_c drop of 50 V at $t = 0.1$ seconds, and a grid voltage short-circuit transient occurring at 0.15 seconds and clearing at 0.18 seconds are shown in Figure 5.25. The slow controller (Design 1) shows a current transient with peak values as large as ±50 A when the fault happens or clears. The controller of Design 2 reduces these peaks to about 30 A. The controllers of Design 3 and Design 4 also reduce the transient to about 17 and 9 A, respectively. If the criterion is not to exceed 20 A over-current, Design 3 and 4 pass this criterion.

Now we intentionally introduce a delay in the control loop (right before forwarding the modulation signal to the PWM) to each of these systems. The faster the system, the shorter the delay that it can tolerate. Figure 5.26 shows their responses for Design 1–4 when a delay of 1, 0.67, 0.4, and 0.22 ms is applied to them, respectively.[9] The responses become highly oscillatory. Moreover, the peak over-current to grid voltage fault increases. The faster the system, the lower the amount of delay it can tolerate. It can easily be computed that the delay margins of these four designs are 1.38, 0.76, 0.43, and 0.24 ms, respectively (see Problem 5.1).

Now we study the impact of adding a filtered-version of the grid voltage as shown in Figure 5.23. We use a filter $H_{\text{LPF}}(s) = \frac{1}{0.001s+1}$. Figure 5.27 shows the responses for Design 1–4 to the same set of disturbances. It is observed that this feed-forward term has been able to reduce the peak of over-current from the original values of 50, 30, 17, 9 A, respectively to 26, 17, 12, 7 A.

Figure 5.28 shows the case where the grid voltage feed-forward is used and the large delays are also intentionally introduced into the control loop. As expected, the responses become oscillatory and the peak over-currents increase. However, compared to Figure 5.26, the grid voltage feed-forward has been able to reduce the peak over-current.

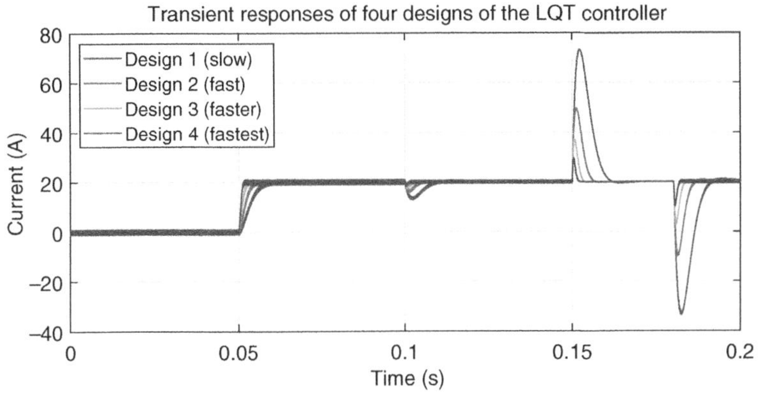

Figure 5.25 Time responses of four LQT-based controllers of Figure 5.24 to a full power command at $t = 0.05$ seconds, a V_c drop of 50 V at $t = 0.1$ seconds, and a grid voltage short-circuit transient occurring at 0.15 seconds and clearing at 0.18 seconds.

8 This is the same design used in Example 5.3.
9 Notice that in these simulations, we use real switch models with PWM. Thus, the delay associated with the PWM is modeled. However, the controller is implemented in continuous-time domain. Thus, the delay associated with the DSP is not considered. One sampling cycle delay at 8 kHz (twice the switching frequency of 4 kHz) corresponds to 0.125 ms.

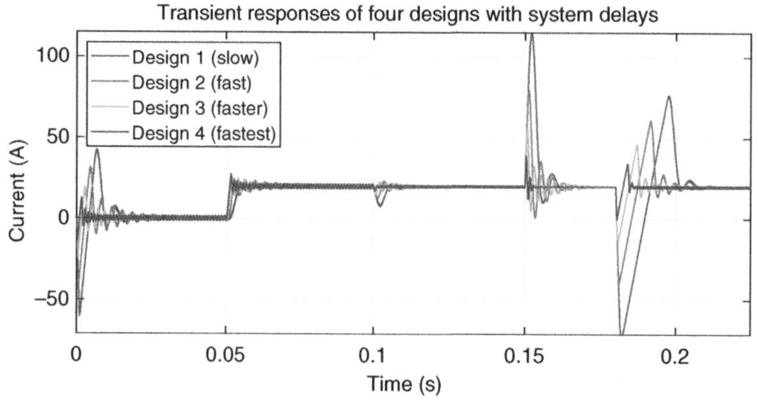

Figure 5.26 Time responses of four LQT-based controllers of Figure 5.24 when a delay of 1, 0.67, 0.4, and 0.22 ms are introduced in Design 1–4, respectively.

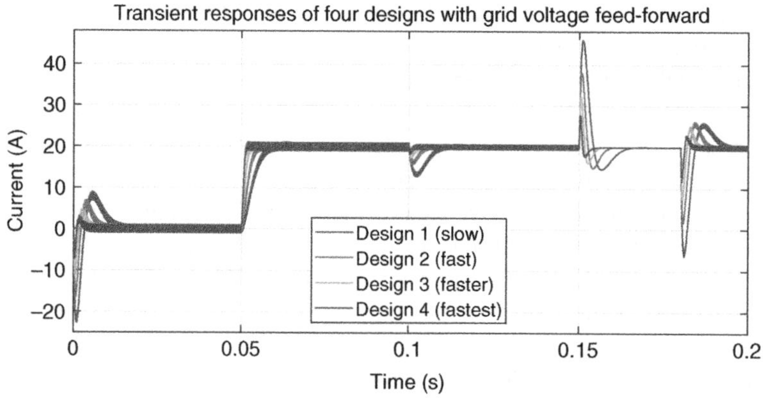

Figure 5.27 Time responses of four LQT-based controllers of Figure 5.24 when a filtered-version ($\tau = 0.001$ seconds) of the grid voltage is feed-forwarded.

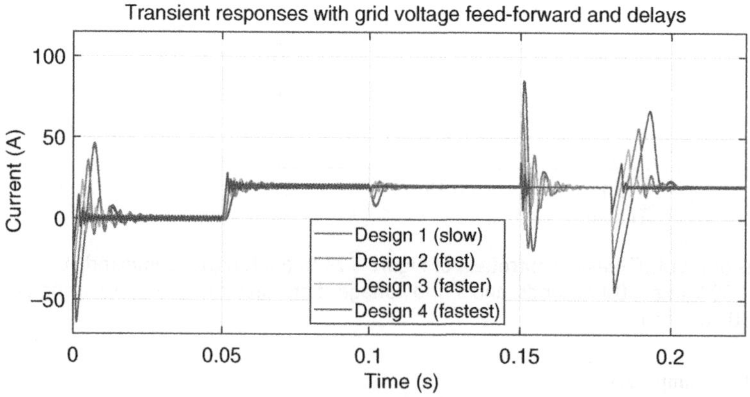

Figure 5.28 Time responses of four LQT-based controllers of Figure 5.24 when a delay of 1, 0.67, 0.4, and 0.22 ms are introduced in Design 1–4, respectively, and the grid voltage feed-forward is also used.

This section and the studied example concludes that in order to successfully limit the current, either the control loop must be very fast (like Design 3 and 4 above) or it must be sufficiently fast and combined with a grid voltage feed-forwarding term (like Design 2). The faster the control loop, the lower their ability to cope with control and system delays and also the measurement and system noises. So, the trade-offs are clear.

5.4 Grid Voltage Support

Section 5.3 discussed the aspects on designing a feedback control system that enables the execution of a power set-point by the converter. It also discussed optimal design of controller gains, soft start, and current limiting aspects. This section delves into the aspect of grid voltage support which is divided into dynamic and static grid supports. The converter is desired to respond in a dynamic way to the quick changes in the grid voltage. Such changes may constantly happen due to the transient load and supply imbalance in the grid. The converter should dynamically support the grid during such transients. The dynamic grid support is closely related to the concept of "inertia." In the static grid support, the converter must respond to grid voltage deviations (offsets) from its desired rated value.

5.4.1 Explanation on Concept of Inertia

In a synchronous generator, the grid frequency is tightly coupled with the rotor speed. Assume that τ_m (p_m) is the mechanical torque (power) supplied by the prime mover, and τ_e (p_e) is the equivalent electrical torque (power) supplied by the generator. The Newton's law indicates that

$$\tau_m - \tau_e = J\dot{\omega}, \quad \text{or} \quad p_m - p_e = J\omega\dot{\omega}, \tag{5.2}$$

where ω is the rotor speed, and J is the moment of inertia of the rotating mass. This equation implies that the output electrical power supplied by the generator is

$$p_e = p_m - J\omega\dot{\omega}$$

and has two components, i.e. p_m: supplied by the prime mechanical source, and $p_{\text{inr}} = -J\omega\dot{\omega}$: supplied from the kinetic energy of the rotor mass. This latter is called the *inertia power* and such a response of the machine is called the inertia response.

The inertia response is grid supportive: when the grid frequency tends to go down/up (↓ / ↑) as a result of a dynamic power shortage/excess in the grid, the inertia power goes up/down (↑ / ↓) and compensate for that power shortage/excess. Moreover, this response is dynamic thanks to being proportional to $\dot{\omega}$. It can even be said it has some "predictive" nature due to the leading angle of derivative function. That is why the inertia of synchronous generators in bulk power system has been very important in damping the fast and dynamic power imbalances. It must be noted that the prime source power p_m cannot respond quickly to fast power dynamics due to its physical and mechanical limitations.

5.4.2 Conflict of Inertia Response and Current Limiting

The standard feedback loop shown in Figure 5.14 can be redrawn in the form of Figure 5.29. Figure 5.29 first illustrates the fact that k_2 feedback term introduces a virtual resistance of $k_2 V_c$

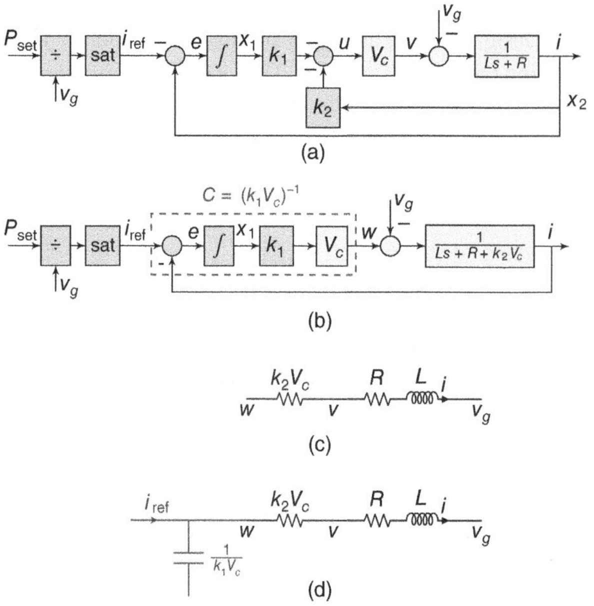

Figure 5.29 (a) Standard LQT-based controller of Figure 5.14, (b) alternative equivalent presentation, (c) equivalent circuit showing the virtual resistor, and (d) complete equivalent circuit showing the virtual resistance and virtual capacitor.

in the loop. Then, the signal $w(t) = v(t) + k_2 V_c i(t)$ represents the voltage at a "virtual" point. It then shows that the k_1 feedback term introduces a virtual capacitance of $C = \frac{1}{k_1 V_c}$ in the control loop.

The differential equation

$$i = i_{\text{ref}} - e = i_{\text{ref}} - \dot{x}_1 = i_{\text{ref}} - \frac{1}{k_1 V_c}\dot{w} = i_{\text{ref}} - J\dot{w} \tag{5.3}$$

leads to

$$p = iv_g = i_{\text{ref}} v_g - J\dot{w}v_g = P_{\text{set}} - Jv_g\dot{w} \tag{5.4}$$

which indicates that the power follows the dynamics of $-Jv_g\dot{w}$ where $J = C = \frac{1}{k_1 V_c}$. This virtual capacitor and its effect may be interpreted as a natural inertia[10] behavior in the following sense: when the grid voltage $v_g(t)$ experiences a sudden or fast change, this change tends to naturally reflect on $w(t)$.[11] However, w is the capacitor voltage which means the capacitor will respond by a fast $i = i_{\text{ref}} - i_c = i_{\text{ref}} - C\dot{w}$ current. More specifically, if the grid voltage tends to rise, the converter current reacts by going down; and vice versa. This is an inertia behavior and is confirmed by Figure 5.25: when the grid voltage suddenly drops at $t = 0.6$ seconds, the current experiences a rise and an upward bump (a hill) that corresponds to a small positive power (and energy) injected to the grid by the converter. Conversely, when the grid voltage suddenly jumps at $t = 0.63$ seconds, the current experiences a fall and a valley that corresponds to power (and energy) drawn from the grid by the converter.

10 Although the word "inertia" is commonly used for a mechanical system but the core concept is the same. Some literature use terms such as "virtual," "synthetic," or "electrostatic" inertia to distinguish this type of inertia from the kinetic inertia [2–5].

11 Because these two are the voltages across an inductive element which resists a sudden current change.

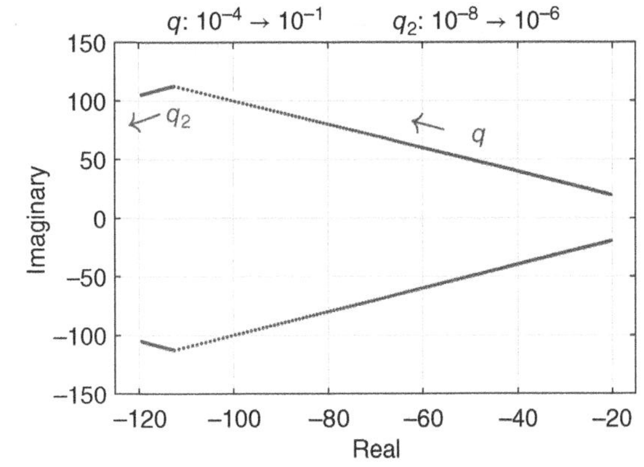

Figure 5.30 LQT-based controller at very slow design to amplify and visualize the inertia response.

This natural inertia behavior is a result of the adopted control system with an integrator. The controller gains can weaken or amplify it (as shown in Figure 5.25: the faster the controller, the lower the inertia) but the nature of this response remains independent from the selection of controller gains.

The "moment of inertia" J (or the virtual capacitance C) is $(k_1 V_c)^{-1}$. Examining the numbers for k_1 calculated in Example 5.5 shows that k_1 changes as 3.16, 10, 31.6, and 100 for Design 1 to Design 4, respectively. This concludes that the moment of inertia and current limiting are in trade-off. When current limiting is achieved, the moment of inertia of this system is practically negligible.

Example 5.6 ***Inertia Response versus Current Limiting***

Consider the same converter system parameters of Table 5.1 and the Design 3 that places the poles at $-1407 \pm j742$ with $K = [31.6 \quad 0.0350]$. Design 0 is shown in Figure 5.30 which places the poles at $-120 \pm j105$ with $K = [0.316 \quad 0.0029]$. This design has been made very slow to increase and visualize the inertia response.

Responses of these two controllers to a grid voltage swing between $t = 0.1$ seconds and $t = 0.5$ seconds, and a sudden voltage drop occurring at 0.6 seconds and clearing at 0.7 seconds are shown in Figure 5.31. The very slow controller (Design 0) shows an inertia response to the voltage swing but it cannot limit the current during sudden voltage drop. The faster controller (Design 3) limits the current at the sudden voltage change but it does not provide an inertia response to the voltage swing.

For this slow design, $J = C = \frac{1}{k_1 V_c} = \frac{1}{0.316 \times 400} = 0.0079$ F $= 7.9$ mF. The grid voltage swing shown in Figure 5.31c has a slope of $\frac{50}{0.1} = 500$ V/s. Thus, from (5.3), the power inertia response will be about $J v_g \dot{w} = 0.0079 \times 300 \times 500 \approx 1.2$ kW. This number is confirmed in Figure 5.31a: the power experiences ± 1.2 kW to the ± 500 V/s voltage swings.

5.4.3 Inertia Response Using Capacitor Emulation

The above discussion and example proved that the inertia response and current limiting are two conflicting objectives in the basic power control approach. In the following, the controller is enhanced to achieve an inertia response while still being able to limit the current.

The first proposed structure is shown in Figure 5.32. Here the current control loop (specified by k_1 and k_2) is fast and thus it allows current limiting. This means that the current waveform i tightly follows its reference i_{ref} and is limited between $I_{\min}$ and $I_{\max}$. The feed-forward term is

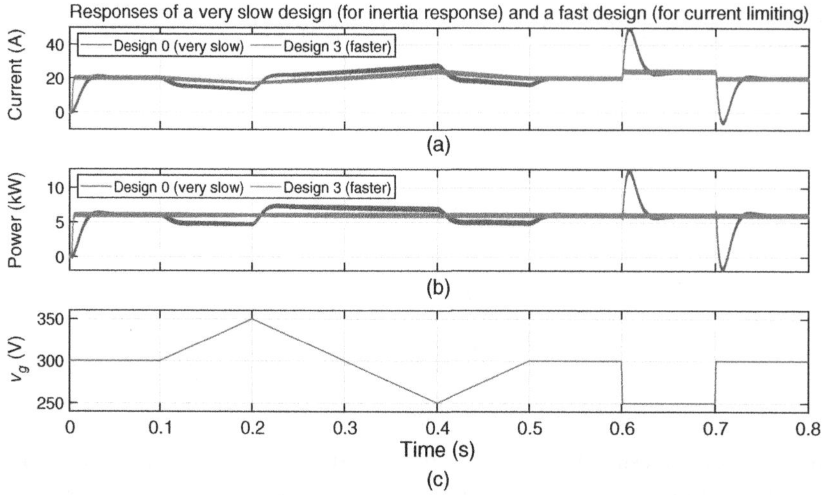

Figure 5.31 Time responses of the very slow and the fast controllers to a grid voltage swing and a sudden grid voltage drop. The slow design provides inertia but cannot limit the current.

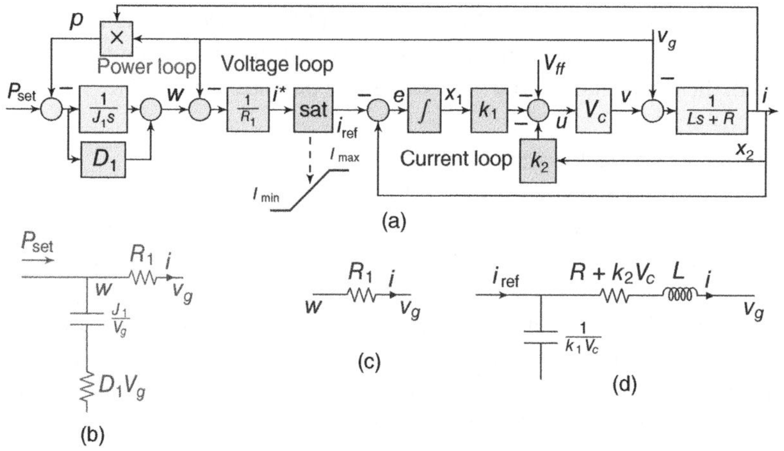

Figure 5.32 (a) Adding a voltage and a power loop around the fast current loop to make the converter emulate a capacitor with inertia response while being able to limit the current. (b) Equivalent circuit of the total converter with emulated capacitor $C = \frac{J1}{V_g}$. (c) Equivalent circuit between w and v_g. (d) Equivalent circuit of fast current loop.

either $V_{ff} = \frac{v_g(0)}{v_c(0)}$ (to give a soft start) or a filtered version of the grid voltage to mitigate transient over-currents as discussed in Section 5.3.5.

In Figure 5.32, an external voltage control loop is closed right after the current control loop. This loop establishes $v_g + R_1 i = w$.[12] And since the current i is continuous, the virtual voltage w will replicate the fast dynamics of the grid voltage v_g.

For $D_1 = 0$, the outer power loop in Figure 5.32 establishes the relationship

$$P_{set} - p = J_1 \dot{w} \tag{5.5}$$

12 Notice that since the current loop is very fast, i_{ref} and i appear the same for the voltage loop.

which indicates that the output power p will follow an inertia dynamics indicated by $-J\dot{\omega}$ like a virtual capacitor. The variable w emulates the voltage across the virtual capacitor. If the voltage drop across R_1 is small (compared with V_g), then w will stay around v_g. Therefore, the capacitor voltage is around V_g.[13] Thus, the emulated capacitance is equal to $C = \frac{J}{V_g}$. This also entails that for a soft start, it is good to initialize this integrator at $w(0) = V_g$. It is now clear to see that D_1 also emulates a virtual resistance $R_d = D_1 V_g$ connected in series with the virtual capacitor.[14]

The conclusion of this discussion is that Figure 5.32 offers an inertia response behavior to the converter (equivalent to a capacitor with size $C = \frac{J_1}{V_g}$) while ensuring that its current is limited during severe disturbances (such as grid voltage short-circuit faults). If the current-limiting property were not a requirement, the variable w/V_c could directly be applied to the control signal u and the converter would mimic a capacitor (without current limiting ability).

Example 5.7 ***Inertia Response Using Capacitor Emulation***
Consider the same converter system parameters of Table 5.1 and the Design 3 that places the poles at $-1407 \pm j742$ with $K = [31.6 \;\; 0.035]$, shown again in Figure 5.33. The limiter bounds are set to $I_{min} = 0$ and $I_{max} = 40$ A. In the system without inertia, $i^* = \frac{P_{set}}{v_g}$ is set as shown in Figure 5.29a. In the one with inertia, the voltage and power loops shown in Figure 5.32 are used. The value of J_1 is set to 1 corresponding to an emulated capacitance of $C = \frac{J_1}{V_g} = 3.3$ mF. The other two control parameters are set as $R_1 = 1$ and $D_1 = 0$.

Responses of the two systems (without and with the inertia loop) to a grid voltage swing between $t = 0.15$ seconds and $t = 0.5$ seconds, and a sudden voltage drop occurring at 0.55 seconds and clearing at 0.65 seconds are shown in Figure 5.31. The converters start at $t = 0$ and their power reference increases from zero to 3 kW at $t = 0.05$ seconds and then to 6 kW at $t = 0.1$ seconds. Here are the observations from this graph: (i) the inertia loop reduces the speed of transient response to power commands; (ii) during the swing of ± 1000 V/s, the converter supplies ± 1 kW power which complies with $J_1 = 1$; (iii) at the instant of sudden grid voltage drop/rise, the controller instantly supplies/absorbs a large inertia power; and (iv) the current is successfully limited between 0 and 40 A.

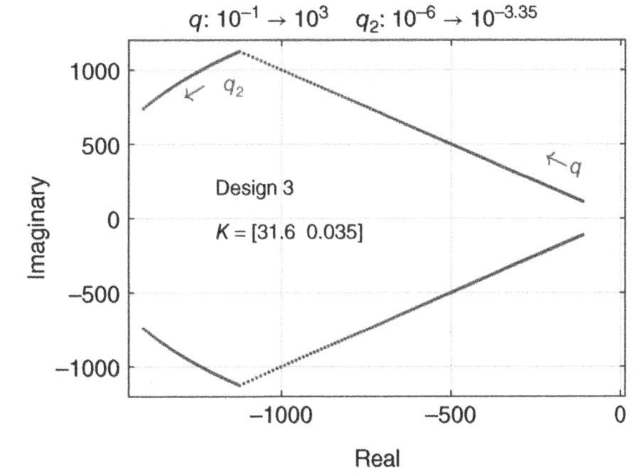

Figure 5.33 LQT-based current controller at fast design to allow current limiting.

13 The term V_g indicates the nominal value of the grid voltage $v_g(t)$.
14 This resistance is not helpful in this simple situation and we have only included here for conceptual and completeness of discussion. It can help dampen the system oscillations in more complex cases such as LCL filters.

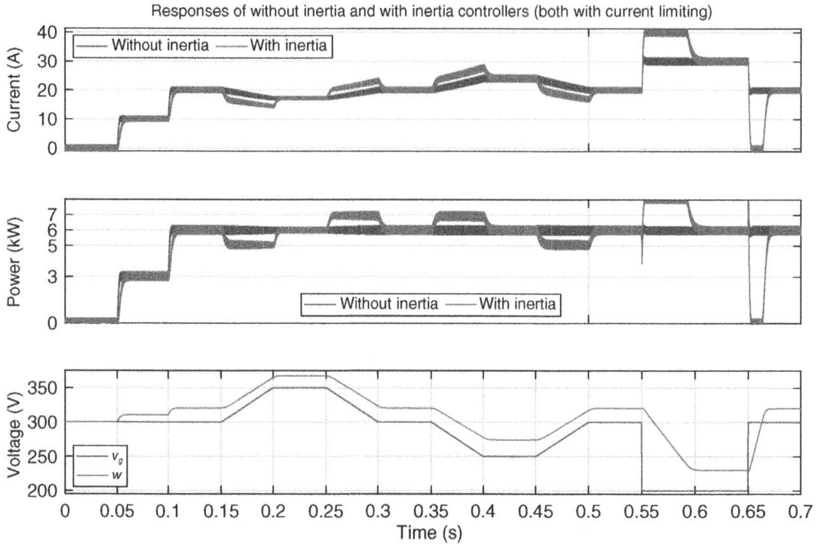

Figure 5.34 Time responses of the fast LQT-based controller without inertia, Figure 5.29, and with an inertia response, Figure 5.32, to power commands (at $t = 0.05$ seconds and $t = 0.1$ seconds) and a grid voltage swing (between $t = 0.15$ and 0.5 seconds) and a sudden grid voltage drop (at $t = 0.55$ seconds) and clearing (at $t = 0.65$ seconds).

It is also worthwhile noticing that the power produced by the controller without inertia is constant and does not experience any noticeable change during the grid voltage transients. The responses of the controller with inertia loop, however, is "grid supportive" in the sense that when the grid voltage is rising/falling, it absorbs/injects power from/to the grid. This may be called a "dynamic voltage support property." It responds to the changes of the grid voltage not to its value (Figure 5.34).

The power loop shown in Figure 5.32 is mathematically equivalent to a capacitor. In a capacitor with capacitance C, the relationship between its net instantaneous power $p_c(t)$ and its voltage $v_c(t)$ is given by

$$p_c(t) = Cv_c(t)\dot{v}_c(t). \tag{5.6}$$

If the capacitor is operating around the nominal voltage V_c, the equation may be approximated by $p_c(t) = CV_c\dot{v}_c(t)$.[15] Thus, this approach basically makes the converter to mimic a capacitor with capacitance $C = \frac{J_1}{V_c}$. Here, we assumed that the dc side of the converter is connected to a voltage source V_c. In most converter applications, this side already has an actual capacitor in itself. This will be discussed in details in this section.

Figure 5.35 shows an improved/refined version of the system of Figure 5.32. (i) The power-loop integrator $\frac{1}{J_1 s}$ has been replaced with $\frac{1}{CV_c s}$ to emphasize the equivalence with a capacitor. (ii) The capacitor voltage increment $\tilde{v}_c = v_c - V_c$ and the grid voltage increment $\tilde{v}_g = v_g - V_g$ are subtracted through a positive factor γ and forwarded to the current loop. (iii) The gain R_1 emulates a resistor in between $\tilde{v}_c - \tilde{v}_g$ and the current increment $\tilde{i} = i - I$ where I is the nominal rated value of current. These relations are shown in the equivalent circuit diagrams illustrated in Figure 5.35. The

15 Or, we may say the first term in its Taylor's series is this. See Problem 5.2.

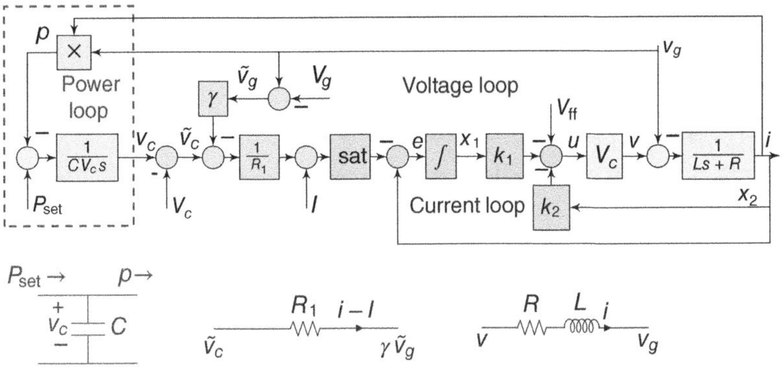

Figure 5.35 Improved/refined version of the capacitor emulation approach for realizing an inertia response.

feed-forward term is $V_{ff} = \frac{V_g(0)}{v_c(0)}$ for actual capacitor and $V_{ff} = \frac{V_g(0)}{V_c}$ for emulated capacitor or can use a filtered version of the grid voltage for improved current limiting.

This discussion results in that the controller will establish a relationship

$$v_c = V_c + \gamma(v_g - V_g) + R_1(i - I) \quad (\text{or}: \quad \tilde{v}_c - \gamma\tilde{v}_g = R_1\tilde{i}) \tag{5.7}$$

between the grid voltage, the capacitor voltage, and the converter current. This equation has the following interesting implications. (i) The capacitor voltage v_c changes with the grid voltage v_g. At a given loading condition, i.e. at a given i, if v_g changes 1 V, v_c will change γ volts. (ii) The capacitor voltage will also change with the current. At a given grid voltage, i.e. at a given v_g, if the converter current changes 1 A, the capacitor voltage will change R_1 volts. (iii) The grid voltage dynamics of $\tilde{v}_g$ will reflect on the capacitor voltage $\tilde{v}_c$ through a factor γ. This implies an inertia response. Moreover, if $\gamma > 1$, this is particularly advantageous when an actual capacitor is in the circuit. This way, the inertia power supplied by the capacitor is equal to $Cv_c\dot{v}_c \approx C\gamma V_c\dot{v}_g$ which indicates an "amplification" for a given physical capacitor size C. Practically, γ cannot be too large so as not to cause large transients on v_c. For $\gamma = 0$, this feature (i.e. providing inertia response to grid voltage variations) will be disabled. In this case, the capacitor voltage will only change with the amount of load. And this change can be adjusted through R_1 and/or I.[16]

Example 5.8 *Inertia from Emulated or Actual Capacitor*

Consider the same converter system parameters of Table 5.1 and the Design 3 that places the poles at $-1407 \pm j742$ with $K = [31.6 \;\; 0.035]$, shown in Figure 5.33. The current limiter bounds are set to $I_{min} = 0$ and $I_{max} = 40$ A. Two systems are considered as follows. System 1 uses the (linearized) mathematical model of a capacitor (the power loop) to emulate the inertia response according to Figure 5.36a. System 2 uses an actual capacitor to emulate the inertia and the controller is shown in Figure 5.36b. In System 1, the dc side of the converter is connected to a 400 V voltage source that supplies all powers, including P_{set} and the inertia. In System 2, it is connected across an actual capacitor (that supplies the inertia) and a source of power (that supplies P_{set}). This source of power is simulated using a controlled current source $i_{set}(t) = \frac{P_{set}}{v_c(t)}$ connected across the capacitor terminals. The same power command P_{set} is applied to both systems. The capacitor size is 2.5 mF which

16 Smaller R_1 means lower range of change of v_c for different loading conditions. If I is set to $\frac{P_{set}}{V_g}$, then the capacitor voltage will remain constant (in the steady state) for all loading conditions.

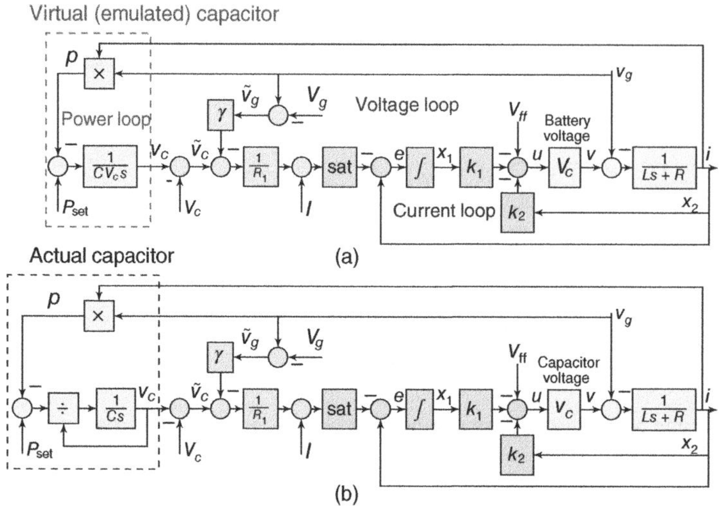

Figure 5.36 Controllers for flexible inertia response and current limiting using (a) emulation of a capacitor, and (b) using the actual capacitor.

corresponds to $\frac{1}{CV_c} = 1$. All other system parameters are set equally in both systems. Specifically, $R_1 = 1$ and $\gamma = 1$.

Responses of the two systems (capacitor-emulated and with actual capacitor) to a grid voltage swing between $t = 0.15$ seconds and $t = 0.5$ seconds, and a sudden voltage drop occurring at 0.55 seconds and clearing at 0.65 seconds (same scenario used in Example 5.7) are shown in Figure 5.37. The converters start at $t = 0$ and their power reference increases from zero to 3 kW at $t = 0.05$ seconds and then to 6 kW at $t = 0.1$ seconds. Here are the observations from this graph: (i) both systems respond almost identically; minor differences is due to the fact that System 1 has constant dc side voltage while System 2 has variable capacitor voltage, (ii) the capacitor voltage follows the grid voltage swings, (iii) the capacitor voltage also varies with the level of power, (iv) the level of inertia power matches the expected level corresponding to the level of grid voltage rate of change, and (v) the current is successfully limited between 0 and 40 A.

In Figure 5.38, two cases of battery and capacitor are shown but this time the grid voltage drop is deeper all the way to 100 V. During this deep voltage drop, the grid cannot absorb the 6 kW power and this will be stored in the capacitor (or battery). During this time, $0.55 < t < 0.65$ seconds, the capacitor voltage rises as much as around 550 V. Once the fault clears at 0.65 seconds, this energy is quickly released to the grid and the capacitor voltage returns to its normal range. The current is limited within the prespecified range for all times. All system parameters are the same for both systems and equal to those of previous simulation.

In Figure 5.39, two cases of battery and capacitor but this time a small capacitance of 0.5 mF is considered. As observed, the system responses show some extended oscillations. Interestingly, the system with actual capacitor shows more stable responses compared to the one with battery.[17] Both systems show inertia responses but at 5 times reduced level compared with the previous case where the capacitor size of 2.5 mF was used. All other system parameters are the same for both systems and equal to those of previous simulation.

---

17 This is likely due to the capacitor voltage being variable and adaptively adjusting to the system conditions while the battery's voltage is constant.

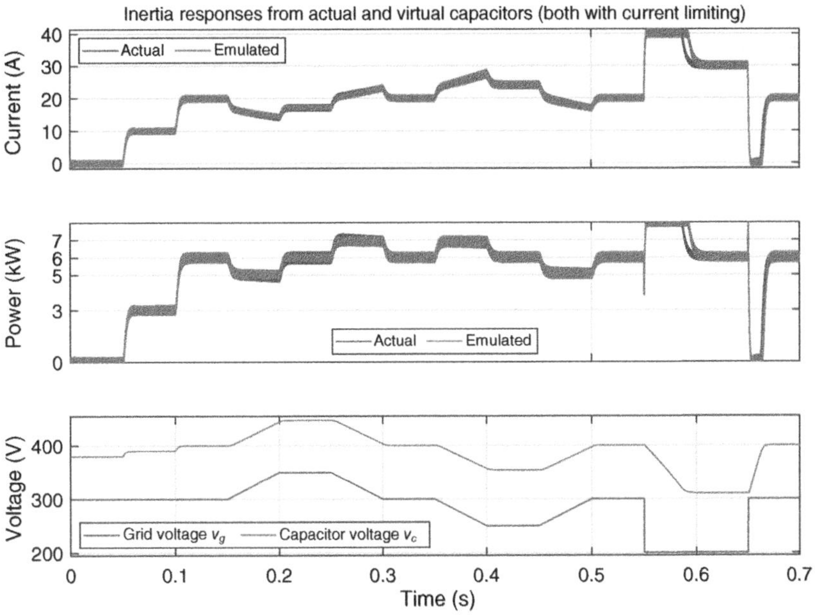

Figure 5.37 Time responses of converters with inertia response for two cases where a battery is used and the inertia is emulated, and an actual capacitor is used. The current controller is design fast using LQT approach and the current limiting is in place. Power commands (at $t = 0.05$ seconds and $t = 0.1$ seconds) and grid voltage swing (between $t = 0.15$ and 0.5 seconds) and a sudden grid voltage drop (at $t = 0.55$ seconds) and clearing (at $t = 0.65$ seconds) are applied.

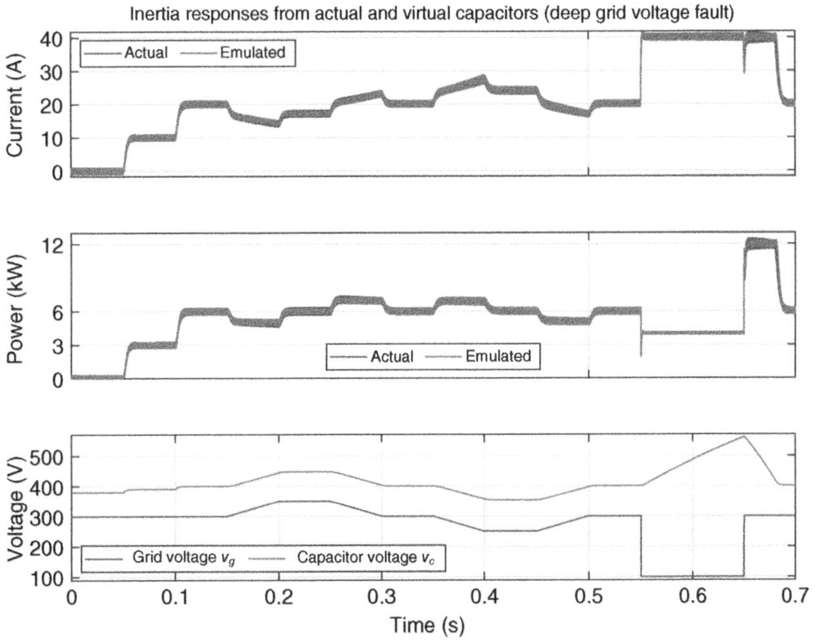

Figure 5.38 Time responses of converters with inertia response at deep grid voltage fault.

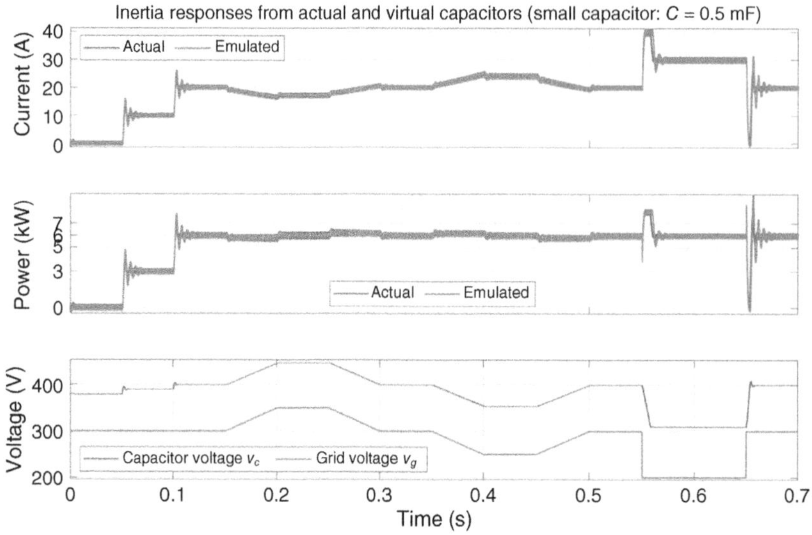

Figure 5.39 Time responses of converters with inertia response at small capacitance of 0.5 mF.

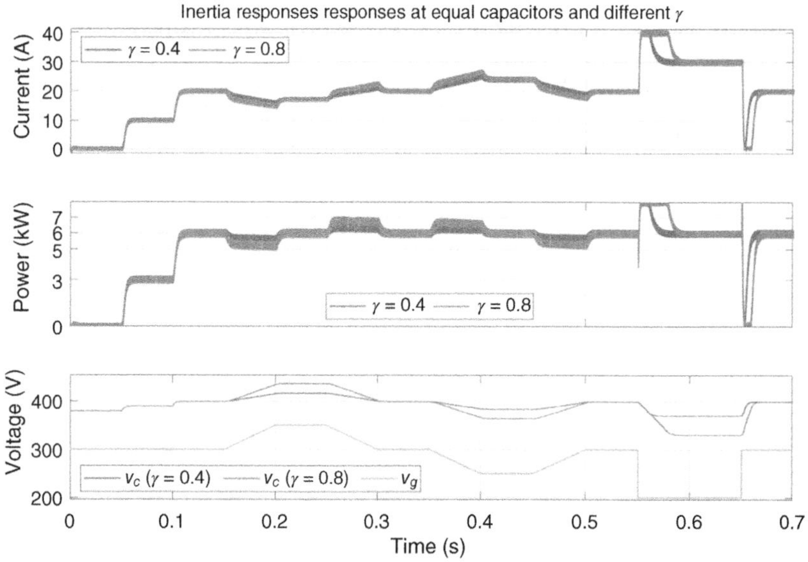

Figure 5.40 Time responses of converters with inertia response for two cases where an equal capacitor size is used and the software gain γ is either 0.4 or 0.8.

In Figure 5.40, two cases of $\gamma = 0.4$ and $\gamma = 0.8$ and the equal capacitor size of 2.5 mF (actual capacitor for both cases) is shown. As expected, doubling γ doubles the level of inertia power response of the system. Smaller γ will lower the range of changes of v_c in response to grid voltage variations.

In Figure 5.41, two cases of $R_1 = 1$ and $R_1 = 0.2$ and the equal capacitor size of 2.5 mF is shown. The constant γ is set to zero to disable direct variations of capacitor voltage caused by grid voltage. As expected, as R_1 increases, the range of deviation of the capacitor voltage from the nominal value of 400 V decreases. For $R_1 = 0.2$, and for the entire operating range of the converter, the voltage v_c

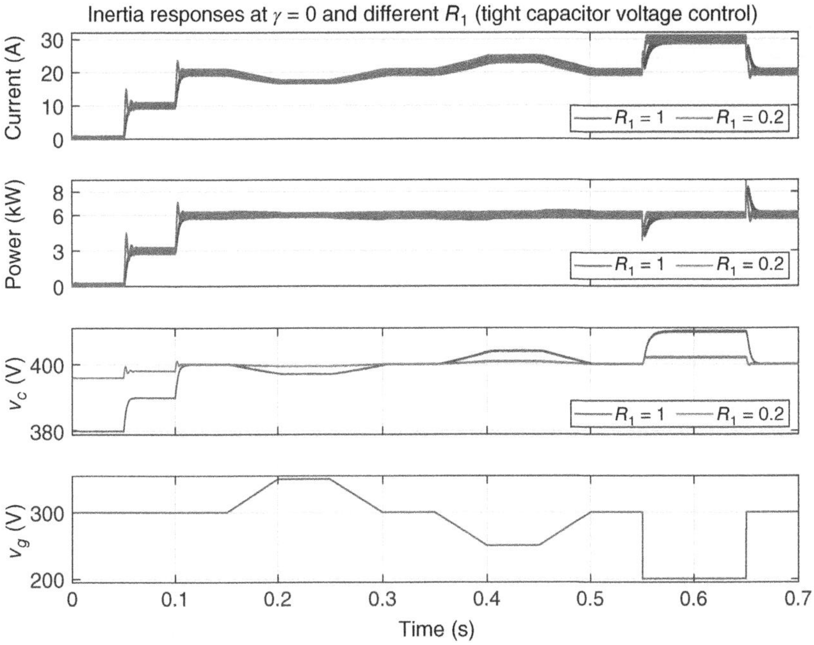

Figure 5.41 Time responses of converters with inertia response for two cases where $R_1 = 1$ and $R_1 = 5$ (at $\gamma = 0$).

changes ±4 V around 400 V, i.e. 1%. This can even further be decreased by decreasing R_1. However, as observed, some oscillations appear in the responses for smaller R_1. We will see in Section 5.4.4 that these oscillations can also brought under control by adding another feedback gain.

5.4.4 Full State Feedback of Power Loop

It was shown in Example 5.8, Figure 5.39, that the system responses tend to become more oscillatory as the capacitor size is reduced. One reason is that the added power loop adds a new state variable which is not considered during the LQT state feedback design of the current controller. It is possible, as shown in Figure 5.42, to add this state variable to the feedback signals of the inner current controller. This section discusses the design of k_1, k_2 and k_3 using the LQT approach with the objective of reducing the response oscillations at small capacitor sizes.

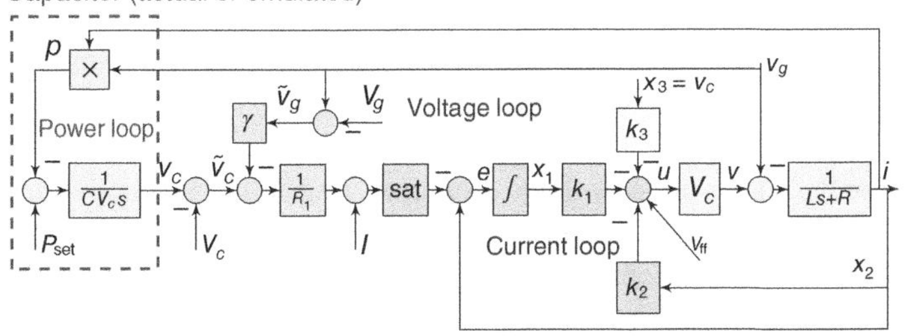

Figure 5.42 Full state feedback of current, and power loops.

The state-space equations of the system are summarized as

$$\begin{aligned}
\dot{x}_1 &= x_2 - \{I + \tfrac{1}{R_1}[(x_3 - V_c) - \gamma(v_g - V_g)]\} \\
\dot{x}_2 &= -\tfrac{R}{L}x_2 + \tfrac{V_c}{L}u - \tfrac{1}{L}v_g \\
\dot{x}_3 &= -\tfrac{1}{CV_c}v_g x_2 + \tfrac{1}{CV_c}P_{\text{set}} \\
u &= -k_1x_1 - k_2x_2 - k_3x_3 + V_{\text{ff}} = -Kx + V_{\text{ff}},
\end{aligned} \tag{5.8}$$

where $V_{\text{ff}} = \frac{v_g(0)}{v_c(0)} + k_3 v_c(0)$ is a constant term for soft start, or the first term may be replaced with a filtered version of the grid voltage for improved current limiting. To proceed with the LQT approach, apply the linear differential operator $\frac{d}{dt}$ to both sides of (5.8) to get

$$\begin{aligned}
\dot{z}_1 &= z_2 - \tfrac{1}{R_1}z_3 \\
\dot{z}_2 &= -\tfrac{R}{L}z_2 + \tfrac{V_c}{L}w \\
\dot{z}_3 &= -\tfrac{1}{CV_c}V_g z_2 \\
w &= -k_1z_1 - k_2z_2 - k_3z_3 = -Kz,
\end{aligned} \tag{5.9}$$

where $z_i = \dot{x}_i$, $w = \dot{u}$, and the derivative of all constants and external inputs are considered zero for the sake of this design. Therefore, the state space representation in terms of z variables is

$$\dot{z}(t) = Az(t) + Bw(t), \text{ where } A = \begin{bmatrix} 0 & 1 & -\frac{1}{R_1} \\ 0 & -\frac{R}{L} & 0 \\ 0 & -\frac{V_g}{CV_c} & 0 \end{bmatrix} \quad B = \begin{bmatrix} 0 \\ \frac{V_c}{L} \\ 0 \end{bmatrix}.$$

To minimize the cost function $J = \int_0^\infty \left[qe^2(t) + q_2z_2^2(t) + q_3z_3^2(t) + w^2(t)\right] dt$ for some $q > 0$, the matrix Q should be chosen as

$$Q = \begin{bmatrix} q & 0 & 0 \\ 0 & q_2 & 0 \\ 0 & 0 & q_3 \end{bmatrix}$$

and the LQR problem is solved in MATLAB: ≫ `K = lqr(A, B, Q, 1)`.

Example 5.9 ***Full State Feedback of Power Loop***

Consider the same converter system parameters of Table 5.1 and the design that places the poles at $-1407 \pm j742$ with $K = [31.6 \;\; 0.0350]$, shown in Figure 5.33. The limiter bounds are set to $I_{\min} = 0$ and $I_{\max} = 40$ A. This design is independent from the capacitor size.

For the rather small capacitor size of $C = 0.5$ mF,[18] a sample of running the above full state feedback LQT design is shown in Figure 5.43. For $q = 10^3$, $q_2 = 10^{-3.35}$ (same as before) and $q_3 = 10^{-3.35}$, the location of closed loop poles are at $-1423 \pm j1123, -1155$, and the controller gains are $K = [k_1 \;\; k_2 \;\; k_3] = [31.6 \;\; 0.05 \;\; -0.0337]$.

Responses of two systems (without the gain k_3 and with k_3) to a grid voltage swing between $t = 0.075$ seconds and $t = 0.25$ seconds, and a sudden voltage drop occurring at 0.275 seconds and clearing at 0.325 seconds are shown in Figure 5.31. The converters start at $t = 0$ and their power reference increases from zero to 3 kW at $t = 0.025$ seconds and then to 6 kW at $t = 0.05$ seconds. Here

18 In this full state feedback method, the capacitor size is used in the design. Thus, the capacitor size can be selected very small. However, the smaller the capacitor size, the lower the amount of inertia response it can supply.

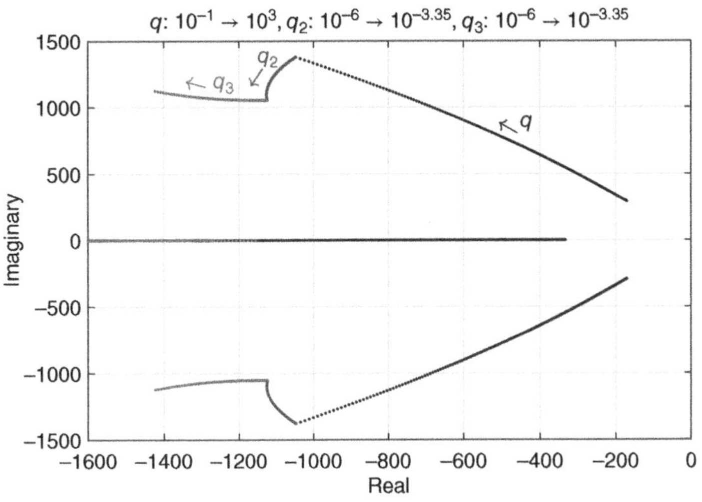

Figure 5.43 LQT-based full state current controller.

are the observations from this graph: (i) the system with full state feedback design exhibits much smoother responses while the other one shows oscillations,[19] (ii) during the swing of ± 1000 V/s, the converter supplies ± 0.2 kW power which confirms the inertia response of a 0.5 mF capacitor that, (iii) at the instant of sudden grid voltage drop/rise, the controller instantly supplies/absorbs a large inertia power,[20] and (iv) the current is successfully limited between 0 and 40 A at all times (Figure 5.44).

One potential issue with the full state feedback of Figure 5.42 is that during the current-limiting mode, where the limiter is saturated, v_c can experience large changes. During this mode, the connection between the outer loop and the inner loop is also severed. This means that the added feedback term $k_3 v_c$ will act as a disturbance to the current loop and it may disturb the current limiting. If this term grows fast, for example when a very deep grid voltage fault occurs which causes most of the power be absorbed by the capacitor, the disturbance is more likely to interfere with current limiting process. Figure 5.45 shows such a scenario where the grid voltage drops deeply to 100 V at $t = 0.275$ seconds and causes large growth of v_c. It is observed that rather than being limited at 40 A, the current is limited around 45 A during the deep fault interval, i.e. $0.275 < t < 0.3$ seconds.[21]

Example 5.10 ***Tight Control of Capacitor Voltage***
In some applications, it might be desirable to regulate the capacitor voltage v_c tightly close to a prespecified value V_c. In this case, the inertia response is disabled (by setting $\gamma = 0$) and the constant R_1 is reduced to a small number. The system block diagram is shown in Figure 5.46. The three controller gains k_1 to k_3 are optimally designed using the full state LQT approach described in this section.

---

19 These oscillations can tend to instability if the capacitor size is further reduced.
20 However, the converter with full state feedback tends to offer a better response by keeping more distance from the current limiting borders.
21 To completely resolve this problem, the added feedback term must be frozen during the current limiting mode as discussed in Example 5.10.

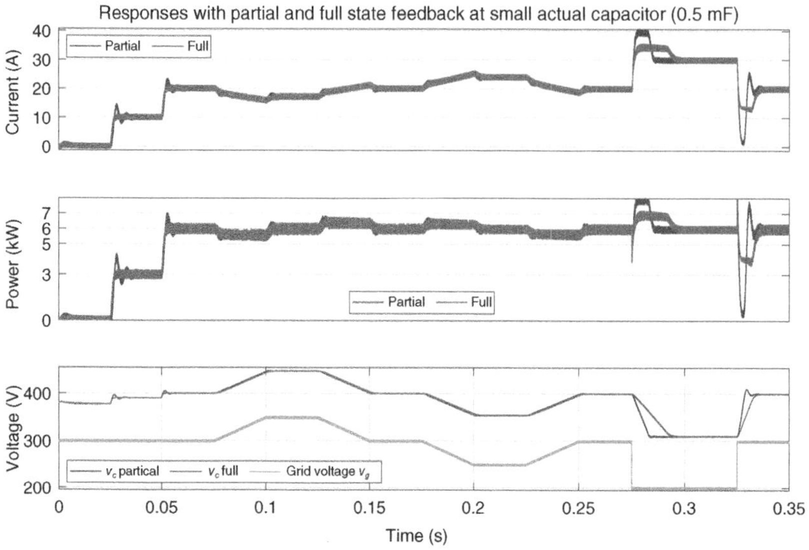

Figure 5.44 Time responses of the partial and full state feedback designs (Figure 5.42 without and with k_3) with small 0.5 mF capacitor, to power commands (at $t = 0.025$ seconds and $t = 0.05$ seconds) and a grid voltage swing (between $t = 0.075$ and 0.25 seconds) and a sudden grid voltage drop (at $t = 0.275$ seconds) and clearing (at $t = 0.325$ seconds).

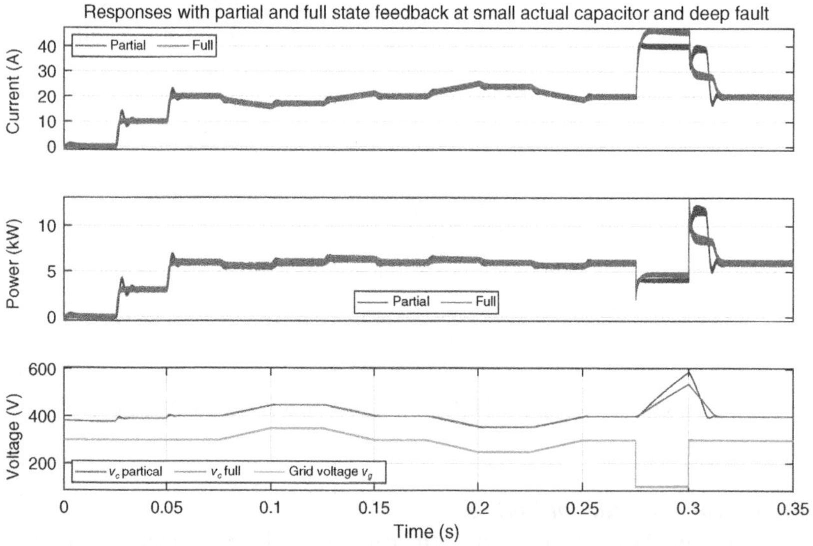

Figure 5.45 Time responses of the partial and full state feedback designs (Figure 5.42 without and with k_3) with small 0.5 mF capacitor, to power commands (at $t = 0.025$ seconds and $t = 0.05$ seconds) and a grid voltage swing (between $t = 0.075$ and 0.25 seconds) and a sudden deep grid voltage drop (at $t = 0.275$ seconds) and clearing (at $t = 0.3$ seconds).

The feedback branch k_3 is frozen during the current-limiting mode to ensure that the current is also tightly limited within the prespecified range $[I_{\min} \;\; I_{\max}]$. One possible freezing/releasing (FR) mechanism is shown in Figure 5.47 where the input and output of saturation block (i_1^* and i_2^*) are used to detect when the system enters into the limiting mode; i.e. when their difference is above a threshold ϵ. Then, the voltage v_c is sampled and held until it again exits the limiting mode.

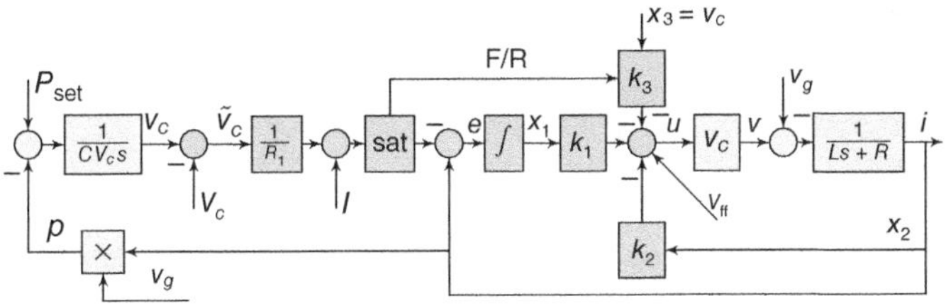

Figure 5.46 Tight capacitor voltage control.

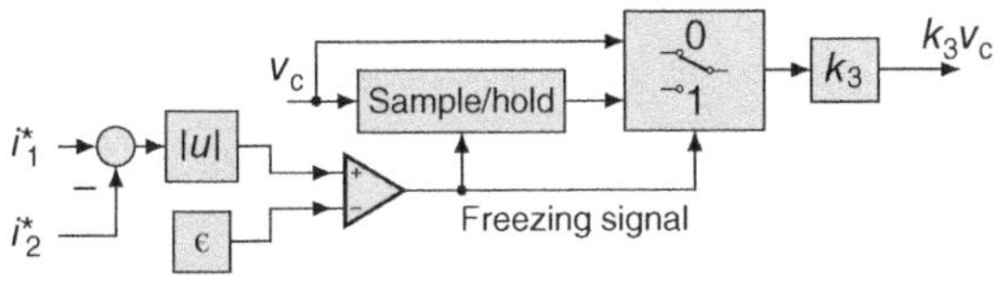

Figure 5.47 Freeze/release (F/R) mechanism during current limiting.

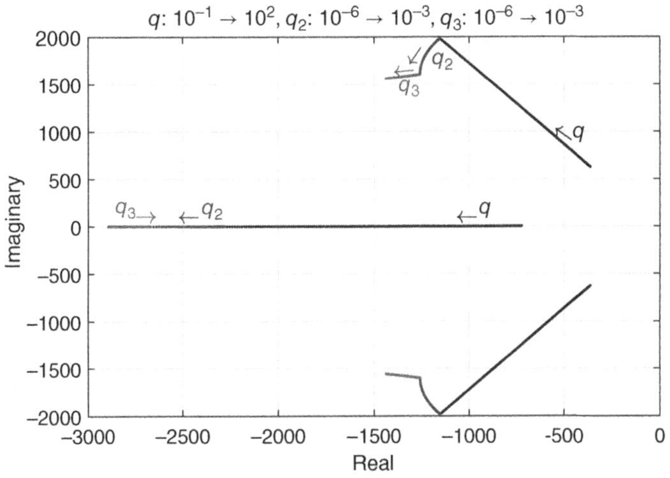

Figure 5.48 LQT-based full state current controller for tight capacitor voltage regulation.

For the same set of converter parameters we have been using, Table 5.1, the capacitor size of $C = 0.5$ mF, and $R_1 = 0.1$ (to achieve tight regulation of capacitor voltage), a sample of running the full state feedback LQT design is shown in Figure 5.48. For $q = 10^2$, $q_2 = 10^{-3} = q_3$, the location of closed loop poles are at $-1438 \pm j1559, -2670$, and the controller gains are $K = [k_1 \ \ k_2 \ \ k_3] = [10 \ \ 0.069 \ \ -0.095]$.

Responses of two systems (i.e. the lose control with inertia designed in Example 5.9, and the tight control of capacitor voltage designed in this example) to a grid voltage swing between $t = 0.075$ seconds and $t = 0.25$ seconds, and a deep sudden voltage drop occurring at 0.275 seconds and clearing at 0.325 seconds are shown in Figure 5.31. The converters start at $t = 0$ and their power reference increases from zero to 3 kW at $t = 0.025$ seconds and then to 6 kW at $t = 0.05$ seconds. Here are the observations from this graph: (i) the system with tight control maintains the voltage at close vicinity of 400 V (with tiny offsets at different loading conditions), and (ii) the current is successfully limited between 0 and 40 A at all times. If the current I is set to $\frac{P_{set}}{V_g}$, those very small offsets at different loading conditions will also be removed (Figure 5.49).

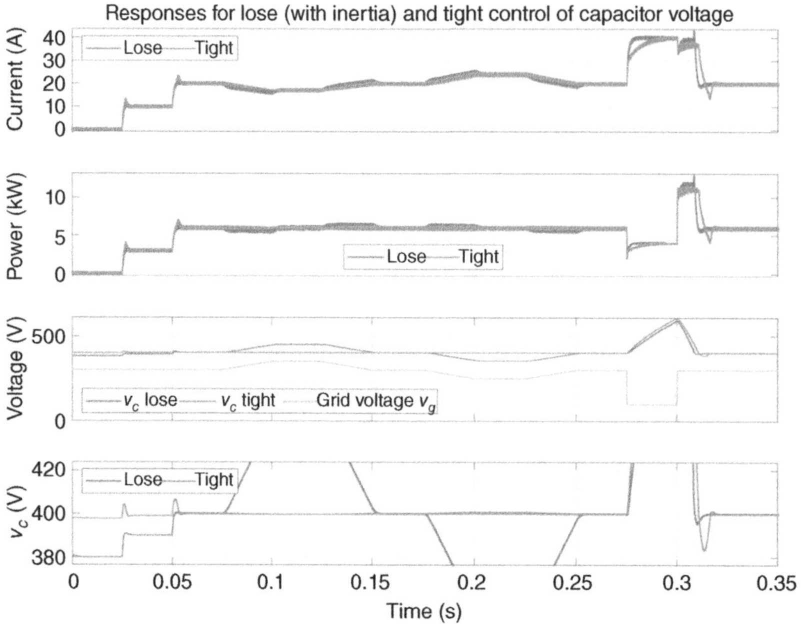

Figure 5.49 Time responses of the system with inertia ($\gamma \neq 0$) and the one with tight control of capacitor voltage ($\gamma = 0$, R_1: small), to power commands (at $t = 0.025$ seconds and $t = 0.05$ seconds) and a grid voltage swing (between $t = 0.075$ and 0.25 seconds) and a sudden deep grid voltage drop (at $t = 0.275$ seconds) and clearing (at $t = 0.325$ seconds).

5.4.5 Static Grid Voltage Support (Droop Function)

The dynamic voltage support approach (or inertia response) was discussed in details in Section 5.4.3. Using either an actual or a virtual capacitor, it adds an increment power which is proportional to the grid voltage derivative: $p_{dyn}(t) = -k\dot{v}_g(t)$, $(k > 0)$. This is a fast and dynamic response to the grid voltage changes and can instantly respond to the grid disturbances, in an inertial way. However, as soon as the changes in grid voltage settle, this power becomes zero.

Another aspect of grid support is "static" support where an increment power proportional to the deviation of the grid voltage from its rated value is generated: $p_{stat}(t) = k[V_g - v_g(t)]$, $k > 0$. This power does not have the strength and swiftness of $p_{dyn}(t)$ in responding to quick (yet small) changes of the grid voltage. It, however, contributes more and more as the grid voltage deviates more from its nominal value of V_g. As the grid voltage tends to settle at a lower (or higher) voltage, and this indicates a shortage (or an excess) of energy in the grid, $p_{stat}(t)$ increases (or decreases) and the converter feeds more (or less) power to the grid to support its shortage (or excess).

To enable this property, assume that P_{set} is the set-point of converter power. Then, the reference of converter power may be modified as

$$P_{ref}(t) = P_{set} + K_v[V_g - v_g(t)], \tag{5.10}$$

where V_g is the nominal value and v_g is the actual (measured) value of the grid voltage, and K_v is a positive constant. This *droop mechanism* indicates that for x volts decrease/increase in the grid voltage, the converter increases/decreases $K_v x$ watts in its power. A limiter sat_p is used to limit the power to the maximum/minimum available power of the resource.

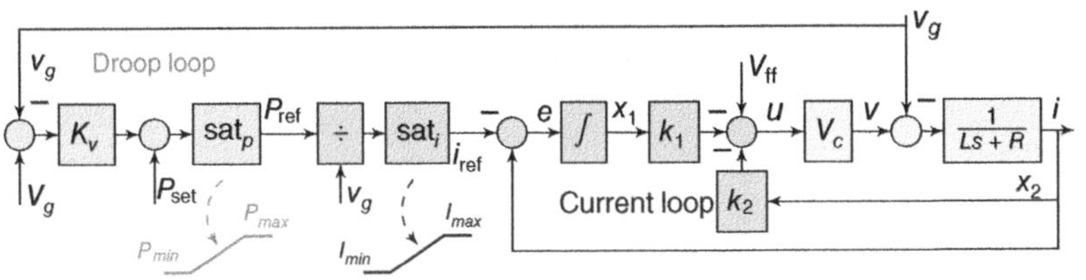

Figure 5.50 Adding a power-voltage droop loop to the power controller to obtain static grid voltage support.

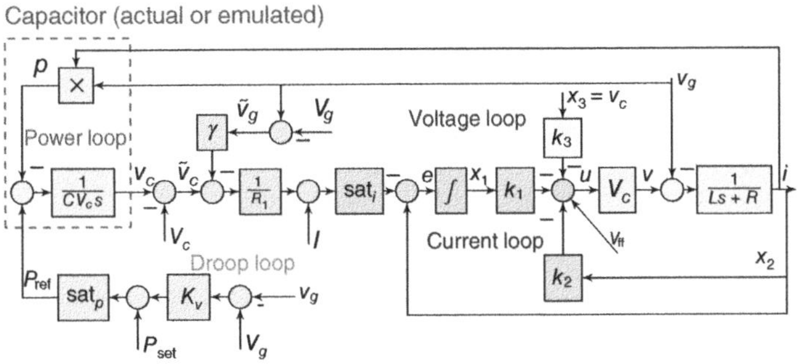

Figure 5.51 Adding a power-voltage droop loop to the power controller with inertia (either using an actual or virtual capacitor) to obtain static grid voltage support.

In the power control without dynamic voltage support (or inertia), the reference current is then directly calculated from $i_{ref} = \frac{P_{ref}}{v_g}$. A limiter sat_i is used to limit the reference current during deep low-voltage transients. This mechanism is added to the power controller and the grid voltage supporting controller is shown in Figure 5.50.

In the controller with dynamic voltage support (or inertia), the set-point power P_{set} is modified properly, as shown in Figure 5.51, to enable static voltage support. This controller thus offers both dynamic and static grid voltage support. In both Figures 5.50 and 5.51, the droop mechanism is enabled by an external control loop (called the droop loop) which determines the power reference based on (5.10). Notice that if an actual capacitor is present, P_{ref} in Figure 5.51 indicates the actual power flowing toward the capacitor from its previous stage.

Example 5.11 ***Comparison of Static and Dynamic Voltage Supports***

Consider the same converter system parameters of Table 5.1 and the LQT current controller design shown in Figure 5.52. For $q = 10^3$ and $q_2 = 10^{-3.35}$, the location of poles is $-1407 \pm j742$ and the controller gain vector is $K = [31.6 \;\; 0.035]$. The feed-forward term is taken from the grid voltage with a low-pass filter with time-constant of 1 ms. The current limiter bounds are set to $I_{min} = 0$ and $I_{max} = 30$ A. This is a partial state feedback design is independent from the capacitor size.[22] The capacitor size is 2.5 mF which corresponds to an inertia power $C\gamma V_c \dot{v}_g = \dot{v}_g$ at $V_c = 400$ V and $\gamma = 1$. The droop gain is $K_v = 50$ which corresponds to 1 kW power for 20 V voltage change. The other parameters are set as before, i.e. $R_1 = 1$, $V_g = 300$ V, $V_c = 400$ V, and $I = 20$ A.

Responses of four systems: (i) without any support, i.e. Figure 5.50 without droop loop, (ii) with static support, i.e. Figure 5.50, (iii) with dynamic support, i.e. Figure 5.51 without droop loop, and

22 For larger values of capacitor (either emulated or actual) which is required anyway to achieve a sufficient level of dynamic support, the control loop is robust and a full state feedback is not required.

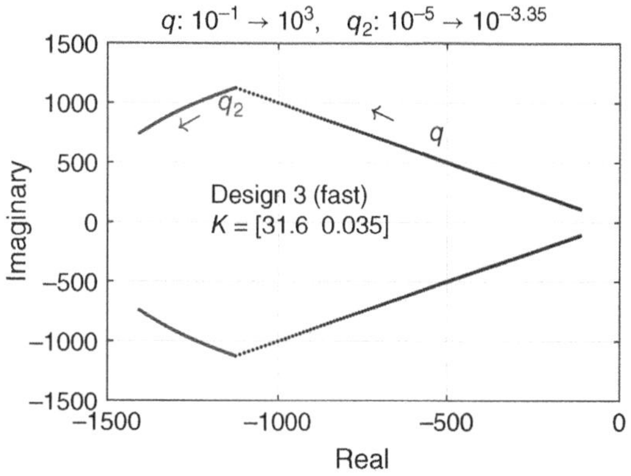

Figure 5.52 LQT-based full state current controller.

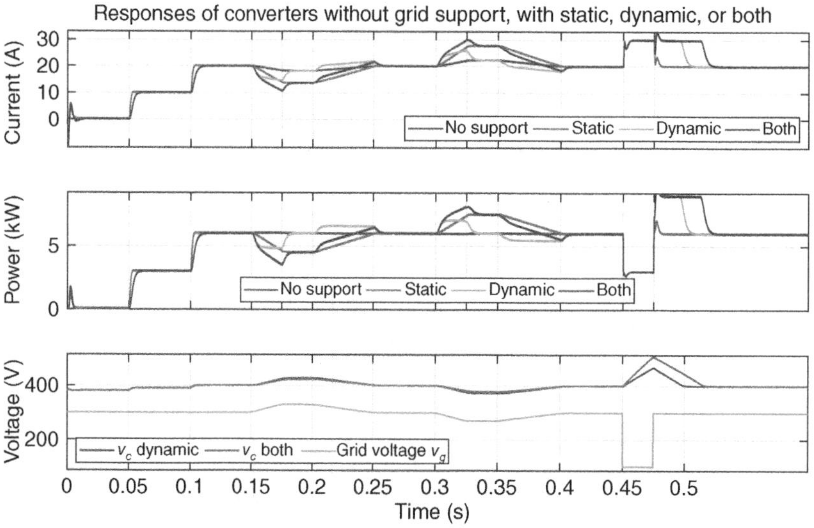

Figure 5.53 Time responses of the systems: (i) without any support, i.e. Figure 5.50 without droop loop, (ii) with static support, i.e. Figure 5.50, (iii) with dynamic support, i.e. Figure 5.51 without droop loop, and (iv) with both static and dynamic support, i.e. Figure 5.51, to set-point power jumps at $t = 0.05$ seconds and $t = 0.1$ seconds, and a grid voltage swing between $t = 0.15$ seconds and $t = 0.4$ seconds, and a sudden voltage drop occurring at 0.45 seconds and clearing at 0.475 seconds.

(iv) with both static and dynamic support, i.e. Figure 5.51, are shown and compared in Figure 5.53. Here first the converters start at zero power. At $t = 0.05$ seconds, the set-point power P_{set} jumps to 3 kW followed by another 3 kW at $t = 0.1$ seconds. The grid voltage experiences a swing between $t = 0.15$ seconds and $t = 0.4$ seconds, and a sudden voltage drop occurring at 0.45 seconds and clearing at 0.475 seconds. Here are the observations from this graph: (i) the system with no support does not give any response to grid voltage swings and continues to supply a constant 6 kW power, (ii) the system with static support follows the grid voltage swing and provides a power opposite to it, (iii) the system with dynamic support responds to changes of grid voltage very quickly but it goes back to zero as soon as the grid voltage settles, (iv) the controller with both supports combines the responses of both supports and responds to both changes and offsets in the grid voltage, and (v) all controllers limit the current successfully during deep grid voltage fault. When the grid voltage restores after the fault, a small over-current is observed which is due to sudden change of the slope in capacitor voltage and the large inertia power change at that moment.

5.4.6 Inertia Power Using Grid Voltage Differentiation

The dynamic (or inertia) support is transient and that is why it can be supplied from a short-term storage element such as a capacitor without requiring a solid source of power and energy, e.g. a battery. The provided inertia power is given by $C\gamma V_c \dot{v}_g$. However, the static (or droop) support is a lasting component of power as long as there is an offset in the grid voltage: $K_v(V_g - v_g)$. Therefore, it must be supplied from a stable source, such as a battery or a PV. Apart from the fact that this power is slow by nature (compared with inertia power), there may be further delays in the source that is supposed to provide this power.[23] Thus, the presence of inertia power is crucial in stabilizing a dynamic system, specially in large ac applications.

If the source of power is guaranteed and does not have delays, a battery or a directly-connected PV array for instance, the inertia power may simply be computed using time-differentiation of the grid voltage and commanded to the power reference as shown in Figure 5.54. Here, the reference power is computed according to

$$P_{ref} = P_{set} + P_{droop} + P_{inertia} = P_{set} + K_v[V_g - v_g(t)] - K_i\dot{v}_g(t) \quad (5.11)$$

where $K_v > 0$ and $K_i > 0$ are the droop and inertia power coefficients. If a capacitor is used, there will be a need for another control loop to ensure that capacitor voltage is maintained within acceptable range.

Example 5.12 ***Direct Inertia Computation Using Voltage Differentiation***
Consider the same system and simulation conditions described in Example 5.11. Now, we compare three cases: (i) the converter with no grid support, (ii) the converter with inertia support where the inertia is calculate using direct grid voltage differentiation, i.e. Figure 5.54, and (iii) the converter using inertia loop on a capacitor, i.e. Figure 5.36b. The droop term is disabled from both systems, $K_v = 0$.

In practice, the grid voltage differentiation needs special care to prevent noise amplification. Here we have used a low-pass filter (LPF), $\frac{1}{\tau s+1}$, first and then a filtered derivative, $\frac{s}{\tau s+1}$, and then a limiter with bounds at −3000 W and 3000 W, as shown in Figure 5.55. The time-constant τ is selected at 0.001 seconds. The constant K_i is set to 1.[24]

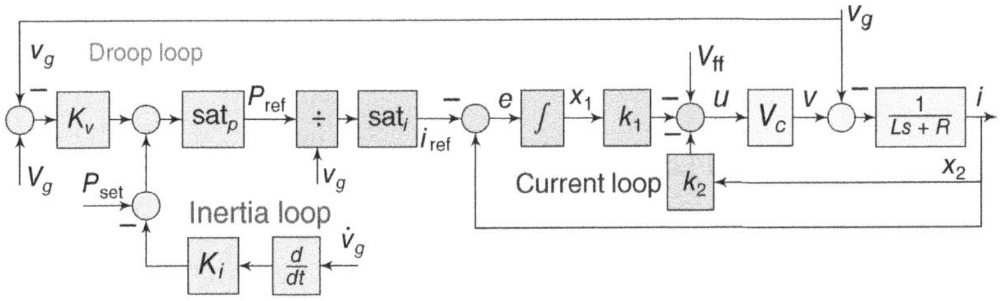

Figure 5.54 Adding the droop and inertia to the power controller to obtain both static and dynamic grid voltage support from a battery source (not capacitor) using direct grid voltage differentiation.

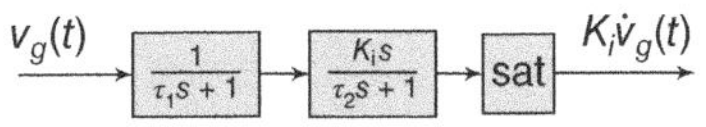

Figure 5.55 Calculating the inertia power using differentiating of grid voltage.

23 For example, the governor control system in a synchronous generator which can have long delays.
24 This will correspond to the same level of inertia supplied from a capacitor of 2.5 mF at the voltage of 400 V.

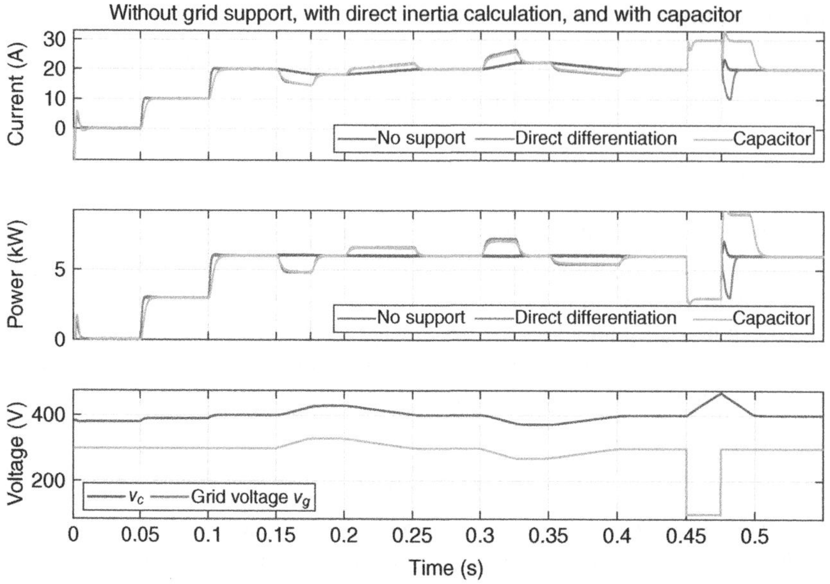

Figure 5.56 Time responses of the systems: (i) the converter with no grid support, i.e. Figure 5.50 without droop loop, (ii) the converter with inertia support where the inertia is calculate using direct grid voltage differentiation, i.e. Figure 5.54, and (iii) the converter using inertia loop on a capacitor, i.e. Figure 5.36b, to set-point power jumps at $t = 0.05$ seconds and $t = 0.1$ seconds, and a grid voltage swing between $t = 0.15$ seconds and $t = 0.4$ seconds, and a sudden voltage drop occurring at 0.45 seconds and clearing at 0.475 seconds.

Figure 5.56 shows the responses of these three systems. The converters start at zero power. At $t =$ 0.05 seconds, the set-point power P_{set} jumps to 3 kW followed by another 3 kW at $t = 0.1$ seconds. The grid voltage experiences a swing between $t = 0.15$ seconds and $t = 0.4$ seconds, and a sudden voltage drop occurring at 0.45 seconds and clearing at 0.475 seconds. Here are the observations from this graph: (i) the system with no support does not give any response to grid voltage swings and continues to supply a constant 6 kW power, (ii) both other systems provide almost equal level of inertia to the grid voltage swings, (iii) the system with direct inertia calculation using grid voltage differentiation does not compromise the speed of responses to power set point changes, and (iv) all controllers limit the current successfully during deep grid voltage fault.

5.4.7 Common Approach: Nested Control Loops

The existing and common dc voltage control systems with current limiting capability use two nested PI control loops as shown in Figure 5.57. The internal loop with PI_i ensures that the fast current tracking is obtained. The external loop with PI_V ensures that the dc voltage is tracked.[25]

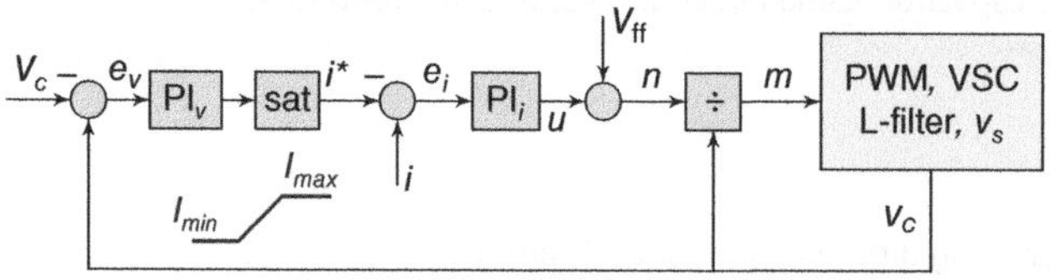

Figure 5.57 Conventional nested-loop dc bus control structure with current limiting.

25 The integrator in PI_V must have an anti-windup mechanism.

The distinct timescale of the two nested loops allows to design the two PI compensators separately. This implies that in the design of the outer loop, the dynamics of the internal loop can be neglected. It is also possible to design these two separate PI compensators using the LQT approach. Moreover, it is possible to enhance their design by including the interactions of the two loops [6].

Our proposed controllers discussed in this chapter have multiple differences and advantages compared with the two-PI nested structure explained as follows. (i) It is compact; it avoids one integrator that reduces the order of system and the system oscillations and increases its stability margins. (ii) The design is in one-shot without a need to separate the timescales of the two nested loops. (iii) It allows a natural yet small swing of the dc voltage in proportion to the converter operating current and the grid voltage variations. This feature enables a stabilizing inertia (or dynamic grid voltage support). Moreover, it can be used for "dc voltage signaling." For instance, if multiple converters are connected to the same dc bus, each converter will be informed of the status of other converters through the dc bus voltage. This can be used for power balancing and also to determine the number of required converters to be online at each time, without requiring a communication among the converters.

5.5 Analysis of Weak Grid Condition

So far, the grid voltage v_g is assumed to be fully determined by the grid side and was not affected by the converter. This is true only if the grid is "strong" or "stiff." In many practical conditions, however, this is not true. The converter current may change the grid voltage v_g. This is called a "weak" grid condition. Mathematically, and from a Thevinin equivalent circuit perspective, the grid side may be modeled by a stiff voltage source behind a grid impedance $L_g s + R_g$. This leads to the equivalent circuit and control diagram shown in Figure 5.58 where v_s is assumed to be a stiff voltage.

If the grid impedance parameters R_g and L_g are known, it is possible to include them in the design stage of the controller gains k_1 and k_2. However, the grid impedance is often unknown and/or may experience variations. Therefore, in this section, we analyze the impact of this impedance on the performance of the controller when the controller is designed without considering it.

The system equations are given by

$$\begin{aligned}\dot{x}_1 &= x_2 - i_{\text{ref}}\\ \dot{x}_2 &= -\tfrac{R}{L}x_2 + \tfrac{V_c}{L}u - \tfrac{1}{L}v_g\\ u &= -k_1x_1 - k_2x_2 + V_{\text{ff}} = -Kx + V_{\text{ff}}\end{aligned} \tag{5.12}$$

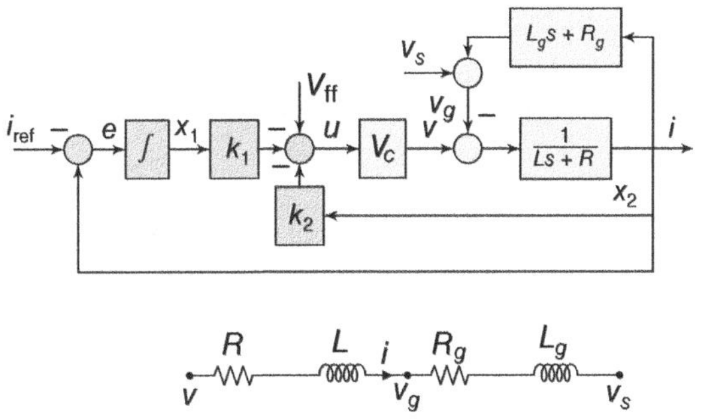

Figure 5.58 Showing impact of grid impedance on the control diagram.

where

$$v_g = v_s + L_g\dot{x}_2 + R_g x_2 = v_s + L_g\left(-\frac{R}{L}x_2 + \frac{V_c}{L}u - \frac{1}{L}v_g\right) + R_g x_2$$

which leads to

$$v_g = \frac{L}{L + L_g}\left[v_s + \left(R_g - \frac{L_g}{L}R\right)x_2 + \frac{L_g V_c}{L}u\right].$$

Substituting the latter in (5.14) results in

$$\begin{aligned}
\dot{x}_1 &= x_2 - i_{\text{ref}}\\
\dot{x}_2 &= -\tfrac{R+R_g}{L+L_g}x_2 + \tfrac{V_c}{L+L_g}u - \tfrac{1}{L+L_g}v_s\\
u &= -k_1x_1 - k_2x_2 + V_{\text{ff}} = -Kx + V_{\text{ff}}.
\end{aligned} \tag{5.13}$$

If the filtered version of the grid voltage is used as the feed forward, $V_{\text{ff}}(s) = \frac{V_c^{-1}}{\tau_f s+1}V_g(s)$, (5.14) changes to

$$\begin{aligned}
\dot{x}_1 &= x_2 - i_{\text{ref}}\\
\dot{x}_2 &= -\tfrac{R+R_g}{L+L_g}x_2 + \tfrac{V_c}{L+L_g}u - \tfrac{1}{L+L_g}v_s\\
\dot{V}_{\text{ff}} &= -\tfrac{1}{\tau_f}V_{\text{ff}} + \tfrac{1}{\tau_f V_c}v_g\\
&= -\tfrac{1}{\tau_f}V_{\text{ff}} + \tfrac{1}{\tau_f V_c}\tfrac{L}{L+L_g}\left[v_s + \left(R_g - \tfrac{L_g}{L}R\right)x_2 + \tfrac{L_g V_c}{L}u\right]\\
u &= -k_1x_1 - k_2x_2 + V_{\text{ff}} = -Kx + V_{\text{ff}}.
\end{aligned} \tag{5.14}$$

Example 5.13 ***Weak Grid Analysis***

To visualize the effect of grid impedance, consider our much used LQT design shown in Figure 5.59 which for $q = 10^3$ and $q_2 = 10^{-3.35}$ leads to $K = [10 \quad 0.035]$ and the closed-loop poles of $-1407 \pm j742$. This is for zero grid impedance, i.e. $R_g = 0$, $L_g = 0$. Now, increase the grid impedance according to $R_g = 10x$ mΩ, $L_g = x$ mH where x varies from 0 to 5. This means that L_g goes up to 5 mH and R_g up to 50 mΩ. As the grid impedance increases, the closed-loop poles shift to the right. At the final point, they are at $-706 \pm j876$ (for constant feed-forward case) and $-610 \pm j1210$, -692 (for grid voltage feed forward through a low-pass filter with time-constant of 0.001 seconds). This means both the speed and the damping of the closed-loop poles decrease when the grid impedance goes up. The grid voltage feed forward tends to have more adverse impact than a constant feed forward term.

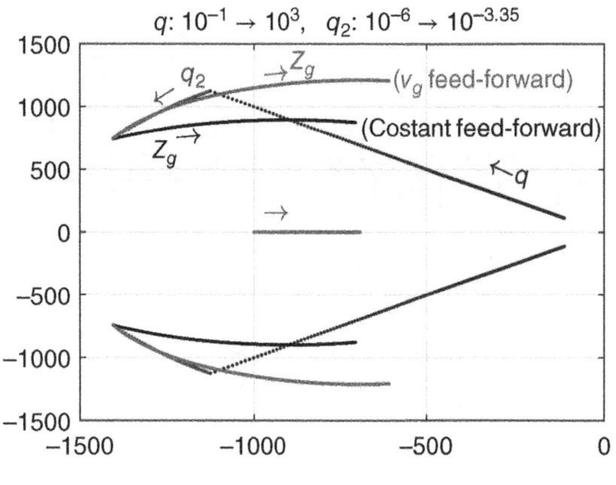

Figure 5.59 Impact of increasing grid impedance on closed-loop poles.

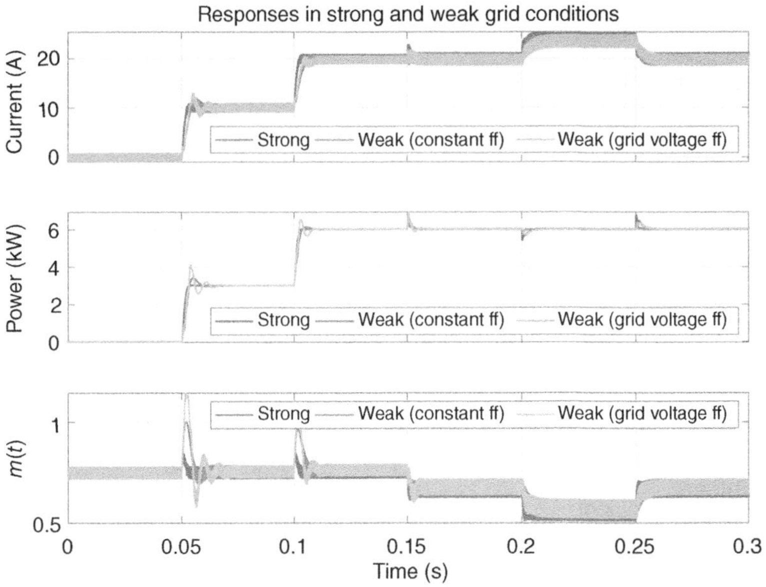

Figure 5.60 Impact of weak grid conditions on transient responses. Two power commands of 3 kW are applied at $t = 0.05$ seconds and $t = 0.1$ seconds, then the dc voltage jumps 50 V at $t = 0.15$ seconds, and finally the grid voltage experiences a 50 V drop during $0.2 \leq t < .25$.

Figure 5.60 shows the responses for three cases: (i) strong grid, (ii) weak grid with constant feed-forward, and (iii) weak grid with grid voltage feed-forward. Two power commands of 3 kW are applied at $t = 0.05$ seconds and $t = 0.1$ seconds, then the dc voltage jumps 50 V at $t = 0.15$ seconds, and finally the grid voltage experiences a 50 V drop during $0.2 \leq t < .25$. The grid impedance is $L_g = 25$ mH and $R_g = 0.25\ \Omega$. It is observed that the weak grid condition causes oscillations. Moreover, the system with grid voltage feed forward is more affected than the constant feed forward. All these comply with the above analysis.

5.6 Load Voltage Control

In Section 5.3, we discussed the power controller. In Section 5.4, we enhanced them to provide dynamic and static grid support. They are also able to control the dc link voltage v_c should such a link is present. In this section, we study the case where a single load is connected at the "grid side" of the converter. The objective will be to control its voltage. Then, we will see that these two sets of control approaches actually merge into one another and we can come up with a unified controller, later called "grid forming" controller, that can maneuver between different modes and address control objectives according to the system operating conditions.

A VSC connected to a dc load with resistance R_g is shown in Figure 5.61 where L and C_1 form the interfacing filter. The capacitor C_1 absorbs the switching ripples of the current i and provides a clean voltage for the load. The control objectives may be stated as follows.

- Control the output voltage v_g at (or closely around) the desired value V_g.[26]
- Maintain this objective robustly despite system uncertainties including those in L, R, C, C_1, v_c, R_g.

26 The same notation as Section 5.3 for this voltage is used as we will combine both methods later.

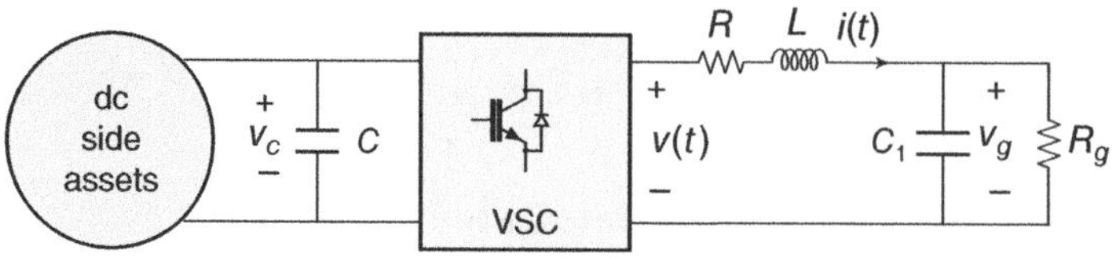

Figure 5.61 VSC connected to a dc load.

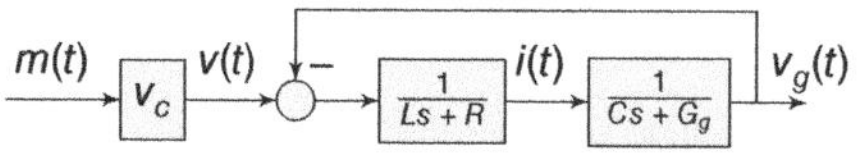

Figure 5.62 Control block diagram of VSC connected to a dc load.

- Respond to load variations in a quick and smooth manner.
- Limit the converter current within $[I_{\min}\ \ I_{\max}]$ during all conditions, e.g. during transient overload or short circuit conditions.

The control block diagram of the system is shown in Figure 5.62 assuming that a full-bridge or a standard buck topology is used. In case of a half-bridge VSC, v_c should be replaced with $0.5v_c$. Inverse of output resistance, called the conductance, is shown by $G_g = R_g^{-1}$.

The systems transfer function is given by

$$\frac{V_g(s)}{M(s)} = \frac{V_c}{LC_1 s^2 + (RC_1 + LG_g)s + RG_g + 1} \tag{5.15}$$

which has two poles.[27] (see Problem 5.3.) For $G_g = 0$ (no-load or open-circuit situation), the two poles are at $\pm j\omega_r = \pm j\frac{1}{\sqrt{LC_1}}$ which indicate the frequency of an undamped resonance between the inductor and capacitor (note that R is close to zero and is neglected here). As we earlier discussed for an LCL filter, there are two ways of damping the resonance phenomenon in such circuits: *Passive damping* is done by adding dissipative components, commonly a resistor in series with capacitor C_1; *Active damping* uses a feedback control to shift the resonance poles to the left without adding a resistor.[28] A full-state feedback control strategy is discussed below where it can flexibly be adjusted to do an active damping or otherwise.

5.6.1 Control Structure and Optimal Design

A full state feedback combined with an integrating output feedback loop is shown in Figure 5.63.[29]

State space differential equations of this system may be expressed as

$$\dot{x}_1(t) = e(t) = x_2(t) - V_g$$

$$\dot{x}_2(t) = -\frac{G_o}{C_1}x_2(t) + \frac{1}{C_1}x_3(t)$$

$$\dot{x}_3(t) = -\frac{1}{L}x_2(t) - \frac{R}{L}x_3(t) + \frac{V_c}{L}m(t)$$

27 We use V_c as the nominal value of v_c.

28 In the load voltage control, however, unlike the grid-connected applications, the capacitor is rather larger which indicates a lower resonance frequency and a more critical role of active damping.

29 In applications where v_c may vary in a wide range, this voltage may be measured and a division block may be added to the controller to linearize its effect. This is also advantageous in ac applications where v_c may have double-frequency harmonics.

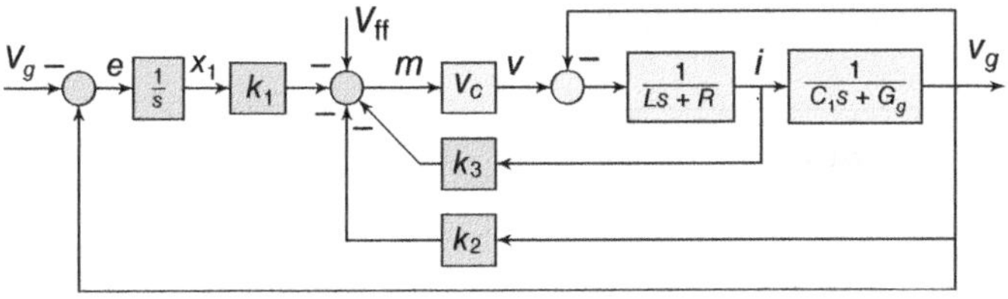

Figure 5.63 Full state feedback control diagram of the VSC connected to a dc load.

where $x_2(t) = v_g(t)$ and $x_3(t) = i(t)$. The control (or modulation) input is

$$m(t) = -k_1x_1(t) - k_2x_2(t) - k_3x_3(t) + V_{ff} = -Kx(t) + V_{ff}$$

where V_{ff} is a feed-forward term for soft-starting and K is the vector of state feedback gains. Using the LQT approach, the operator $\frac{d}{dt}$ is applied to arrive at

$$\dot{z}_1(t) = z_2(t)$$

$$\dot{z}_2(t) = -\frac{G_g}{C_1}z_2(t) + \frac{1}{C_1}z_3(t)$$

$$\dot{z}_3(t) = -\frac{1}{L}z_2(t) - \frac{R}{L}z_3(t) + \frac{V_c}{L}n(t)$$

where

$$z_i(t) = \dot{x}_i(t), \ i = 1, 2, 3, \quad n(t) = \dot{m}(t).$$

Notice particularly that

$$z_1(t) = \dot{x}_1(t) = e(t) = v_g(t) - V_g$$

is the voltage tracking error. In order to minimize the cost function

$$J = \int_0^\infty \left[qe(t)^2 + q_2z_2(t)^2 + q_3z_3(t)^2 + n(t)^2\right] dt,$$

we choose

$$A = \begin{bmatrix} 0 & 1 & 0 \\ 0 & -\frac{G_g}{C_1} & \frac{1}{C_1} \\ 0 & -\frac{1}{L} & -\frac{R}{L} \end{bmatrix}, \ B = \begin{bmatrix} 0 \\ 0 \\ \frac{V_c}{L} \end{bmatrix}, \ Q = \begin{bmatrix} q & 0 & 0 \\ 0 & q_2 & 0 \\ 0 & 0 & q_3 \end{bmatrix}$$

and solve in MATLAB: ≫ `K = lqr(A, B, Q, 1)`.

Example 5.14 ***dc Load Voltage Control***

The converter system parameters used for the examples in this section are given in Table 5.2. Figure 5.64 shows the locus of closed-loop poles when q, q_2, and q_3 vary. As expected, q does the main work while q_2 and q_3 finely increase the damping of oscillatory mode. For the final values of $q = 10^3$, $q_2 = 10^{-3.5}$, and $q_3 = 10^{-3.5}$, the closed-loop poles are at $-1563, -1292 \pm j1542$. This indicates a low-frequency time-constant of below 1 ms combined with an under-damped mode with a damping around 0.64.[30] The controller gain vector is $K = [31.6 \ \ 0.037 \ \ 0.051]$.

30 This seems to be around the fastest response for a switching frequency of 4 kHz. If we use a unipolar PWM that effectively doubles the switching frequency, we can push the poles further more to the left. Recall, anyway, that at least a few switching cycles must fit within a time-constant of the system responses. Finally, the capacitor size C_1 (as well as the inductance L) can be reduced for higher switching frequencies that allow faster responses.

Table 5.2 Converter system parameters used for the examples in this section

Parameter	Symbol	Value	Unit
dc side voltage	V_c	400	V
Load voltage	V_g	300	V
Load resistance	R_g	90	Ω
Load power	P_g	1000	W
Inductance	L	5	mH
Capacitance	C_1	400	μF
Parasitic resistance	R	30	mΩ
Switching frequency	f_{sw}	4	kHz

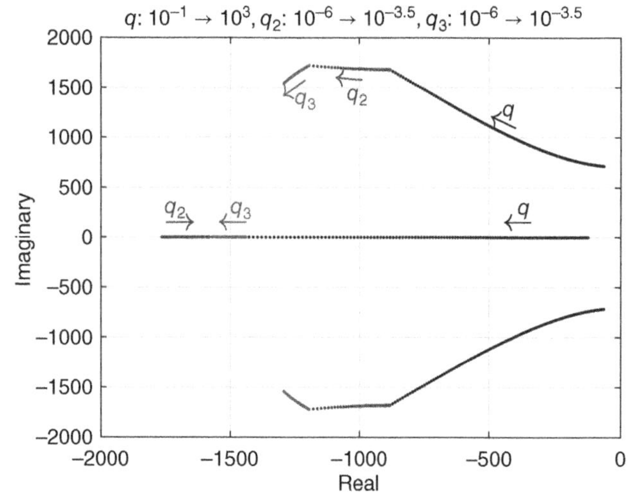

Figure 5.64 Locus of closed-loop poles of load voltage control system when elements of matrix Q vary.

A sample of responses of this control system is shown in Figure 5.65. The converter starts at $t = 0$ while the load of 1 kW is connected. The load-side capacitor is initially charged to 300 V and the feed-forward term $V_{ff} = 300/400 + 0.037 \times 300$ is added for soft starting.[31] As expected, the converter starts smoothly. The output voltage reference V_g jumps to 320 V at $t = 0.01$ seconds and turns back to 300 V at $t = 0.02$ seconds. The controller quickly and smoothly tracks this command. At $t = 0.03$ seconds, the dc voltage drops to 370 V and then restores to 400 V at $t = 0.04$ seconds. The output voltage experiences a small transient glitch of about ± 1 V. At $t = 0.05$ seconds, another 1 kW load is switched on, and it then is switched off at $t = 0.06$ seconds. A similar thing is repeated with a large load of 3 kW at $t = 0.07$ seconds and $t = 0.08$ seconds. The output voltage experiences a small transient glitch at the instants of sudden load changes that quickly die out within a couple of milliseconds.

31 Otherwise, the converter will be exposed to large over-current during the start. This issue is addressed by the current-limiting controller discussed in Section 5.6.2.

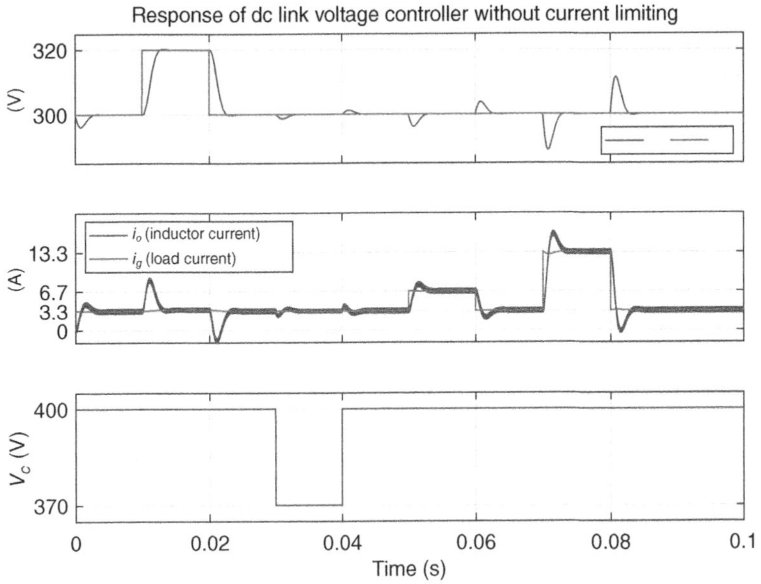

Figure 5.65 Responses of dc load voltage controller to command change at $t = 0.01$ seconds and $t = 0.02$ seconds, to dc voltage disturbance at $t = 0.03$ to $t = 0.04$ seconds, and to two sudden load changes during $t = 0.05$–0.06 seconds and $t = 0.07$–0.08 seconds.

5.6.2 Current Limiting

Section 5.6.1 and Example 5.14 show that the converter output current can become very large when transient low-voltage incidents happen on the load side either caused by a large load change or by a transient short-circuit or low-voltage fault on the load side. Since the power electronic switches have small over-current capacity, it is necessary that the controller has current limiting capability in response to such incidents. One proposed structure is shown in Figure 5.66.[32]

In the approach of Figure 5.66, the controller relaxes the output voltage by removing the integrating function on its error. The reference for the current is then generated according to

$$i^* = I + k_o(V_g - v_g), \tag{5.16}$$

where I is the rated (or a set-point) current. Equation (5.16) indicates a *droop relationship* between the output voltage and the converter current: the output voltage is reduced as the current grows. When a transient low-voltage happens at v_g, then i^* can become very large. The block denoted by "sat" saturates or limits the current within the prespecified values I_{min} and I_{max}.

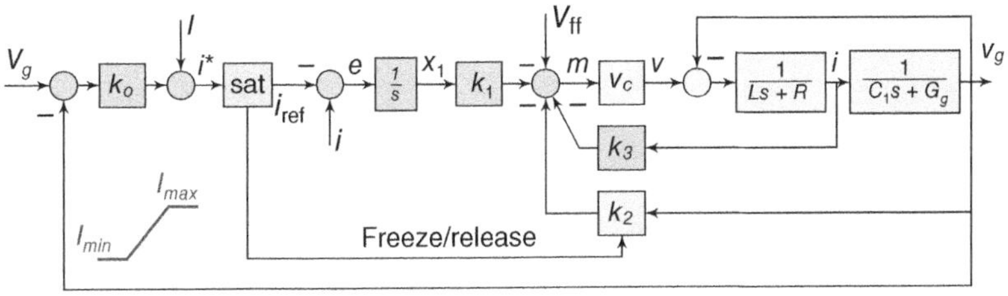

Figure 5.66 Full state feedback of dc load voltage control with current limiting.

32 This is different from the common approach of using two nested PI controllers.

The droop gain k_o must be properly selected such that during entire linear region operation of the converter, the output voltage changes within an acceptable range, e.g. 1–2% around V_g. Let I be the rated current of the converter. Within the linear region, (5.16) indicates that the relationship between output voltage and converter current is $v_g = V_g + k_o^{-1}(I - i)$. So, $v_g = V_g$ at $i = I$. And $v_g = V_g + k_o^{-1}I$ when $i = 0$. So, if for example 2% voltage deviation is allowed for this range of current, then $k_o^{-1}I = 0.02V_g$ which leads to $k_o = \frac{I}{0.02V_g}$.

When the controller is in current-limiting mode, i.e. the saturation block is in its nonlinear region, the load voltage may deviate possibly much beyond its normal values. During such periods, the feedback term of v_g through the gain k_2 should be frozen. Once the controller is back to its normal linear operation, this term is also released to its normal operation. One freezing/releasing scheme is shown in Figure 5.67 using the standard Matlab/Simulink blocks. The two inputs of this scheme are the input and output of the sat block. Its output will be forwarded to the k_2 gain. This scheme simply (i) detects the instance t_1 that the sat block enters into saturation and holds $v_g(t)$ at $v_g(t_1)$ and outputs zero, and (ii) detects the instance t_2 where the sat returns into linear region and smoothly outputs $v_g(t) - v_g(t_2)$.

State space differential equations of this system may be expressed as

$$\begin{aligned}
\dot{x}_1(t) &= e(t) = k_o(x_2 - V_g) - I^* + x_3 \\
\dot{x}_2(t) &= -\frac{G_g}{C_1}x_2(t) + \frac{1}{C_1}x_3(t) \\
\dot{x}_3(t) &= -\frac{1}{L}x_2(t) - \frac{R}{L}x_3(t) + \frac{V_c}{L}m(t)
\end{aligned}$$

where $x_2(t) = v_g(t)$ and $x_3(t) = i(t)$. The control (or modulation) input is

$$m(t) = -k_1x_1(t) - k_2x_2(t) - k_3x_3(t) = -Kx(t)$$

which is in the standard form of a linear full state feedback law. Using the LQT approach, the operator $\frac{d}{dt}$ is applied to arrive at

$$\begin{aligned}
\dot{z}_1(t) &= k_o z_2(t) + z_3(t) \\
\dot{z}_2(t) &= -\frac{G_g}{C_1}z_2(t) + \frac{1}{C_1}z_3(t) \\
\dot{z}_3(t) &= -\frac{1}{L}z_2(t) - \frac{R}{L}z_3(t) + \frac{V_c}{L}n(t)
\end{aligned}$$

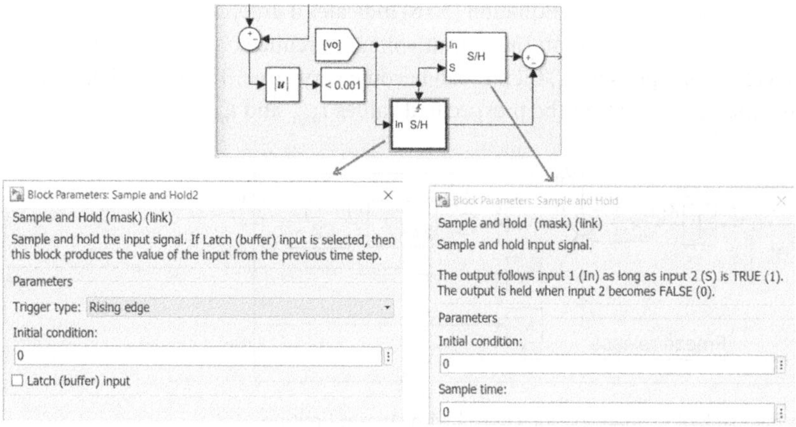

Figure 5.67 A Matlab/Simulink Mechanism to Freeze/Release k_2. Source: The MathWorks, Inc.

where

$$z_i(t) = \dot{x}_i(t), \ i = 1, 2, 3, \quad n(t) = \dot{m}(t).$$

Notice particularly that

$$z_1(t) = \dot{x}_1(t) = e(t)$$

is the tracking error. In order to minimize the cost function

$$J = \int_0^\infty [qe(t)^2 + q_2 z_2(t)^2 + q_3 z_3(t)^2 + n(t)^2]dt,$$

we choose

$$A = \begin{bmatrix} 0 & k_o & 1 \\ 0 & -\frac{G_g}{C_1} & \frac{1}{C_1} \\ 0 & -\frac{1}{L} & -\frac{R}{L} \end{bmatrix}, \ B = \begin{bmatrix} 0 \\ 0 \\ \frac{V_c}{L} \end{bmatrix}, \ Q = \begin{bmatrix} q & 0 & 0 \\ 0 & q_2 & 0 \\ 0 & 0 & q_3 \end{bmatrix}$$

and solve in MATLAB: ≫ `K = lqr(A, B, Q, 1)`.

Example 5.15 ***dc Load Voltage Control with Current Limiting***
Consider the system parameter values of $V_c = 400$ V, $R = 30$ mΩ, $L = 5$ mH, $C_1 = 400\ \mu$F, $V_g = 300$ V, $R_g = 90\ \Omega$ (1 kW load at 300 V). Figure 5.68 shows the locus of closed-loop poles when q, q_2, and q_3 vary. As expected, q moves the dominant low-frequency pole of the system while q_2 and specially q_3 damp the resonance mode. For the final values of $q = 10^3$, $q_2 = 10^{-3.5}$, and $q_3 = 10^{-3.5}$, the closed-loop poles are at −1393, $-1438 \pm j1589$. This indicates a low-frequency time-constant of below 1 ms combined with an under-damped mode with a damping around 0.65. The controller gain vector is $K = [31.6 \quad 0.027 \quad 0.053]$.

A sample of responses of this control system is shown in Figure 5.69. The V_{ff} term is set to a low-pass filtered version of v_g as $V_{\text{ff}} = \frac{1/400}{0.001s+1}$ to help with current limiting as discussed in Section 5.3.5. The current limits (in the "sat" block) are set to ±8 A. The rated current is

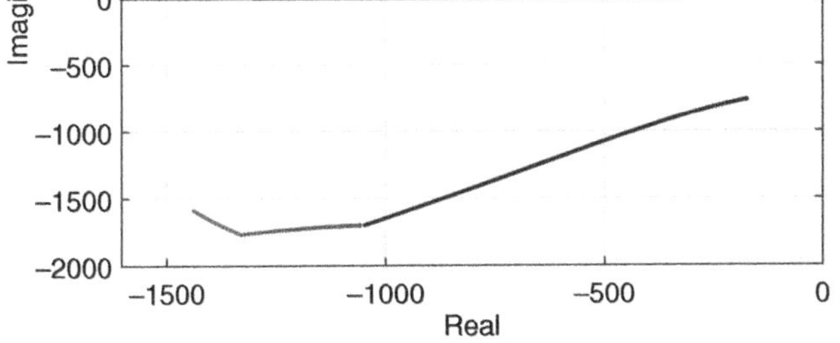

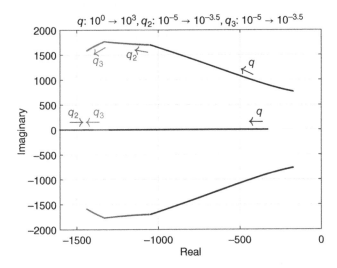

Figure 5.68 Locus of closed-loop poles of load voltage control system with current limiting when elements of matrix Q vary.

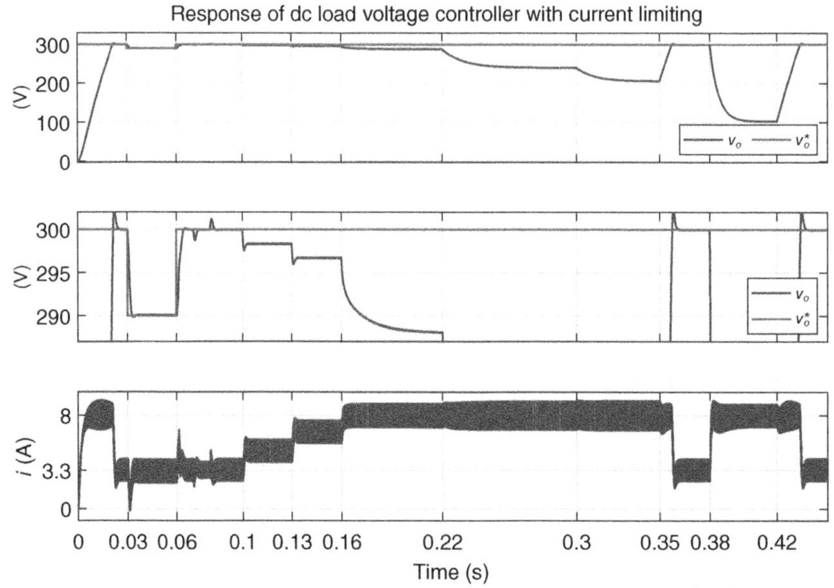

Figure 5.69 Responses of dc load voltage controller with current limiting to command change at $t = 0.03$ seconds and $t = 0.06$ seconds, to dc voltage disturbance at $t = 0.07$ to $t = 0.08$ seconds, to consecutive load additions at $t = 0.1,\ 0.13,\ 0.16,\ 0.22$, and 0.3 seconds, to removing of all these added loads at $t = 0.35$ seconds, to a deep voltage sag at $t = 0.38$ seconds, and to clearing of the voltage sag at $t = 0.42$ seconds.

$\frac{1000}{300} = 3.33$ A. The converter starts at $t = 0$ while the load of 1 kW is connected. The converter smoothly (linearly) increases the load voltage while the converter current is at the maximum of 8 A. It takes about 20 ms for the voltage to build to 300 A. The output voltage reference jumps to 290 V at $t = 0.03$ seconds and turns back to 300 V at $t = 0.06$ seconds. The controller quickly and smoothly tracks this command. At $t = 0.07$ seconds, the dc voltage drops to 370 V and then restores to 400 V at $t = 0.08$ seconds. The output voltage experiences almost no change. At every time instant $t = 0.1,\ 0.13,\ 0.16,\ 0.22,\ 0.30$ seconds, a 0.5 kW load is switched on. This causes the output voltage to slightly drop at first, and as the converter current saturates after $t = 0.16$ seconds load, it will drop at larger steps to make up for the balance of power. The current remains limited at 8 A. At $t = 0.35$ seconds, all the added loads are removed and the converter output voltage returns to 300 V and its current to the rated 3.33 A (as fast as within almost 10 ms). At $t = 0.38$ seconds, a deep voltage sag occurs that reduces the voltage to about 100 V and is cleared at $t = 0.42$ seconds. The output voltage is nicely controlled and the current is limited at all times.

5.7 Grid-Forming Converter Controls

The dc load voltage controller with current limiting just discussed in Section 5.6.2, Figure 5.66, is much similar to the power controller with static grid support discussed in Section 5.4.5, Figure 5.50. They are practically equivalent with only some minor differences. This fact indicates that the same controller can operate the converter in both grid-connected and load-connected (also called stand-alone, isolated, and islanded) conditions. In other words, such controllers can control the power flow when connected to the grid and can control the voltage (and form a grid)

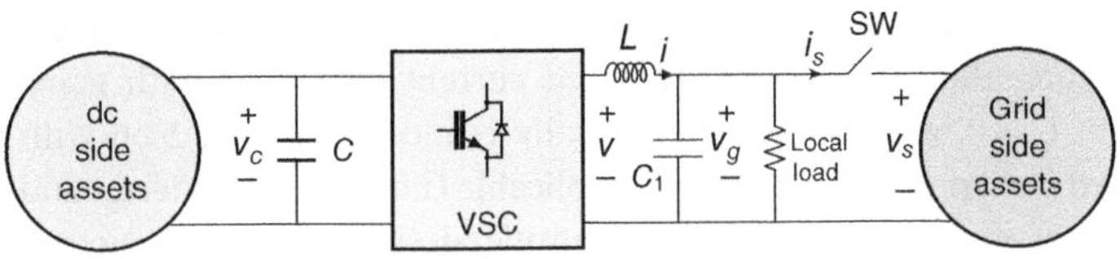

Figure 5.70 Converter with grid-connected and standalone operation.

when connected to a load. These type of controllers may be called *grid-forming controllers*. On the other hand, the power controllers without grid voltage support discussed in Section 5.3 cannot establish (or form) a voltage if the grid is not available. These are commonly called *grid-following controllers*.

This section discusses further details on grid-forming controllers for a dc converter. Such converters are able to operate in both grid-connected and islanded conditions. They can also seamlessly transfer from one mode of operation to the other. The grid-forming converters are the desired option for building the grids and microgrids which may have a number of converters and are to function together within multiple operating scenarios.

Figure 5.70 shows the circuit diagram of the converter which has a local load and is also connected to the grid through a switch. The switch may be closed or open corresponding to grid-connected or stanadlone operation of the converter, respectively. The converter output voltage (at the load terminals) rated value is V_g and its instantaneous value is $v_g(t)$. The capacitor C, which may or may not exist depending on the converter topology, has the voltage $v_c(t)$ and its rated value is V_c. The control objectives are as follows. (i) Maintain the load voltage close to its rated value. (ii) Limit the converter current $i(t)$ within $[I_{\min}\ \ I_{\max}]$. (iii) Maintain the capacitor voltage v_c close to its rated value V_c. (iv) Have seamless transition between grid-connected and standalone modes of operation. Seamless transition means smooth and with an acceptable level of transients. (v) Offer grid-supportive responses. This feature is desired for high penetration of such converters in the grid.

5.7.1 Grid-Forming Control Without a dc-Side Capacitor

Figure 5.71 shows a grid-forming control strategy without a capacitor C or without emulating it in the controller. This capacitor, as we have seen in Section 5.4, can provide dynamic voltage support (or inertia response). In the absence of such capacitor, and if the inertia response is desired, the method of grid voltage differentiation of Section 5.4.6 can be used as shown in Figure 5.71.

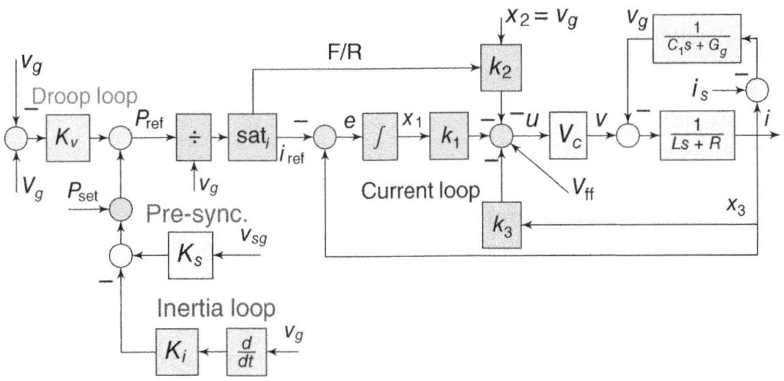

Figure 5.71 Grid-forming control without a capacitor or an emulated capacitor. The inertia can be provided using direct differentiation of the grid voltage.

The controller of Figure 5.71 is basically a refined version of the previously studied controllers (Figures 5.50 and 5.66). At the core, there is a fast LQT-based current controller with gains k_1 to k_3. In standalone operation, $i_s = 0$, and the system basically reduces to Figure 5.66 with $K_v = V_g k_0$ and the same LQT-based design of Section 5.6.2 is applicable (note that the inertia and pre-synchronization branches are not engaged in the current controller design). In grid-connected operation, i_s is determined by the grid dynamics. If the grid model is known, it can be used to further enhance the controller.[33] This, however, is not necessary, and if the controller is robust enough, it can handle the variations in i_s.

The feed-forward term V_{ff} may be constant at $\frac{v_g(0)}{V_c}$ to have a soft start or may be a feedback term from v_g, e.g. $V_{ff}(s) = \frac{1/V_c}{\tau s+1} V_g(s)$, to assist with better current limiting. The saturation block sat maintains the converter current within $[I_{min} \;\; I_{max}]$. The constant K_v is the droop coefficient and determines how much the voltage v_g may change when its power changes.[34] Its relationship with k_0 used in Figure 5.66 is $K_v = V_g k_0$. The power set-point is P_{set}. Finally, the pre-synchronization term $K_s v_{sg} = K_s(v_s - v_g)$ is enabled before closing the switch in order to bring v_g close to v_s and to achieve a soft transition. The gain K_s determines how strong the converter pushes v_g to v_s within the power limitations of the converter.

Example 5.16 ***Grid-forming Controller without a Capacitor on dc Side***
Consider the converter parameters given in Table 5.2. No dc-side capacitor is used or emulated. The grid-forming controller of Figure 5.71 is used with the same current controller we used in Example 5.15, i.e. $K = [31.6 \;\; 0.027 \;\; 0.053]$. The current limit is set to 8 A. The droop coefficient $K_v = 300$ is used meaning that for every 3.3 V voltage change, the converter changes 1000 W power to support the voltage. This is also consistent with the value of $k_0 = 1$ used in the design stage in Example 5.15. The synchronization gain $K_s = 250$ is used meaning that the converter supplies 1000 W for a 4 V difference between v_g and v_s. The inertia term is not used, i.e. $K_i = 0$.

Figure 5.72 shows a sample of the converter responses. Initially, the switch is open, and everything is at rest while the rated load of 1 kW is connected. At $t = 0$, the converter starts and quickly builds the voltage across the load to 300 V in about 20 ms. At $t = 0.05$ seconds, an additional 1000 W load is switched on. As a result, the converter increases its power. The output voltage drops to about 296.8 V. At $t = 0.1$ seconds, the synchronization branch is activated. As a result, the output voltage v_g is pushed close to 298.3 V which is close to the grid voltage v_s. At $t = 0.15$ seconds, the switch closes and connects the converter to the grid. The transition takes place seamlessly and a current of about 2 A flows from the grid to the load. The converter adjusts accordingly. At $t = 0.2$ seconds, the grid voltage v_s drops to 295 V. The converter responds accordingly by slightly increasing its current. At $t = 0.25$ seconds, the switch opens and islands the converters and load from the grid. The transition takes place seamlessly. The converter current is successfully limited below 8 A at all times.

A few remarks may be stated as follows. (i) The synchronization block brings v_g close to v_s for smooth reconnection to grid. This term may not be much crucial in dc DER applications as a small dc voltage offset may not cause much current transients specially if a grid impedance is also present.

---

33 For instance, if a Thevinin equivalence of grid network is used, then $I_s(s) = \frac{V_g(s)-V_s(s)}{L_s s+R_s}$ and then an additional feedback branch from i_s can be added and optimally designed. This is much similar to the optimal controller design for an LCL filter discussed in Section 5.9.

34 For instance, if the voltage is only allowed to change 5% when the converter power changes over the full range of P_{rated}, then $K_v = \frac{P_{rated}}{0.05V_g}$.

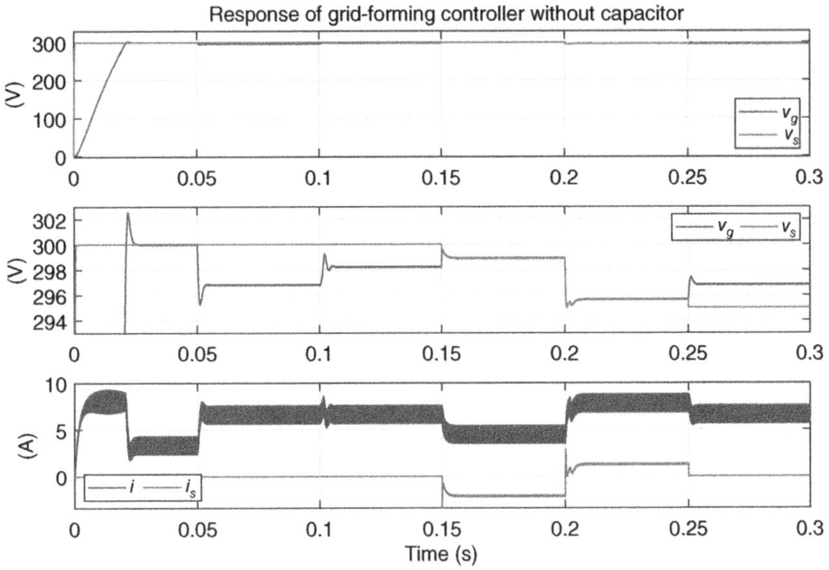

Figure 5.72 Responses of the grid-forming controller of Figure 5.71. $0 < t < 0.15$ seconds and $t > 0.25$ seconds: standalone; $0.15 < t < 0.25$: grid-connected. $t = 0.05$ seconds: load increase; $t = 0.1$ seconds: synchronization activated; $t = 0.15$: switch closes; $t = 0.2$ seconds: grid voltage drops; $t = 0.25$ seconds: switch opens.

However, this concept is much useful in ac applications. (ii) The feedback term k_2 taken from the grid voltage is not absolutely necessary. Specially, if a passive resistor is connected in series with C_1, this term may safely be removed.

5.7.2 Grid-Forming Controller with a dc-Side Capacitor

A grid-forming controller with an emulated dc-side capacitor and one with an actual dc-side capacitor are shown in Figures 5.73 and 5.74, respectively. These are basically the same as those discussed

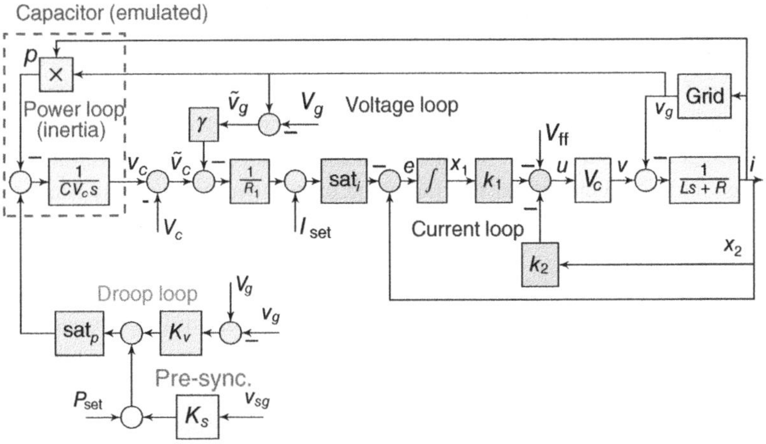

Figure 5.73 Grid-forming controller with an emulated dc-side capacitor providing dynamic grid voltage support (inertia response.)

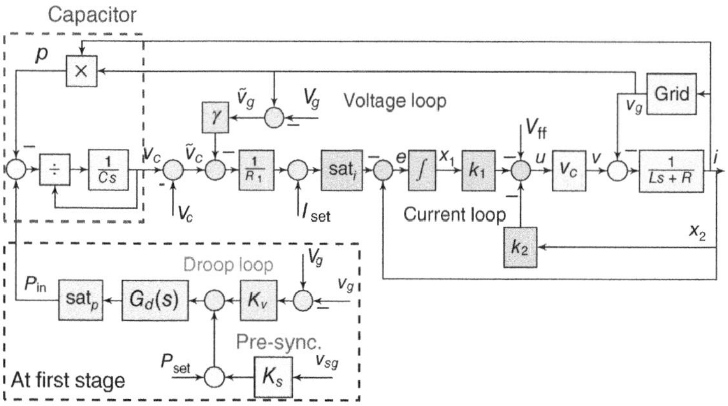

Figure 5.74 Grid-forming controller with an actual dc-side capacitor providing dynamic grid voltage support (inertia response.)

in Sections 5.4 and 5.4.5. Note that Figure 5.73 is directly implementable by a converter connected across a battery. However, Figure 5.74 implies that a first stage supplies the power to the capacitor and the blocks denoted within "at first stage" are implemented at the first stage. The block sat_p denotes the limits of power that can be supplied, either by the battery in Figure 5.73 or by the first stage source in Figure 5.74. The block $G_d(s)$ models the delay (or similar filtering) dynamics of the first stage.

The most internal loop is an LQT-based current controller with current limiting. The voltage loop ensures that the capacitor power is transmitted to the output as an inertia power. The droop loop ensures that the converter responds to grid voltage offsets. Finally, the pre-synchronization branch forces the load voltage v_g to move close to the grid voltage v_s for smooth transition from standalone to grid-connected operation.

In Figures 5.73 and 5.74, $v_g(t)$ and $v_c(t)$ denote the converter output and the capacitor voltages with the rated values of V_g and V_c; $v_s(t)$ shows the grid side voltage, on the right side of the switch, and $v_{sg}(t) = v_s(t) - v_g(t)$; P_{set} and I_{set} are the power and current set-points.[35] The controller establishes the equation

$$R_1[i(t) - I_{\text{set}}] = [v_c(t) - V_c] - \gamma[v_g(t) - V_g], \quad Cv_c(t)\dot{v}_c(t) = P_{\text{ref}} - p(t), \tag{5.17}$$

which since the inductor current $i(t)$ is continuous, it implies that fast transients of v_c and v_g are proportional via the factor γ, and this implies a dynamic grid voltage support or inertia response. To explain more, when the grid voltage experiences a fast transient, it will be reflected on the capacitor voltage with a factor γ. It then generates a power $-Cv_c(t)\dot{v}_c(t)$ at the output power $p(t)$ even if P_{set} does not experience a change. This is much similar to what happens in the rotor of a synchronous generator when it supplies kinetic inertia.

Example 5.17 ***Grid-forming Controller with dc-Side Capacitor***
Consider the converter parameters given in Table 5.2. A dc-side capacitor of 1 mF is connected. The grid-forming controller of Figure 5.74 is used with the same current controller we used before, i.e. $K = [31.6 \quad 0.035]$. The power $P_{\text{ref}} = P_{\text{in}}$ (in practice coming from the prime source) is simulated using a controlled current source connected across the capacitor.[36] The current limit is set to 8 A and

35 If we set $P_{\text{set}} = V_g I_{\text{set}}$, then the capacitor voltage becomes independent from the loading level.
36 The index "in" is used in this system to emphasize that this power comes from the input side.

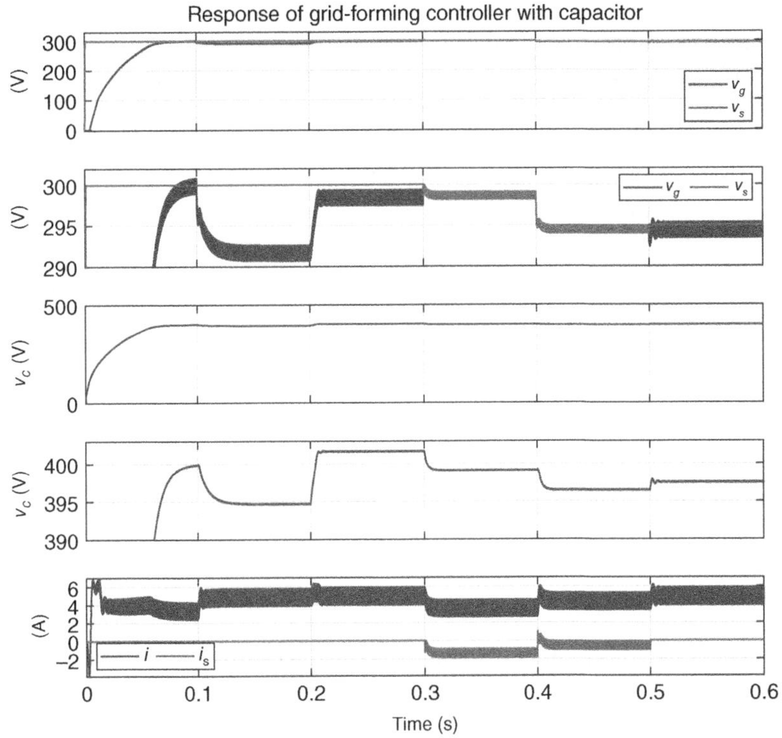

Figure 5.75 Responses of the grid-forming controller of Figure 5.74. $0 < t < 0.3$ seconds and $t > 0.5$ seconds: standalone; $0.3 < t < 0.5$: grid-connected. $t = 0.1$ seconds: load increase; $t = 0.2$ seconds: synchronization activated; $t = 0.3$ seconds: switch closes; $t = 0.4$ seconds: grid voltage drops; $t = 0.5$ seconds: switch opens.

the power limit to 2 kW (the rated values are 3.33 A and 1 kW, respectively). The droop coefficient $K_v = 50$ is used meaning that for every 20 V voltage change, the converter changes 1000 W power to support the voltage. The synchronization gain $K_s = 250$ is used meaning that the converter supplies 1000 W for 4 V difference between v_g and v_s. The inertia "amplification gain" γ is set to 1. Also, $R_1 = 2$ is used and $P_{set} = 1000$. The power limit in sat_p is set to 2 kW.

Figure 5.75 shows a sample of the converter responses. Initially, the switch is open, and everything is at rest while the rated load of 1 kW is connected. At $t = 0$, the converter starts and quickly builds the voltages both across the load to 300 V and across the capacitor to 400 V, in less than 70 ms. At $t = 0.1$ seconds, an additional 500 W load is switched on. As a result, the converter instantly increases its current. The output voltage drops to about 292 V. The capacitor voltage also drops to about 395 V with a very similar dynamic to the load voltage (i.e. an inertia behavior).[37] At $t = 0.2$ seconds, the synchronization branch is activated. As a result, the output voltage v_g is pushed close to 299 V which is close to the grid voltage v_s. At $t = 0.3$ seconds, the switch closes and connects the converter to the grid. The transition takes place seamlessly and a small current of about 1 A flows from the grid to the load. The converter adjusts accordingly. At $t = 0.4$ seconds, the

37 When the grid voltage drops, the inertia response drops the capacitor voltage. However, the droop term increases the power coming from the input. This latter power increases the capacitor voltage. The actual change on the capacitor voltage is the sum of these two effects. That is why 8 V drop in v_g causes only 5 V drop in v_c despite the fact that $\gamma = 1$ indicates equal level of drop in both as far as the inertia alone is concerned.

grid voltage v_s drops to 295 V. The converter responds accordingly by slightly increasing its current. The capacitor voltage dynamically supports the grid voltage by reducing its value almost with the same dynamic of the load voltage. At $t = 0.5$ seconds, the switch opens and islands the converters and load from the grid. The transition takes place seamlessly. The converter current is successfully limited below 8 A at all times.

To study the impact of first stage dynamics, the transfer function $G_d(s) = \frac{1}{\tau_d s+1}$ in Figure 5.74 with two values of $\tau_d = 10$ ms (corresponding to a small delay) and $\tau_d = 100$ ms (corresponding to a large delay) is considered. Figure 5.76 shows a sample of the converter responses for these two cases. Initially, the switch is open, and everything is at rest while the rated load of 1 kW is connected. At $t = 0$, the converter starts. At $t = 0.5$ seconds, an additional 500 W load is switched on. At $t = 1$ second, the synchronization branch is activated. At $t = 1.5$ seconds, the switch closes and connects the converter to the grid. At $t = 2$ seconds, the grid voltage v_s drops to 295 V. At $t = 2.5$ seconds, the switch opens and islands the converters and load from the grid. It is observed that a delay of 100 ms (modeled by the first-order transfer function $G_d(s)$) has caused more oscillations into the system responses. This is because the static support (droop term) cannot immediately respond to the grid side disturbances. The system has, however, been able to stabilize the responses thanks to the robustness of the controller and also the inertia support provided by the capacitor.

A few remarks may be stated here regarding the above grid-forming controllers with a dc-side (emulated or actual) capacitor. The internal current controller is a partial state feedback. If the

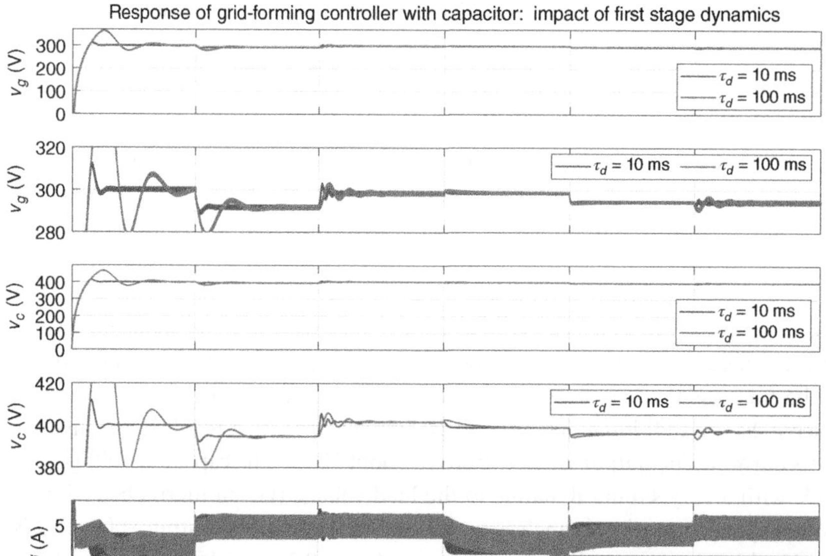

Figure 5.76 Responses of the grid-forming controller of Figure 5.74 considering the first-stage dynamics modeled by $G_d(s) = \frac{1}{\tau_d s+1}$. $0 < t < 1.5$ seconds and $t > 2.5$ seconds: standalone; $1.5 < t < 2.5$ seconds: grid-connected. $t = 0.5$ seconds: load increase; $t = 1$ second: synchronization activated; $t = 1.5$ seconds: switch closes; $t = 2$ seconds: grid voltage drops; $t = 2.5$ seconds: switch opens.

circuit and controller parameters are selected at borderlines (for example a too small dc-side capacitor, or a too large inertia amplification gain γ, or a too small R_1) the dynamic responses may exhibit oscillations. Specifically, high-frequency resonance between L and C_1 may be excited. In such cases, the following actions can resolve the issue. (i) Readjust the circuit and controller parameters properly if possible. (ii) Use a more complete state feedback that uses the capacitor voltage as well, as discussed in Section 5.4.4, or the grid voltage, as discussed in Section 5.6.2. A true full state feedback must engage both the capacitor voltage v_c and the grid voltage v_g as discussed in Section 5.7.2.1. (iii) Use passive damping. In the above example, a small resistor of $R_d = 1\ \Omega$ is connected in series with the filter capacitance C_1 to introduce damping of the resonance poles. The power losses caused by this resistor is in the order of 0.1% of the rated converter power which is negligible.

5.7.2.1 Full State Feedback

A full state feedback of the grid-forming controller with an actual dc capacitor is shown in Figure 5.77. The two other variables, i.e. v_c and v_g, are also included in addition to those used in the previous version of Figure 5.74. Notice that both newly added feedback branches, via k_3 and k_4, must be frozen (as shown in Figure 5.67) during the current-limiting mode. The optimal LQT-based design of the controller gains is discussed below.

The state space equations of the whole system is given by

$$\begin{aligned}
\dot{x}_1 &= (x_2 - I_{\text{set}}) - \tfrac{1}{R_1}(x_3 - V_c) + \tfrac{\gamma}{R_1}(x_4 - V_g)\\
\dot{x}_2 &= -\tfrac{R}{L}x_2 - \tfrac{1}{L}x_4 + \tfrac{V_c}{L}u\\
\dot{x}_3 &= \tfrac{1}{CV_c}[K_v(V_g - x_4) - I_{\text{set}}(x_4 - V_g) - V_g(x_2 - I_{\text{set}})]\\
\dot{x}_4 &= \tfrac{1}{C_1}x_2 - \tfrac{G_g}{C_1}x_4 - \tfrac{1}{C_1}i_s\\
u &= -k_1x_1 - k_2x_2 - k_3x_3 - k_4x_4 + V_{\text{ff}},
\end{aligned} \tag{5.18}$$

where $x_2 = i, x_3 = v_c, x_4 = v_g$. Notice that the third equation has originally been nonlinear, $Cx_3\dot{x}_3 = P_{\text{set}} + K_v(V_g - x_4) - x_2x_4$, and has been linearized around the rated operating conditions of $x_2 = I_{\text{set}}$, $x_3 = V_c$ and $x_4 = V_g$.

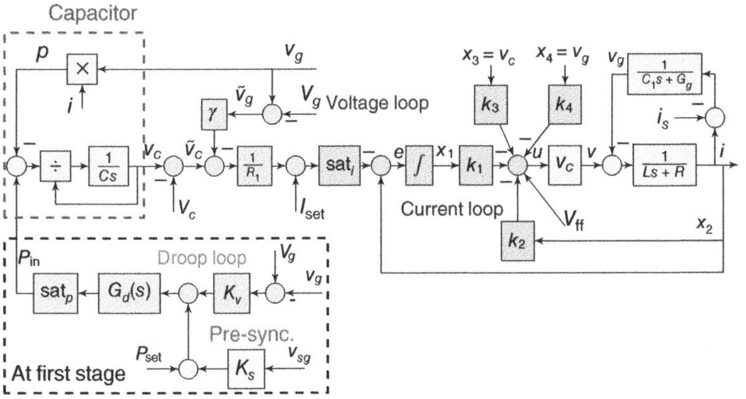

Figure 5.77 Grid-forming controller with an actual dc-side capacitor providing dynamic grid voltage support (inertia response): "full state feedback."

To proceed with the LQT approach, we apply the linear differential operator $\frac{d}{dt}$ to both sides of (5.18) to arrive at

$$\begin{aligned}
\dot{z}_1 &= z_2 - \tfrac{1}{R_1} z_3 + \tfrac{\gamma}{R_1} z_4 \\
\dot{z}_2 &= -\tfrac{R}{L} z_2 - \tfrac{1}{L} z_4 + \tfrac{V_c}{L} w \\
\dot{z}_3 &= -\tfrac{V_g}{CV_c} z_2 - \tfrac{K_v + I_{set}}{CV_c} z_4 \\
\dot{z}_4 &= \tfrac{1}{C_1} z_2 - \tfrac{G_g}{C_1} z_4 \\
w &= -k_1 z_1 - k_2 z_2 - k_3 z_3 - k_4 z_4 = -Kz,
\end{aligned} \tag{5.19}$$

where $z_i = \dot{x}_i$ and $w = \dot{u}$. Specifically, $z_1 = \dot{x}_1 = e$ is the current tracking error. Equation set (5.19) is in the standard LQR format $\dot{z} = Az + Bw, \quad w = -Kz$ with

$$A = \begin{bmatrix} 0 & 1 & -\dfrac{1}{R_1} & \dfrac{\gamma}{R_1} \\ 0 & -\dfrac{R}{L} & 0 & -\frac{1}{L} \\ 0 & -\dfrac{V_g}{CV_c} & 0 & -\dfrac{K_v + I_{set}}{CV_c} \\ 0 & \dfrac{1}{C_1} & 0 & -\dfrac{G_g}{C_1} \end{bmatrix}, \quad B = \begin{bmatrix} 0 \\ \dfrac{V_c}{L} \\ 0 \\ 0 \end{bmatrix}.$$

This will readily allow to minimize the cost function

$$J = \int_0^\infty \left(qe^2 + q_2 z_2^2 + q_3 z_3^2 + q_4 z_4^2 + w^2 \right) dt$$

using the command MATLAB: ≫ `K = lqr(A, B, Q, 1)` where Q is a diagonal matrix with elements $q > 0, q_2 \geq 0, q_3 \geq 0, q_4 \geq 0$.

During the current-limiting mode, the closed loop is characterized by the reduced-order set of equations

$$\begin{aligned}
\dot{x}_1 &= x_2 - I_{max} \\
\dot{x}_2 &= -\tfrac{R}{L} x_2 - \tfrac{1}{L} x_4 + \tfrac{V_c}{L} u \\
u &= -k_1 x_1 - k_2 x_2
\end{aligned} \tag{5.20}$$

whose poles are the eigenvalues of

$$A_1 = \begin{bmatrix} 0 & 1 \\ -\frac{V_c}{L} k_1 & -\frac{R}{L} - \frac{V_c}{L} k_2 \end{bmatrix}.$$

The following example shows some numerical results.

Example 5.18 ***Grid-forming Full State Feedback Controller with dc-Side Capacitor***

For the same set of system parameters used in Example 5.17, Figure 5.68 shows the locus of the closed-loop poles of the grid-forming controller with full state feedback controller when q, q_2, q_3, and q_4 vary. As expected, q has the dominant effect and there is practically no need to even change the others. For the final values of $q = 10^{3.5}$, $q_2 = 10^{-3.25}$, $q_3 = 10^{-3.25}$, and $q_4 = 10^{-4.25}$, the closed-loop poles are at $-115, -1430, -1602 \pm j1515$. The controller gain vector is $K = [56.2 \ \ 0.059 \ \ -0.021 \ \ 0.018]$. The poles of current controller during the current-limiting mode, i.e. eigenvalues of matrix A_1 defined above, are at -1325 and -3395 which are fast enough to enable tight limiting (Figure 5.78).

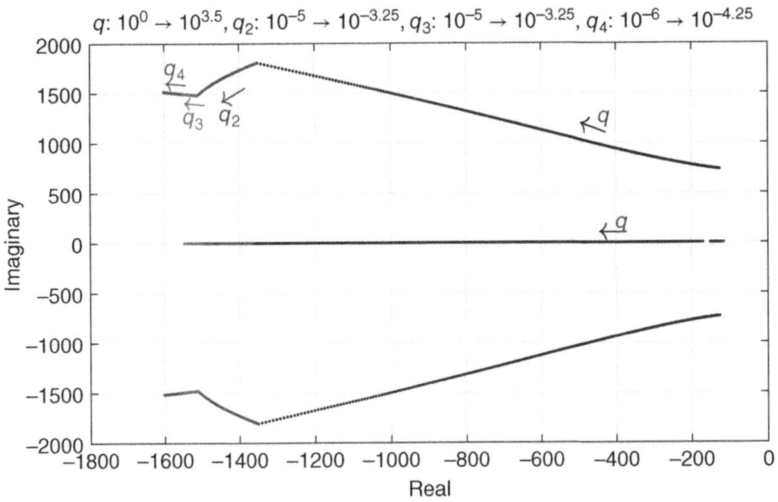

Figure 5.78 Locus of closed-loop poles of grid-forming full state feedback controller when elements of matrix Q vary.

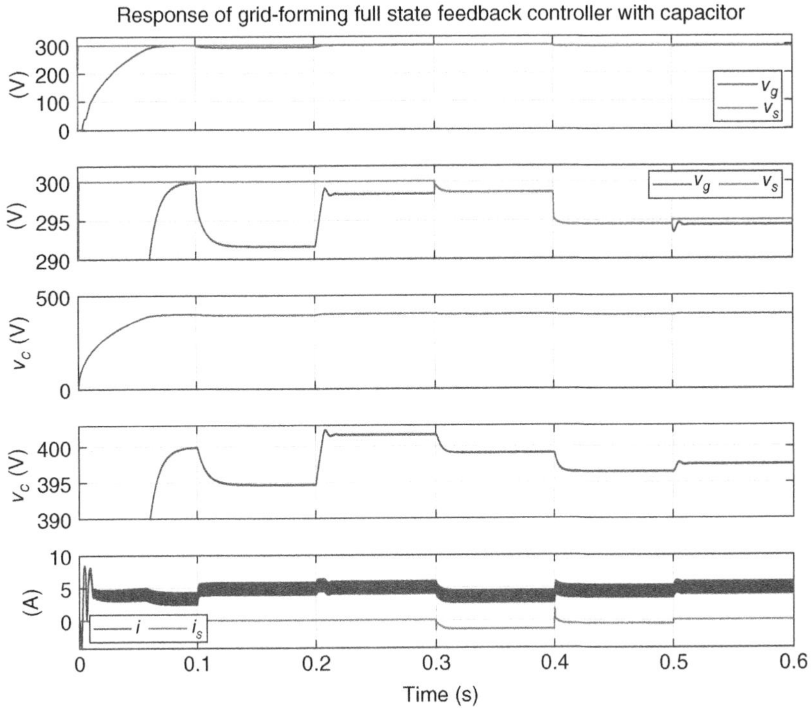

Figure 5.79 Responses of the full state feedback grid-forming controller of Figure 5.77. $0 < t < 0.3$ seconds and $t > 0.5$ seconds: standalone; $0.3 < t < 0.5$ seconds: grid-connected. $t = 0.1$ seconds: load increase; $t = 0.2$ seconds: synchronization activated; $t = 0.3$ seconds: switch closes; $t = 0.4$ seconds: grid voltage drops; $t = 0.5$ seconds: switch opens.

Figure 5.79 shows a sample of the converter responses. The same simulation scenario of Figure 5.76 is used. The converter responses are smooth and robust. Notice that no resistor is connected in series with C_1 in this example. In other words, the controller itself introduces active damping to the resonance mode. In Example 5.17, a small resistor was placed in series with C_1 to passively damp the resonance ringings.

5.8 Control Scenarios in a PV Converter

In a PV system, multiple control scenarios may be considered to harvest the proper amount of power. MPPT is a common approach where a feedback control loop is devised to continuously extract the maximum available power from the PV system. As the PV system penetration is increasing, local over-generation may occur. Therefore, solar power curtailment (where the PV system operates below its maximum power capacity) or storage scenarios (where the excess energy is stored) become necessary. This section discusses some feedback control systems to control the extraction and flow of power in a PV system.

5.8.1 PV Voltage Control

The electric power supplied by a PV system is determined by the voltage across its panels. The relationship between the PV power and voltage, however, is nonlinear, see Figure 5.80. When the voltage is below the voltage of the MPP V_{mp}, i.e. left side of the power voltage characteristics, the power increases by an increase in the voltage. On the right side, however, the power decreases when the voltage goes up.

It is worthwhile noting that, in the absence of a feedback control on the PV voltage, the right side of the curve tends to be stable and its left side is unstable. To understand this, assume that a capacitor is connected across the panels as shown in Figure 5.81. In a steady operation, the power supplied by the PV panels P_{in} is equal to the output power P_{out}, and the voltage v_c is constant. Assume that the PV panel is operating at a point on the right side of maximum power voltage, i.e. $V_{mp} \leq v_c \leq V_{oc}$. If the output power P_{out} goes up (as a result of a load switching on the grid side), this momentarily discharges the capacitor and the voltage v_c goes down, and the PV power P_{in} goes up. The PV will find a new steady point at a lower voltage. If the output power P_{out} goes down (as a result of a load switching off on the grid side), the capacitor voltage rises, and this lowers the PV power P_{in}. The system will find a new steady operation at a higher voltage v_c. In either case, a naturally correct and stabilizing response is provided.

Now, assume that the PV is operating at a voltage below the maximum power voltage, i.e. $0 \leq v_c \leq V_{mp}$, and the output power P_{out} goes up, this will lower v_c which in turn lowers P_{in}. The voltage

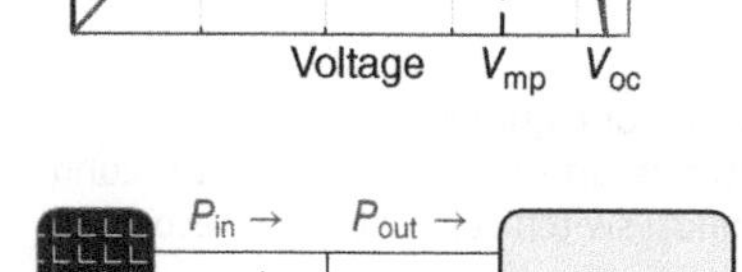

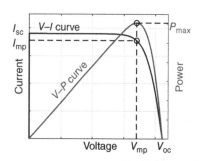

Figure 5.80 *I–V* and *P–V* characteristics of a solar module.

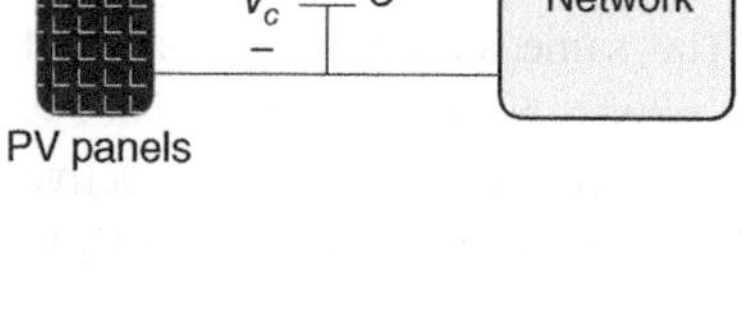

Figure 5.81 General PV system.

v_c keeps going down. If, on the other hand, the output power P_{out} goes down, the voltage v_c goes up and increases P_{in}. The voltage v_c keeps going up. This is an unstable response.

When the controller has the ability of controlling the PV voltage, the PV array can be biased at any voltage between 0 to V_{oc} as long as it does not violate the physical limitations of the converter system.

Example 5.19 ***PV Voltage Control***

Consider the PV system shown in Figure 5.82. The PV array comprises 2 parallel strings each including 14 series-connected modules of 1Soltech 1STH-215-P model. Each module has a maximum power of 213.15 W, voltage of maximum power of 29 V, and open-circuit voltage of 36.3 V (at 25° and 1000 W/m^2). The array will have a maximum power of 5968.2 W at a maximum power voltage of 406 V and open-circuit voltage of 508.2 V as shown in Figure 5.83.

The tight dc voltage controller of Figure 5.46, Example 5.10, redrawn in Figure 5.84 is used. The converter parameters are the same as used in Example 5.10, i.e. $C = 0.5$ mF, $L = 5$ mH, $V_g = 300$ V, $f_{sw} = 4$ kHz. For $R_1 = 0.1$ (to achieve tight regulation of capacitor voltage), and selection of $q = 10^2$, $q_2 = 10^{-3} = q_3$, the location of closed loop poles are at $-1438 \pm j1559, -2670$, and the controller gains are $K = [k_1\ \ k_2\ \ k_3] = [10\ \ 0.069\ \ -0.095]$, see Figure 5.48. The feedback branch k_3 is frozen during the current-limiting mode to ensure that the current is also tightly limited within the prespecified range $[I_{min}\ \ I_{max}] = [-30\ \ 30]$ A. The freezing/releasing (FR) mechanism is shown in Figure 5.47. Note that the rated current is $I = \frac{P}{V_g} = \frac{6000}{300} = 20$ A. The feed forward

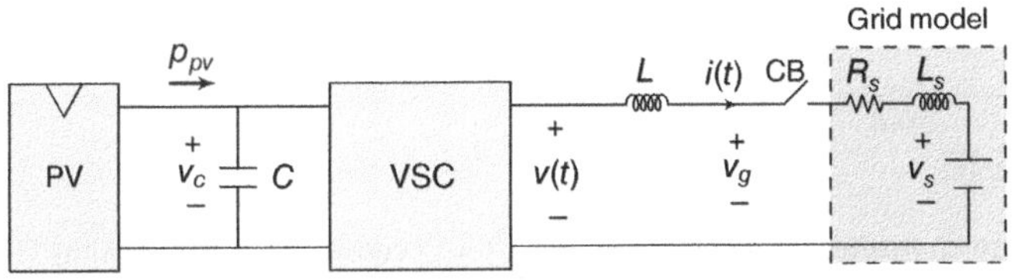

Figure 5.82 Circuit diagram for a converter interfacing a PV array to a dc grid.

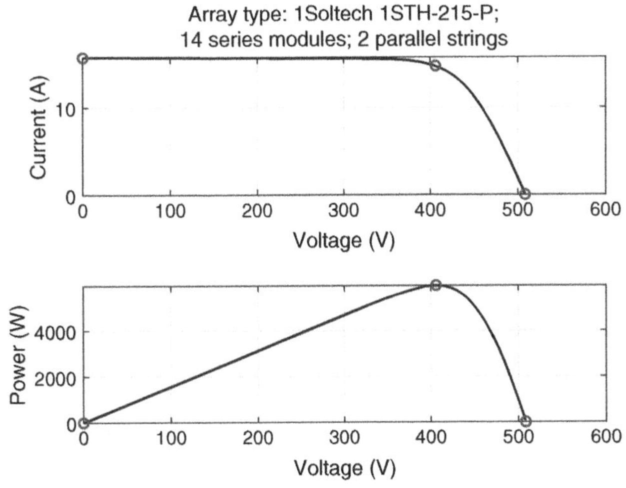

Figure 5.83 PV array characteristics at 25° and 1000 W/m^2.

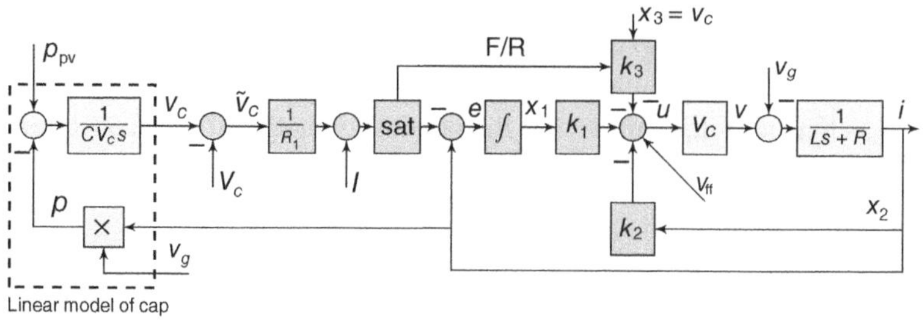

Figure 5.84 Tight capacitor voltage control, see Example 5.10, used for direct control of PV voltage.

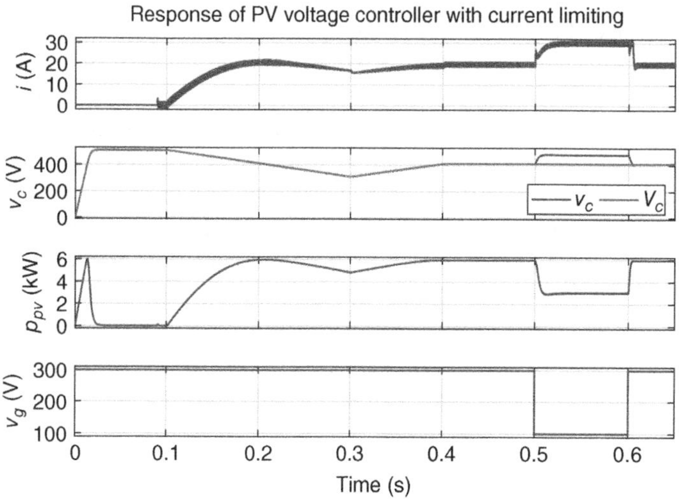

Figure 5.85 Responses of the PV voltage controller with current limiting. $t = 0$: converter is not working and capacitor is discharged; $t = 0.07$ seconds: circuit breaker closes; $t = 0.09$ seconds: converter starts; $0.1 < t < 0.3$ seconds: the reference voltage is linearly decreased at the rate of -1000 V/s; $0.3 < t < 0.4$ seconds: the reference voltage is linearly increased at the rate of 1000 V/s; $t = 0.5$ seconds: deep grid voltage fault; $t = 0.6$ seconds: fault clears.

term $V_{ff} = k_3 v_c(t_o) + \frac{1/400}{0.001s+1} v_g$ is chosen, where $v_c(t_o)$ is the capacitor voltage at the instant where converter starts switching, for soft start and current limiting enhancement.

The reference to the voltage controller, V_c, is provided manually. This determines the voltage that the PV is commanded to work at. A sample of system responses is shown in Figure 5.85 where the converter output current i, the PV voltage v_c and its command V_c, the PV power p_{pv}, and the grid voltage v_g are depicted. In this simulation, the PV array is connected to the converter at $t = 0$ seconds while the converter is not switching, its capacitor voltage is initially zero, and is not connected to the grid. The PV charges the capacitor voltage to its open-circuit voltage of 508.2 V in around 20 ms. During this interval, the PV scans its characteristics from zero voltage to open-circuit voltage. At $t = 0.07$ seconds, the circuit breaker connects the converter to the grid and, at $t = 0.09$ seconds, the control system is activated.

The reference voltage is linearly decreased at a rate of -1000 V/s starting from $t = 0.1$ seconds and ending at $t = 0.3$ seconds. Figure 5.85 shows that the converter starts smoothly and follows the voltage command faithfully. As the voltage decreases, the operating point moves from the open-circuit voltage all the way to the maximum power voltage and then moves below it to about 300 V. Then, the reference voltage is increased during [0.3 0.4] seconds at the rate 1000 V/s up to about 400 V. The controller executes the command very closely and the power increases. At $t = 0.5$ seconds, a deep low voltage fault occurs that drops v_g to 100 V. The PV immediately decreases the power and the controller limits the current at 30 A. At $t = 0.6$ seconds, the fault clears and the system returns to normal condition quickly after that.

Remark on nonlinearity of PV characteristics: In Figure 5.84, the PV power p_{pv} is a nonlinear function of v_c according to the PV power versus voltage characteristics. This characteristics is a function of solar irradiance, temperature, partial shading, PV panels' manufacturing brand, and even some other ambient conditions, and aging of panels. So, much uncertainties and changes are involved. Our approach has been to simply ignore and not model any of these in the control model! The robustness of the proposed optimally-designed controller, verified by simulating the system within various operating conditions and severe disturbances, indicates that such an approach is sound and better than trying to explicitly address those nonlinearities.

5.8.2 MPPT via PV Voltage Control

Unlike the voltage control, the power control (meaning to adjust the power to a given or intended value) in an intermittent PV resource is harder due to random and uncertain nature of intermittencies. One approach is that the power control loop adjust the reference, i.e. V_c, for the control system using an additional feedback control loop. This section talks about the control system to achieve a particular yet common power control scenario, i.e. MPPT. The controller is responsible for operating the panels at the maximum power voltage V_{mp}.

Circuit diagram of a converter interfacing a PV array to a dc bus is shown in Figure 5.82. The voltage v_{pv} directly determines the amount of current and power drawn from the PV module as shown in Figure 5.80.[38] Figure 5.80 also clearly shows that there is one single voltage at which the solar module generates maximum power. This point depends on solar light intensity, ambient temperature and other characteristics of the solar module including circuit uncertainties and aging. Therefore, generally speaking, a feedback mechanism is required to find and track that point if it is desired to extract the maximum power at all times. Such an algorithm is called MPPT algorithm and control.

The block diagram of the common control system to achieve the MPPT is shown in Figure 5.86. The PV side measurements (including the PV voltage and current) are used by MPPT Algorithm and Control to generate the reference value for the voltage, v^*_{pv}. The internal loop (PV Voltage Controller) performs the PV voltage control and current limiting. This part (PV Voltage Controller) is the same as what was earlier discussed its multiple versions in details and used one of them in Example 5.19.

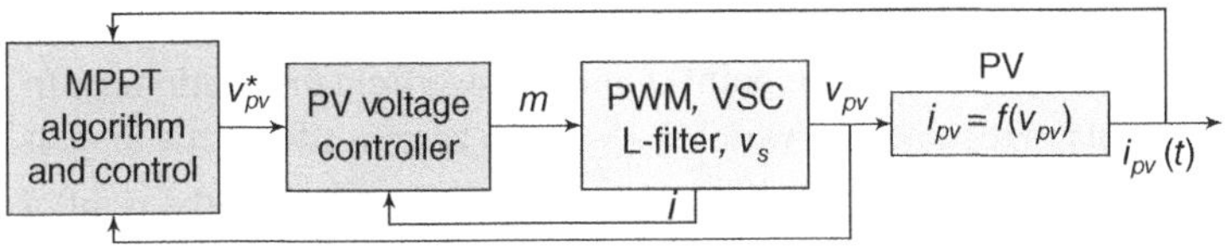

Figure 5.86 Block diagram of control system to achieve MPPT.

38 In this discussion v_{pv} is the same voltage we denoted by v_c. They can be used interchangeably. Similarly, the reference for this voltage is denoted by v^*_{pv} which is the same as what we denoted by V_c in earlier discussions.

The *MPPT Algorithm and Control* conceptually operates based on perturbing the reference and observing the response; called **perturb-and-observe (P&O)** method. It may be summarized in the following steps.

Step 1: Select the time step-size T and the voltage step-size ΔV. Choose an initial value for the voltage reference, $v^*_{pv}(0)$.

Apply this to the system and let the PV system reach a steady operation. Calculate the power $p_{pv}(0) = v_{pv}(0)i_{pv}(0)$ where i_{pv} is the current of PV module.

Set $\alpha = -1$ or $\alpha = 1$ to decrease and increase from the initial value of voltage.[39]

For all discrete instants of time, $n = 1, 2, 3, \cdots$, corresponding to actual time $t = nT$, perform the following steps.

Step 2: Set $v^*_{pv}(n) = v^*_{pv}(n-1) + \alpha\Delta V$.

Step 3: Calculate the PV power $p_{pv}(n) = v_{pv}(n)i_{pv}(n)$ and the PV power increment $\Delta p_{pv}(n) = p_{pv}(n) - p_{pv}(n-1)$.

If $[\Delta p_{pv}(n) > 0$ and $\alpha > 0]$ OR $[\Delta p_{pv}(n) < 0$ and $\alpha < 0]$; then, set $\alpha = 1$ and go to Step 2.

If $[\Delta p_{pv}(n) < 0$ and $\alpha > 0]$ OR $[\Delta p_{pv}(n) > 0$ and $\alpha < 0]$; then set $\alpha = -1$ and go to Step 2.

Remark 5.1 The time step-size T should be sufficiently large to ensure that the internal control system has completed its response to the command of voltage. Meanwhile, it should be sufficiently small to ensure that continuous tracking is performed and maximum power is extracted in a stable way over a course of time where the maximum power experiences quick changes.

Remark 5.2 The voltage step-size ΔV should be sufficiently large to ensure large steps are taken when moving toward the maximum power and the MPPT is achieved quickly. Meanwhile it should be sufficiently small to prevent large oscillation and chattering around the MPP when it is reached. In a more sohpisticated way, ΔV can be adjusted adaptively: when $|\Delta p_{pv}(n)| = |p_{pv}(n) - p_{pv}(n-1)|$ is large (meaning operating distant from the maximum power), then ΔV can be large and as $|\Delta p_{pv}(n)|$ becomes smaller then ΔV can also become smaller. In other words, $\Delta V = c_0|\Delta p_{pv}(n)|$ where c_0 is a positive constant.

The control system of Figure 5.86 is clearly a nonlinear system. It comprises the internal PV Voltage Control system and the outer MPPT algorithm and control. The MPPT is characterized by two parameters ΔV and T and they should be adjusted properly. In practice, and to simplify the design and analysis, the internal voltage control loop is designed to be "much" faster than the external MPPT control loop. In this way, the two control systems may be considered decoupled and their designs can be done separately. It is possible to derive a mathematically more rigorous way of analysis and design of the MPPT controller as discussed below.

5.8.3 Mathematical Modeling of MPPT Algorithm

Mathematically, the MPPT algorithm searches for the point that the power remains stationary. In other words, the MPPT loop attempts to regulate the variable $\frac{d}{dv_{pv}}p_{pv}$ to zero in a feedback loop. If the operating point is the on the left of the MPP, it increases the voltage; and if it is on the right of it, it decreases the voltage. This concept can be generalized and visualized using the block diagram

---

39 When a PV system is at rest and the panels are connected across the input terminals of the converter (without the inverter operating), the voltage across the capacitor may increase all the way to the open-circuit voltage of the panels. This way, $v_{pv}(0)$ is at highest and $\alpha = -1$ must be chosen.

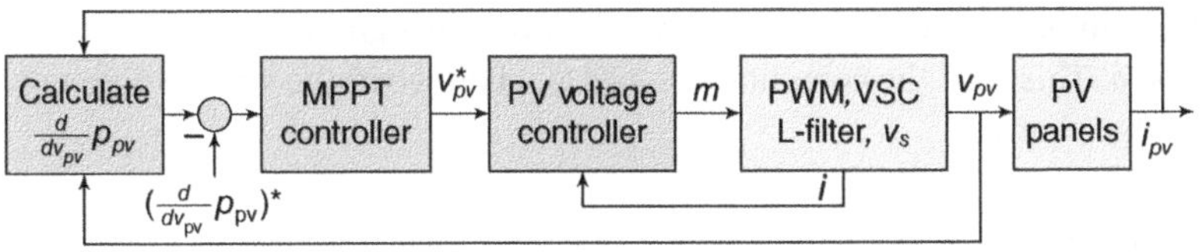

Figure 5.87 Detailed MPPT mechanism including $\frac{d}{dv_{pv}}p_{pv}$ calculation and MPPT controller.

of Figure 5.87. Thus, an MPPT algorithm comprises a way of calculating the voltage-derivative of the power, and an MPPT controller which may be a simple PI controller to regulate it to zero, i.e. in this case $(\frac{d}{dv_{pv}}p_{pv})^* = 0$.

Calculating the derivative is practically a challenge due to presence of noise, switching noise and oscillations. Low-frequency oscillations, dominantly the double-frequency, exist in single-phase ac and unbalanced three-phase systems. High-frequency oscillations are caused by distorted grid voltage and/or nonlinear loads. Two methods for calculating the derivative term is explained as follows.

5.8.3.1 Calculation of $\frac{d}{dv_{pv}}p_{pv}$: Method 1

In this approach, $\frac{d}{dt}p_{pv}$ and $\frac{d}{dt}v_{pv}$ are calculated and divided as

$$\frac{d}{dv_{pv}}p_{pv} = \frac{\frac{d}{dt}p_{pv}}{\frac{d}{dt}v_{pv}}. \tag{5.21}$$

To compute the differentiation of a noisy/oscillatory/switching variable, proper filtering stages must be used. Thus, a low-pass filter, e.g. $G_{\text{LPF}}(s) = \frac{1}{\tau_1 s+1}$, combined with a filtered derivative, e.g. $G_{\text{D}}(s) = \frac{s}{\tau_2 s+1}$, may be used, as shown in Figure 5.88. The time-constants τ_1 and τ_2 are selected large enough to adequately filter the noise without much delay. These must be larger than the time-constant of the voltage control loop.

One issue with computing the division in (5.21) is that either the denominator or both the numerator and denominator can approach zero, particularly when the operating point moves toward the MPP or when the controller operates the PV system at an equilibrium point. Thus, computation errors may arise. To overcome this problem, the relationship can be modified as

$$\frac{d}{dv_{pv}}p_{pv} = \frac{\frac{d}{dt}p_{pv}}{\frac{d}{dt}v_{pv}} = \left\{ \begin{array}{ll} \frac{d}{dt}p_{pv}\dfrac{\text{sign}\left(\dfrac{d}{dt}v_{pv}\right)}{\left|\dfrac{d}{dt}v_{pv}\right|+\epsilon}, & \left|\dfrac{d}{dt}v_{pv}\right| \geq \delta; \\ \hline \overline{} \\ \frac{d}{dt}p_{pv}\dfrac{\overline{\text{sign}\left(\dfrac{d}{dt}v_{pv}\right)}}{\left|\dfrac{d}{dt}v_{pv}\right|+\epsilon}, & \left|\dfrac{d}{dt}v_{pv}\right| < \delta \end{array} \right. \tag{5.22}$$

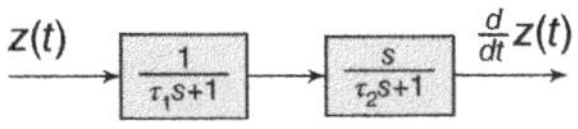

Figure 5.88 Differentiating a noisy/oscillatory/switching variable $z(t)$.

where ϵ is a small positive number to avoid divide-by-zero. The over-bar sign means that the variable is held when $|\frac{d}{dt}v_{\mathrm{pv}}| < \delta$. This way, the computation goes to a halt when the voltage does not experience a sufficiently large change.

5.8.3.2 Calculation of $\frac{d}{dv_{pv}}p_{\mathrm{pv}}$: Method 2

If fast and accurate sampling of the PV voltage and current measurements is possible, synchronized sampling of them can directly remove the switching noises. The waveforms of the converter are shown in Figure 5.89 where $i(t)$ is the inductor current and $v_{\mathrm{pv}}(t)$ is the PV-module voltage.

It is observed from Figure 5.89 that when $i(t)$ is rising, $v_{\mathrm{pv}}(t)$ is falling and vice-versa. Consider the time instants t_1, t_2, and t_3 as shown on this graph. The voltage-derivative of PV power may be approximated by

$$\frac{dp_{\mathrm{pv}}}{dv_{\mathrm{pv}}} \approx \frac{p_{\mathrm{pv}}(t_2) - p_{\mathrm{pv}}(t_1)}{v_{\mathrm{pv}}(t_2) - v_{\mathrm{pv}}(t_1)} = \frac{v_{\mathrm{pv}}(t_2)i_{\mathrm{pv}}(t_2) - v_{\mathrm{pv}}(t_1)i_{\mathrm{pv}}(t_1)}{v_{\mathrm{pv}}(t_2) - v_{\mathrm{pv}}(t_1)}. \tag{5.23}$$

Instead of t_1 and t_2, t_2 and t_3 may also be used. Or for a better accuracy, both may be calculated and averaged over a full cycle. The way of calculating the derivative function as descried here is very fast. A low-pass filter may also be used to further remove other possible noises.

Example 5.20 ***MPPT Via PV Voltage Control***

Consider the same PV system of Example 5.19 and the same voltage controller with current limiting. Here an MPPT control loop is added. The reference to the voltage controller is provided by the MPPT controller which is a pure integrator with gain 10, i.e. the transfer function of $\frac{10}{s}$. Method 1 described above is used to compute the derivative of power with respect to voltage, as shown in Figure 5.90. Here $\tau_1 = \tau_2 = 0.005$ seconds is used, $\delta = 0.1$, $\epsilon = 0.001$, and a limiter [-100, 50] and another similar low-pass filter is used to ensure that the power to voltage derivative is smooth. The PV side capacitor size C is increased to 1 mF (from the original value of 0.5 mF used in Example 5.19) to allow smoother dp/dv computation. However, we did not redesign the controller and used the same design of Example 5.19. The desired responses indicate the robustness of the controller.

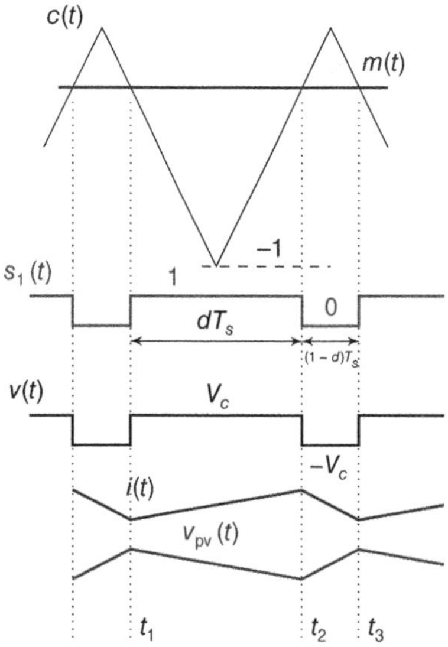

Figure 5.89 Synchronized sampling for fast and smooth $\frac{dP}{dV}$ calculation.

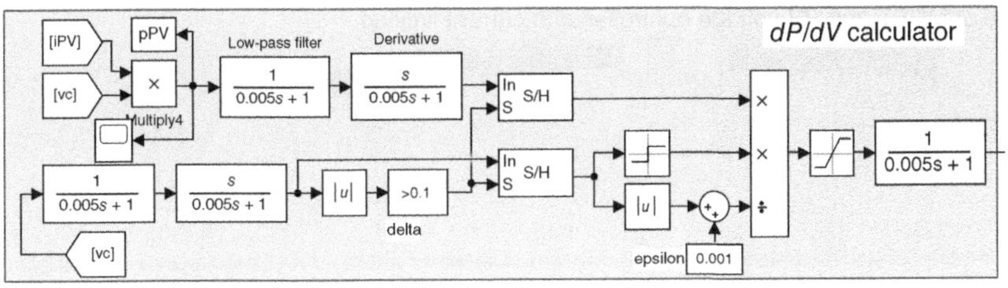

Figure 5.90 The filtering steps used to compute $\frac{dp}{dv}$.

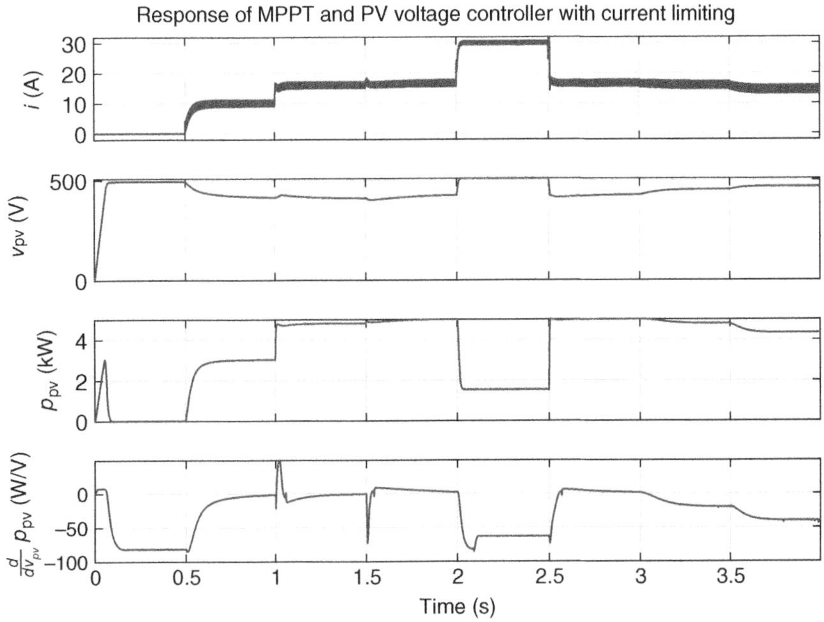

Figure 5.91 Responses of the MPPT and PV voltage controller with current limiting. $t = 0$: converter is not working and capacitor is fully discharged; $t = 0.5$ seconds: converter starts; $t = 1$ second: sudden 300 W/m^2 solar irradiance rise; $t = 1.5$ seconds: sudden 10° temperature fall; $t = 2$ seconds: sudden grid voltage fault down to 50 V (current limiting at 30 A); $t = 2.5$ seconds: fault clears; $t = 3$ seconds: $\left(\frac{d}{dv_{pv}}p_{pv}\right)^*$ jumps from 0 to −20 (shifting from MPPT to power curtailing); $t = 3$ seconds: $\left(\frac{d}{dv_{pv}}p_{pv}\right)^*$ jumps from −20 to −40.

A sample of system responses is shown in Figure 5.91 where the converter output current i, the PV voltage v_{pv}, the PV power p_{pv}, and the variations of $\frac{dp_{pv}}{dv_{pv}}$ are depicted. In this simulation, the PV array is connected to the converter at $t = 0$ seconds while the converter is not working, its capacitor voltage is initially zero, and is not connected to the grid. The PV charges the capacitor voltage to close to 500 V in less than one second.[40] At $t = 0.5$ seconds, the circuit breaker connects the converter to the grid and the control system is activated. The controller drives the PV toward its MPP quickly and practically gets there within half a second. At $t = 1$ second, the solar irradiance experience an abrupt jump of 300 W/m^2 and the controller follows it. At $t = 1.5$ seconds, the panels' temperature experiences a sudden drop of 10° Celsius. The controller adjusts to it as a result of which the power slightly increases. At $t = 2$ seconds, the grid voltage drops all the way to 50 V.

40 During this interval, the PV scans its characteristics from zero voltage to open-circuit voltage.

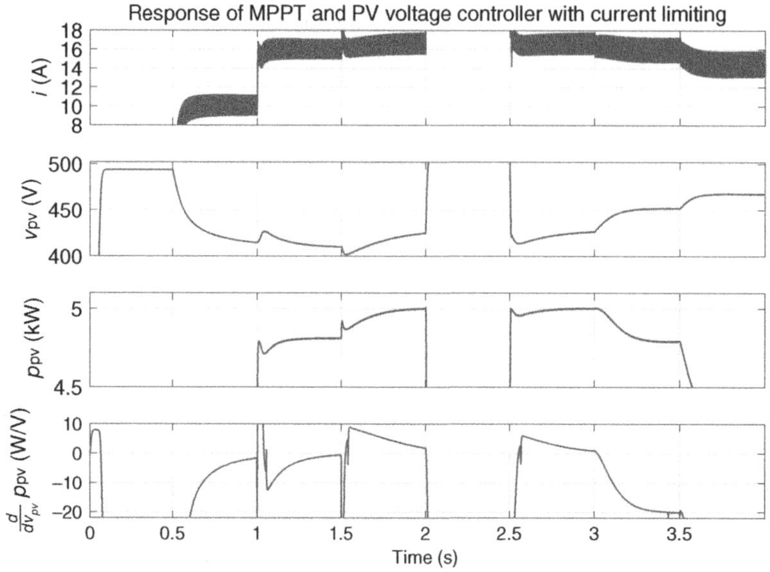

Figure 5.92 Zoomed version of Figure 5.91 for better visualization.

The current limiting gets activated and limits the converter current at the preset value of 30 A. At $t = 2.5$ seconds, the grid voltage becomes normal and the controller adjusts to it smoothly. At $t = 3$ seconds, the reference $(\frac{d}{dv_{pv}}p_{pv})^*$ is changed from 0 to -20 W/V and to -40 at $t = 3.5$ seconds. The controller follows it and as a result, the PV voltage is increased and its power is decreased (i.e. the operation is shifted from MPPT to power curtailing or "deloading"). A zoomed version of Figure 5.91 is shown in Figure 5.92 for better visualization.

5.8.4 PV Power Control

Increasingly, and with higher deployment of these systems, it is desirable to have a "direct and fast control" over the extracted power from a PV system.[41] As we explained before, if the PV voltage is biased above the maximum power voltage, it has an inherent stability. This way, directly controlling the power will automatically adjust the PV voltage. So, there is no need to an explicit voltage control loop as long as it remains on the right side of the maximum power voltage point, i.e. as long as the requested power does not go above the maximum available power of the PV.

The power controller was introduced and discussed in details in Section 5.3. It then was improved by adding a static grid voltage support, or droop mechanism, in Section 5.4.5. The dynamic voltage support, or inertia response, was added to it in Section 5.4.6. Finally, it was upgraded to a grid-forming controller in Section 5.7.1. The diagram of this most complete structure is further enhanced and shown in Figure 5.93. Notice that the core structure is very simple (depicted and marked by "core blocks") and is just a current control loop which is designed optimally using the LQT approach. The other blocks are explained as follows.

The droop loop and the inertia loop provide static and dynamic grid voltage support as explained in Sections 5.4.5, and 5.4.6. The two blocks sat_i and sat_p indicate the limits of current and power,

41 In the control scenario studied so far, the power is "indirectly" controlled via the voltage. Voltage, in its turn, may be determined through another loop such as the MPPT. These multiple control loops plus the required increased size of capacitor (to facilitate good control over voltage, and also to facilitate smooth MPPT) conflict with a fast and "instantaneous" change of the PV power.

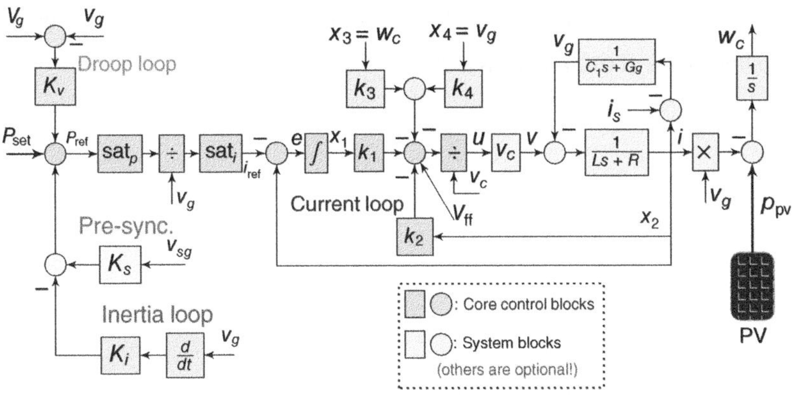

Figure 5.93 Power control of PV system. The power command is applied at P_{set}. Core control blocks and system blocks are identified. Other control blocks are optional and may be added to achieve further objectives.

respectively. The two feedback branches with gains k_3 and k_4 complete the controller to a "full" state feedback and they may be used to improve the dynamic behavior of the control system in extreme and disturbed conditions. These two branches may be frozen during the current limiting mode. The branch with gain K_s helps with smooth transition from standalone to grid-connected mode when applicable.

There are two other optional changes introduced in this diagram. (i) A division block is added right after the control signal is calculated. This linearizes the converter multiplication into v_c. (ii) Rather than using the capacitor voltage v_c as a feedback signal, the capacitor energy $w_c = \frac{1}{2}Cv_c^2$ is used. This also helps having a linear behavior over a wider range of v_c variations.

The controller design stage is much similar to what was explained in Section 5.7.2.1 and is adjusted and presented as follows. The state space equations of the whole system, linearized around $v_g = V_g$, is given by[42]

$$\begin{aligned}
\dot{x}_1 &= (x_2 - I_{\text{set}}) + \frac{K_v}{V_g}(x_4 - V_g) + \frac{P_{\text{set}}}{V_g^2}(x_4 - V_g) \\
\dot{x}_2 &= -\frac{R}{L}x_2 - \frac{1}{L}x_4 + \frac{1}{L}v \\
\dot{x}_3 &= P_{\text{pv}} - V_g I_{\text{set}} - V_g(x_2 - I_{\text{set}}) - I_{\text{set}}(x_4 - V_g) \\
\dot{x}_4 &= \frac{1}{C_1}x_2 - \frac{G_g}{C_1}x_4 - \frac{1}{C_1}i_s \\
v &= -k_1x_1 - k_2x_2 - k_3x_3 - k_4x_4 + V_{\text{ff}}.
\end{aligned} \tag{5.24}$$

---

42 The term $\frac{1}{v_g}$ is linearized (using Taylor's series expansion) and approximated as

$$\frac{1}{v_g} \approx \frac{1}{V_g} - \frac{1}{V_g^2}(v_g - V_g) = \frac{2}{V_g} - \frac{1}{V_g^2}v_g.$$

Also the term $\frac{v_g - V_g}{v_g}$ is approximated by

$$\frac{v_g - V_g}{v_g} \approx 0 + \frac{V_g}{V_g^2}(v_g - V_g) = \frac{1}{V_g}(v_g - V_g).$$

Similarly,

$$iv_g \approx I_{\text{set}}V_g + V_g(i - I_{\text{set}}) + I_{\text{set}}(v_g - V_g).$$

where $x_2 = i, x_3 = v_c, x_4 = v_g, I_{set} = \frac{P_{set}}{V_g}$.[43] To proceed with the LQT approach, we apply the linear differential operator $\frac{d}{dt}$ to both sides of (5.24) to arrive at

$$\begin{aligned}
\dot{z}_1 &= z_2 + \frac{K_v V_g + P_{set}}{V_g^2} z_4 \\
\dot{z}_2 &= -\frac{R}{L} z_2 - \frac{1}{L} z_4 + \frac{1}{L} w \\
\dot{z}_3 &= -V_g z_2 - I_{set} z_4 \\
\dot{z}_4 &= \frac{1}{C_1} z_2 - \frac{G_g}{C_1} z_4 \\
w &= -k_1 z_1 - k_2 z_2 - k_3 z_3 - k_4 z_4 = -Kz,
\end{aligned} \tag{5.25}$$

where $z_i = \dot{x}_i$ and $w = \dot{u}$. Specifically, $z_1 = \dot{x}_1 = e$ is the current tracking error. Equation set (5.25) is in the standard LQR format $\dot{z} = Az + Bw, \quad w = -Kz$ with

$$A = \begin{bmatrix} 0 & 1 & 0 & \frac{K_v V_g + P_{set}}{V_g^2} \\ 0 & -\frac{R}{L} & 0 & -\frac{1}{L} \\ 0 & -V_g & 0 & -I_{set} \\ 0 & \frac{1}{C_1} & 0 & -\frac{G_g}{C_1} \end{bmatrix}, \quad B = \begin{bmatrix} 0 \\ \frac{1}{L} \\ 0 \\ 0 \end{bmatrix}.$$

This will readily allow to minimize the cost function

$$J = \int_0^\infty (qe^2 + q_2 z_2^2 + q_3 z_3^2 + q_4 z_4^2 + w^2) dt$$

using the command MATLAB: ≫ `K = lqr(A, B, Q, 1)` where Q is a diagonal matrix with elements $q > 0, q_2 \geq 0, q_3 \geq 0, q_4 \geq 0$.

During the current-limiting mode, the closed loop is characterized by the reduced-order set of equations

$$\begin{aligned}
\dot{x}_1 &= x_2 - I_{max} \\
\dot{x}_2 &= -\frac{R}{L} x_2 - \frac{1}{L} x_4 + \frac{1}{L} v \\
u &= -k_1 x_1 - k_2 x_2 + V_{ff}
\end{aligned} \tag{5.26}$$

whose poles are the eigenvalues of

$$A_1 = \begin{bmatrix} 0 & 1 \\ -\frac{1}{L} k_1 & -\frac{R}{L} - \frac{1}{L} k_2 \end{bmatrix}.$$

Example 5.21 ***Power Control in PV System***

For a simplified case where the two feedback branches associated with k_3 and k_4 are not used, and the droop loop is not present either, the state equations for the system reduce to

$$\begin{aligned}
\dot{x}_1(t) &= e(t) = i(t) - i_{ref} = x_2(t) - i_{ref} \\
\dot{x}_2(t) &= -\frac{R}{L} x_2(t) + \frac{1}{L} v(t) + \frac{1}{L}(V_{ff} - v_g),
\end{aligned}$$

43 The inertia and pre-synchronization loops are not considered in this formulation. The latter only intervenes during short periods prior to grid connection. The former may easily be included by substituting for $\dot{v}_g$ from the fourth equation of (5.24) but we decided to make it simple here. See Problem 5.4.

where $x_2(t) = i(t)$ and V_{ff} is selected as $\frac{1}{\tau s+1}v_g$ (with $\tau = 0.001$ in this example) for soft start and enhanced current limiting. Apply the operator $\frac{d}{dt}$ to both sides of the two above equations to obtain

$$\dot{z}_1(t) = z_2(t)$$
$$\dot{z}_2(t) = -\tfrac{R}{L}z_2(t) + \tfrac{1}{L}w(t),$$

where $z_1(t) = \dot{x}_1(t) = e(t), z_2(t) = \dot{x}_2(t)$ and $w(t) = \dot{u}(t)$. Notice that we assume $\frac{d}{dt}i_{\text{ref}} = 0$. Also notice that

$$w(t) = \dot{u}(t) = -k_1\dot{x}_1(t) - k_2\dot{x}_2(t) = -k_1z_1(t) - k_2z_2(t) = -Kz(t),$$

where $K = [k_1\ k_2]$ and $z(t) = \begin{bmatrix} z_1(t) \\ z_2(t) \end{bmatrix}$. Therefore, the state space representation in terms of z variables is

$$\dot{z}(t) = Az(t) + Bw(t), \text{ where } A = \begin{bmatrix} 0 & 1 \\ 0 & -\frac{R}{L} \end{bmatrix} B = \begin{bmatrix} 0 \\ \frac{1}{L} \end{bmatrix}.$$

It is worthwhile mentioning that all this derivation is not necessary and can simply be written by just setting the rows and columns asociated with z_3 and z_4 to zero in (5.24) and (5.25). Anyway, to minimize the cost function $J = \int_0^\infty [qe^2(t) + q_2z_2^2 + w^2(t)]dt$ for some $q > 0, q_2 \geq 0$, the matrix Q should be chosen as

$$Q = \begin{bmatrix} q & 0 \\ 0 & q_2 \end{bmatrix}$$

and the LQR problem is solved in MATLAB: $\gg$ `K = lqr(A, B, Q, 1)`.

Figure 5.94 shows the locus of closed-loop poles of the current control system when q varies up to 10^8 and q_2 varies up to $10^{1.25}$. At the final location, poles are at $-1085 \pm j907$, and the controller gains are $K = [10\ 000\ \ 10.8]$.[44]

Figure 5.95 shows a sample of responses of this control system. Here, the PV, converter, and grid parameters are the same as Example 5.20. The circuit breaker (CB) closes at $t = 0.07$ seconds and the converter and controller start at $t = 0.09$ seconds. The reference power P_{set} ramps from 0 to 3 kW during $t = 0.1$ seconds to $t = 0.2$ seconds, then jumps to 4.5 kW at $t = 0.3$ seconds, to 5.4 kW at $t = 0.4$ seconds, and finally jumps back to 3 kW at $t = 0.5$ seconds. At $t = 0.6$ seconds, the solar irradiance drops 200 W/m^2. At $t = 0.65, 0.7$, and 0.75 seconds, v_g experiences jumps of +50 V, −300 V and +250 V, respectively.

Figure 5.94 Variation of closed-loop poles of the current control system.

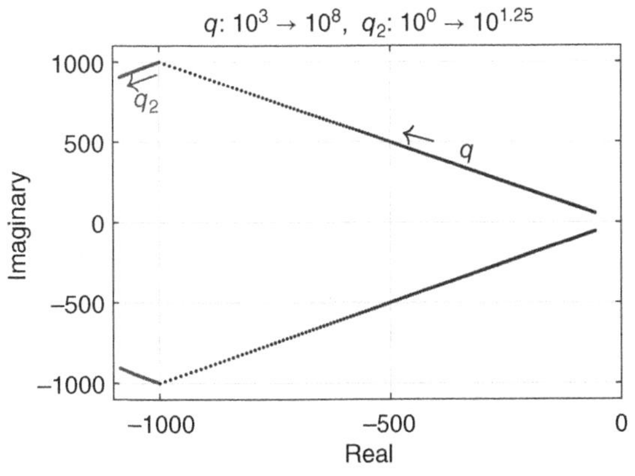

44 We design the current control loop to be quite fast in order to have inherent current limiting in response to sharp grid voltage transient faults.

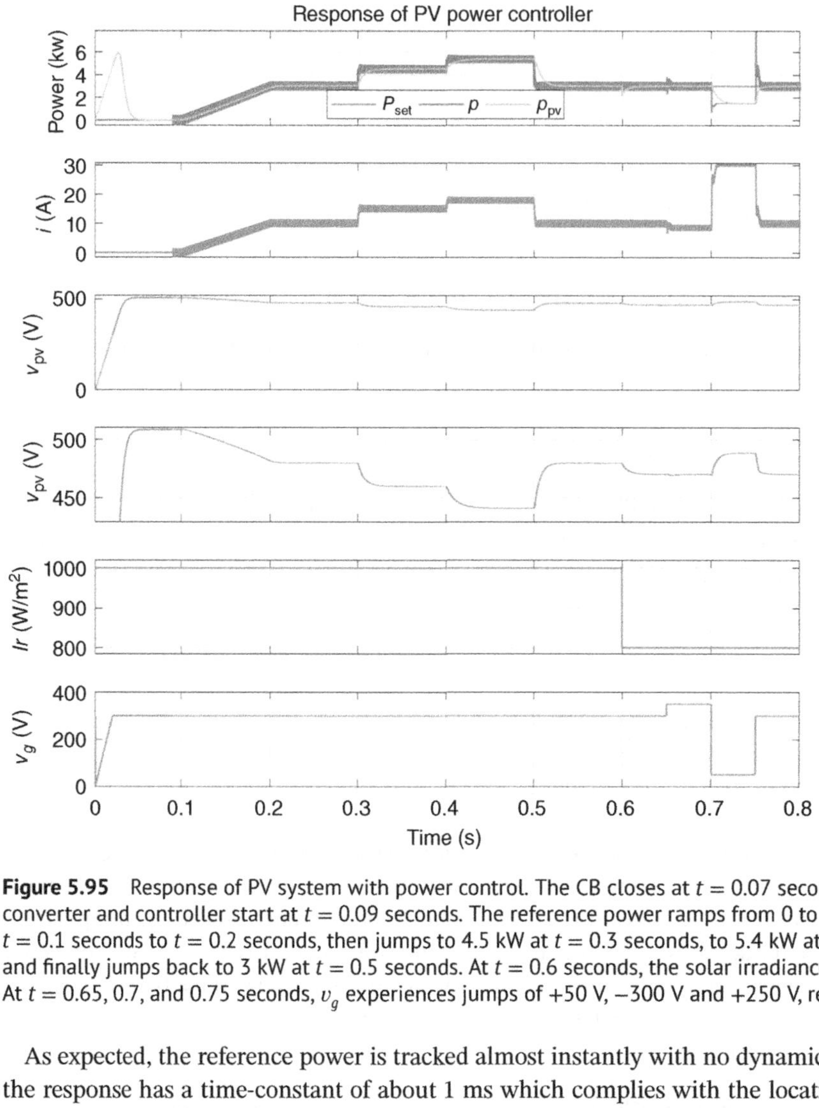

Figure 5.95 Response of PV system with power control. The CB closes at $t = 0.07$ seconds and the converter and controller start at $t = 0.09$ seconds. The reference power ramps from 0 to 3 kW during $t = 0.1$ seconds to $t = 0.2$ seconds, then jumps to 4.5 kW at $t = 0.3$ seconds, to 5.4 kW at $t = 0.4$ seconds, and finally jumps back to 3 kW at $t = 0.5$ seconds. At $t = 0.6$ seconds, the solar irradiance drops 200 W/m^2. At $t = 0.65$, 0.7, and 0.75 seconds, v_g experiences jumps of +50 V, −300 V and +250 V, respectively.

As expected, the reference power is tracked almost instantly with no dynamics. More precisely, the response has a time-constant of about 1 ms which complies with the location of closed-loop poles being around −1000 distance from the imaginary axis. The PV automatically adjusts its voltage in response to the changes in the current. The irradiance change does not cause any change in the converter power. The fast current controller ensures that the current stays limited even when the grid experiences a deep low voltage fault.

In another simulation study, we are going to compare four systems described as follows. *System 1*: the same one just designed and studied above. *System 2*: enable the static grid voltage support term, i.e. the droop term $K_v(V_g - v_g)$, with $K_v = 100$. Thus, for every 10 V of voltage change, the converter will supply 1000 W of supportive power upon its availability and upon the current staying within the limits. *System 3*: enable the dynamic grid voltage support term, i.e. the differentiation term $-K_i\dot{v}_g$, with $K_i = 10$.[45] Thus, for every 100 V/s of voltage change rate, the converter will supply 1000 W of supportive power upon its availability and upon the current staying within the limits. *System 4*: both above static and dynamic grid support terms are enabled. The power limiter is set at 6.5 kW[46]

45 The derivative of v_g is computed by passing it through the transfer function $\frac{s}{0.001s+1}$.

46 The PV panels can only generate close to 6 kW but during transients, as shown in the simulations, the power can go above 6 kW due to the inertia provided by the capacitor.

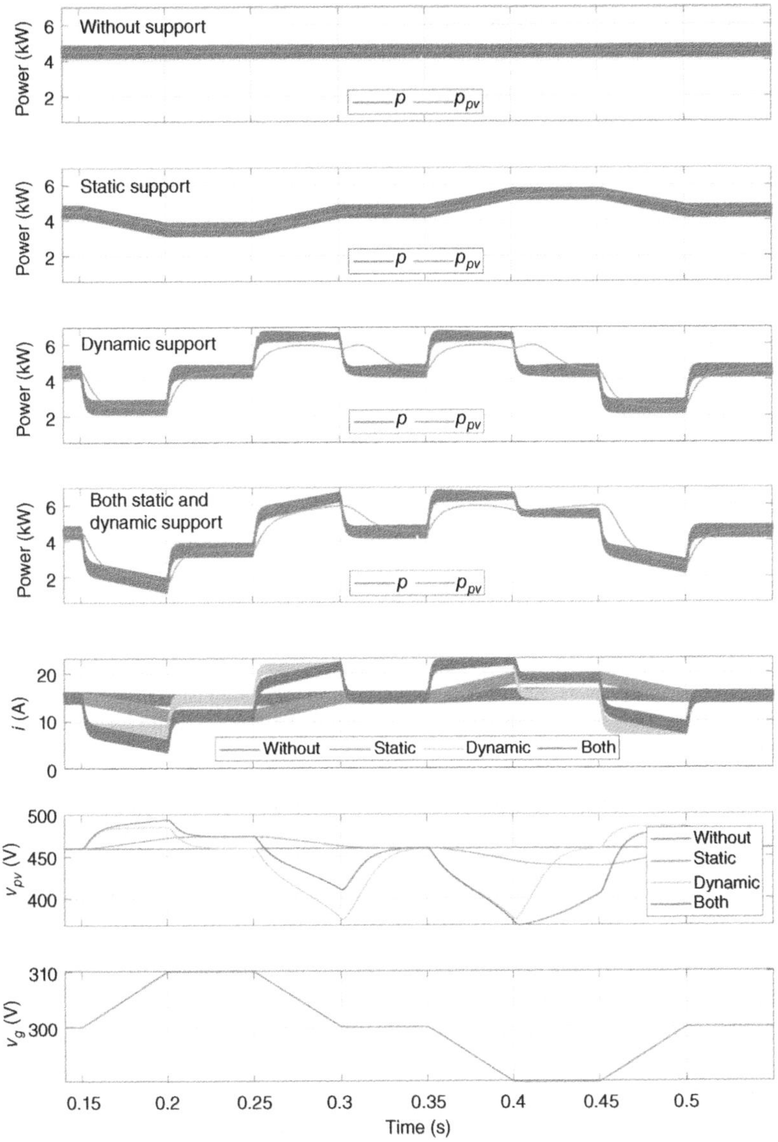

Figure 5.96 Response of PV system with power control. Four systems are considered: no support, static support, dynamic support, and both supports. The CB closes at $t = 0.07$ seconds and the converter and controller start at $t = 0.09$ seconds. The grid voltage v_g follows the shown pattern. The large inertia pushes the PV temporarily into the unstable region but it manages to recover.

and the current limiter at 30 A. The power set-point P_{set} is set at 4500 W. All other controller gains are the same as explained above. The controller is not redesigned. The successful results shown below illustrate the robust performance of the controller.

Figure 5.96 shows a sample of responses of these four control systems. The CB closes at $t = 0.07$ seconds and the converter and controller start at $t = 0.09$ seconds. The grid voltage v_g ramps to 310 V during $0.15 < t < 0.2$ seconds (rate of 200 V/s), then stays at 310 V and then ramps down to 300 V. Later, it ramps down to 290 V, stays at 290 V and again ramps up to 300 V. The system with no support provides a constant power regardless of grid voltage variations. The system with static grid support provides a grid-supporting response by simply copying the grid voltage pattern

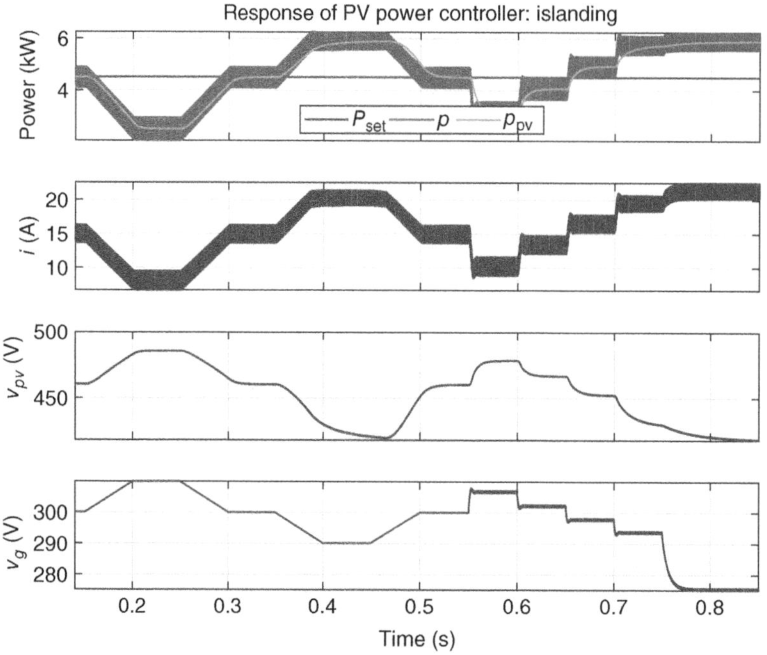

Figure 5.97 Response of PV system with power control. The converter and local load island at $t = 0.55$ seconds. Multiple local loads are switched one consecutively at $t = 0.6$, 0.65, 0.7, and 0.75 seconds. The converter maintains the output voltage.

in an inverse form (and amplified by 100). The system with dynamic grid support responds to the "changes" in the grid voltage, i.e. during the ramp up and ramp down times. The system with noth supports simply combines both effects.

Interestingly, the output power goes above the maximum available PV power (6 kW) around and right before $t = 0.3$ seconds and also $t = 0.4$ seconds. We have intentionally chosen a large inertia factor K_i to illustrate this effect. The PV voltage goes (rather deep into the left side) below the maximum power voltage of 406 V but it manages to return back to the stable region.[47]

The ability of this controller to maintain the output voltage even when the grid becomes unavailable and isolated is shown in Figure 5.97. In this simulation all the controller gains are similar to the previous case except K_v is increased to 200 (for a better output voltage regulation) and $K_i = 0$ to disable the inertia response. The PV array is capable of generating close to 6 kW and the power limiter, sat_p, is set to 5900. At $t = 0.55$ seconds, the grid switch opens and islands the converter from the grid. A local load with resistance of 30 Ω (corresponding to 3 kW at 300 V) is present.[48] Since this load is smaller than the power set-point, $P_{\text{set}} = 4.5$ kW, the voltage slightly goes up to about 306.5 V. At $t = 0.6$ seconds, an additional load with resistance 90 Ω (corresponding to 1 kW at 300 V) is switched on. The output voltage drops to 302.5 V. Another similar load is connected at $t = 0.65$ seconds. Now, the total load goes above the set-point and, as a result, the voltage goes slightly below 300 V, to 298 V. A third similar load is switched on at $t = 0.7$ seconds and the voltage drops to 293.5 V. Now, the PV is operating almost at its maximum available power. Then, another similar load is switched on at $t = 0.75$ seconds. This time, the voltage drops a larger step to 276 V because the PV cannot supply more power.

47 This behavior is much similar to the well-known stability analysis of a synchronous generator (single machine infinite bus -SMIB- analysis) where the rotor can stabilize after a grid fault as long as it satisfies the equal-area criterion [7].

48 Note that we have also added a capacitor of 1 mF in parallel with the local load to absorb the switching ripples.

The above example illustrated that the PV power controller with grid support features is also grid forming in the sense that it can establish the output voltage at (or close to) the given voltage V_g even if the grid is unavailable. The static and dynamic support gains, K_v and K_i, must be properly selected for a desirable steady and dynamic voltage responses, respectively. However, there must be in a mechanism in place to adjust the limit in the power limiter block, sat_p, to the maximum available power of the PV array. This, can be a challenge as this power is constantly changing with the weather conditions and must be properly estimated.

5.9 LCL Filter*

Using a first-order inductive filter is the simplest way of reducing the switching ripples on the converter current. However, higher order filters may also be advantageously used. One common approach is to use an inductive-capacitive-inductive (LCL) filter, as shown in Figure 5.98. A capacitor C is added to absorb the switching ripples of i_1 and make the output current nice and clean. In this way, the inductance L_1 can be made smaller, allowing large ripples in i_1, while the output current i_2 still remains clean even with a small L_2.

An LCL topology allows the total inductance $L_1 + L_2$ to be much smaller than the inductance of an L topology. Apart from the physical size aspect, the LCL topology also leads to reduced core loss (due to lower size of the magnetic cores). Although the reduced inductance will have a smaller parasitic resistance, its copper loss may not be noticeably reduced because its current ripple increases in L_1.

In order to mathematically analyze the effect of the LCL filter, we notice that the transfer functions from the switching voltage v to the the inductors' currents in Figure 5.98 are given by[49]

$$\frac{I_2(s)}{V(s)} = \frac{1}{(L_1 + L_2)s(\frac{s^2}{\omega_r^2} + 1)}, \quad \frac{I_1(s)}{V(s)} = \frac{L_2Cs^2 + 1}{(L_1 + L_2)s(\frac{s^2}{\omega_r^2} + 1)}, \tag{5.27}$$

where $\omega_r = \frac{1}{\sqrt{L_{\text{eq}}C}}$ is called the resonance frequency, and $L_{\text{eq}} = L_1||L_2 = \frac{L_1L_2}{L_1+L_2}$. The corresponding transfer function for the L filter is given for reference as

$$G_L(s) = \frac{I(s)}{V(s)} = \frac{1}{Ls}. \tag{5.28}$$

Comparing the L and LCL filter TFs reveals the following facts. (i) The L filter magnitude frequency response has a first-order characteristics, i.e. −20 dB/dec slope, for all frequencies. (ii) The LCL filter frequency response of i_2 has a first-order characteristics for low frequencies, i.e. those below the resonance frequency, and has a third-order characteristics, i.e. −60 dB/dec, for large frequencies. This means that the switching ripples on i_2 may be strongly attenuated compared to the L filter. (iii) The switching ripples on i_1 are still large, similar to a first-order TF. (iv) The LCL filter has a resonance mode, i.e. the infinite magnitude, at the frequency of ω_r.

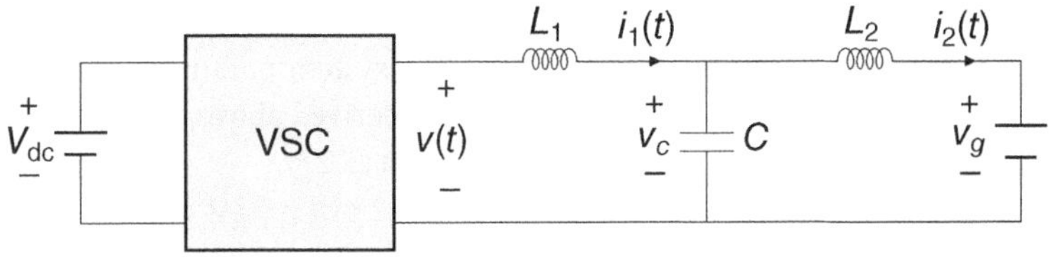

Figure 5.98 VSC with an LCL output filter.

49 See Problem 5.5.

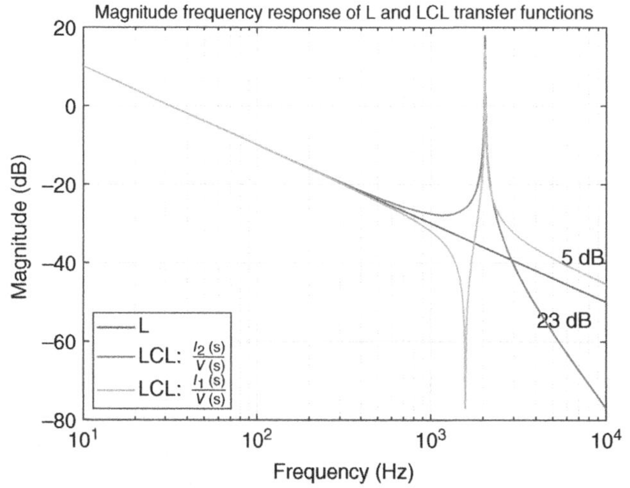

Figure 5.99 Magnitude frequency responses of the L-filter and LCL-filter transfer functions (5.28) and (5.27). ($L = 5$ mH, $L_1 = 3$ mH, $L_2 = 2$ mH, $C = 5$ μF)

Magnitude frequency responses of these transfer functions are depicted in Figure 5.99 for a set of parameters given as $L = 5$ mH, $L_1 = 3$ mH, $L_2 = 2$ mH, $C = 5$ μF. The resonance frequency is around 2 kHz. It is observed that while $L_1 + L_2$ is equal to L, the LCL can offer 23 dB more attenuation at the frequency of 8 kHz as far as the output current i_2 is concerned. While this is a great advantage, it should be noticed that the current flowing in inductor L_1 has higher level of ripples. For this set of numbers, i_1 has about 5 dB lower level of attenuation compared to the L filter current i.

The conclusion is that the LCL filter can offer a much higher attenuation of the switching noises at the output current i_2 if the switching frequency is sufficiently larger than the resonance frequency. The three circuit elements L_1, L_2, and C must be designed properly to satisfy this condition and also to maintain the ripples of i_1 within an acceptable range.

The resonance mode of the LCL filter poses a challenge and can easily lead to ringings in the responses and instabilities. There are two ways of dealing with the resonance phenomenon called (i) *the passive damping*: adding some resistance to the circuit, and (ii) *the active damping*: damping the resonance using feedback. The common passive damping approach is to insert a resistor in series with the capacitor. The switching ripples that are absorbed by the capacitor will flow into the resistance and cause additional ohmic losses. The active damping has the advantage of not causing such additional power losses. However, it requires measurement of required signals and design of a fast and strong feedback controller. These two approaches are discussed in this section.

The differential equations of the LCL filter in Figure 5.98 are[50]

$$L_1 \frac{d}{dt} i_1(t) = v(t) - v_c(t),\; C\frac{d}{dt} v_c(t) = i_1(t) - i_2(t),\; L_2 \frac{d}{dt} i_2(t) = v_c(t) - v_g \tag{5.29}$$

which can be shown in a block diagram form as shown in Figure 5.100. The system is of order three, i.e. it has three state variables, i_1, v_c, and i_2. The control input is the modulation index m. The variable v_g is treated as a disturbance; and the dc voltage V_{dc} as a system parameter in this representation. This diagram, or alternatively the state-space equations derived above, can readily be used to obtain the expressions for different TFs, such as those given in (5.27).

50 Note that the parasitic resistances of circuit components are not considered in this set of equations. Such parasitic terms do not have any impact on the optimal control designs discussed in this text.

5.9.1 Passive Damping of Resonance Mode

The common and most effective approach for passive damping of resonnace mode in an LCL filter is to add a resistor in series with the capacitor, as shown in Figure5.101.

The equivalent control diagram including the effect of additional resistor R_d is shown in Figure 5.102. Using this diagram, the transfer function to the output current can readily be written as

$$\frac{I_2(s)}{V(s)} = \frac{\frac{1}{L_1 s}\left(\frac{1}{Cs}+R_d\right)\frac{1}{L_2 s}}{1+\frac{1}{L_1 s}\left(\frac{1}{Cs}+R_d\right)+\frac{1}{L_2 s}\left(\frac{1}{Cs}+R_d\right)} = \frac{R_d Cs+1}{(L_1+L_2)s\left(\frac{s^2}{\omega_r^2}+R_d Cs+1\right)}. \tag{5.30}$$

This means that R_d directly introduces a damping to the resonance mode. The damped resonance poles are now at

$$-\beta \pm j\sqrt{\omega_r^2-\beta^2}, \quad \beta = 0.5R_d C\omega_r^2. \tag{5.31}$$

The relationship between the introduced damping ratio and the resistance R_d is

$$\zeta = 0.5R_d C\omega_r = 0.5\sqrt{\frac{C}{L_{\text{eq}}}}R_d \Rightarrow \boxed{R_d = 2\zeta\sqrt{\frac{L_{\text{eq}}}{C}}}. \tag{5.32}$$

For the above numerical values ($L_1 = 3$ mH, $L_2 = 2$ mH, $C = 5\ \mu$F), equation (5.32) reduces to $R_d = 31\zeta$. To have a damping ratio of 0.1 or 0.2, the required R_d is about 3 Ω and 6 Ω, respectively. The larger values of R_d yield to higher damping but the ohmic power loss of

$$p_d = R_d i_c^2 = R_d(i_1 - i_2)^2 \approx R_d i_{\text{ripple}}^2 \tag{5.33}$$

can compromise the efficiency of the converter as will be shown in the forthcoming studies. In (5.33), i_{ripple} is the switching ripple component of current i_1. Moreover, as observed from the magnitude frequency responses shown in Figure 5.103, larger values of R_d will also reduce the level of switching noise attenuation. Figure 5.103 shows that the attenuation level has decreased about

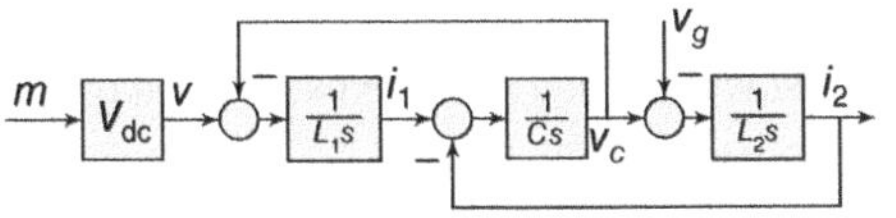

Figure 5.100 Control block diagram of FB-VSC with LCL filter.

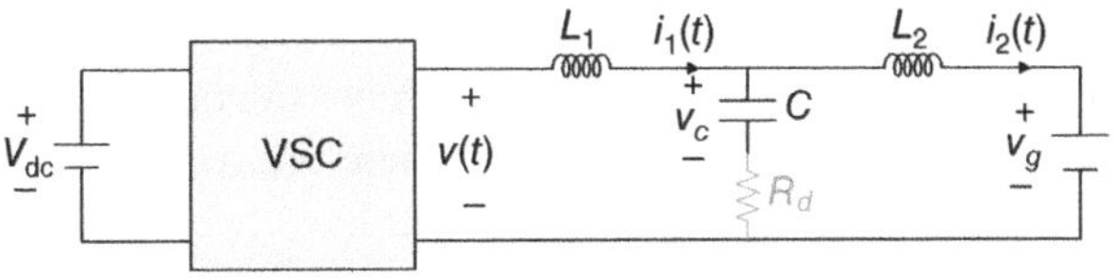

Figure 5.101 Passive damping of resonance mode using R_d.

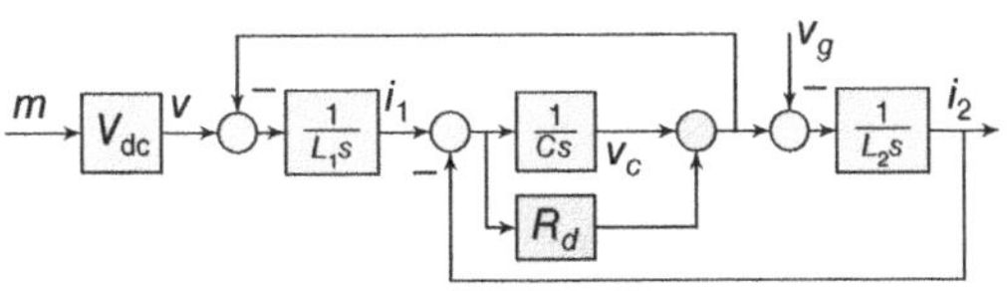

Figure 5.102 Control block diagram of VSC with LCL filter including passive damping.

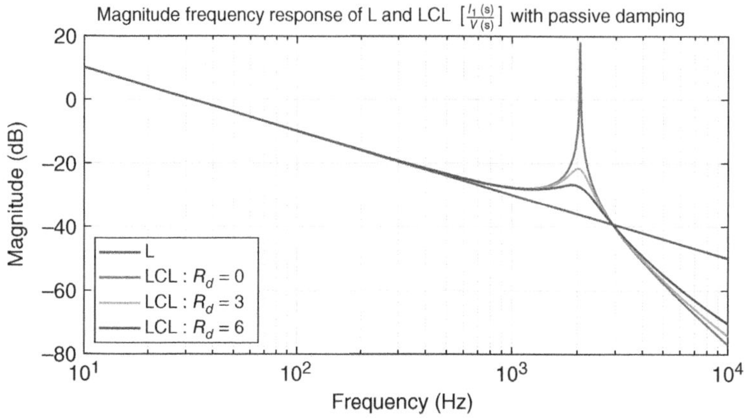

Figure 5.103 Magnitude frequency responses of the L-filter and LCL-filter without and with the passive damping. ($L = 5$ mH, $L_1 = 3$ mH, $L_2 = 2$ mH, $C = 5$ μF).

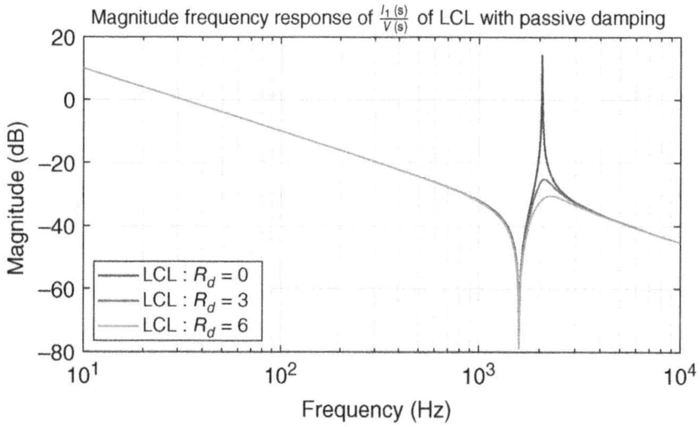

Figure 5.104 Magnitude frequency response of $\frac{I_1(s)}{V(s)}$ in an LCL filter without and with the passive damping. ($L = 5$ mH, $L_1 = 3$ mH, $L_2 = 2$ mH, $C = 5$ μF)

2 dB for $R_d = 3\ \Omega$ and about 5 dB for $R_d = 6\ \Omega$. Despite this, the attenuation level of the damped LCL filter is still remains much higher than that of the L filter.

If an effective passive damping approach is adopted in an LCL filter, its transfer function $\frac{I_1(s)}{V(s)}$ is much similar to a pure L filter with total inductance of $L_1 + L_2$. Figure 5.104 shows the magnitude frequency response of this transfer function for the above numerical values. Therefore, a controller similar to what was proposed for the L filter may be used. However, we will have two options, as whether to regard i_1 as the output or i_2 as the output. These two perspectives are shown in Figure 5.105. The actual output of the converter that interacts with the grid is i_2. However, we care about limiting i_1 because this is the current flowing into the switches. Thus, choosing i_1 as the output has both advantages, i.e. reducing the design to a simple L filter and allowing direct converter current limiting.

Example 5.22 ***Passive Damping of Resonance Mode in LCL Filter***

In Example 5.3, an optimal controller was designed for the L filter converter. The system parameters $V_{dc} = 400$ V, $V_g = 300$ V, $f_{sw} = 4$ kHz, $L = 5$ mH, $R = 5$ mΩ. The LQT controller placed the poles

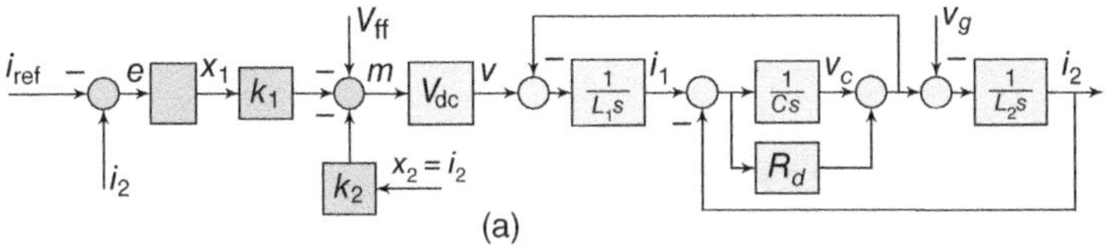

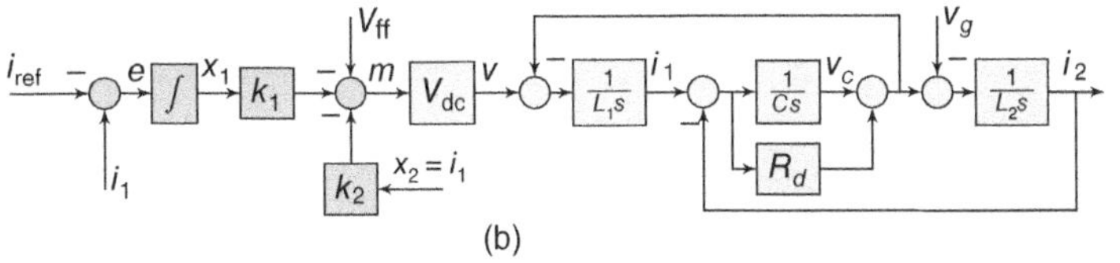

Figure 5.105 Full control diagram of VSC with LCL filter including passive damping: (a) closing the feedback loop on i_2, (b) closing the feedback loop on i_1.

of the closed loop system around $-1407 \pm j742$ with the gains $K = [31.6 \;\; 0.035]$. In this example, the L filter is replaced with an LCL filter with the values $L_1 = 3$ mH, $C = 5$ μF, $L_2 = 2$ mH. The resonance frequency is around 2 kHz. If a unipolar PWM is used, the effective switching frequency will be 8 kHz. Figure 5.99 confirms that at this frequency, the switching ripples of i_2 are over 20 dB (10 times) smaller compared to those of the L filter. The ripples in i_1, however, are about 5 dB (1.8 times) larger compared to those of the L filter.

(a) According to (5.32), a resistance of $R_d = 6\ \Omega$ in series with the capacitor, will introduce a damping of about $\zeta = 0.2$ into the resonance poles. Let us pick this value.

(b) The resonance mode is now passively (and sufficiently)[51] damped. Let us treat the LCL filter as an L filter and use the controller of Figure 5.105b. Since $L_1 + L_2 = L = 5$ mH, to have the same location of closed-loop poles, the controller gains are the same as $K = [31.6 \;\; 0.035]$.

(c) The time responses of the LCL filter and L filter converters are shown in Figure 5.106. The converters start at zero current. At $t = 0.01$ seconds, a step command of 10 A is applied and followed by a step increase to 20 A at $t = 0.02$ seconds. At $t = 0.03$ seconds, the voltage V_{dc} is dropped to 380 V. At $t = 0.04$ seconds, the voltage v_g is stepped down to 280 V. The LCL filter currents i_2 and i_1 are shown on the top two panels. On the next two panels, the currents of two L filter and the power loss in resistor R_d are shown. Following observations are made.

(1) The LCL filter's output current i_2 is much cleaner than the L filter's current.
(2) The current i_1, however, has more ripples than the associated L filter current.
(3) The resonance mode is well damped and is only slightly excited when the jump in v_g occurs. For the sake of reference, Figure 5.107 shows the responses of the system when R_d is reduced to 0.6 Ω. The resonance ringings are clearly visible.
(4) The ripple current in i_1 is almost entirely absorbed by the capacitor and also flows into the damping resistor R_d causing an ohmic power loss of about 6 W which is about 0.1% of the full converter power of 6 kW. However, when the converter operates at low power, e.g. at 0.6 kW, this will amount to about 1% which is noticeable. Therefore, although the LCL offers a clean output current (with a low percentage ripple even at low currents), the relative power loss is higher at low powers. Therefore, the advantage of this method is more pronounced at high power.

51 A damping ratio of 0.2 for a high-frequency resonance mode is normally sufficient.

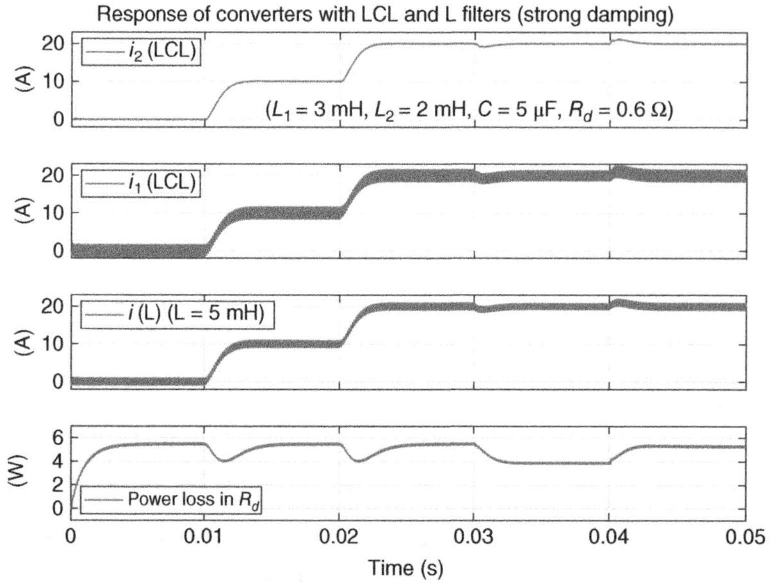

Figure 5.106 Responses of the LCL and L filter converters at strong passive damping of resonance mode. At $t = 0.01$ seconds, a step command of 10 A is applied and followed by a step increase to 20 A at $t = 0.02$ seconds. At $t = 0.03$ seconds, the voltage V_{dc} is dropped to 380 V. At $t = 0.04$ seconds, the voltage v_g is stepped down to 280 V. From top to bottom: output current (i_2) of LCL, i_1 of LCL, i of L, and the power loss in R_d.

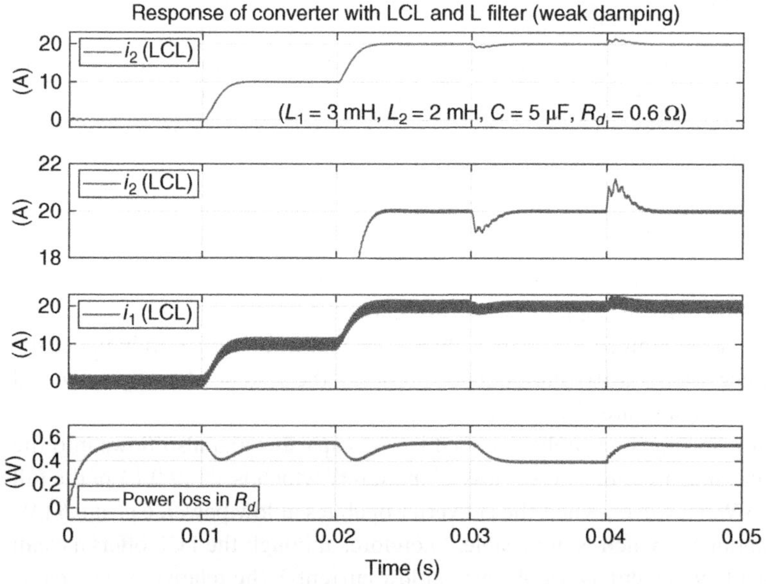

Figure 5.107 Responses of the LCL and L filter converters at weak passive damping of resonance mode. From top to bottom: output current (i_2) of LCL, its zoomed version, i_1 of LCL, and the power loss in R_d.

The LCL filter approach allows reduction of inductor sizes. This, however, will cause larger fluctuations of the converter current in response to disturbances such as grid voltage faults. Therefore, its current controller needs to be sufficiently faster to be able to limit the current when grid voltage (deep or short-circuit) faults occur. A filtered version of the grid voltage (or capacitor voltage) may be used to enhance the current limiting, as discussed in details for L filter in Section 5.3.5. Therefore, to achieve current limiting in LCL with smaller inductances, faster controller is required. This may require to increase the switching frequency to allow the fast controller to function properly.

5.9.2 Full-State Feedback with Active Damping

Section 5.9.1 demonstrated that when a passive damping is used, the LCL filter can be treated as an L filter and only a feedback from i_1 is sufficient to perform the current control. When the losses associated with the passive damping are of concern, an active damping approach must be used. To this purpose, this section presents the structure and design of a full-state feedback controller for the VSC with LCL filter. The closed-loop feedback structure is shown in Figure 5.108 where it comprises an output integrating loop, to remove the steady offset errors from the output current, incorporated in a full state feedback. The state variable of the integrator is denoted by x_1 and the state variables of the LCL filter converter by $x_2 = i_1, x_3 = v_c$, and $x_4 = i_2$.

Notice that we have the option to closed the outer loop on i_1 and i_2, as also discussed in the section on passive damping. Closing on i_1 has the advantage of having direct control on it and being able to limit it directly. The feed-forward term $V_{\rm ff}$ is used to achieve soft starting of the converter and also to enhance current limiting. Its value must be such that the converter supplies a voltage v equal to v_c at the start. So, if v_c is charged to $vc(0)$ at the start, then $(-k_3 v_c(0) + V_{\rm ff})V_{\rm dc} = v_c(0)$ or $V_{\rm ff} = \frac{v_c(0)}{V_{\rm dc}} + k_3 v_c(0)$. If the first term is replaced with a filtered-version of v_g, i.e. v_g passed through $\frac{1/V_{\rm dc}}{\tau s+1}$, it can help with current limiting during grid fault conditions.

The entire state space equations of the system are summarized as

$$\begin{aligned}
\dot{x}_1(t) &= e(t) = x_2(t) - i_{\rm ref} \\
\dot{x}_2(t) &= -\frac{1}{L_1}x_3(t) + \frac{V_{\rm dc}}{L_1}u(t) + \frac{V_{\rm dc}V_{\rm ff}}{L_1} \\
\dot{x}_3(t) &= \frac{1}{C}x_2(t) - \frac{1}{C}x_4(t) \\
\dot{x}_4(t) &= \frac{1}{L_2}x_3(t) - \frac{v_g}{L_2}
\end{aligned} \tag{5.34}$$

and the control law is

$$u(t) = -k_1x_1(t) - k_2x_2(t) - k_3x_3(t) - k_4x_4(t) = -[k_1\ k_2\ k_3\ k_4]\begin{bmatrix} x_1(t) \\ x_2(t) \\ x_3(t) \\ x_4(t) \end{bmatrix} = -Kx(t).$$

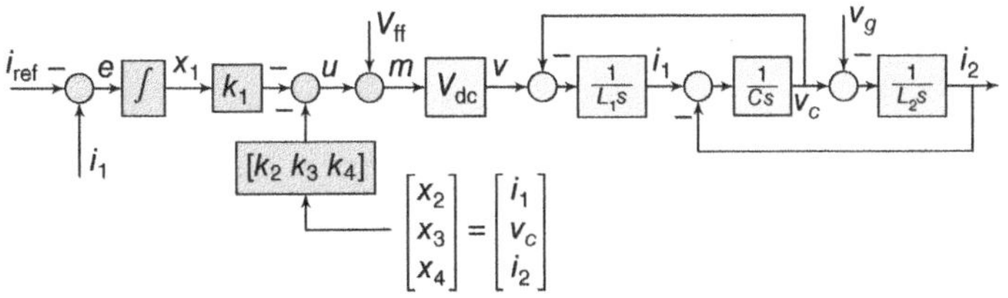

Figure 5.108 Full-state feedback control with active damping for LCL filter.

Applying the linear operator $\frac{d}{dt}$ to the both sides of these two equations lead to

$$\begin{aligned}
\dot{z}_1(t) &= z_2(t) \\
\dot{z}_2(t) &= -\frac{1}{L_1}z_3(t) + \frac{V_{dc}}{L_1}w(t) \\
\dot{z}_3(t) &= \frac{1}{C}z_2(t) - \frac{1}{C}z_4(t) \\
\dot{z}_4(t) &= \frac{1}{L_2}z_3(t) \\
w(t) &= -Kz(t)
\end{aligned} \tag{5.35}$$

where $z_i = \dot{x}_i$, $i = 1, 2, 3, 4$ and $w = \dot{u}$. Particulrely, it is noticed that $z_1 = \dot{x}_1 = e(t)$ is the tracking error. Using the LQR approach, defining the cost function

$$J = \int_0^{\infty} \left(qe^2 + q_2 z_2^2 + q_3 z_3^2 + q_4 z_4^2 + w^2\right) dt$$

with the matrices

$$A = \begin{bmatrix} 0 & 1 & 0 & 0 \\ 0 & 0 & -\frac{1}{L_1} & 0 \\ 0 & \frac{1}{C} & 0 & -\frac{1}{C} \\ 0 & 0 & \frac{1}{L_2} & 0 \end{bmatrix}, \quad B = \begin{bmatrix} 0 \\ \frac{V_{dc}}{L_1} \\ 0 \\ 0 \end{bmatrix}, \quad A = \begin{bmatrix} q & 0 & 0 & 0 \\ 0 & q_2 & 0 & 0 \\ 0 & 0 & q_3 & 0 \\ 0 & 0 & 0 & q_4 \end{bmatrix}$$

and running in MATLAB: ≫ `K = lqr(A, B, Q, 1)` gives the controller gain vector K.

Example 5.23 ***Active Damping of Resonance Mode in LCL Filter***
Figure 5.109 shows the evolution of closed-loop poles of the system when the elements of matrix Q vary. There are two pairs of complex poles where only one of each is shown in Figure 5.109 for more clarity. As expected, it is observed that q is the major element that impacts the low-frequency poles, and q_3 is the major element that introduces damping into the resonance poles. The two elements q_2 and q_4 need not be used unless more damping of the low-frequency poles are desired. Here, they are shown only to illustrate how they vary the poles. The final location of poles shown in Figure 5.109 are $-1412 \pm j540$ (low-frequency mode) and $-3245 \pm j13\ 187$ (resonance mode). The damping of resonance poles is just over $\zeta = 0.2$. The controller gain vector is $K = [31.6\ \ 0.070\ \ 0.0013\ \ -0.030]$.

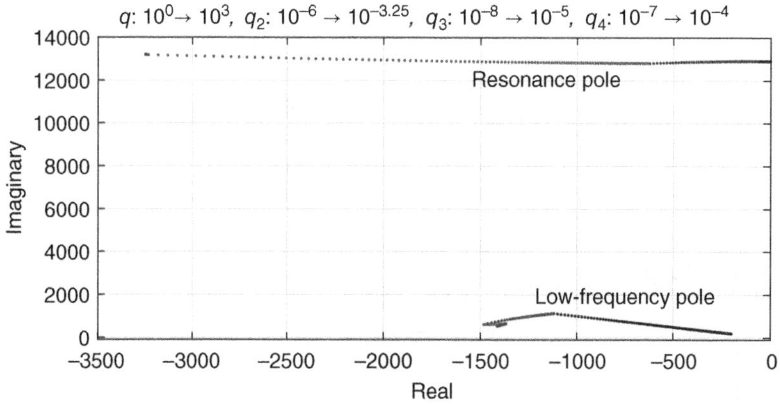

Figure 5.109 Closed loop poles of the converter with LCL filter using active damping. ($L_1 = 3$ mH, $L_2 = 2$ mH, $C = 5\ \mu$F)

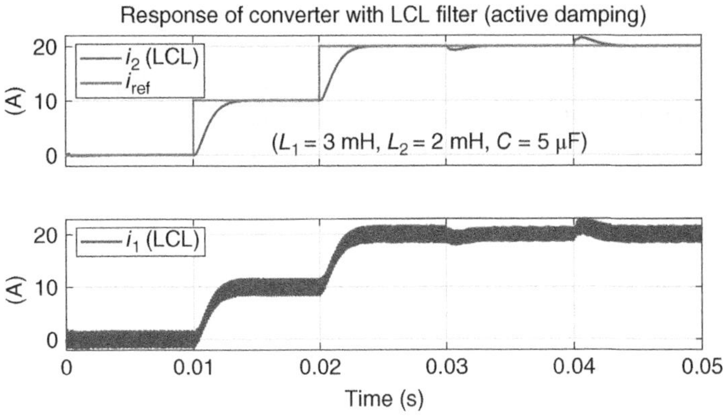

Figure 5.110 Response of the converter with LCL filter using active damping.

Figure 5.110 shows the time responses of the converter with LCL filter and the above-designed controller with active damping. The converter starts at zero current then two step commands of 10 A are applied at $t = 0.01$ and 0.02 seconds. At $t = 0.03$ seconds, V_{dc} steps down from 400 V to 380 V. At $t = 0.04$ seconds, V_s steps up from 300 to 320 V. (Same scenario used in Example 5.22.) The responses show successful performance of the controller to actively damping the resonance mode and to executing the commands.

The active damping controller has a side effect: reduced delay margin (DM) of the control loop. In practice, there are delays in control loops, such as the one-sample delay in digital controllers that use a DSP, and the delay effect of PWM. Such delays can quickly degrade and even completely ruin the performance of the active damping controller. Figure 5.111 shows the performance of the controller when 0.7 times the sampling-time delay is manually introduced in the control loop.[52]

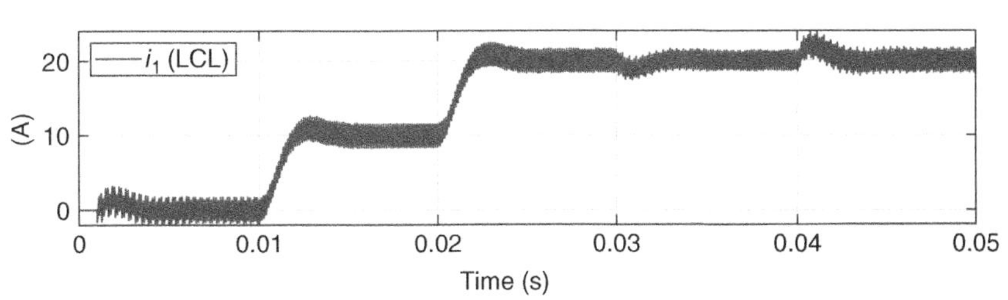

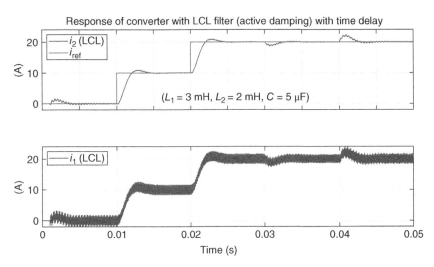

Figure 5.111 Response of the converter with LCL filter using active damping when a time-delay of 0.7 the sampling cycle, i.e. 0.0875 ms, is introduced in the control loop.

52 The controller itself is implemented in continuous-time domain. Thus, this delay can partly model the effect of digital implementation of the controller. In our simulations, we use switches and apply PWM. Therefore, the effect of PWM is presented.

As observed, this delay has significantly caused the ringings in the response. The controller cannot damp the resonance due to this delay. If the delay is increased above this value, the system quickly becomes unstable. A simple calculation shows that the delay margin of the above-designed LCL filter controller is about 0.24 ms. The DM is about 0.44 when the resonance is passively damped (See problem 5.6). In Section 5.9.3, a systematic method to address this issue is studied.

5.9.3 Delay Compensation Technique Using LQT Approach

The objective of this section is to design a full-state optimal controller that includes the effect of system delays, e.g. the delay caused by digital implementation, into consideration. The method is general but we explain it in the context of converter with LCL filter.

A time delay operation of T seconds is mathematically represented by $G_d(s) = e^{-Ts}$. This is an infinite-dimensional rational transfer function but can be best approximated at different levels according to the Pade approximation. The first- and second-order Pade approximations of the delay function are

$$-\frac{s-\frac{2}{T}}{s+\frac{2}{T}}, \quad \frac{s^2-\frac{6}{T}s+\frac{12}{T^2}}{s^2+\frac{6}{T}s+\frac{12}{T^2}}. \tag{5.36}$$

The MATLAB command: ≫ [p, q] = pade(T, n) returns the numerator and denominator coefficients for the nth-order Pade approximation of the delay of e^{-Ts}. We use this approach to convert the delay TF to a rational TF in the analysis in this section.

The proposed full-state feedback control loop is shown in Figure 5.112, the block denoted by D represents the actual system delays including digital control and other delays. A new block is added in the controller denoted by $\tilde{D}$ which is the Pade approximation of the delay function. The amount of delay used in this function does not have to be equal to the conceived delays of the system. By incorporating this block in its structure, the controller includes the state variables of the delay function in its formulation.

The entire state space equations of the system are summarized as

$$\begin{aligned}
\dot{x}_1(t) &= e(t) = x_2(t) - i_{\text{ref}} \\
\dot{x}_d(t) &= A_d x_d(t) + B_d u(t) + B_d V_{\text{ff}} \\
\dot{x}_2(t) &= -\tfrac{1}{L_1}x_3(t) + \tfrac{V_{\text{dc}}}{L_1}C_d x_d(t) + \tfrac{V_{\text{dc}}}{L_1}D_d u(t) + \tfrac{V_{\text{dc}}}{L_1}D_d V_{\text{ff}} \\
\dot{x}_3(t) &= \tfrac{1}{C}x_2(t) - \tfrac{1}{C}x_4(t) \\
\dot{x}_4(t) &= \tfrac{1}{L_2}x_3(t) - \tfrac{v_g}{L_2}
\end{aligned} \tag{5.37}$$

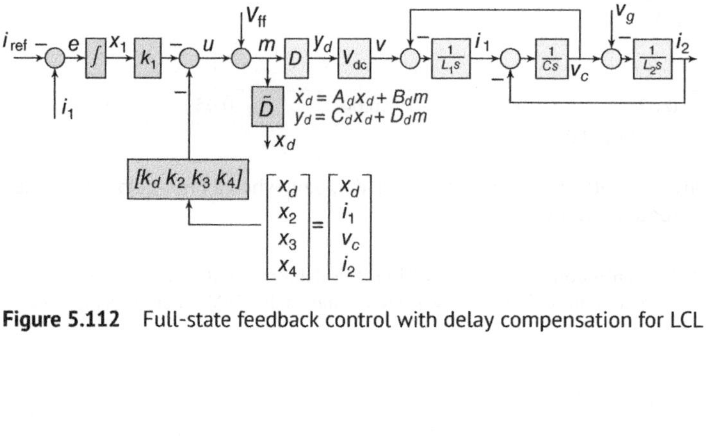

Figure 5.112 Full-state feedback control with delay compensation for LCL filter.

where A_d and B_d are state space matrices of the Pade approximation of the delay function.[53] The control law is

$$u(t) = -k_1x_1(t) - k_dx_d(t) - k_2x_2(t) - k_3x_3(t) - k_4x_4(t) = -Kx(t),$$

where $K = -[k_1\ k_d\ k_2\ k_3\ k_4]$ and $x(t) = \left[x_1(t)\ x_d^T(t)\ x_2(t)\ x_3(t)\ x_4(t)\right]^T$. Applying the linear operator $\frac{d}{dt}$ to the both sides of these equations lead to

$$\begin{aligned}
\dot{z}_1(t) &= z_2(t) \\
\dot{z}_d(t) &= A_dz_d(t) + B_dw(t) \\
\dot{z}_2(t) &= -\tfrac{1}{L_1}z_3(t) + \tfrac{V_{dc}}{L_1}C_dz_d(t) + \tfrac{V_{dc}}{L_1}D_dw(t) \\
\dot{z}_3(t) &= \tfrac{1}{C}z_2(t) - \tfrac{1}{C}z_4(t) \\
\dot{z}_4(t) &= \tfrac{1}{L_2}z_3(t) \\
w(t) &= -Kz(t),
\end{aligned} \tag{5.38}$$

where $z_i = \dot{x}_i$, $i = 1, d, 2, 3, 4$ and $w = \dot{u}$. We particularly note that $z_1 = \dot{x}_1 = e$ is the tracking error.

Using the LQR approach, defining the cost function

$$J = \int_0^\infty \left(qe^2 + q_dz_d^2 + q_2z_2^2 + q_3z_3^2 + q_4z_4^2 + w^2\right) dt$$

with the matrices

$$A = \begin{bmatrix} 0 & 0_{1\times r} & 1 & 0 & 0 \\ 0_{r\times 1} & A_d & 0_{r\times 1} & 0_{r\times 1} & 0_{r\times 1} \\ 0 & \frac{V_{dc}}{L_1}C_d & 0 & -\frac{1}{L_1} & 0 \\ 0 & 0_{1\times r} & \frac{1}{C} & 0 & -\frac{1}{C} \\ 0 & 0_{1\times r} & 0 & \frac{1}{L_2} & 0 \end{bmatrix},\quad B = \begin{bmatrix} 0 \\ B_d \\ \frac{V_{dc}}{L_1}D_d \\ 0 \\ 0 \end{bmatrix},\quad Q = \begin{bmatrix} q & 0 & 0 & 0 & 0 \\ 0 & q_d & 0 & 0 & 0 \\ 0 & 0 & q_2 & 0 & 0 \\ 0 & 0 & 0 & q_3 & 0 \\ 0 & 0 & 0 & 0 & q_4 \end{bmatrix}$$

and running in MATLAB: `>> K = lqr(A, B, Q, 1)` leads to the controller gains K.

Example 5.24 ***Active Damping of Resonance with Delay Compensation***
Figure 5.113 shows the time responses of the converter with LCL filter and active damping when a time-delay of half the switching cycle (or one sampling cycle)[54], i.e. $T_d = \frac{1}{8000} = 125\ \mu s$, is present in the control loop and is compensated by the controller using the above approach. For this design, the Pade approximation of order one of the delay function is used. The elements of Q matrix in the LQT design are varied the same as those in LCL filter without delay, i.e. those mentioned in Example 5.23. The controller gains are $K = [31.6\ \ 16\,850\ \ 0.057\ \ -0.0027\ \ -0.0124]$. The location of closed-loop poles are also almost the same as those shown in Example 5.23 except that there is

---

53 For a delay of e^{-T_ds} to be Pade-approximated of order r, simply type in Matlab:
`>> [Pd, Qd] = pade(Td, r); [Ad, Bd, Cd, Dd] = tf2ss(Pd, Qd);`

54 In digital implementation of a single-phase full-bridge VSC using unipolar PWM, the sampling frequency is selected at twice the switching frequency.

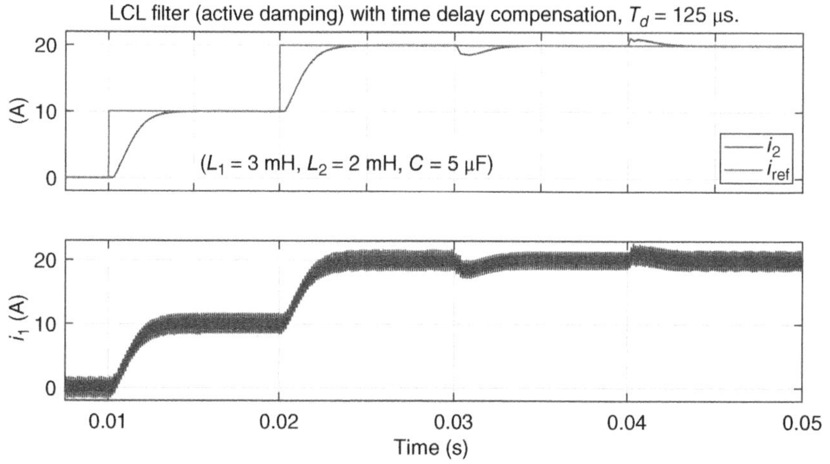

Figure 5.113 Responses of the converter with LCL filter using active damping when a time-delay of one sampling cycle, i.e. 0.125 ms, is introduced in the control loop, and the delay is compensated in the controller.

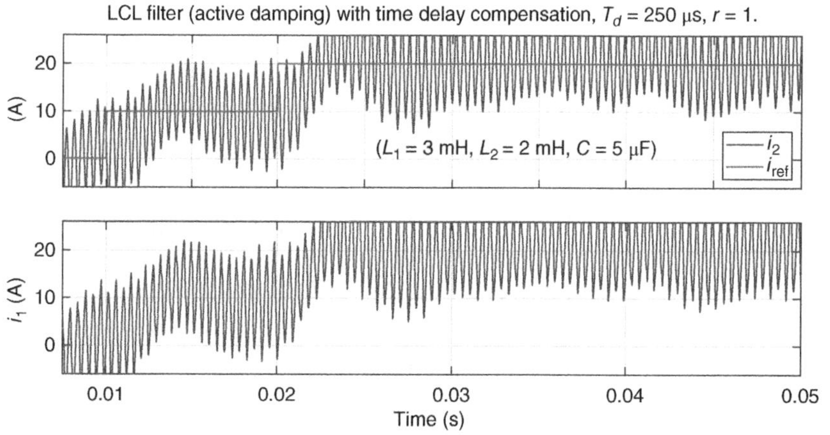

Figure 5.114 Responses of the converter with LCL filter using active damping when a *time-delay of two sampling cycle*, i.e. 0.250 ms, is introduced in the control loop, and the delay is compensated in the controller. The Pade approximation of *order 1* is used.

an additional pole at $s = -\frac{2}{T_d} = -16\ 000$. The responses in Figure 5.113 confirm that the resonance mode is successfully damped despite the presence of delay.

Figure 5.114 shows the time responses of the converter with LCL filter and active damping when a time-delay of *two* sampling cycles, i.e. $T_d = \frac{1}{4000} = 250\ \mu s$, is present in the control loop and is compensated by the controller using the above approach. For this design, still the Pade approximation of order *one* of the delay function is used. The control loop cannot maintain its stability. Figure 5.115 shows the responses when the Pade approximation of order 2 is used. The controller stabilized the loop. The controller gains are $K = [31.6\ \ 1.1 \times 10^4\ \ 2.66 \times 10^8\ \ -0.013\ \ -1.9 \times 10^{-3}\ \ 0.061]$ where the second and third elements constitute k_d (for the delay unit).

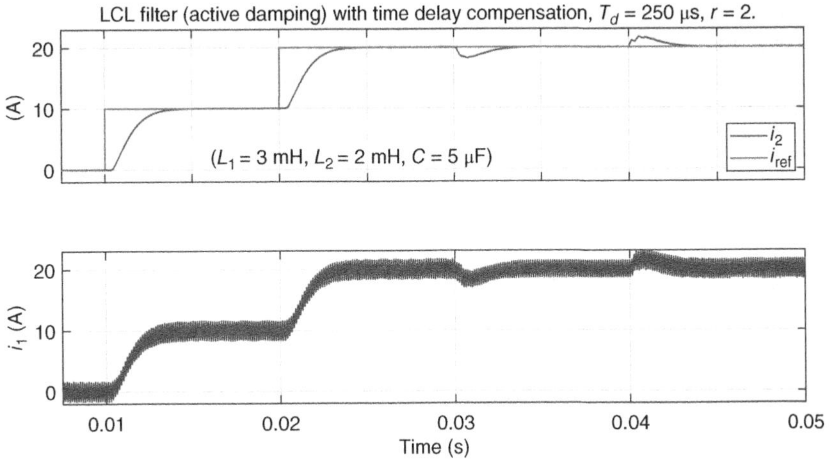

Figure 5.115 Responses of the converter with LCL filter using active damping when a time-delay of two sampling cycle, i.e. 0.250 ms, is introduced in the control loop, and the delay is compensated in the controller. The Pade approximation of *order 2*is used.

5.10 Summary and Conclusion

This chapter reviewed various control aspects pertaining to the integration of a dc DER. Their principles are also largely applicable to three-phase converters controlled in synchronous reference frame (or *dq* frame) as well. Solar PV application was given particular attention and issues such as MPPT of PV generators was thoroughly studied. The techniques discussed here are applicable to wide variety of applications where a dc converter (or a three-phase converter controlled in *dq* frame) is adopted including but not limited to solar PV and battery charging applications. Special attention is paid to efficient modeling of the systems for control designs.

We have also paid attention to the ability of the controllers to limit their currents quickly when a severe disturbance (such as a grid side short circuit) occurs. Furthermore, aspects of grid voltage support including dynamic support (or inertia response) and static support (or droop function) are studied. Grid-forming versus grid-following controllers are discussed and their interrelations are revealed. And finally, some challenging issues with using higher order filters (such as the LCL filter) are explained and addressed. Many of these aspects can be properly adjusted and applied to ac DERs which are discussed in Chapters 6 and 7 of this book.

Problems

5.1 Delay Margin Analysis of Basic PI and LQT Designs

(a) Consider the four LQT designs used in Example 5.5. Using the MATLAB command *margin*, find the phase margins and gain crossover frequency of each design. Compute the delay margin for each design by dividing the phase margin (in radians) to the gain crossover frequency (in rad/s) and show that they are respectively around 1.38, 0.78, 0.43, and 0.24 ms.

(b) Consider the PI design used in Example 5.1. Similar to above, show that its delay margin is about 0.32 ms. Thus, the corresponding LQT design (Design 2 with delay margin of 0.78 ms) has a delay margin over two times larger than the PI design.

5.2 Consider the capacitor power equation $p_c(t) = Cv_c(t)\dot{v}_c(t)$. This may be considered as a non-linear function of x and y ac $p_c = Cxy$ where $x = v_c$ and $y = \dot{v}_c$. Using the Taylor's series expansion of this function around the point $(x_o, y_o) = (V_c, 0)$ prove that its linearization will be $p_c = CV_c y = CV_c\dot{v}_c(t)$.

5.3 Derive (5.15) for the system of Figure 5.62.

5.4 Derive a more complete version of (5.24) by including the inertia loop in the formulation. Hint: use $K_i\dot{v}_g = K_i\dot{x}_4 = K_i(\frac{1}{C_1}x_2 - \frac{G_g}{C_1}x_4 - \frac{1}{C_1}i_s)$.

5.5 For the converter with LCL filter, shown in Figure 5.98,
(a) derive (5.27) using direct manipulation of transfer functions or using the Mason's rule;
(b) derive similar equations for $\frac{I_2(s)}{V_g(s)}$ and $\frac{I_1(s)}{V_g(s)}$.

5.6 Delay Margin Analysis of Passive/Active Damping in LCL
(a) Consider the controller with passive damping for LCL filter used in Example 5.22. Using the MATLAB command *margin*, find the phase margin and gain crossover frequency of this design. Compute the delay margin by dividing the phase margin (in radians) to the gain crossover frequency (in rad/s) and show that it is around 0.44 ms.
(b) Consider the controller with active damping of LCL resonance mode used in Example 5.23. Similar to above, show that its delay margin is about 0.24 ms. Thus, active damping can compromise the DM seriously.

References

1 Wikipedia page for solar irradiance. https://en.wikipedia.org/wiki/Solar_irradiance. Accessed: 2020-02-08.

2 Mihai Sanduleac, Lucian Toma, Mircea Eremia, Irina Ciornei, Constantin Bulac, Ion Tristiu, Andreea Iantoc, Jo ao F Martins, and Vitor F Pires. On the electrostatic inertia in microgrids with inverter-based generation only: an analysis on dynamic stability. *Energies*, 12(17):3274, 2019.

3 Robert Eriksson, Niklas Modig, and Katherine Elkington. Synthetic inertia versus fast frequency response: a definition. *IET Renewable Power Generation*, 12(5):507–514, 2018.

4 Zhongda Chu, Uros Markovic, Gabriela Hug, and Fei Teng. Towards optimal system scheduling with synthetic inertia provision from wind turbines. *IEEE Transactions on Power Systems*, 35(5):4056–4066, 2020.

5 Mohammadreza Fakhari Moghaddam Arani and Ehab F El-Saadany. Implementing virtual inertia in DFIG-based wind power generation. *IEEE Transactions on Power Systems*, 28(2):1373–1384, 2012.

6 Nima Amouzegar Ashtiani, S Ali Khajehoddin, and Masoud Karimi-Ghartemani. Optimal design of nested current and voltage loops in grid-connected inverters. In *2020 IEEE Applied Power Electronics Conference and Exposition (APEC)*, pages 2397–2402. IEEE, 2020.

7 Prabha S Kundur, Neal J Balu, and Mark G Lauby. Power system dynamics and stability. *Power System Stability and Control*, 3: 827–950, 2017.

6

Single-Phase Alternating-Current (ac) DERs

In a single-phase ac distributed energy resource (DER) application, the power electronic converter is used to exchange the power between a dc side (either originally dc, e.g. in a PV system, or rectified to dc, e.g. in a wind energy system) and an ac side. The ac side can be a load, a grid, or a microgrid. This chapter studies the control system objectives and challenges pertaining to such a system. Various scenarios are treated while more and more of control objectives are taken into consideration and addressed.

(i) The power control via a current feedback loop (CFL) – also called the grid-following control – is presented. The CFL is properly structured to allow application of an optimal control approach. The CFL is invariably used in most applications due to the need for current limiting. (ii) The phase-locked loop (PLL) concept is discussed. Particularly, the single-phase enhanced phase-locked loop (ePLL) is studied and its analysis and design are presented. (iii) Subsequently, the concept of "grid support" is studied and is added to the power control to derive the virtual synchronous machine (VSM). It provides grid support functions as well as a "grid-forming" property. (iv) The dc voltage control and support is also presented. Particularly, obtaining inertia from the dc link capacitor is studied. (v) The ac load voltage control is discussed. It is shown how the VSM and grid-forming controller can be derived from the load voltage control by adding a CFL and relaxing its reference voltage.

6.1 Power Balance in a dc/ac System

A converter (e.g. a VSC) interfacing a dc side with an ac side is shown in Figure 6.1 where $v_g(t)$ is the ac grid side voltage with peak value of V_g and frequency ω rad/s or f Hz, i.e. $\omega = 2\pi f$. The interface filter L suppresses the switching ripples of the converter ac-side current $i(t)$.[1]

Assume that $v_g(t) = V_g\cos(\omega t)$ and $i(t) = I\cos(\omega t + \phi)$. Then, through simple trigonometric relations, the instantaneous ac side power is expressed as[2]

$$p(t) = v_g(t)i(t) = P[1 + \cos(2\omega t)] + Q\sin(2\omega t) = P + S\cos(2\omega t + \phi), \tag{6.1}$$

where

$$P = \frac{1}{2}V_g I\cos\phi, \quad Q = -\frac{1}{2}V_g I\sin\phi, \quad S = \frac{1}{2}V_g I \tag{6.2}$$

1 Unlike the dc case, the switching ripple levels change over a full cycle of an ac signal.
2 See Problem 6.1.

Modeling and Control of Modern Electrical Energy Systems, First Edition. Masoud Karimi-Ghartemani.

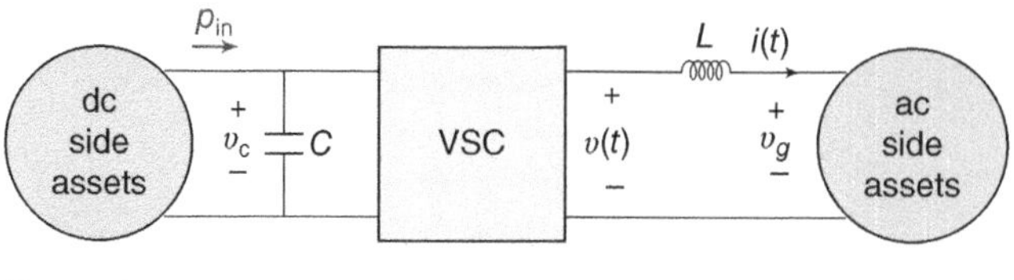

Figure 6.1 Interfacing a dc side and an ac side: dc/ac conversion.

are, respectively, the real power (in watts: W), the reactive power (in volt amperes reactive: VAr), and the apparent power (in volt amperes: VA). The instantaneous power may thus be written as $p(t) = P + p_2(t)$ where $p_2(t)$ is the double-frequency power.

Assume that p_{in} is constant. For instance, a first-stage converter is employed to tightly operate the PV at its maximum power. Therefore, the double-frequency component of the power needs to be supplied by the capacitor C. Therefore, the equation $v_c(t)i_c(t) = p_2(t)$ is held which means[3]

$$Cv_c(t)\dot{v}_c(t) = p_2(t) = S\cos(2\omega t + \phi) \Rightarrow \frac{1}{2}Cv_c^2(t) = \frac{S}{2\omega}\sin(2\omega t + \phi) + K,$$

where K is the constant of integration. While v_c^2 only has a dc and a double-frequency component, $v_c(t)$ will have infinite number of harmonics. However, its dc component and double-frequency component are dominant. Therefore, the capacitor voltage may be approximately written as

$$v_c(t) = V_c + v_{c_2}(t) = V_c + V_{c_2}\sin(2\omega t + \phi).$$

The current of capacitor will then be approximately equal to

$$i_c(t) = C\dot{v}_c(t) = 2C\omega V_{c_2}\cos(2\omega t + \phi).$$

The capacitor's 2-f power is then expresses as

$$p_{c_2}(t) = 2C\omega V_c V_{c_2}\cos(2\omega t + \phi)$$

which should balance $p_2(t)$. Therefore, $S = 2C\omega V_c V_{c_2}$, which leads to

$$\boxed{V_{c_2} = \frac{S}{2C\omega V_c}.} \tag{6.3}$$

For a 10 kVA power, 1 mF capacitor, 500 V dc bus, and 60 Hz frequency,

$$V_{c_2} = \frac{S}{2C\omega V_c} = \frac{10\,000}{2 \times 0.001 \times 377 \times 500} \approx 26.5\ V.$$

If the capacitance is decreased to 250 μF, the voltage will increase to about 106 V. This analysis simply concludes that rather a large capacitor is required to ensure reasonable level of oscillations across this capacitor.

In a *single-stage* PV converter, the capacitor should be large enough to prevent large double-frequency oscillations because these oscillations appear right across the PV panel and they disturb the maximum power extraction process.

In a two-stage converter, larger oscillations may be allowed across the intermediary link capacitor as long as the voltage oscillations do not violate the converter limitations. For example, if a VSC is used, the oscillations should not cause the capacitor voltage to become smaller than the grid voltage. The capacitor voltage has the form $v_c(t) = V_c + V_{c_2}\sin(2\omega t + \phi)$. Figure 6.2 shows the grid voltage and the capacitor voltage for these numerical values: 240 V (rms), 60 Hz grid, $C = 400$ μF, $V_c = 400$ V, $S = 10$ kVA, and two values of $\phi = 0$ and $\phi = -60°$. The peak double-frequency ripple

3 In this analysis, we ignore the converter losses and the power in inductor.

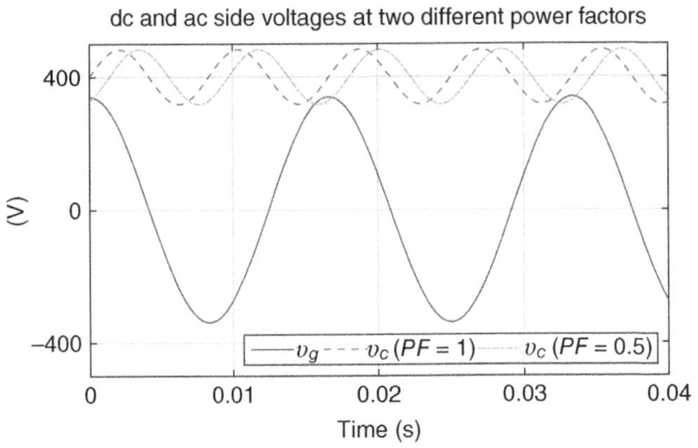

Figure 6.2 Grid and capacitor voltages for two cases of unity and 0.5 PF lagging.

is about 83 V (peak-to-peak 166 V). It is observed that the converter dc side voltage violates the limitation and becomes smaller than the grid voltage when the power factor decreases.

6.1.1 Power Decoupling

The process of preventing the ac (double-frequency component, or ac ripples of) power from penetrating to the PV panels is called the *power decoupling*. It is done in three ways explained below.

- Use a sufficiently large capacitor C to reduce the ripples below a certain level. This often requires using high-density technologies such as the *electrolytic* technology. Electrolytic capacitors have a short life-time before failure especially for outdoor applications.
- Use a double-stage topology which enables distributing the ripples between the PV side and intermediary link. The intermediary link can be allowed to have large level of voltage ripples and its possible impact on the ac side can be addressed using appropriate controls.
- Add auxiliary power electronic circuitry to provide local paths (to other small capacitors/inductors) for the double-frequency power preventing it from penetrating to the PV terminals.

The second approach is more common. It allows the two stages to address control objectives more effectively. In this method, a reasonably large level of oscillations is allowed across the intermediary link capacitor. However, the oscillations across the PV panel are tightly controlled to avoid interfering with the power extraction.

6.2 Power Control Method via Current Feedback Loop (CFL)

On the ac (or grid) side, the power has a real and a reactive component. It is common to use a *CFL* approach to control these two components of the power via a feedback loop closed on the converter current. This feedback loop allows further shaping of the converter current (such as removing its harmonic distortions) and limiting it during grid sudden low-voltage disturbances. These aspects are discussed in details in this section. In all the discussions, we assume that a single-phase full-bridge VSC is used to interface the dc side with the ac side.

6.2.1 Input Linearization and Feedforward Compensation

Simplified control (i.e. averaged over a switching cycle) diagram of the system is shown in Figure 6.3, see Section 2.7. In this diagram, v_c is the dc side (capacitor) voltage, v_g is the ac side (grid) voltage, $v(t)$ is the voltage of the converter, $i(t)$ is the converter current, L is the filter inductance and R models the sum of parasitic resistance of the inductor and conducting resistance of switches.

The dc side voltage $v_c(t)$ comprises a dc component V_c and a double frequency term $v_{c_2}(t)$. Assuming that the modulation signal $m(t)$ is a pure sinusoidal signal at the fundamental frequency, then the product $m(t)v_c(t)$ produced by the converter has a term $m(t)v_{c_2}(t)$ which will generate a third harmonic in the converter current. One way to overcome this problem is to use an input *feedback linearization* in the form of a division by the capacitor voltage as shown in Figure 6.4. The capacitor voltage is measured and is used to cancel out the product effect.

A feedforward of the grid voltage is also often used to achieve pre-synchronization and soft-start performance of the converter. A PLL provides a smooth version of the grid voltage while tracking its frequency swings. Assume that the PLL is started first and reaches its steady-state, then the inverter starts. This will lead to a smooth and soft start because the inverter will immediately generate a voltage which is synchronous with the grid voltage (assuming that $u(t)$ is initialized at zero). There will not be a starting problem in terms of possible disturbances to the grid or converter over-current.[4]

Figure 6.5 shows the resulted system after the input feedback linearization of the bus voltage and the feedforward compensation of the grid voltage. The function $h(t)$ lumps together all possible high frequency terms that have not been compensated such as the high frequency harmonics of the grid voltage. The signal $u(t)$ now serves as the new control input that we design. Once this is designed, the actual modulation index $m(t)$ is calculated according to

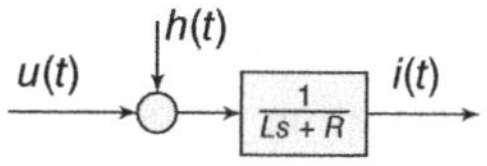

Figure 6.5 Resulted diagram after feedback linearization and feedforward compensation.

$$\boxed{m(t) = \frac{u(t)+v_{ff}(t)}{v_c(t)}.} \tag{6.4}$$

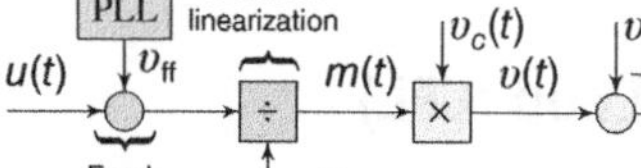

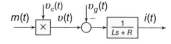

Figure 6.3 Block diagram model of the FB-VSC.

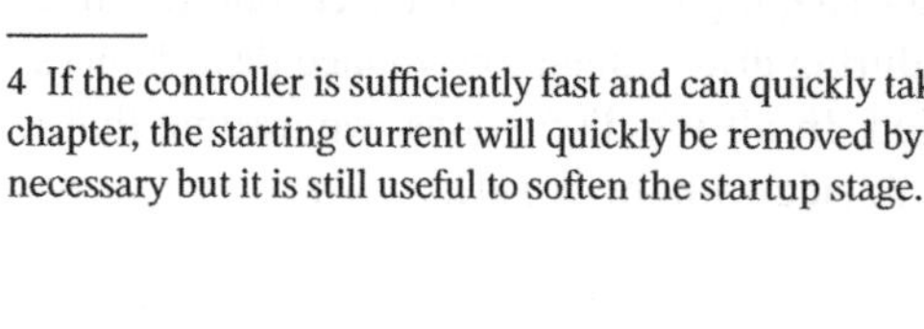

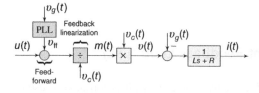

Figure 6.4 Feedback linearization and feedforward compensation.

4 If the controller is sufficiently fast and can quickly take charge, as we will design later in the examples in this chapter, the starting current will quickly be removed by the controller. In such cases, the feed-forward term is not necessary but it is still useful to soften the startup stage.

6.2.2 Control Structure

The general control objectives may be stated as follows. (i) Regulate the real power at the set-point P_{set} and the reactive power at the set-point Q_{set}. (ii) Keep the current $i(t)$ clean and sinusoidal.[5] The aspect of dc bus controller is studied later.

The optimal control structure (inspired from the linear quadratic tracker [LQT] concept) is shown in Figure 6.6. In Figure 6.6, "Resonant" stands for a second-order generalized integrator type of controller characterized by $s^2 + \omega_o^2$ in its transfer function, where ω_o is the rated value of frequency in rad/s, e.g. 120π rad/s. The function $i_{ref}(t)$ is the reference or desired value for the current to be generated by the inverter. The tracking error is $e(t) = i(t) - i_{ref}(t)$. The resonant controller supplies an infinite (practically very large) gain into the feedback loop at the frequency ω_o. This will cause the steady state error to become zero.[6]

6.2.3 Calculating and Limiting Reference Current

The reference current $i_{ref}(t)$ must be calculated such that the real and reactive powers become equal to their set-points P_{set} and Q_{set}. Employing the resonance controller implies that $i(t)$ will become exactly equal to $i_{ref}(t)$ in the steady condition. Without loss of generality, assume that $v_g(t) = V_g \cos(\omega t)$. Express the current as

$$i(t) = I\cos(\omega t + \phi) = I\cos\phi\ \cos(\omega t) - I\ \sin\ \phi\sin(\omega t) = I_d\ \cos(\omega t) - I_q\ \sin(\omega t).$$

The component $I_d \cos(\omega t)$ which is "parallel" (or direct) with the grid voltage is active (or real) component of the current and is responsible for real power transfer. The component $-I_q \sin(\omega t)$ which is "perpendicular" or quadrature to the grid voltage is the reactive component of the current and is responsible for reactive power exchange. More specifically,

$$P = \frac{1}{2}V_g I\cos\phi = \frac{1}{2}V_g I_d \Rightarrow I_d = \frac{2P}{V_g},\ Q = -\frac{1}{2}V_g I\sin\phi = -\frac{1}{2}V_g I_q \Rightarrow I_q = -\frac{2Q}{V_g}.$$

The functions $s_d(t) = \cos(\omega t)$ and $s_q(t) = \sin(\omega t)$ are in-phase and quadrature-phase (90°-delayed) unit vectors of the grid voltage.[7] To summarize, the reference current is computed from

$$\boxed{i_{ref}(t) = I_d s_d(t) - I_q s_q(t),\ I_d = \frac{2P_{set}}{V_g},\ I_q = -\frac{2Q_{set}}{V_g},} \tag{6.5}$$

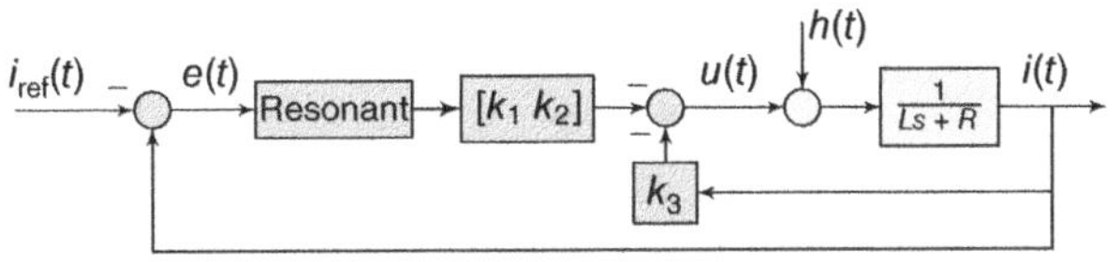

Figure 6.6 Feedback control structure using resonant controller.

5 There are various industry standards such as IEEE-1547 that provide regulatory requirements of a grid-connected DER in terms of the levels of its current harmonics [1].

6 If we use a simple P controller instead of the resonant controller, and choose its gain large to achieve tight tracking of the current, the error can become small but will not be completely removed. See Problem 6.2 for a study of the extent that such an approach can work.

7 Note that if we formulate the grid voltage as a sine function, i.e. $v_g(t) = V_g \sin(\omega t)$, then its in-phase and quadrature-phase unit vectors are $s_d(t) = \sin(\omega t)$ and $s_q(t) = -\cos(\omega t)$, respectively. In this textbook, we consistently adopt a cosine function for the grid voltage.

where V_g is the peak grid voltage, and $s_d(t)$ and $s_q(t)$ are the unity magnitude sinusoidal functions which are respectively parallel and 90° delayed with respect to the grid voltage. These pieces of grid information are provided by a PLL as shown in Figure 6.7.

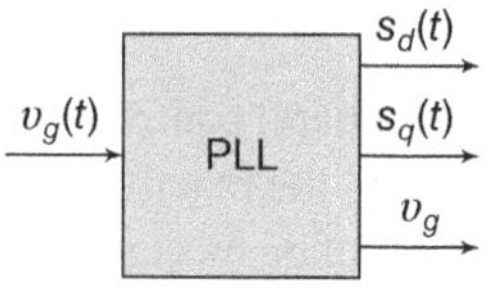

Figure 6.7 PLL providing the unity magnitude synchronized sinusoids and the magnitude.

During the transient low-voltage (or short-circuit) grid faults, V_g drops and as a result, a large current tends to flow to balance the power. To protect the converter, this current must be limited. One approach, shown in Figure 6.8, limits the total current $I = \sqrt{I_d^2 + I_q^2}$. In this diagram, the block denoted by "sat" is a linear saturation block that limits its input within $[0 \;\; I_{max}]$ as defined by[8]

$$\text{sat}(I_1) = \begin{cases} I_1, & 0 \le I_1 \le I_{max} \\ I_{max}, & I_1 \ge I_{max} \end{cases}. \tag{6.6}$$

If the controller is sufficiently fast, it will keep the converter current close to the reference current at all times, and thus, limits it.

6.2.4 Single-Phase ePLL

Structural block diagram of the ePLL is shown in Figure 6.9 [2]. Its input is the grid voltage $v_g(t)$ and it provides several outputs such as the direct and quadrature unit vectors $s_d(t)$ and $s_q(t)$, the magnitude A (which estimates V_g), the frequency ω (in rad/s), and the phase angle ϕ in the expression $v_g(t) = A\cos\phi$.[9,10]

The ePLL has three controlling parameters: μ_1, μ_2, and μ_3. The constants ω_o and A_o indicate the rated frequency and magnitude of the input signal. With these two added outside the integrators, the initial conditions of the integrators are set to zero.[11] In the following, some analysis of the equations of the ePLL are performed in order to find how to design these parameters.

6.2.4.1 Linear Analysis of ePLL

Let $\boxed{\mu_1 = \mu_2 = \mu}$.[12] The ePLL two main equations from Figure 6.9 are derived by inspection as

$$\dot{A} = \mu e \cos\phi, \quad \dot{\phi} = \omega - \mu \frac{e}{A} \sin\phi,$$

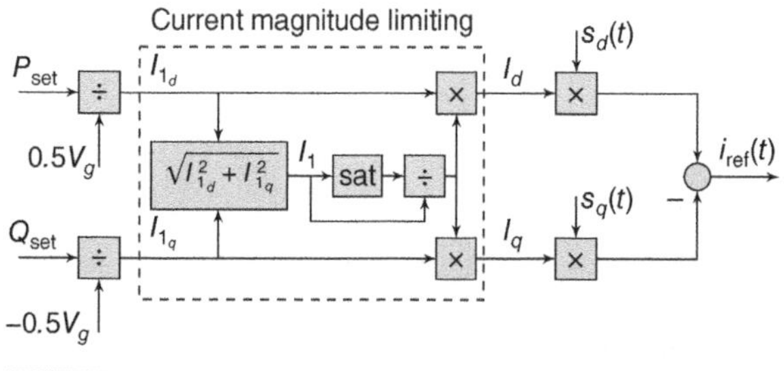

Figure 6.8 Generating and limiting the reference current.

8 Notice that I_1 in Figure 6.8 represents the current magnitude and is non-negative.
9 If the grid voltage is expressed in terms of a sine function, i.e. $v_g(t) = A\sin\phi$, the block "cos" is replaced with "sin" and the block "− sin" is replaced with "cos," in the ePLL diagram of Figure 6.9.
10 The division by A may be replaced with the division by $|A| + \epsilon$ (for a small positive ϵ, such as $0.001A_o$ where A_o is the nominal value of voltage magnitude) to avoid a possible divide-by-zero. This way also guarantees that the estimated magnitude remains always positive; see Problem 6.3.
11 Most existing literature on ePLL do not consider the addition by A_o at the output of its integrator. Therefore, the initial condition of this integrator may be set to A_o for smoother/faster starting.
12 This does not cause loss of generality for the type of applications we encounter here.

Figure 6.9 The ePLL structure.

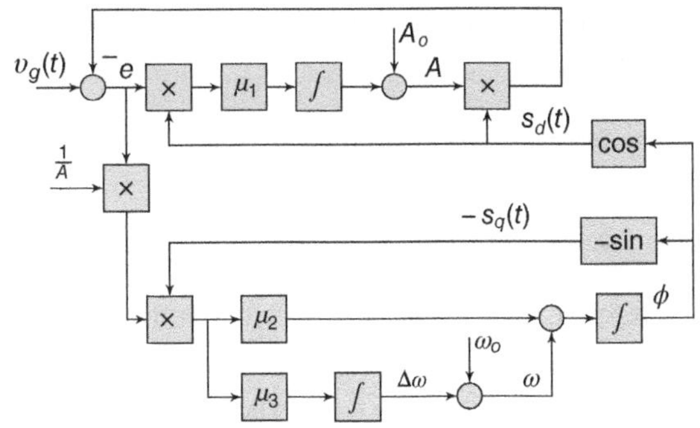

where $e = v_g - A\cos\phi$. If we define $x_1 = A\cos\phi = As_d$ and $x_2 = A\sin\phi = As_q$, it is readily seen that

$$\dot{x}_1 = \dot{A}\cos\phi - A\dot{\phi}\sin\phi = \mu e \cos^2\phi - A\left(\omega - \mu\frac{e}{A}\sin\phi\right)\sin\phi = -\omega x_2 + \mu e$$

$$\dot{x}_2 = \dot{A}\sin\phi + A\dot{\phi}\cos\phi = \mu e\sin\phi\cos\phi + A\left(\omega - \mu\frac{e}{A}\sin\phi\right)\cos\phi = \omega x_1.$$

These two equations show a set of LTI equations assuming the dynamics of ω are ignored which is a reasonable assumption for power system applications.

Assuming $\omega = \omega_o$ being constant, and taking Laplace transform from the above two equation results in

$$X_1(s) = \frac{\mu s}{s^2 + \omega_o^2}E(s), \quad X_2(s) = \frac{\mu\omega_o}{s^2 + \omega_o^2}E(s)$$

and since $E(s) = V_g(s) - X_1(s)$, we get

$$X_1(s) = \frac{\mu s}{s^2 + \mu s + \omega_o^2}V_g(s), \quad X_2(s) = \frac{\mu\omega_o}{s^2 + \mu s + \omega_o^2}V_g(s).$$

This means that the "amplitude/phase" dynamics of the ePLL is characterized by $s^2 + \mu s + \omega_o^2 = 0$. Choosing $\boxed{\mu = 2\zeta\omega_o}$, the poles are at $-\zeta\omega_o \pm j\zeta\omega_o\sqrt{\frac{1}{\zeta^2} - 1}$. For instance, for $\zeta = \frac{\sqrt{2}}{2}$, $\mu = \sqrt{2}\omega_o$ and the poles are at $-\frac{\sqrt{2}}{2}\omega_o \pm j\frac{\sqrt{2}}{2}\omega_o$.[13]

The gain μ_3 determines the bandwidth of the frequency estimation loop. In order to derive a linear model for the "phase/frequency" dynamics of the ePLL, assume that $v_g(t) = V_g\cos\phi_g$. Then, $e = v_g - A\cos\phi = V_g\cos\phi_g - A\cos\phi$. then,

$$\dot{\Delta\omega} = -\mu_3\frac{e}{A}\sin\phi = -\frac{\mu_3}{A}[V_g\cos\phi_g - A\cos\phi]\sin\phi$$

$$= \frac{\mu_3 V_g}{2A}\sin(\phi_g - \phi) - \frac{\mu_3 V_g}{2A}\sin(\phi_g + \phi) + \frac{\mu_3}{2}\sin(2\phi) \approx \frac{\mu_3 V_g}{2A}\sin(\phi_g - \phi).$$

The term $-\frac{\mu_3 V_g}{2A}\sin(\phi_g + \phi) + \frac{\mu_3}{2}\sin(2\phi)$ is a high-frequency term (as opposed to $\frac{\mu_3 V_g}{2A}\sin(\phi_g - \phi)$ which is a low-frequency term) and also goes to zero as A and ϕ approach V_g and ϕ_g. Therefore, discarding it in this simple analysis is justifiable.

Similarly, for the angle, we get

$$\dot{\phi} = \omega_o + \Delta\omega - \frac{\mu}{A}[V_g\cos\phi_g - A\cos\phi]\sin\phi \approx \omega_o + \Delta\omega + \frac{\mu V_g}{2A}\sin(\phi_g - \phi).$$

13 For the 60 Hz ac frequency, these are around $-267 \pm j267$ which indicate a time-constant of around 3.8 ms. This seems to be the fastest response possible with this PLL.

Now, let us assume $\phi_g = \omega_o t$ and $\phi = \omega_o t + \Delta\phi$, then the above two differential equations may be written as

$$\dot{\Delta\omega} = -\frac{\mu_3 V_g}{2A}\sin(\Delta\phi) \approx -\frac{\mu_3}{2}\Delta\phi$$

$$\dot{\Delta\phi} = \Delta\omega - \frac{\mu V_g}{2A}\sin(\Delta\phi) \approx \Delta\omega - \frac{\mu}{2}\Delta\phi.$$

Therefore, the phase/frequency dynamics is characterized by $s^2 + 0.5\mu s + 0.5\mu_3 = 0$ which has two poles at $-\frac{\mu}{4} \pm \frac{\mu}{4}\sqrt{1 - \frac{8\mu_3}{\mu^2}}$. Since the frequency does not experience very fast changes in a power system, these poles do not need to be very fast. Indeed, the responses will be more robust if the poles are real. For instance, for $\boxed{\mu_3 = \frac{3\mu^2}{32}}$, the poles are at $-\frac{3}{8}\mu$ and $-\frac{1}{8}\mu$.[14]

6.2.4.2 Two Modifications to the ePLL

It is very important that the ePLL must continue to provide an accurate estimate of s_d and s_q during the severe conditions, such as the abrupt transients and disturbances of the grid voltage. Moreover, it can often happen that the input to the PLL has a small dc offset which may have been present in the original V_g or have been created during the measurement and signal conditioning stage. Figure 6.10 shows two modifications that are made to the ePLL to improve these two aspects of its performance as explained below.

To enhance the performance during sever grid voltage transients, two modifications are made. First, a limiter (saturation) block is added to keep the frequency deviation between $[-2\pi\Delta f_{min} \;\; 2\pi\Delta f_{max}]$. For instance, if the frequency range is [55 65] Hz, the limiter bounds are $[-10\pi \;\; 10\pi]$. This will prevent the frequency from experiencing swings outside this band during severe conditions (such short-circuit faults, etc.). Second, the gain μ_3 is made adaptive to $\frac{\mu_3}{1+\lambda\frac{|e|}{|A|}}$. The denominator becomes large subsequent to an abrupt disturbance. This, again, will prevent the frequency from shifting too much during an abrupt disturbance. The constant $\lambda > 0$ determines how much this operation takes place. For $\lambda = 0$, this feature is disabled. A value of $\boxed{10 < \lambda < 50}$ can have much impact on improving the ePLL responses during harsh conditions.

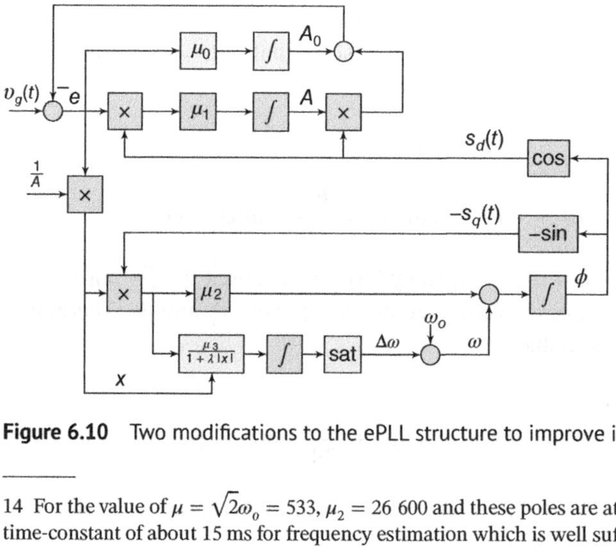

Figure 6.10 Two modifications to the ePLL structure to improve its disturbance response.

14 For the value of $\mu = \sqrt{2}\omega_o = 533$, $\mu_2 = 26\,600$ and these poles are at −200 and −67. This indicates a time-constant of about 15 ms for frequency estimation which is well sufficient.

To deal with the possible dc offset, a new branch (including a gain μ_0 and an integrator) is added. Here, A_0 will estimate the offset of the input.[15] To design μ_0, we notice that the loop characteristic equation is $1+\frac{\mu_0}{s}+\frac{\mu s}{s^2+\omega_o^2}=0$ which reduces to $s^3+(\mu+\mu_0)s^2+\omega_o^2 s+\mu_0\omega_o^2=(s^2+2\zeta\omega_n s+\omega_n^2)(s+\beta)=0$. One choice (that we use in this chapter) is $\zeta=0.5$ and $\beta=\zeta\omega_n$ which leads to $\boxed{\mu=0.95\omega_o,\ \mu_0=0.27\omega_o}$.[16]

6.2.5 Controller Formulation and LQT Design

Returning to the CFL of Figure 6.6, the generalized integrator (or resonant controller) has two state variables. Define those state variables as

$$X_1(s)=\frac{1}{s^2+\omega_o^2}E(s),\quad X_2(s)=\frac{s}{s^2+\omega_o^2}E(s), \tag{6.7}$$

where $E(s)$ is the Laplace transform of the tracking error $e(t)$. This means

$$sX_1(s)=X_2(s)\Rightarrow \dot{x}_1(t)=x_2(t)$$

and

$$sX_2(s)=\frac{s^2}{s^2+\omega_o^2}E(s)=\frac{s^2+\omega_o^2-\omega_o^2}{s^2+\omega_o^2}E(s)=E(s)-\omega_o^2 X_1(s)$$

which means

$$\dot{x}_2(t)=-\omega_o^2 x_1(t)+e(t).$$

Thus, the state-space model of the resonant controller is

$$\begin{bmatrix}\dot{x}_1\\ \dot{x}_2\end{bmatrix}=\begin{bmatrix}0 & 1\\ -\omega_o^2 & 0\end{bmatrix}\begin{bmatrix}x_1\\ x_2\end{bmatrix}+\begin{bmatrix}0\\ 1\end{bmatrix}e. \tag{6.8}$$

This is the controllable canonical form. Also notice from (6.7) that

$$(s^2+\omega_o^2)X_1(s)=E(s),\quad (s^2+\omega_o^2)X_2(s)=sE(s). \tag{6.9}$$

The entire state-space description of the system (including the resonant controller and the converter model) is summarized as

$$\begin{aligned}\dot{x}_1(t)&=x_2(t)\\ \dot{x}_2(t)&=-\omega_o^2 x_1(t)+x_3(t)-i_{\text{ref}}(t)\\ \dot{x}_3(t)&=-\tfrac{R}{L}x_3(t)+\tfrac{1}{L}u(t)+\tfrac{1}{L}h(t),\end{aligned}$$

where $x_3(t)=i(t)$ is the inverter current. In matrix form, it is

$$\dot{x}=Ax+Bu+B_{\text{ref}}i_{\text{ref}}+B_g h,\quad y=Cx, \tag{6.10}$$

where $y=x_3$ and the matrices/vectors are

$$A=\begin{bmatrix}0 & 1 & 0\\ -\omega_o^2 & 0 & 1\\ 0 & 0 & -\frac{R}{L}\end{bmatrix},\ B=B_g=\begin{bmatrix}0\\ 0\\ \frac{1}{L}\end{bmatrix},\ B_{\text{ref}}=\begin{bmatrix}0\\ -1\\ 0\end{bmatrix},\ C^T=\begin{bmatrix}0\\ 0\\ 1\end{bmatrix}. \tag{6.11}$$

15 Notice the notations: A_o for the rated magnitude, A_0 for the estimated dc.

16 For a 60 Hz system, $\omega_o=377$ rad/s, this leads to $\mu=360$ and $\mu_0=100$. The ePLL poles will be at -154, $-154\pm j267$. And μ_3 will be $\mu_3=\frac{3}{32}\mu^2=12\,000$.

The control law is

$$u(t) = -k_1x_1(t) - k_2x_2(t) - k_3x_3(t) = -Kx(t), \quad K = [k_1 \;\; k_2 \;\; k_3]$$

which is a full linear state feedback.

In order to proceed with the LQT method, assume $h(t) = 0$ for simplicity, and apply the operator $\frac{d^2}{dt^2} + \omega_o^2$ to both sides of state space equations to obtain

$$\begin{aligned}\dot{z}_1(t) &= z_2(t)\\ \dot{z}_2(t) &= -\omega_o^2 z_1(t) + z_3(t)\\ \dot{z}_3(t) &= -\tfrac{R}{L} z_3(t) + \tfrac{1}{L} w(t),\end{aligned}$$

where $z_i(t) = \ddot{x}_i(t) + \omega_o^2 x_i(t)$ for $i = 1, 2, 3$ and $w(t) = \ddot{u}(t) + \omega_o^2 u(t)$. Notice that $\frac{d^2}{dt^2} i_{\text{ref}}(t) + \omega_o^2 i_{\text{ref}}(t) = 0$ because the current reference is a pure sinusoidal at ω_o. Notice also that

$$w(t) = -k_1z_1(t) - k_2z_2(t) - k_3z_3(t) = -Kz(t)$$

which is a full linear state feedback. Also notice that from (6.9) we can immediately conclude that

$$\boxed{z_1(t) = e(t), \quad z_2(t) = \dot{e}(t)}$$

which are the tracking error and its derivative.

Therefore, using the LQT technique, the cost function

$$J = \int_0^{\infty} \left[qe(t)^2 + q_2\dot{e}(t)^2 + q_3z_3(t)^2 + w(t)^2\right] dt$$

may be minimized using Matlab: $\gg$ `K = lqr(A, B, Q, 1)` where `Q = diag([q q2 q3])`.

Example 6.1 ***Power Controller with Optimal CFL***

The system parameters used in this example, and subsequent ones in this chapter, are given in Table 6.1. According to $S = \frac{1}{2}VI$, the maximum peak current is $\frac{2S}{V} = \frac{2\times 2000\sqrt{2}}{240\sqrt{2}} = 16.7$ A. We set the current magnitude limit in the "sat" block to 22 A.

For this set of parameters, a snapshot of the closed-loop root-locus is shown in Figure 6.11 when q varies from 10^{10} to 10^{14} (blue); q_2 varies from 10^5 to 10^8 (red); and q_3 varies from 10^{-1} to $10^{1.5}$ (magenta). At final point, the poles are located at -1035, $-973 \pm j996$. These poles correspond to

Table 6.1 Converter system parameters used for the examples in this chapter.

Parameter	Symbol	Value	Unit
dc side voltage	V_{dc}	500	V
Grid side voltage (rms)	V_g	240	V
Grid side frequency	f_o (ω_o)	60 (120π)	Hz (rad/s)
Inductance	L	5	mH
Parasitic resistance	R	30	mΩ
Switching frequency	f_{sw}	4	kHz
Power rating	P_{rate}	2	kW
Reactive power rating	Q_{rate}	± 2	kVAr
Current limits	$I_{max,\ min}$	± 22	A

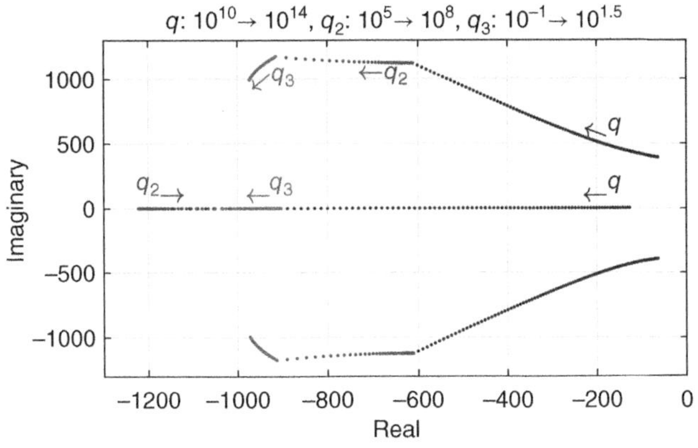

Figure 6.11 Poles of the closed-loop system using LQT approach.

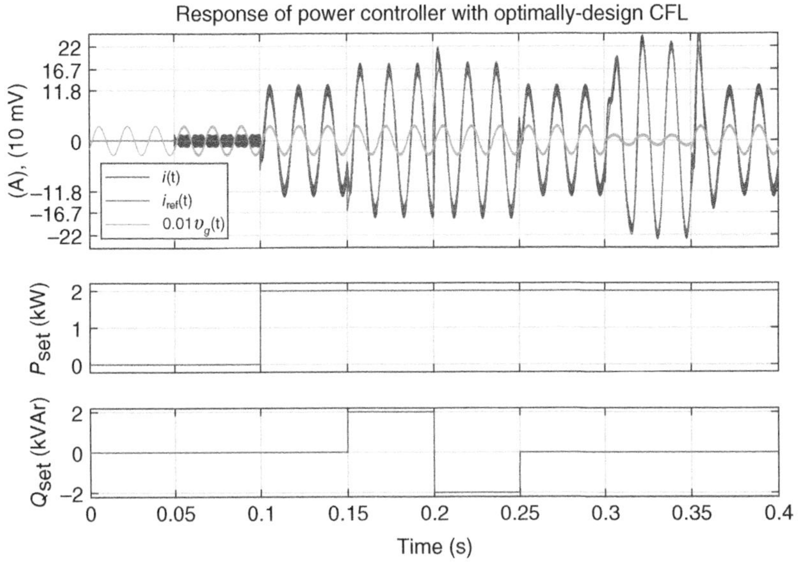

Figure 6.12 Responses of inverter with optimal controller. $t = 0$: ePLL starts, $t = 0.05$ seconds: converter starts, $t = 0.1$, 0.15, 0.20, 0.25 seconds: real and reactive power commands applied, $t = 0.3$ seconds: grid fault (voltage sag down to 30%), $t = 0.35$ seconds: fault is cleared.

a rather fast response with about 1 ms time constant to allow tight current control and limiting.[17] The two complex poles have a damping ratio of about 0.7. At the final point, the controller gains are equal to $K = [k_1 \;\; k_2 \;\; k_3] = [7.9 \times 10^6 \;\; 19.1 \times 10^3 \;\; 14.9]$.

A snapshot of the inverter responses is shown in Figure 6.12. At $t = 0.05$ seconds, the gating commands are applied to the switches. Prior to this time, the ePLL is functioning to prepare a pre-synchronization reference for soft-start.[18] The inverter starts at $t = 0.05$ seconds.

17 We use a unipolar PWM which means the effective switching frequency is 8 kHz. Thus, a time-constant of 1 ms for the control system is very reasonable. It can even be further increased if need be.

18 The parameters of the ePLL are set at $\mu = 360$, $\mu_0 = 100$, and $\mu_3 = 12\,000$ as designed before. The PLL signal $V_g s_d(t)$ is used for the feed-forward term $v_{\rm ff}$ in the controller.

At $t = 0.1$ seconds, the real power reference of 2 kW (and zero reactive power, i.e. unity power factor operation) is applied. The converter immediately generates a current of about 11.8 A peak and in-phase with the grid voltage. At $t = 0.15$ seconds, the reactive power set-point of 2 kVAr is applied (i.e. the power factor of 0.7 lagging). The current peak, expectedly, increases to about 16.7 A, and it lags the grid voltage 45°. The change of reactive power to −2 kVAr (i.e. the power factor of 0.7 leading) at $t = 0.2$ seconds quickly (and smoothly) shifts its angle to 45° leading. The reactive power is restored to 0 at $t = 0.25$ seconds. At $t = 0.3$ seconds, the grid voltage experiences a deep drop of 70% (down to 72 V rms) and restores to normal at $t = 0.35$ seconds. The converter responds by limiting the current at 22 A. The controller succeeds to (i) yield a soft-start, (ii) execute real and reactive power commands quickly and smoothly, (iii) generate sinusoidal current, and (iv) limit the converter current.

6.2.6 Impact of Grid Voltage Harmonics

In a practical system, the grid voltage $v_g(t)$ is noisy and also has numerous harmonics. These are caused by nonlinear components, such as nonlinear loads, and are present on the grid voltage. In this section, we study the impact of such distortions on the operation of an inverter tied to such a grid. We also enhance the controller to effectively operate in such conditions.

Returning to our basic closed-loop feedback control diagram of Figure 6.6, the transfer function from $h(t)$ (which represents v_g) to $i(t)$ is[19]

$$
\begin{aligned}
G(s) &= \frac{I(s)}{V_g(s)} = \frac{\frac{1}{Ls+R}}{1+\frac{k_3}{Ls+R}+\frac{k_2 s+k_1}{\left(s^2+\omega_o^2\right)(Ls+R)}} \\
&= \frac{s^2+\omega_o^2}{\left(s^2+\omega_o^2\right)(Ls+R)+k_3\left(s^2+\omega_o^2\right)+k_2 s+k_1}.
\end{aligned}
\tag{6.12}
$$

Magnitude of this transfer function for frequencies [10 3000] Hz is shown in Figure 6.13 for the design values used in Example 6.1. The magnitude is zero or $-\infty$ dB at the grid frequency. Without feedback, this transfer function is simply $\frac{1}{Ls+R}$ and is also shown for reference. It is observed that the

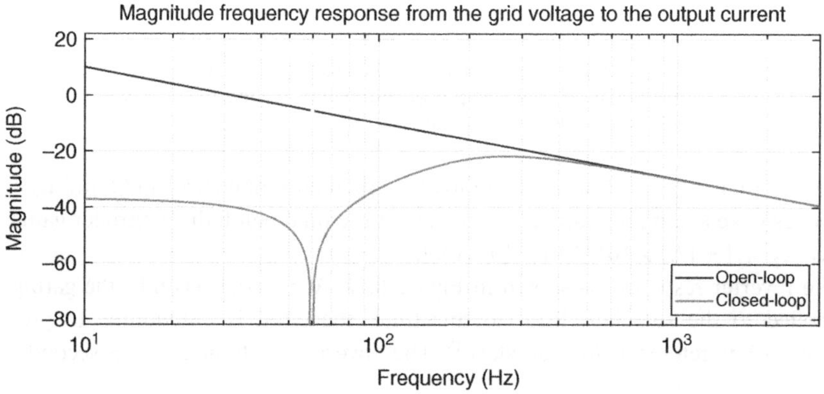

Figure 6.13 Magnitude transfer function of (6.12) for the system of Figure 6.6.

19 This is the system represented by the state space model $ss(A - BK, B_g, C, 0)$ with matrices defined in (6.11).

feedback does not make a noticeable contribution in reducing the impact of harmonics above the third harmonic. Both the closed-loop and open-loop magnitudes are practically identical above the frequency of 300 Hz (that is the fifth harmonic). This means that, with this design, the only way to reduce harmonics is to increase L which is a practically demanding solution: doubling the inductor only introduces $20\log(2) = 6$ dB attenuation.

Figure 6.13 also shows how far each harmonic is suppressed by the system. For instance, a magnitude of −20 dB (happening around harmonics 5 and 7), means a gain factor of 0.1. In other words, a fifth harmonic in the grid voltage gets attenuated with a factor of 0.1 before being transferred to the current. Let us assume that a 5th harmonic as much as $4.8\sqrt{2}\cos(600\pi t)$ (which is 2% of a 240 V grid) is added at the point associated with $h(t)$. Then, the grid current will be $i(t) = I\cos(120\pi t) + I_5\cos(600\pi t + \theta_5)$ where $I_5 = 0.1 \times 4.8 \times \sqrt{2} = 0.57$ A. The total harmonic distortion (THD) of the current will be $\frac{0.57}{I} \times 100$. If the inverter operates at its rated ($S = 3$ kVA), $I = 17.7$ A, its current THD is about 3.2%. If the inverter operates at its third power (1 kVA), the THD goes up to about 10%.

At the zero frequency, dc, the magnitude is about −37 dB which is about 0.014. If 1% dc component is present on the grid voltage ($2.4\sqrt{2} = 3.4$ V), it will cause a dc current of $0.014 \times 3.4 = 48$ mA. This is about 0.5% of a 10 A, and 1% of a 5 A current. The utility grids have highly restricted levels on the amount of harmonics and dc that a grid-connected inverter can feed to the grid. They do not allow current THDs above 5% or so with limits on individual harmonics and dc as well [3].

The transfer function from $i_{ref}(t)$ to $i(t)$ of Figure 6.6 is[20]

$$G(s) = \frac{I(s)}{I_{ref}(s)} = \frac{\dfrac{1}{Ls+R}\dfrac{k_2 s + k_1}{s^2+\omega_o^2}}{1 + \dfrac{k_3}{Ls+R} + \dfrac{k_2 s + k_1}{\left(s^2+\omega_o^2\right)(Ls+R)}} = \frac{k_2 s + k_1}{\left(s^2+\omega_o^2\right)(Ls+R) + k_3\left(s^2+\omega_o^2\right) + k_2 s + k_1}. \tag{6.13}$$

Magnitude of this transfer function is shown in Figure 6.14.[21] This shows how far these harmonics get suppressed by the system. For instance, a magnitude of −10 dB (happening around harmonics

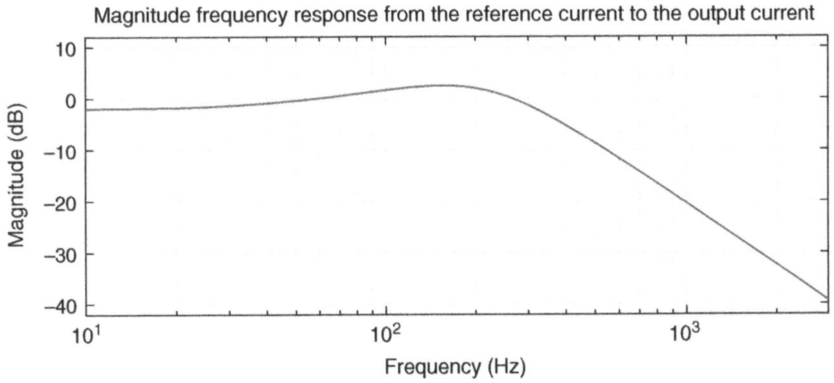

Figure 6.14 Magnitude transfer function from i_{ref} to i for Figure 6.6.

20 This is the system represented by the state space model $ss(A - BK, B_{ref}, C, 0)$.
21 Note that this function is unity (0 dB) at the grid frequency ω_o regardless of design parameters.

7 and 11), means a gain factor of 0.3. In other words, a seventh harmonic in the reference current gets attenuated with a factor of 0.3 before being transferred to the current. Let us assume a current reference is $i^*(t) = I_1^* \cos(120\pi t) + 0.02I_1^* \cos(840\pi t)$ which shows a seventh harmonic of 2%. Then, the grid current will be $i(t) = I_1^* \cos(120\pi t) + 0.006I_1^* \cos(840\pi t + \theta_7)$. The THD generated by this will be 0.6%.

This analysis shows that the impact of reference current harmonics is much less than the grid voltage harmonics (for the range of power and voltages we are working with in single-phase ac DER applications). Therefore, the impacts of the grid voltage harmonics on the inverter current must be mitigated. In Section 6.2.7, a systematic way of addressing this problem is presented.

6.2.7 Harmonics and dc Control Units

In order to attenuate the inverter current harmonics and decrease its THD, the controller is modified as shown in Figure 6.15. In Figure 6.15, to show the concept by way of an example, for rejecting the nth harmonic at the frequency $n\omega_o$, a harmonic resonant controller which has $s^2 + n^2\omega_o^2$ in its denominator is used, and for rejecting the dc component, the regular integrator with s in the denominator is used.

There are two alternatives as to the input connection of the harmonic and dc controllers. If it is connected to the the error signal $e(t)$, the controller will increase the magnitude transfer function of the reference current to unity at those frequencies! In other words, the controller treats the reference current harmonics also as the components of interest and follows them as opposed to rejecting them. This option may be used only if the reference current is clean and distortion-free. To avoid this problem, the input point of the harmonic and dc controller units must be connected to the current $i(t)$. This way, the controllers remove the harmonics and the dc from the output current regardless of the source of those harmonics and dc.

The main tracking resonant controller has two state variables as

$$X_1(s) = \frac{1}{s^2 + \omega_o^2} E(s),\ X_2(s) = \frac{s}{s^2 + \omega_o^2} E(s),$$

where $E(s)$ is the Laplace transform of the tracking error function $e(t)$. Thus,

$$sX_1(s) = X_2(s) \Rightarrow \dot{x}_1(t) = x_2(t)$$

$$sX_2(s) = \frac{s^2}{s^2 + \omega_o^2} E(s) = E(s) - \omega_o^2 X_1(s) \Rightarrow \dot{x}_2(t) = -\omega_o^2 x_1(t) + e(t).$$

The harmonic resonant controller has two state variables as

$$X_3(s) = \frac{1}{s^2 + n^2\omega_o^2} I(s),\ X_4(s) = \frac{s}{s^2 + n^2\omega_o^2} I(s),$$

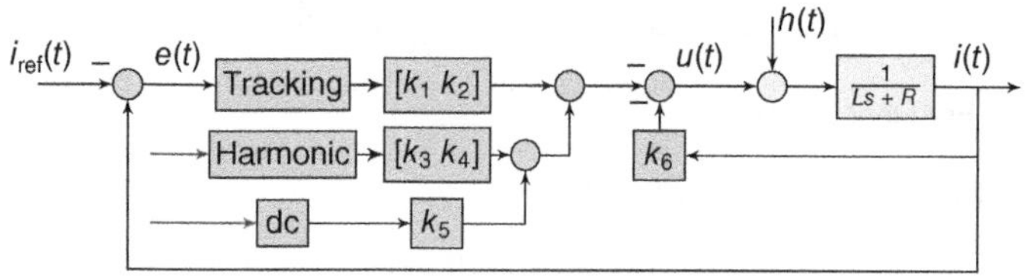

Figure 6.15 Feedback control structure using to reject harmonics and dc.

where $I(s)$ is the Laplace transform of the current $i(t)$. This means that

$$sX_3(s) = X_4(s) \Rightarrow \dot{x}_3(t) = x_4(t)$$
$$sX_4(s) = \frac{s^2}{s^2 + n^2\omega_0^2} I(s) = I(s) - n^2\omega_0^2 X_3(s) \Rightarrow \dot{x}_4(t) = -n^2\omega_0^2 x_3(t) + i(t).$$

The dc controller has one state variable as

$$X_5(s) = \frac{1}{s} I(s) \Rightarrow \dot{x}_5(t) = i(t).$$

Finally, the state-space description of the entire system is summarized as

$$\begin{aligned}
\dot{x}_1(t) &= x_2(t)\\
\dot{x}_2(t) &= -\omega_0^2 x_1(t) + x_6(t) - i_{\text{ref}}(t)\\
\dot{x}_3(t) &= x_4(t)\\
\dot{x}_4(t) &= -n^2\omega_0^2 x_3(t) + x_6(t)\\
\dot{x}_5(t) &= x_6(t)\\
\dot{x}_6(t) &= -\tfrac{R}{L} x_6(t) + \tfrac{1}{L} u(t) + \tfrac{1}{L} h(t),
\end{aligned}$$

where $x_6(t) = i(t)$ is the inverter current, and $h(t)$ represents the grid voltage harmonics. In matrix form, it is

$$\dot{x} = Ax + Bu + B_{\text{ref}} i_{\text{ref}} + B_h h, \quad y = Cx, \tag{6.14}$$

where $y = x_6$ and the matrices/vectors are

$$A = \begin{bmatrix} 0 & 1 & 0 & 0 & 0 & 0\\ -\omega_0^2 & 0 & 0 & 0 & 0 & 1\\ 0 & 0 & 0 & 1 & 0 & 0\\ 0 & 0 & -n^2\omega_0^2 & 0 & 0 & 1\\ 0 & 0 & 0 & 0 & 0 & 1\\ 0 & 0 & 0 & 0 & 0 & -\frac{R}{L} \end{bmatrix}, B = B_h = \begin{bmatrix} 0\\ 0\\ 0\\ 0\\ 0\\ \frac{1}{L} \end{bmatrix}, B_{\text{ref}} = \begin{bmatrix} 0\\ -1\\ 0\\ 0\\ 0\\ 0 \end{bmatrix}, C^T = \begin{bmatrix} 0\\ 0\\ 0\\ 0\\ 0\\ 1 \end{bmatrix}. \tag{6.15}$$

The control law is

$$u(t) = -k_1x_1(t) - k_2x_2(t) - k_3x_3(t) - k_4x_4(t) - k_5x_5(t) - k_6x_6(t) = -Kx(t)$$

which is a full linear state feedback.

In order to proceed with the LQT method, we apply the operator $\left(\frac{d^2}{dt^2} + \omega_0^2\right)\left(\frac{d^2}{dt^2} + n^2\omega_0^2\right)\frac{d}{dt}$ to both sides of state space equations to obtain:[22]

$$\begin{aligned}
\dot{z}_1(t) &= z_2(t)\\
\dot{z}_2(t) &= -\omega_0^2 z_1(t) + z_6(t)\\
\dot{z}_3(t) &= z_4(t)\\
\dot{z}_4(t) &= -n^2\omega_0^2 z_3(t) + z_6(t)\\
\dot{z}_5(t) &= z_6(t)\\
\dot{z}_6(t) &= -\tfrac{R}{L} z_6(t) + \tfrac{1}{L} w(t),
\end{aligned}$$

where $z_i(t) = \left(\frac{d^2}{dt^2} + \omega_0^2\right)\left(\frac{d^2}{dt^2} + n^2\omega_0^2\right)\frac{d}{dt}x_i(t)$ for $i = 1, 2, 3, 4, 5, 6$ and $w(t) = \left(\frac{d^2}{dt^2} + \omega_0^2\right)\left(\frac{d^2}{dt^2} + n^2\omega_0^2\right)\frac{d}{dt}u(t)$. Therefore, using the LQT technique, solve in

MATLAB: ≫ `K = lqr(A, B, Q, 1)` where $Q = \text{diag}([q \;\; q_2 \;\; q_3 \;\; q_4 \;\; q_5 \;\; q_6])$.

22 We assume that $h(t)$ can only have a dc and an nth-order harmonic since we designed our controller to reject these two components only. This can be generalized to reject more number of harmonics if needed.

Note that in this case, the tracking error does not have to become zero in steady state because the reference current may be polluted but the output current must be clean. Thus, the tracking error can have a dc and/or an nth harmonic. If we define the modified tracking error as

$$E_m(s) = s(s^2 + n^2\omega_0^2)E(s)$$

then $e_m(t)$ will go to zero. It is now easy to show that

$$\boxed{z_1(t) = e_m(t),\ z_2(t) = \dot{e}_m(t).}$$

The signal $z_1 = e_m$, thus, represents the (low-frequency) transients of the tracking error without engaging its steady-state value. The signal $z_2 = \dot{e}_m$ represents some higher frequency oscillations of that error. Similarly, the state variables z_3 and z_4 represent the low-frequency and high-frequency components associated with the transient response of the harmonic controller. The variable z_5 represents the dynamics of the dc controller. Finally, z_6 represents the dynamics of the output current.

The associated LQT cost function will be

$$J = \int_0^{\infty} \left[q e_m^2(t) + q_1 \dot{e}_m^2(t) + \sum_{i=3}^{6} q_i z_i^2(t) + w^2(t) \right] dt.$$

The design stage may be done by adjusting q first and then moving to the other gains while closely watching the closed-loop eigenvalues. The following example illustrates the process and shows some results.

Example 6.2 ***Harmonics and dc Control Units***
Figure 6.16 shows the evolution of closed-loop poles when q and subsequent q_i's are adjusted. The controller has three parts: a resonance at the fundamental frequency, a resonance at the fifth harmonic, and a dc. Thus, the controller is of order 5 and the entire system has an order of 6 as explained above. Final location of poles is given by $-1054 \pm j1031$, $-800 \pm j1867$, -653, and -172. The controller gain vector is $K = [8.7 \times 10^6,\ 16 \times 10^3,\ 10^7,\ 19 \times 10^3,\ 10^4,\ 22.6]$.

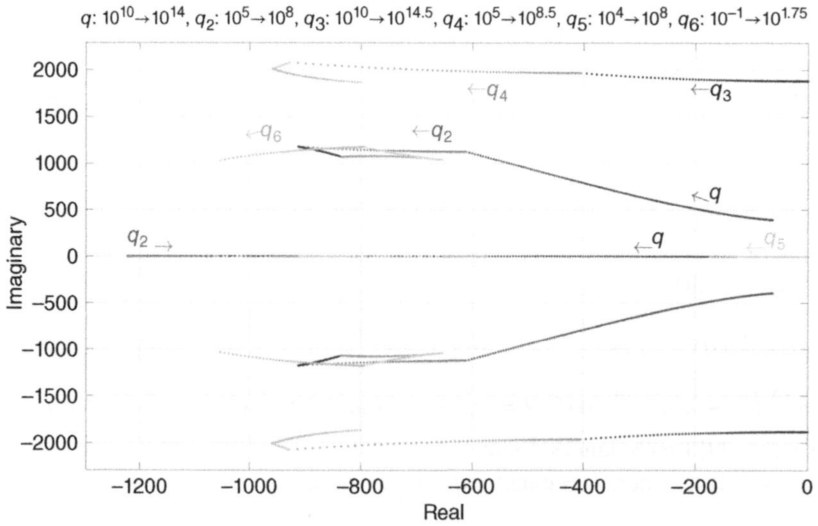

Figure 6.16 Evolution of the poles of the closed-loop system when elements of Q vary.

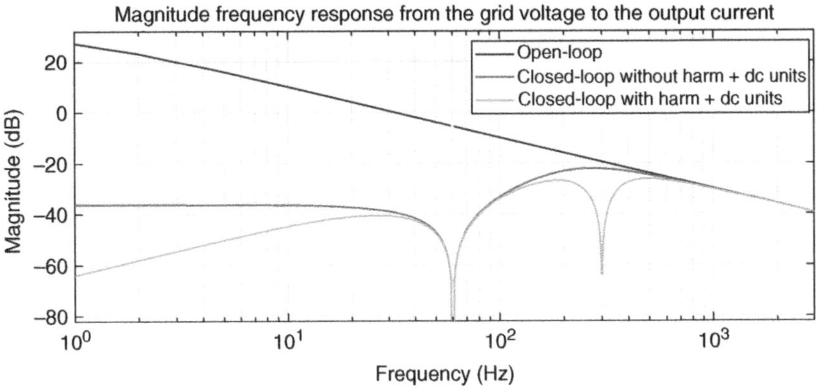

Figure 6.17 Magnitude frequency response of $\frac{I(s)}{V_g(s)}$.

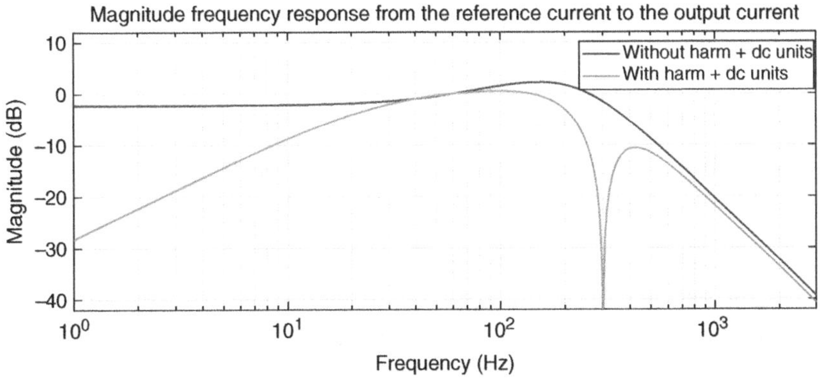

Figure 6.18 Magnitude frequency response of $\frac{I(s)}{I_{ref}(s)}$.

Magnitude frequency response of the transfer function $\frac{I(s)}{V_g(s)}$ is shown in Figure 6.17.[23] It is observed that adding the harmonic and dc units cause this function to block those components, i.e. the fifth harmonic at 300 Hz and the dc at frequency of zero. Figure 6.18 shows the magnitude frequency response of the transfer function $\frac{I(s)}{I_{ref}(s)}$.[24] The same observation is made.

A simulation scenario is defined as follows: the inverter is supposed to inject 2 kW real power at unity power factor to the grid. The PLL starts at $t = 0$ seconds. The inverter gating signals are applied at $t = 0.05$ seconds and the controller is enabled. At $t = 0.1$ seconds, the command of 2 kW is applied. At $t = 0.2$ seconds, the fifth harmonic rejection unit in the controller is enabled. At $t = 0.3$ seconds, the dc rejection unit in the controller is enabled. The grid voltage has 5% of the fifth harmonic and a large dc offset of about −55 V. The dc offset is chosen too large to make the current offset visible. In fact, the controller without dc rejection unit already suppresses the grid voltage offsets quite well and the need to a dc unit is not urgent. Figure 6.18 shows that the controller responds correctly and removes the harmonic and the offset from the inverter current.

23 This is the system represented by the state space model $ss(A - BK, B_h, C, 0)$.
24 This is the system represented by the state space model $ss(A - BK, B_{ref}, C, 0)$.

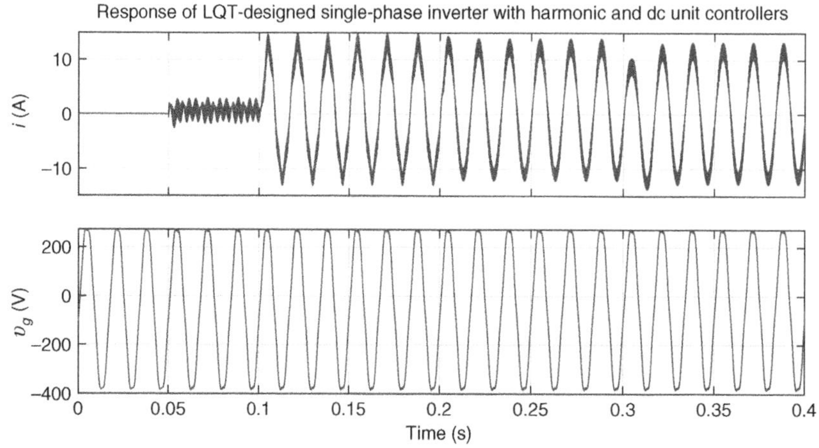

Figure 6.19 The inverter responses in highly distorted (5% of the fifth harmonic and 15% dc offset) grid voltage conditions. At $t = 0.2$ seconds, the harmonic unit is enabled. At $t = 0.3$ seconds, the dc unit is enabled.

6.2.8 Weak Grid Condition and PLL Impact*

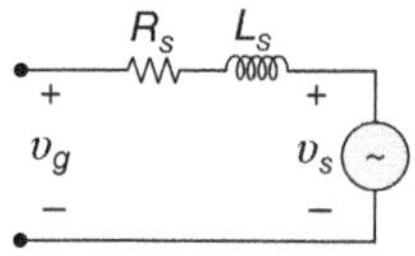

Figure 6.20 Weak grid condition modeling as a voltage source behind an impedance.

Weak grid is a condition where the grid voltage $v_g(t)$, as seen from the converter output, is weak. This means that the converter operation can have an effect on this voltage (Figure 6.19). This condition happens, commonly, when the point of connection of the converter to the grid is distant from the main grid lines. It may be that the grid is weak in nature, e.g. is built using small generators which may not be able to regulate the voltage tightly. At any case, and mathematically, the weak grid condition may be modeled using a voltage source behind an impedance as shown in Figure 6.20. In weak grid studies, often the voltage source $v_s(t)$ is considered as a fixed voltage and fixed frequency source. It may also be replaced by the model of a synchronous generator and a load to better represent the power system voltage and frequency dynamics.

The weak grid condition may degrade the performance of the controller. When a disturbance occurs, the grid voltage v_g may experience fluctuations. Such fluctuations go into the PLL and enter in the control loop. The previous design in Section 6.2.2 does not consider this effect. An enhanced controller that includes the weak grid conditions into consideration with ePLL modeling is shown in Figure 6.21 and will be studied in this section.

6.2.8.1 Short-Circuit Ratio (SCR)

Commonly, to quantify the level of grid weakness, an index called the short-circuit current ratio (SCR) is used. For the circuit of Figure 6.20, the SCR is defined as the ratio of short-circuit current to the inverter current. The short-circuit current is the current that flows in the grid impedance when v_g is short-circuited. Thus,

$$\mathrm{SCR} = \frac{I_{\mathrm{sc}}}{I_{\mathrm{rated}}} = \frac{\dfrac{V_{\mathrm{rated}}}{\sqrt{R_s^2 + (L_s\omega)^2}}}{\dfrac{S_{\mathrm{rated}}}{V_{\mathrm{rated}}}} = \frac{V_{\mathrm{rated}}^2}{S_{\mathrm{rated}}\sqrt{R_s^2 + (L_s\omega)^2}}, \tag{6.16}$$

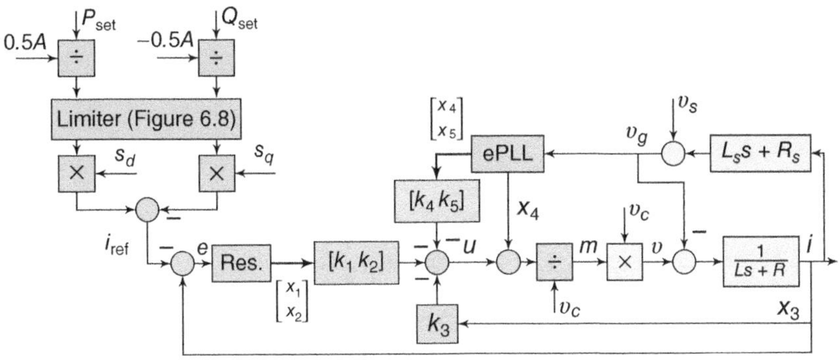

Figure 6.21 Full power controller including the PLL dynamics for weak grid conditions.

where V_{rated} is the rated voltage (rms) of the system and S_{rated} is the rated VA of the inverter. The larger the SCR, the stiffer the grid. As the SCR decreases (e.g. goes below 5), the grid terminals become weaker.

6.2.8.2 LTI Model of Reference Current Generation

Assume $x_4 = As_d$ and $x_5 = As_q$ are the state variables of the ePLL. The reference current is computed according to Figure 6.8 which (ignoring the limiter block) can be written as

$$i_{ref} = \frac{2P_{set}}{A}s_d + \frac{2Q_{set}}{A}s_q = 2\frac{P_{set}x_4 + Q_{set}x_5}{x_4^2 + x_5^2} \approx \frac{2}{V_g^2}(P_{set}x_4 + Q_{set}x_5), \tag{6.17}$$

where in the last term, V_g indicates the nominal value of the grid voltage. In other words, we have approximated the nonlinear term in the denominator around its nominal value.

6.2.8.3 Controller and Its Design

The full diagram of the control system is shown in Figure 6.21. As observed, the two state variables of the ePLL, i.e. x_4 and x_5, are also used in the feedback controller. In the following we write down the entire state equations of this system and design the gains using the LQT approach.

The state-space description of the system is summarized as

$$\begin{aligned}
\dot{x}_1(t) &= x_2(t)\\
\dot{x}_2(t) &= -\omega_o^2 x_1(t) + x_3(t) - \tfrac{2P_{set}}{V_g^2}x_4 - \tfrac{2Q_{set}}{V_g^2}x_5\\
\dot{x}_3(t) &= -\tfrac{R}{L}x_3(t) + \tfrac{1}{L}u(t) + \tfrac{1}{L}x_4 - \tfrac{1}{L}v_g\\
\dot{x}_4(t) &= -\omega_o x_5(t) - \mu x_4(t) + \mu v_g\\
\dot{x}_5(t) &= \omega_o x_4(t),
\end{aligned} \tag{6.18}$$

where $x_3(t) = i(t)$ is the inverter current, x_4 and x_5 are the PLL state variables, and the constant μ is the ePLL gain. The grid voltage v_s satisfies

$$v_g = v_s + L_s\dot{x}_3 + R_s x_3$$

which means

$$v_g = v_s + L_s\left(-\frac{R}{L}x_3 + \frac{1}{L}u + \frac{1}{L}x_4 - \frac{1}{L}v_g\right) + R_s x_3.$$

This results

$$v_g = \frac{L}{L+L_s}\left[v_s + \left(R_s - R\frac{L_s}{L}\right)x_3 + \frac{L_s}{L}x_4 + \frac{L_s}{L}u\right].$$

Substituting in (6.18) leads to

$$\begin{aligned}
\dot{x}_1 &= x_2 \\
\dot{x}_2 &= -\omega_o^2 x_1 + x_3(t) - \frac{2P_{\text{set}}}{V_g^2} x_4 - \frac{2Q_{\text{set}}}{V_g^2} x_5 \\
\dot{x}_3 &= -\frac{R+R_s}{L+L_s} x_3 + \frac{1}{L+L_s} x_4 + \frac{1}{L+L_s} u - \frac{1}{L+L_s} v_s \\
\dot{x}_4 &= -\mu \frac{RL_s - LR_s}{L+L_s} x_3 - \mu \frac{L}{L+L_s} x_4 - \omega_o x_5 + \mu \frac{L_s}{L+L_s} u + \mu \frac{L}{L+L_s} v_s \\
\dot{x}_5 &= \omega_o x_4.
\end{aligned} \tag{6.19}$$

The control law is

$$u(t) = -k_1 x_1(t) - k_2 x_2(t) - k_3 x_3(t) - k_4 x_4(t) - k_5 x_5(t) = -Kx(t)$$

which is a full linear state feedback.

In order to proceed with the LQT method, we assume v_s is a pure sinusoidal functions at ω_o. Apply the operator $\frac{d^2}{dt^2} + \omega_o^2$ to both sides of (6.19) to obtain

$$\begin{aligned}
\dot{z}_1 &= z_2 \\
\dot{z}_2 &= -\omega_o^2 z_1 + z_3 - \frac{2P_{\text{set}}}{V_g^2} z_4 - \frac{2Q_{\text{set}}}{V_g^2} z_5 \\
\dot{z}_3 &= -\frac{R+R_s}{L+L_s} z_3 + \frac{1}{L+L_s} z_4 + \frac{1}{L+L_s} w \\
\dot{z}_4 &= -\mu \frac{RL_s - LR_s}{L+L_s} z_3 - \mu \frac{L}{L+L_s} z_4 - \omega_o z_5 + \mu \frac{L_s}{L+L_s} w \\
\dot{z}_5 &= \omega_o z_4,
\end{aligned}$$

where $z_i(t) = \ddot{x}_i(t) + \omega_o^2 x_i(t)$ for $i = 1, 2, 3, 4, 5$ and $w(t) = \ddot{u}(t) + \omega_o^2 u(t)$. Notice that $\frac{d^2}{dt^2} v_s + \omega_o^2 v_s = 0$. Notice also that

$$w(t) = -k_1 z_1(t) - k_2 z_2(t) - k_3 z_3(t) - k_4 z_4(t) - k_5 z_5(t) = -Kz(t)$$

which is a full linear state feedback. Also notice that

$$z_1(t) = e(t), \quad z_2(t) = \dot{e}(t)$$

which are the tracking error and its derivative.

Therefore, using the LQT technique, the cost function

$$J = \int_0^\infty \left[qe(t)^2 + q_2 \dot{e}(t)^2 + q_3 z_3(t)^2 + q_4 z_4^2 + q_5 z_5^2 + w(t)^2 \right] dt$$

may be minimized using Matlab: ≫ `K = lqr(A, B, Q, 1)` where

$$A = \begin{bmatrix} 0 & 1 & 0 & 0 & 0 \\ -\omega_o^2 & 0 & 1 & -\frac{2P_{\text{set}}}{V_g^2} & -\frac{2Q_{\text{set}}}{V_g^2} \\ 0 & 0 & -\frac{R+R_s}{L+L_s} & \frac{1}{L+L_s} & 0 \\ 0 & 0 & -\mu \frac{RL_s - LR_s}{L+L_s} & -\mu \frac{L}{L+L_s} & -\omega_o \\ 0 & 0 & 0 & \omega_o & 0 \end{bmatrix}, \quad B = \begin{bmatrix} 0 \\ 0 \\ \frac{1}{L+L_s} \\ \mu \frac{L_s}{L+L_s} \\ 0 \end{bmatrix},$$

and $Q = \text{diag}([q,\ q_2,\ q_3,\ q_4,\ q_5])$.

Example 6.3 ***Weak Grid and PLL Modeling***
The same inverter and its parameters used in Example 6.1 is used and its controller is further enhanced to include two feedback branches from the ePLL as discussed in this section. For the grid impedance values, the nominal values of $L_s = 5$ mH and $R_s = 0.1\ \Omega$ are used. The power set-points P_{set} and Q_{set} are set to zero.[25] A snapshot of the root-locus is shown in Figure 6.22.

In Figure 6.22, q varies from 10^{10} to 10^{15} (blue); q_2 varies from 10^5 to $10^{8.5}$ (red); q_3 varies from 10^{-1} to 10^2 (magenta); q_4 varies from 10^{-1} to 10^1 (black); and finally q_5 varies from 10^{-1} to 10^2 (cyan). At final point, the closed-loop poles are located at $-1383, -977 \pm j1141, -269 \pm j269$. The controller gains are equal to $k_1 = 2.7 \times 10^7$, $k_2 = 48 \times 10^3$, $k_3 = 33.7$, $k_4 = 0.84$, and $k_5 = 0.28$.

The controller defined above integrates the state variables of the PLL, i.e. x_4 and x_5, as well as an indication of the grid impedance, i.e. R_s and L_s. It is therefore more robust to grid disturbances. Several simulation studies are presented in this section to illustrate this. Performance of the system of Example 6.1 is also shown for reference. Figure 6.23 shows their performances when the grid impedance is quite small ($L_s = 0.5$ mH, $R_s = 10$ mΩ), thus the grid is stiff.[26] As expected, both controllers perform much similarly. There is a small starting transient caused by the newly added feedback terms from the ePLL.

Figure 6.24 shows their performances when the grid impedance is increased to $L_s = 5$ mH, $R_s = 100$ mΩ. This reduces the SCR to around 10. Both controllers perform much similarly. The one with PLL feedback gains tend to show slightly more robust performance during the grid fault conditions.

Figure 6.25 shows their performances when the grid impedance is further increased to $L_s = 10$ mH, $R_s = 200$ mΩ. This reduces the SCR to around 5. The controller with the PLL feedback gains visibly shows more robust performances. In this study, the grid fault level is reduced to 50% because the controllers have much hard time maintaining the stability for deeper grid faults.

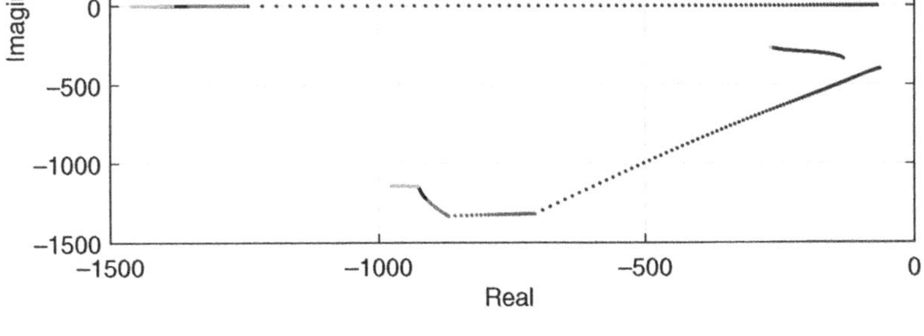

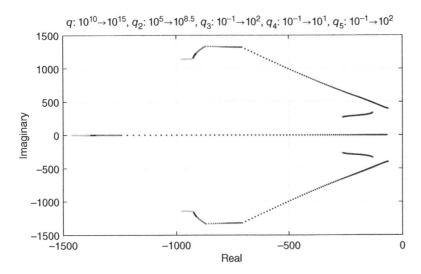

Figure 6.22 Poles of the closed-loop system for weak grid conditions and with PLL modeling, using LQT approach.

25 These two parameters will not have much impact on the results. See Problem 6.4.
26 $\text{SCR} = \frac{240^2}{3000\sqrt{0.01^2+(0.0005\times377)^2}} \approx 100.$

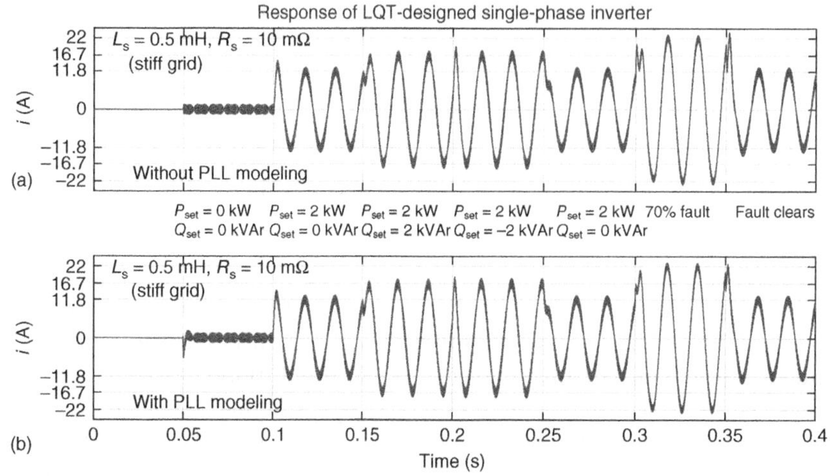

Figure 6.23 Performances of controllers without and with PLL modeling in stiff grid conditions. (a) Without and (b) with.

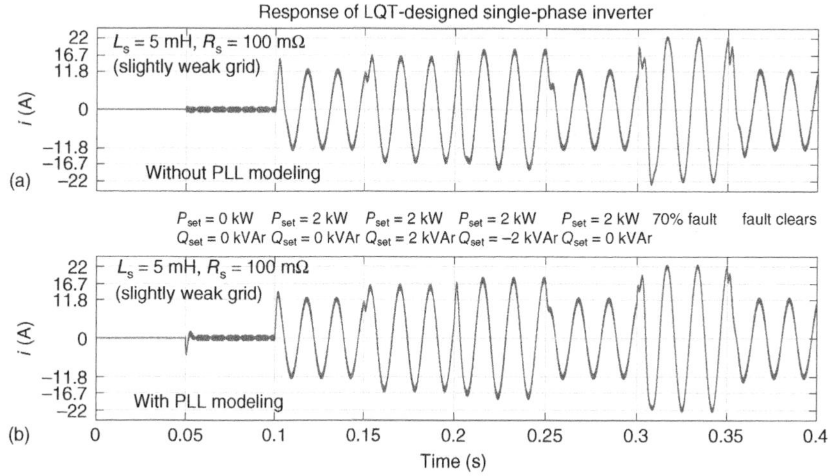

Figure 6.24 Performances of controllers without and with PLL modeling in slightly weak grid conditions.

Figure 6.26 shows their performances when the grid impedance is even further increased to $L_s = 15$ mH, $R_s = 300$ mΩ. This reduces the SCR to around 3.4. The system without PLL feedback gain is unstable. The instability is much more pronounced in the region where the reactive power command is negative. During this time, the grid voltage drops (due to the flow of reactive power into the inverter which causes a voltage drop across the grid inductance), and causes the current to increase (in order to make up for the commands of power). Meanwhile, the current limiter prevents the current from going up. This aggravates the instability during this time. The controller with the PLL feedback gains still shows relatively robust responses.

Figure 6.27 shows their performances when the grid impedance is $L_s = 10$ mH, $R_s = 200$ mΩ ("weak" grid). Here the grid voltage drop is made deeper than the previous case, i.e. deep into 60%.

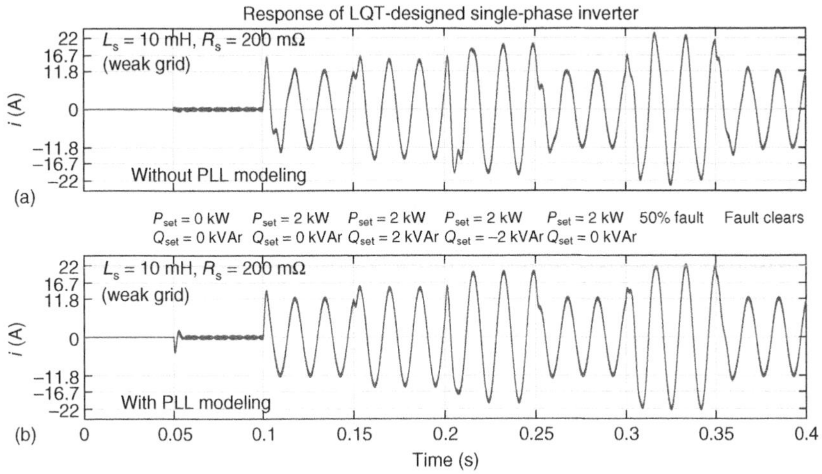

Figure 6.25 Performances of controllers without and with PLL modeling in weak grid conditions.

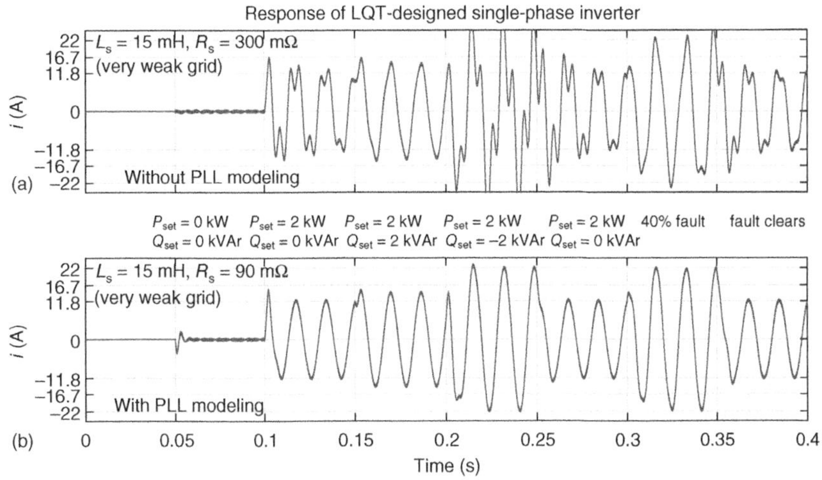

Figure 6.26 Performances of controllers without and with PLL modeling in very weak grid conditions.

This made the controller without the PLL feedback gains unstable. Therefore, we reduced its power from 2 to 1.6 kW and it became stable. This means that in order to be able to withstand a grid voltage disturbance of this magnitude, the inverter power must be limited to 1.6 kW. The controller with the PLL feedback gains, however, was able to withstand this level of disturbance at the rated power of 2 kW. We even increased the power to as much as 3 kW, as shown in Figure 6.27b, and it was still providing a robust and stable response.

Equation (6.16) indicates that the converter power S is multiplied into the grid impedance. This explains why at larger grid impedance values (weaker grids), the converter may not be able to transfer higher power because this corresponds to a much reduced SCR.

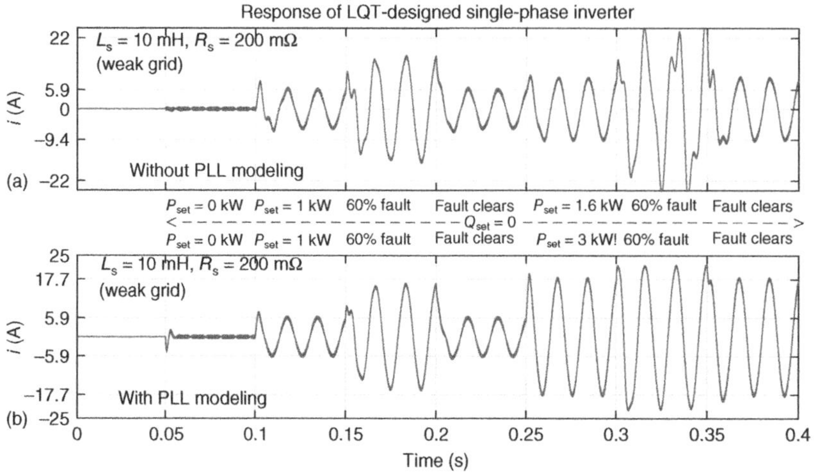

Figure 6.27 Performances of controllers without and with PLL modeling in weak grid conditions and large grid voltage disturbance (60%): the controller without PLL modeling can only transfer up to 1.6 kW while the one with PLL modeling can go much higher.

6.3 Grid-Supportive Controls

An inverter that exchanges the fixed powers P_{set} and Q_{set} and does not adjust them according to the grid conditions does not support the grid. Of course, the set-points P_{set} and Q_{set} may be adjusted by a central supervisory controller (also called the secondary controller) to account for grid conditions. This process, however, engages the optimization and communication stages and cannot provide an immediate and quick support. In this section, we discuss scenarios and policies where the inverter can provide quick grid support within its local (or primary) control system. The secondary controller will still be able to adjust the power set-points at a slower timescale to account for system-level optimization.

When the grid is in a condition where it needs support, this condition reflects on the voltage v_g where the inverter is connected to. The inverter must be able to identify such condition and take a "supportive" action. Contrary to a dc system, however, the voltage v_g is in the most fundamental case characterized by a *magnitude* and a *frequency*. The angle information may also be used.

In the existing ac power grid, where the bulk of power is generated by synchronous generators, it is the frequency of the system that is primarily an indication of the real power balance in the system. When it is below 60 Hz, for example, it means that the real power demand is larger than its supply. On the other hand, and again in the existing ac system with dominantly inductive transmission lines, it is the voltage magnitude that primarily indicates the reactive power balance in the system, When it is below the specified value, for example, it means that the demand of reactive power is more than its supply.

There are, however, a number of reservations to the above discussion. (i) The correspondence between real power and frequency on one hand and the reactive power and voltage magnitude on the other is not 100%, so to speak. Even in a large ac system dominated by synchronous generators and inductive lines, the voltage magnitude is impacted by the real power balance. The frequency is also impacted by the reactive power balance but to a much smaller degree. (ii) At the distribution level, where the (non-synchronous) DERs are often interconnected, and the transmission lines have larger R/X-ratio, the aforementioned correspondences will even be less valid. This whole discussion entails that the DER will have a harder time to identify the type of support that it needs to provide to the grid in order to really and effectively improve its condition.

6.3.1 Static (or Steady-State) Support

Assume that ω_o and V_o are the expected frequency and voltage at the local grid terminals. The actual values are measured (for example by an ePLL) and denoted by ω_g and V_g.[27] Then, the static incremental real and reactive powers may be formed as

$$\begin{bmatrix} \Delta P_s \\ \Delta Q_s \end{bmatrix} = \begin{bmatrix} \cos\theta_s & \sin\theta_s \\ -\sin\theta_s & \cos\theta_s \end{bmatrix} \begin{bmatrix} K_{fs}(\omega_o - \omega_g) \\ K_{vs}(V_o - V_g) \end{bmatrix}, \tag{6.20}$$

where θ_s is a given angle between 0° and 90°. For $\theta_s = 0$, we have the well-known droop terms (compatible with an ac system dominated by synchronous generators and inductive lines), i.e. $\Delta P_s = K_{fs}(\omega_o - \omega_g)$ and $\Delta Q_s = K_{vs}(V_o - V_g)$. For $\theta_s = 90°$, the "reverse" droop terms $\Delta P_s = K_{fs}(V_o - V_g)$ and $\Delta Q_s = -K_{vs}(\omega_o - \omega_g)$ are obtained that seem to be more advantageous for distribution systems with non-synchronous assets and resistive lines [4]. Thus, by using the angle θ_s, (6.20) allows a single point of freedom to combine the two to make a trade-off between the benefits of both terms. Moreover, since the matrix has a unity norm, the magnitude of the power vector, that is the apparent power S, remains unchanged for all values of angle. This is useful to prevent over-loading of the inverter.

6.3.2 Dynamic (or Inertia) Support

This support includes incremental real and reactive powers which are proportional to the derivatives of the frequency and voltage. They may be formed as

$$\begin{bmatrix} \Delta P_d \\ \Delta Q_d \end{bmatrix} = \begin{bmatrix} \cos\theta_d & \sin\theta_d \\ -\sin\theta_d & \cos\theta_d \end{bmatrix} \begin{bmatrix} -K_{fd}\dot{\omega}_g \\ -K_{vd}\dot{V}_g \end{bmatrix}, \tag{6.21}$$

where θ_d is a given angle between 0° and 90°. For $\theta_d = 0$, we have the well-known inertia terms (compatible with the kinetic inertia and magnetic field inertia in a synchronous generator), i.e. $\Delta P_d = -K_{fd}\dot{\omega}_g$ and $\Delta Q_d = -K_{vd}\dot{V}_g$. By using the angle θ_d, (6.21) also generalizes and allows a single point of freedom to allow a cross-inertia response from voltage to the real power and from frequency to the reactive power. The frequency and voltage derivatives, $\dot{V}_g$ and $\dot{\omega}_g$, may be tapped from an ePLL as shown in Figure 6.28 and using some filtering stages.

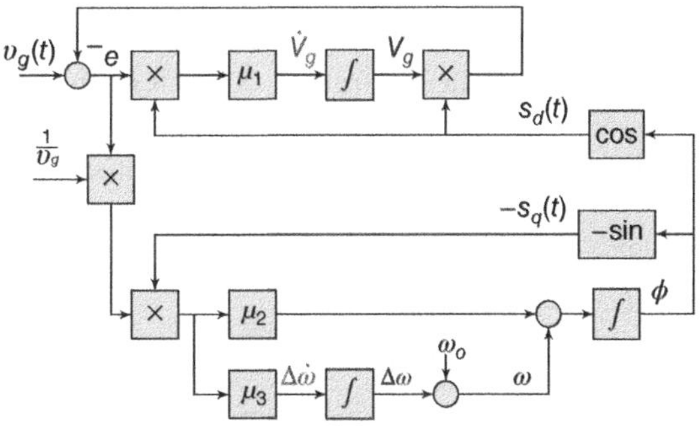

Figure 6.28 Derivative of voltage and frequency in the ePLL.

27 During this discussion, we assume that the measurement device (ePLL, in this case) is able to estimate these two variables accurately and quickly.

6.3.3 Power Controller with Grid Support

The same power control approach discussed in Section 6.2 can be upgraded to provide grid support. The real and reactive power references are generated from their set-points and the grid-support increments as

$$\begin{bmatrix} P_{\text{ref}} \\ Q_{\text{ref}} \end{bmatrix} = \begin{bmatrix} P_{\text{set}} \\ Q_{\text{set}} \end{bmatrix} + R^{-1}(\theta_s) \begin{bmatrix} K_{fs}(\omega_o - \omega_g) \\ K_{vs}(V_o - V_g) \end{bmatrix} + R^{-1}(\theta_d) \begin{bmatrix} -K_{fd}\dot{\omega}_g \\ -K_{vd}\dot{V}_g \end{bmatrix}, \tag{6.22}$$

where $R^{-1}(\theta)$ is the inverse rotation matrix defined in (6.20) and (6.21). In this approach, the grid voltage magnitude and frequency, and their rate of change (or derivative), must be measured and used to compute the power references. It is thus very important to have a fast and smooth estimation of these four variables.

Finally, to prevent overloading of the converter (that could happen during transient but longer-term drops in the grid voltage magnitude and/or frequency), the power reference vector of (6.22) may be limited such that its magnitude remains under the apparent power rating of the converter.

6.3.4 Virtual Synchronous Machine (VSM)

The method of providing grid support using the power control discussed above requires measurement of grid voltage attitudes (voltage magnitude, frequency, and their derivatives) to generate the reference for the powers as expressed in (6.22). An alternative approach is to internally synthesize a voltage that its attitudes copy (or track) the attitudes of the grid voltage. To start the discussion, we want the inverter to produce the real and reactive powers equal to the right side of (6.22). At this desired operating condition, the inverter must satisfy

$$\begin{bmatrix} K_{fd}\dot{\omega}_g \\ K_{vd}\dot{V}_g \end{bmatrix} = R(\theta_d) \begin{bmatrix} P_{\text{set}} - P \\ Q_{\text{set}} - Q \end{bmatrix} + R(\theta_d)R^{-1}(\theta_s) \begin{bmatrix} K_{fs}(\omega_o - \omega_g) \\ K_{vs}(V_o - V_g) \end{bmatrix}. \tag{6.23}$$

This is shown in the diagram of Figure 6.29 where the constant parameters are renamed according to $J\omega_o = K_{fd}$, $M = K_{vd}$, $K_f = K_{fs}$, $K_v = K_{vs}$ to simplify the notations and also use some common notations in the literature. Specifically, J has the meaning of "moment of inertia." We have also assumed $\theta_s = \theta_d = \theta$.

In Figure 6.29, the synthesized voltage $v(t)$ represents the inverter voltage and can be forwarded to the PWM unit to generate the gating signals for the VSC. The frequency ω_g is the grid voltage which will be copied (within its bandwidth) by the VSM frequency ω. Thus, it is possible to replace ω_g by ω (in the frequency droop term). Although V also mimics the dynamics of V_g but its value can be significantly different from V_g. Therefore, the grid voltage magnitude V_g needs to be measured and used in the voltage droop term. The mechanism described above and shown in Figure 6.29 to generate an internal voltage $v(t)$ which mimics the grid voltage is conceptually and in principle very

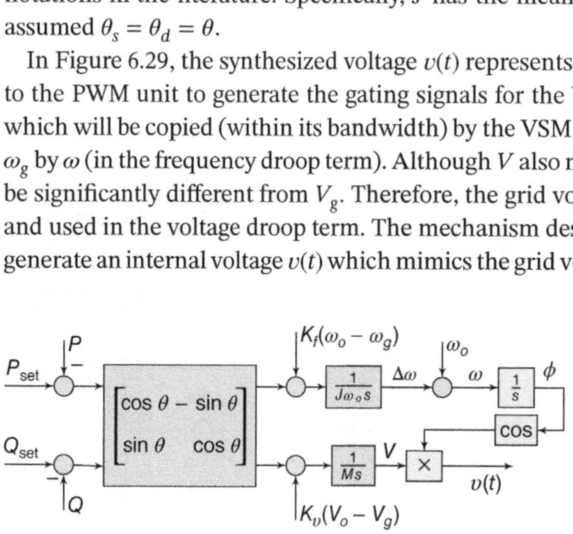

Figure 6.29 Virtual synchronous machine (VSM) concept.

similar to what is done in an actual synchronous machine. Hence, the name virtual synchronous machine (VSM), among some other similar names, have been used for this approach.

The VSM of Figure 6.29 has two limitations: (i) it lacks a damping strategy (similar to what is used in damper windings of synchronous generators)[28], and (ii) it does not do current limiting.[29] Figure 6.30 shows a modified structure where the current reference i_{ref} has been generated and limited to prevent over-current transients. This signal is then forwarded to the CFL.

To further understand the operation of Figure 6.30, we notice that since the current control loop is very fast, i_{ref} is close to i, and the voltage $v_1(t)$ emulates the voltage at a point that is connected through the virtual impedance $Z_v(s) = L_v s + R_v$ to the grid voltage and the same current i flows in it, as shown by the circuit branch in Figure 6.30. Therefore, the resistance R_v also functions as another damper.

For $\theta = 0$, the VSM is much similar to a conventional synchronous generator where the real power and frequency, and reactive power and voltage, are coupled. In an SG, the synchronous reactance $X = L_v\omega_o$ is much larger than R_v. For $\theta = 90°$, on the other hand, it appears that R_v must be dominant, and L_v may even be completely removed. This system will expectedly have much stable responses although it is not compatible with the existing SGs. It could be the basis for a new ac power grid where such inverters constitute a large share compared with the SGs.

The current-limiting approach shown in Figure 6.30, however, has a problem: since i_{ref} is a sinusoidal function, the limiter clamps it into a distorted function. This will distort the current during the current-limiting mode. Moreover, the linear current controller (CC) will have hard time tracking a nonlinear (clipped sinusoidal) reference. To address, these problems, this section of the controller (i.e. the current-limiting function) is implemented in synchronous frame as shown in Figure 6.31. The synchronous frame is defined using the VSM angle ϕ: $x_{dq} = R(-\phi)x_{\alpha\beta}$ where $R(\cdot)$ is the 2×2 rotation matrix, as explained below.

Define $x_{dq} = x_d + jx_q = e^{-j\phi}x_{\alpha\beta} = (\cos\phi - j\sin\phi)(x_\alpha + jx_\beta) = (\cos\phi x_\alpha + \sin\phi x_\beta) + j(-\sin\phi x_\alpha + \cos\phi x_\beta)$. Then, the equation $L_v\frac{d}{dt}i_1 + R_v i_1 = v_1 - v_g$ can be extended to the β component where this component is obtained by applying a simple filter $\frac{\omega_o - s}{\omega_o + s}$ as shown in Figure 6.32. This filter is an all-pass filter, unity magnitude at all frequencies, and introduces 90° delay at the frequency of ω_o. Adding the α and β components results in $L_v\frac{d}{dt}i_{1_{\alpha\beta}} + R_v i_{1_{\alpha\beta}} = v_{1_{\alpha\beta}} - v_{g_{\alpha\beta}}$ which can be written

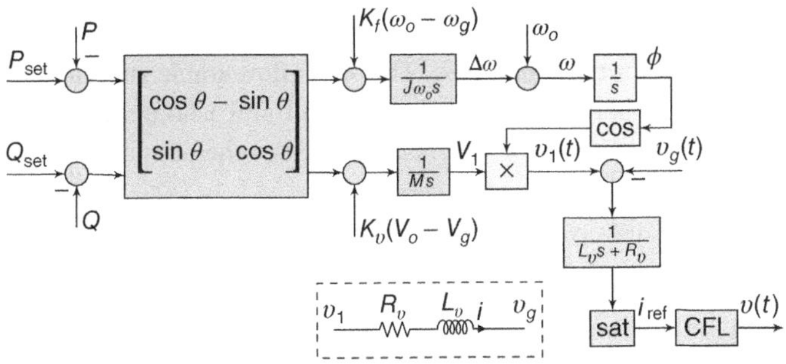

Figure 6.30 Adding a virtual impedance to allow current limiting in VSM.

28 Technically, the term $K_f(\omega_o - \omega)$ is a damper but it may not be strong enough depending on the size of K_f.

29 Overloading may be addressed by readjusting the power set-points as well as limiting the droop terms. Note that we distinguish between "overloading" and "over-current." Overloading is when the converter supplies more power than its power rated capacity. Over-current condition is when the converter current crosses the limit of current imposed by the switches. Over-current condition must be prevented quickly to protect the switches. However, overloading is a condition that must not continue over an extended time but can happen during shorter intervals.

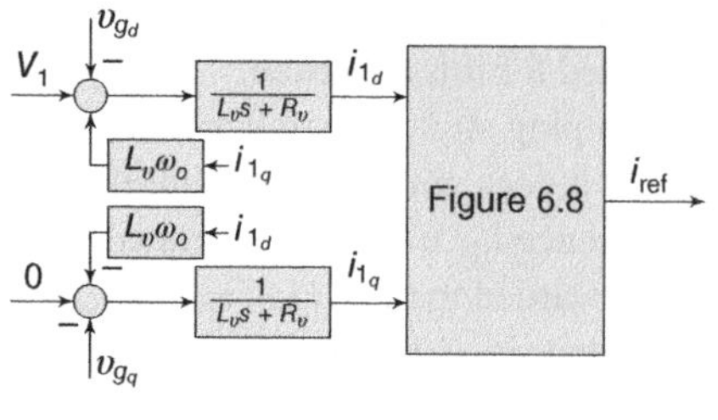

Figure 6.31 Implementation of current-limiting mechanism in synchronous frame.

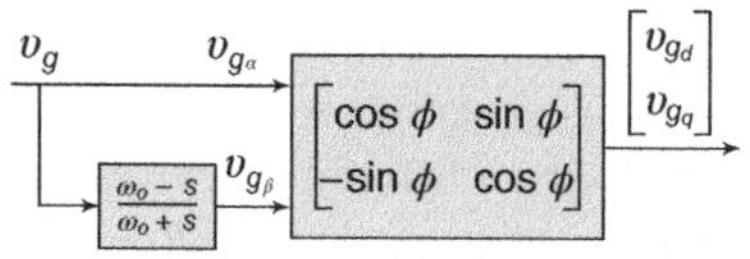

Figure 6.32 Calculating *dq* components.

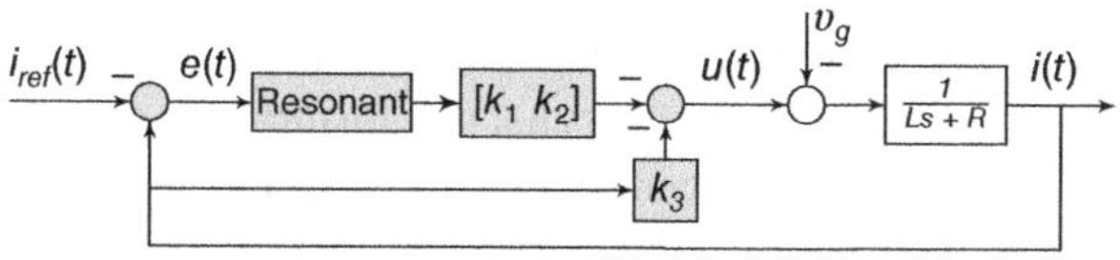

Figure 6.33 The current feedback loop (CFL) for VSM: no feed-forward term.

as $L_v \frac{d}{dt}(i_{1_{dq}} e^{j\phi}) + R_v i_{1_{dq}} e^{j\phi} = v_{1_{dq}} e^{j\phi} - v_{g_{dq}} e^{j\phi}$. This leads to $L_v \frac{d}{dt} i_{1_{dq}} + L_v j\omega_o i_{1_{dq}} + R_v i_{1_{dq}} = v_{1_{dq}} - v_{g_{dq}}$ where we approximated $\dot{\phi}$ with ω_o which is justifiable. Finally,

$$
\begin{aligned}
L_v \frac{d}{dt} i_{1_d} + R_v i_d &= V_1 - v_{g_d} + L_v \omega_o i_{1_q} \\
L_v \frac{d}{dt} i_{1_q} + R_v i_q &= 0 - v_{g_d} - L_v \omega_o i_{1_d},
\end{aligned}
\tag{6.24}
$$

where $v_{1d} = V_1$ and $v_{1q} = 0$ because its angle is used for the transformation.

The CFL used in VSM is shown in Figure 6.33 which is the common structure we have introduced before. However, here the feed-forward term is *not* used. This will allow stable standalone (islanded) operation and seamless transition between the grid-connected and islanded modes. Notice that the modulation index is $m(t) = \frac{u(t)}{v_c(t)}$ and $u(t)$ internally replicates the actual inverter voltage $v(t)$.

In addition to allowing the current limiting, this synchronous (or *dq*) frame calculation has some other advantages as described below.

1. The grid voltage magnitude V_g used in the droop term $K_v(V_o - V_g)$ can be calculated using

$$
\boxed{V_g = \sqrt{v_g^2 + v_{g_\beta}^2}.} \tag{6.25}
$$

2. The real and reactive powers P and Q used in the VSM loop can be calculated using[30]

$$
\boxed{P = 0.5 V_1 i_{1_d}, \quad Q = -0.5 V_1 i_{1_q}.} \tag{6.26}
$$

30 Notice that $P = 0.5(v_{1d} i_{1_d} + v_{1q} i_{1_q})$ and $Q = 0.5(v_{1q} i_{1_d} - v_{1d} i_{1_q})$ but $v_{1d} = V_1$ and $v_{1q} = 0$ which leads to (6.26).

This means that the powers are calculated using the "internal" VSM current and voltage rather than actual converter current and grid voltage.

These equations result in a fast and smooth calculation of these three variables.

Remark Equation (6.26) indicates that the powers are calculated at the virtual terminal associated with $v_1(t)$. As the equivalent circuit of Figure 6.30 shows, the virtual impedance $L_v s + R_v$ stands between this point and the actual terminal associated with the grid voltage v_g. Therefore, if the power set-points P_{set} and Q_{set} are meant to be the powers at the grid terminals, they must be properly adjusted to count for the "virtual" real and reactive powers consumed in this impedance. Thus, the following terms must be added to them, respectively.

$$P_{v,set} = 0.5R_v I_{set}^2, \quad Q_{v,set} = 0.5L_v \omega_o I_{set}^2, \quad I_{set} = \frac{\sqrt{P_{set}^2 + Q_{set}^2}}{0.5V_o}. \tag{6.27}$$

6.3.4.1 Stability Analysis and Design of VSM

It is clear from Figure 6.30 that the VSM resembles a PLL. This control system, indeed, integrates the function of PLL within itself and does not need a PLL to operate. Therefore, it is possible to use the available ePLL theory to perform a stability analysis and design. In order to establish even a closer analogy between the VSM and the ePLL, an additional damping branch, denoted by the block D in Figure 6.34, is added in this section.

The angle and magnitude differential equations may be written from Figure 6.34

$$\begin{bmatrix} \dot{\phi} \\ \dot{V} \end{bmatrix} = \begin{bmatrix} \omega + DK_f(\omega_o - \omega) \\ \frac{K_v}{M}(V_o - V_g) \end{bmatrix} + \begin{bmatrix} D & 0 \\ 0 & \frac{1}{M} \end{bmatrix} \begin{bmatrix} \cos\theta & -\sin\theta \\ \sin\theta & \cos\theta \end{bmatrix} \begin{bmatrix} P_{set} - P \\ Q_{set} - Q \end{bmatrix}, \tag{6.28}$$

where V_1 is denoted by V to simplify the notations. The powers are calculated from (6.26), which can be expressed as

$$\begin{bmatrix} P \\ Q \end{bmatrix} = 0.5V \begin{bmatrix} i_{1_d} \\ -i_{1_q} \end{bmatrix} = 0.5V \begin{bmatrix} \cos\phi & \sin\phi \\ \sin\phi & -\cos\phi \end{bmatrix} \begin{bmatrix} 1 \\ \frac{\omega_o - s}{\omega_o + s} \end{bmatrix} i_{ref}.$$

Define $x = [x_1 \ x_2]^T = [V\cos\phi \ V\sin\phi]^T$, then

$$\dot{x} = \dot{V} \begin{bmatrix} \cos\phi \\ \sin\phi \end{bmatrix} + V\dot{\phi} \begin{bmatrix} -\sin\phi \\ \cos\phi \end{bmatrix} = \begin{bmatrix} -x_2 & x_1 \\ x_1 & x_2 \end{bmatrix} \begin{bmatrix} \dot{\phi} \\ \frac{\dot{V}}{V} \end{bmatrix}. \tag{6.29}$$

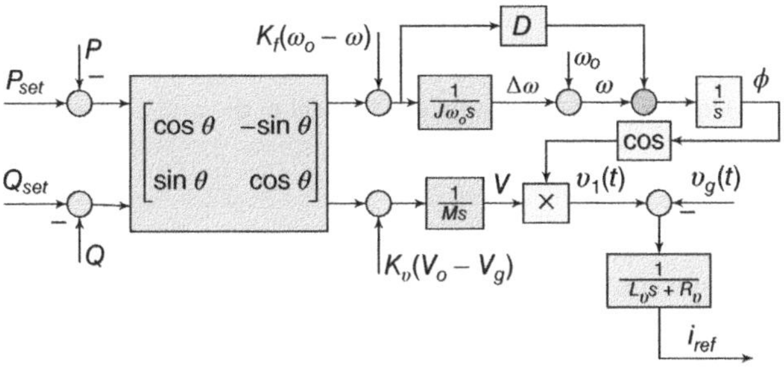

Figure 6.34 Adding the damping term D to establish closer analogy with ePLL.

At this level of stability analysis, assume that the droop terms are neglected (which is a justified assumption for grid-connected operation), then

$$\dot{x} = \begin{bmatrix} -\omega x_2 \\ \omega x_1 \end{bmatrix} + \begin{bmatrix} -x_2 & x_1 \\ x_1 & x_2 \end{bmatrix} \begin{bmatrix} D & 0 \\ 0 & \frac{1}{MV} \end{bmatrix} \begin{bmatrix} \cos\theta & -\sin\theta \\ \sin\theta & \cos\theta \end{bmatrix} \begin{bmatrix} P_{set} - P \\ Q_{set} - Q \end{bmatrix}. \tag{6.30}$$

Let us assume

$$\boxed{D = \frac{1}{MV} = \frac{\mu}{V^2}, \quad \mu > 0.} \tag{6.31}$$

Then, it is not difficult to see that (see Problem 6.6),

$$\begin{bmatrix} -x_2 & x_1 \\ x_1 & x_2 \end{bmatrix} \begin{bmatrix} \cos\theta & -\sin\theta \\ \sin\theta & \cos\theta \end{bmatrix} \begin{bmatrix} \cos\phi & \sin\phi \\ \sin\phi & -\cos\phi \end{bmatrix} = V \begin{bmatrix} \sin\theta & -\cos\theta \\ \cos\theta & \sin\theta \end{bmatrix}, \tag{6.32}$$

$$\begin{bmatrix} -x_2 & x_1 \\ x_1 & x_2 \end{bmatrix} \begin{bmatrix} \cos\theta & -\sin\theta \\ \sin\theta & \cos\theta \end{bmatrix} \begin{bmatrix} P_{set} \\ Q_{set} \end{bmatrix} = \begin{bmatrix} \sin\theta & \cos\theta \\ \cos\theta & -\sin\theta \end{bmatrix} \begin{bmatrix} P_{set} & Q_{set} \\ Q_{set} & -P_{set} \end{bmatrix} \begin{bmatrix} x_1 \\ x_2 \end{bmatrix}. \tag{6.33}$$

Combining these results in[31]

$$\begin{aligned} \dot{x} = & \begin{bmatrix} 0 & -\omega \\ \omega & 0 \end{bmatrix} \begin{bmatrix} x_1 \\ x_2 \end{bmatrix} + \frac{\mu}{V^2} \begin{bmatrix} \sin\theta & \cos\theta \\ \cos\theta & -\sin\theta \end{bmatrix} \begin{bmatrix} P_{set} & Q_{set} \\ Q_{set} & -P_{set} \end{bmatrix} \begin{bmatrix} x_1 \\ x_2 \end{bmatrix} \\ & -\frac{\mu}{2} \begin{bmatrix} \sin\theta & -\cos\theta \\ \cos\theta & \sin\theta \end{bmatrix} \begin{bmatrix} 1 \\ \frac{\omega_o - s}{\omega_o + s} \end{bmatrix} i_{ref}. \end{aligned} \tag{6.34}$$

Assuming zero power set-points,[32] and constant frequency condition $\omega = \omega_o$, this is an LTI equation. The transfer function from i to $x_1 = v_1$ is equal to $\frac{X_1(s)}{I(s)} = -\mu G(s)$ where[33]

$$G(s) = \frac{\sin\theta + \cos\theta}{2} \frac{(s - z_1)(s - z_2)}{(s^2 + \omega_o^2)(s + \omega_o)}, \tag{6.35}$$

and

$$z_1 = \omega_o \frac{\cos\theta - \sin\theta - \sqrt{2}}{\cos\theta + \sin\theta}, \quad z_2 = \omega_o \frac{\cos\theta - \sin\theta + \sqrt{2}}{\cos\theta + \sin\theta}. \tag{6.36}$$

The transfer function control block diagram of the system may be represented as in Figure 6.35.

The characteristics equation of the loop will be

$$1 + \mu \frac{1}{L_v s + R_v} G(s) = 1 + \mu \frac{\sin\theta + \cos\theta}{2L_v} \frac{(s - z_1)(s - z_2)}{(s^2 + \omega_o^2)(s + \omega_o)(s + \frac{R_v}{L_v})} = 0. \tag{6.37}$$

To study the roots of (6.37), the root-locus approach is considered as described below. First we notice that the zeros of $G(s)$ are real for all values of θ between zero to 90°. Their variations with respect to θ is shown in Figure 6.36.

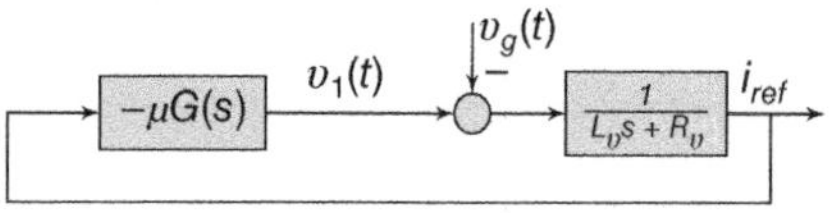

Figure 6.35 Approximate LTI model of the grid-connected VSM.

31 Note that (6.34) has slightly abused the notations as it mixes the time-domain and Laplace-domain variables. But this simplifies the discussion and should not cause confusion to the reader.

32 Case of nonzero power set-points can also be readily studied by linearizing the term $\frac{1}{V^2}$ around its operating point. See Problem 6.7.

33 See Problem 6.8.

Figure 6.36 Variations of zeros of $G(s)$ when θ changes.

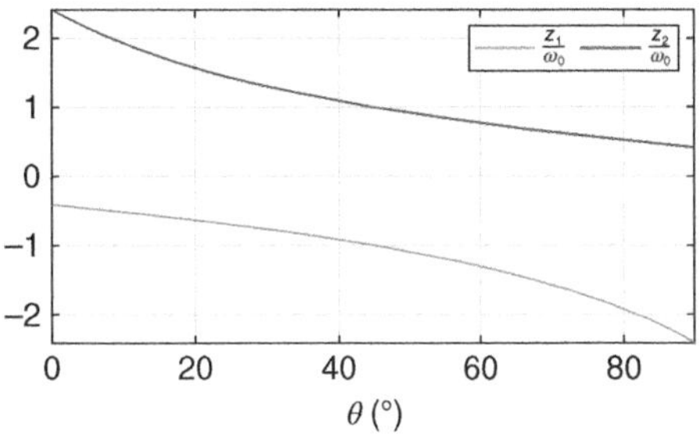

Figure 6.37 Locus of the roots of (6.38) when μ varies from 0 to (6.39).

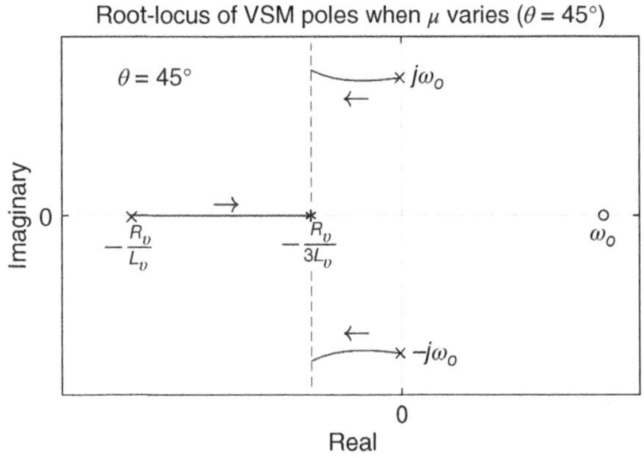

Therefore, to simplify the design, we suggest to choose $\theta = 45°$. With this choice, the characteristic equation reduces to

$$1 + \mu \frac{1}{\sqrt{2}L_v} \frac{(s - \omega_o)}{(s^2 + \omega_o^2)\left(s + \frac{R_v}{L_v}\right)} = 0, \tag{6.38}$$

which has three roots. When μ (as a design parameter) increases, the roots vary according to the sketch shown in Figure 6.37. The real root moves to the right. The complex poles move to the left and go up. It is easy to prove that for

$$\boxed{\mu = \frac{2\sqrt{2}R_v\left(R_v^2 + 9L_v^2\omega_o^2\right)}{9L_v(R_v + 3L_v\omega_o)}} \tag{6.39}$$

the three roots are vertically aligned at the real part of $-\frac{R_v}{3L_v}$.[34] As for the design of R_v and L_v, following is the suggestion. (i) Choose the magnitude of this impedance, i.e. $Z_v = \sqrt{R_v^2 + L_v^2\omega_o^2}$, around the reasonable value of 0.2 pu. (ii) Choose $\frac{R_v}{L_v}$ at $\sqrt{2}\omega_o$ which is the midway of z_1 for entire θ. This will lead to

$$\boxed{L_v = \frac{0.2Z_b}{\sqrt{3}\omega_o}, \quad R_v = \sqrt{2}\omega_o L_v,} \tag{6.40}$$

where $Z_b = \frac{V_b^2}{S_b}$ and S_b and V_b are the base values of the power and voltage of the inverter.

34 See Problem 6.9.

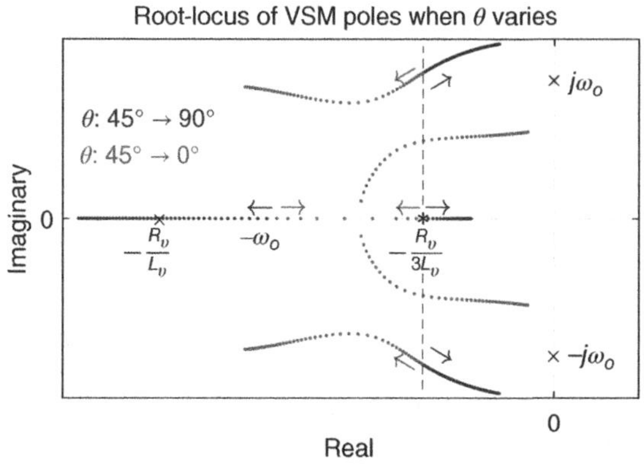

Figure 6.38 Locus of the roots of (6.37) when θ varies, at μ of (6.39), R_v and L_v are selected from (6.40).

Figure 6.37 shows the locus of the roots of (6.38) when μ varies from 0 to the suggested value given in (6.39) which confirms that the the three poles are aligned at $-\frac{R_v}{3L_v}$ at this value of μ. For the suggested value of $\frac{R_v}{L_v} = \sqrt{2}\omega_o = 533$, the three poles are aligned at the real value of -178.

Figure 6.38 shows the locus of the roots of (6.37) when θ varies, while μ is fixed at the value of (6.39), and R_v and L_v are fixed at the values of (6.40). For both cases where θ approaches zero (corresponding to an SG scenario), and when θ goes above 45°, the damping of the poles go down. This is understandable because (6.37) and (6.40) are designed for $\theta = 45°$. However, this graph and also the simulation results shown later confirm that this design is good enough for the entire range of θ. For a specific θ, the design of μ, R_v, and L_v can be further adjusted and tuned.

Discussion: The case $\theta = 0$ represents the common VSM (which emulates a synchronous generator). The case $\theta = 90°$ signifies a more stable and damped control system. The problem with this VSM, however, is that it is not compatible with the existing power grid which is dominantly powered by synchronous generators. Therefore, it will tend to function incoherent with them. As a result, even though this VSM is in itself more stable, when operating in parallel with many conventional SGs, it may degrade the total system's stability [4]. A value of θ in between the two seems to be a more practically desired choice.

Design of J: The parameter J represents "the moment of inertia" of the machine. In conventional large synchronous generators, the inertia constant $H = \frac{0.5J\omega_o^2}{S}$ is between 1 to 10, larger for larger machines. This can immediately lead to a selection for J that is compatible with the conventional machines as $J = \frac{2HS}{\omega_o^2}$ (where S is the DER VA rating). Alternatively, and based on the ePLL theory, we notice that $\mu = DV^2 = \frac{V}{M}$ play the role of $\mu = \mu_1 = \mu_2$. Thus, $\frac{V^2}{J\omega_o}$ will play the role of μ_3. Thus, for instance, a selection of $\mu_3 = \frac{3}{32}\mu^2$ will lead to $\frac{V^2}{J\omega_o} = \frac{3}{32}\mu^2$ or $J = \frac{32V^2}{3\omega_o\mu^2}$.

Comment on Adaptive Parameters: The above analysis and design approach yielded to the VSM parameters D, M, and J that are functions of V. This means that, strictly speaking, they are adaptive. While there is no challenge with this and it can be implemented easily in practice, it is also possible to replace the variable V with its nominal value V_o which is the nominal value of grid voltage. This does not seem to introduce a noticeable impact on the VSM performance except possibly during the start-up and synchronization of the converter. During the start-up synchronization, the magnitude V may experience some swings. Therefore, using adaptive parameters can lead to faster and stronger pre-synchronization. It should also be noted that the division by V must be replaced by division $|V| + \epsilon$ to avoid an unlikely but possible divide-by-zero during starting synchronization. The positive constant ϵ may be set at 0.1% of the nominal voltage magnitude.

6.3.4.2 Start-up Synchronization

One great advantage of the VSM discussed in this section is that it can readily synchronize to the grid prior to the start of applying the gating signals to the VSC. Assume that P_{set} and Q_{set} are initially set to zero. Then, during the pre-synchronization time, we virtually close the feedback loop (without operating the converter) by emulating the filter and the grid as shown in Figure 6.39.

During this time, the converter is not switching. Therefore, the current i is zero. The virtual current i_v emulates the converter current. By regulating this current to zero (setting $i_r = 0$), the converter output voltage v (or u) will be synchronized with the grid voltage v_g. When i_v is sufficiently small (indicating that the synchronization is complete), the current i_v is replaced by the actual current i, the reference point i_r is replaced with the i_{ref} (generated by the VSM controller), and the converter starts. This process is expressed by

$$i_x = \begin{cases} i_v, & t < t_{start}; \\ i(t) \text{ [measured current]}, & t \geq t_{start}. \end{cases}$$
$$i_r = \begin{cases} 0, & t < t_{start}; \\ i_{ref}(t) \text{ [calculated by VSM controller]}, & t \geq t_{start}. \end{cases} \tag{6.41}$$

and shown in Figure 6.39. Note that the modulating signal is obtained from $u(t) = m(t)v_c$ where v_c is the dc side voltage.

6.3.4.3 Grid-Connection Synchronization

The VSM is a grid-forming controller and can operate the inverter in both standalone and grid-connected modes. To seamlessly transition from standalone (isolated) operation to grid-connected mode, prior to closing the switch SW in Figure 6.40, the converter side voltage v_g must be sufficiently synchronized with the grid side voltage v_s to avoid possible voltage disturbances for the load and current transients for the converter.

A synchronization mechanism is shown in Figure 6.41 where the difference between the converter-side voltage v_g and the grid-side voltage v_s, projected on the two axes d and q, is used to manipulate the converter powers in a feedback loop. The positive constants K_{s_d} and K_{s_q} determine how strongly the synchronization pushes the voltage v_g close to v_s. A reasonable selection for these parameters is $K_{s_d} = K_{s_q} = K_v$, where K_v is the droop coefficient of the voltage versus reactive power loop. Notice that by increasing the real power, the converters increases v_{g_d} and brings it closer to v_{s_d}. However, it will need to decrease its reactive power to increase v_{g_q}. The equation

Figure 6.39 Mechanism for achieving synchronization prior to starting the inverter.

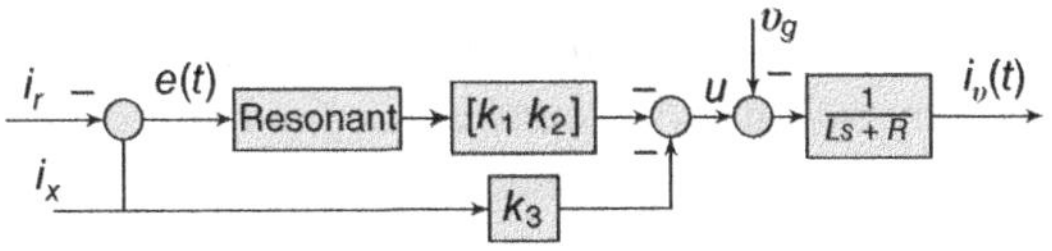

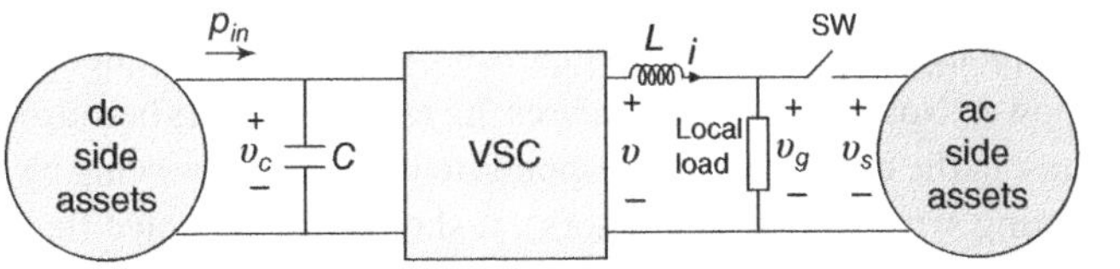

Figure 6.40 Grid-connected and standalone modes of operation.

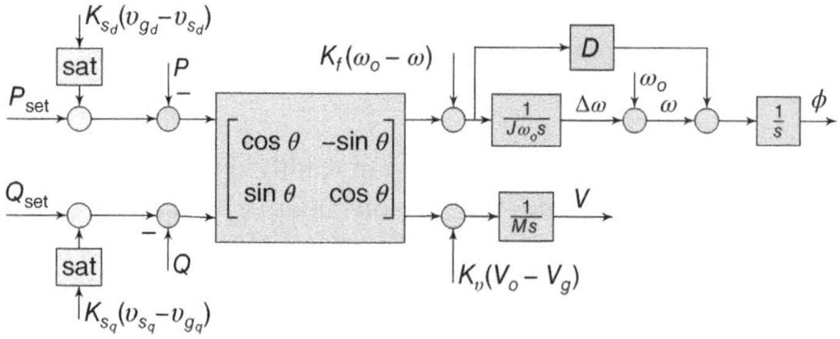

Figure 6.41 Mechanism to achieve synchronization prior to transition from standalone to grid-connected operation.

$P = 0.5Vi_d$ and $Q = -0.5Vi_q$) explain this: increasing real power directly increases V (leading to an increase in v_{g_d}) but decreasing Q increases i_q (leading to an increase in v_{g_q}). A limiter block, denoted by "sat" in Figure 6.41, is used to limit the synchronization power within $\pm S$ to prevent possible transients subsequent to enabling the synchronization block.

Example 6.4 ***Single-Phase VSM***
Consider the inverter and system shown in Figure 6.40 with the same parameters given in Table 6.1. The local load is 1.5 kVA at 0.8 PF lagging. A small capacitor of 2 μF is connected across the load terminals to filter the switching noises during standalone operation, and a resistance of 1 Ω is connected in series with it to passively damp the resonance. In this example, we design a complete VSM-based controller for this inverter and study its performance.

The controller parameters are selected as follows. (i) The virtual impedance parameters are selected according to (6.40) at $L_v = 6.2$ mH, $R_v = 3.3$ Ω. (ii) The parameter θ is chosen at 45°. (iii) The controller parameter $\mu = 982$ is selected according to (6.39) to place all three poles on a vertical line at -178 (one third of $\frac{R_v}{L_v}$). (iv) The moment of inertia J is selected at $J = 0.05$. This will correspond to an inertia constant of about $H = 1.2$ seconds. (v) The droop coefficients are selected at $K_v = \frac{S}{0.2V_o} = 41.7$ VAr/V, $K_f = \frac{S}{0.05\omega_o} = 150$ W s/rad. (vi) The grid-connection synchronization gain is selected at $K_s = K_v$. 7) The same LQT-based current controller of Example 6.1 is used.

Figure 6.42 shows the *starting, power set-points tracking, and grid low-voltage ride through responses* of the controller. During $0 \le t < 0.1$ seconds, the controller synchronizes to the grid voltage according to Figure 6.39. At $t = 0.1$ seconds, the inverter starts operating. The real and reactive power set-points are faithfully tracked. The current is limited during the grid voltage fault. During the fault, the inverter naturally reduces the real power and increases its reactive power. This is a correct fault ride-through operation. The internal VSM frequency and voltage variables (i.e. $f = \frac{\omega}{2\pi}$ and V) are also shown in Figure 6.42.

Figure 6.43 shows the *transition from grid-connected to standalone* (at $t = 0.8$ seconds), activation of *synchronization mechanism* of Figure 6.41 at $t = 1.2$ seconds, and *reconnection to grid* at $t = 1.5$ seconds. It is observed that all transitions are made seamlessly. The synchronization is achieved smoothly and quickly. Once it is enabled, it pushes v_{g_d} and v_{g_q} to the vicinity of v_{s_d} and v_{s_q}.

Responses of the controller to a *grid frequency swing* is shown in Figure 6.44. As the grid frequency goes up, the inverter decreases its real power. Note that it also increases the reactive powers because we have chosen $\theta = 45°$. If $\theta = 0°$ is chosen, the converter only responds to the frequency swing by changing its real power (similar to existing synchronous generators). It should also be noted that the power swings include both the dynamic (inertia) and the static (droop) components. However, since the inertia component is relatively small, only the droop component is visible in Figure 6.44.

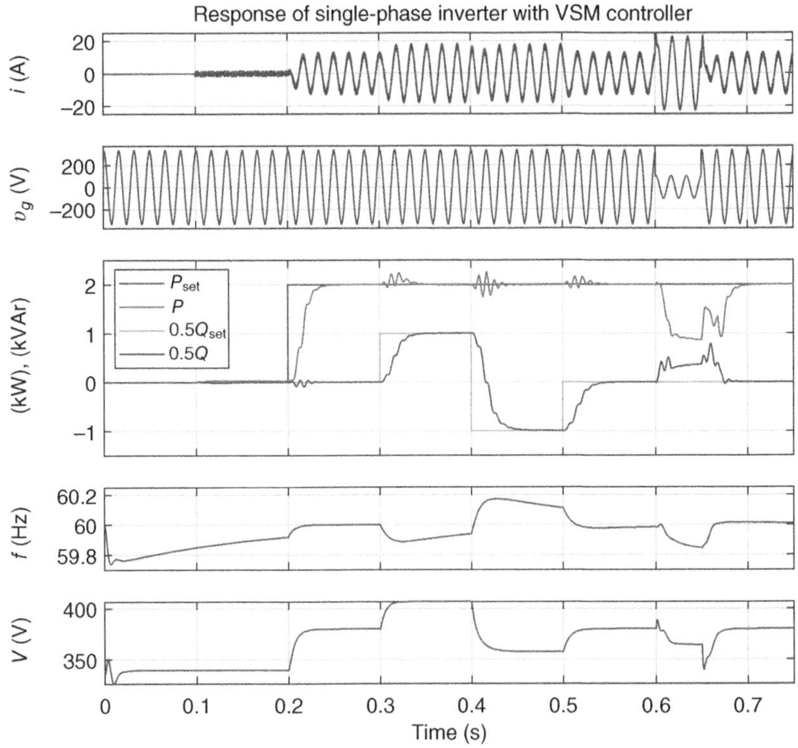

Figure 6.42 Responses of VSM controller: starting synchronization during $0 \leq t < 0.1$ seconds; inverter starts at $t = 0.1$ seconds; real and reactive power set-points at $t = 0.2, 0.3, 0.4$, and 0.5 seconds; low-voltage grid fault at $t = 0.6$ seconds; and fault clears at $t = 0.65$ seconds.

Responses of the controller to a *grid voltage magnitude swing* is shown in Figure 6.45. As the grid voltage magnitude goes down, the inverter increases its power. It increases both the real and reactive powers because we have chosen $\theta = 45°$. If $\theta = 0°$ is chosen, the converter will only respond to the voltage magnitude swing by changing its reactive power (similar to existing synchronous generators). It should also be noted that the power swings include both the dynamic (inertia) and the static (droop) components. However, since the inertia component is relatively small, only the droop component is visible in Figure 6.45.

Example 6.5 ***Power Controller with Grid Support***

This example simulates the power controller with the grid support, as described in (6.22). The two angles θ_s and θ_d are selected equally at 45° to make it similar to the VSM of Example 6.4. The grid frequency, magnitude, and their derivatives are computed from the ePLL as shown in Figure 6.28. The derivatives are passed through a first-order low-pass filter with the transfer function $\frac{1}{0.1s+1}$ as these two variables are very noisy. The droop and inertia gains are selected equivalent to those of the VSM of Example 6.4 for comparison. All other circuit parameters and also the current controller gains are unchanged.

Figure 6.46 shows the *starting, power set-points tracking, and grid low-voltage ride through responses* of the controller. During $0 \leq t < 0.1$ seconds, the controller synchronizes to the grid voltage and at $t = 0.1$ seconds, the inverter starts operating. The real and reactive power set-points are faithfully tracked. The current is limited during the grid voltage fault. During the fault, the

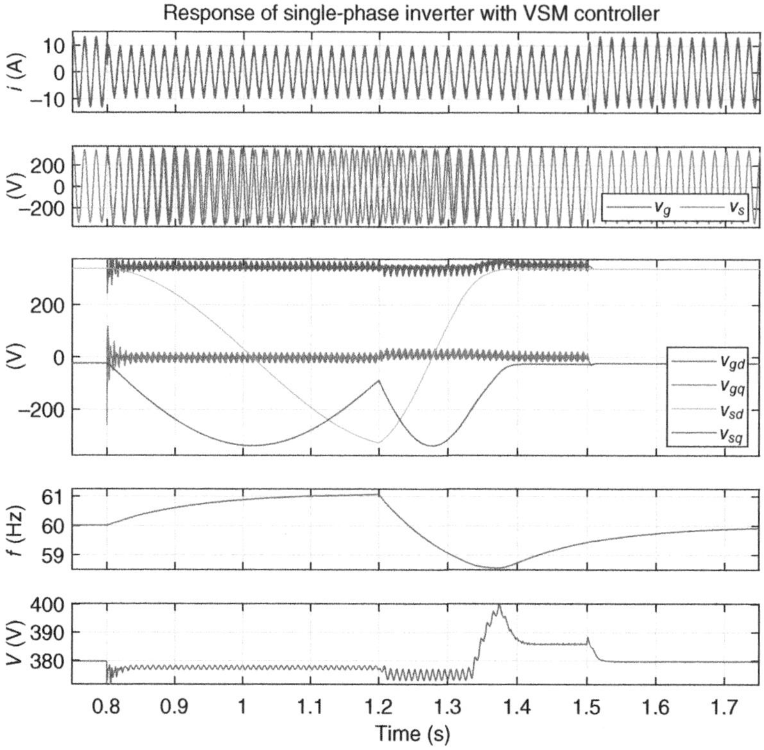

Figure 6.43 Responses of VSM controller: transition from grid-connected to standalone at $t = 0.8$ seconds; enabling the synchronization mechanism at $t = 1.2$ seconds; and reconnecting to the grid at $t = 1.5$ seconds.

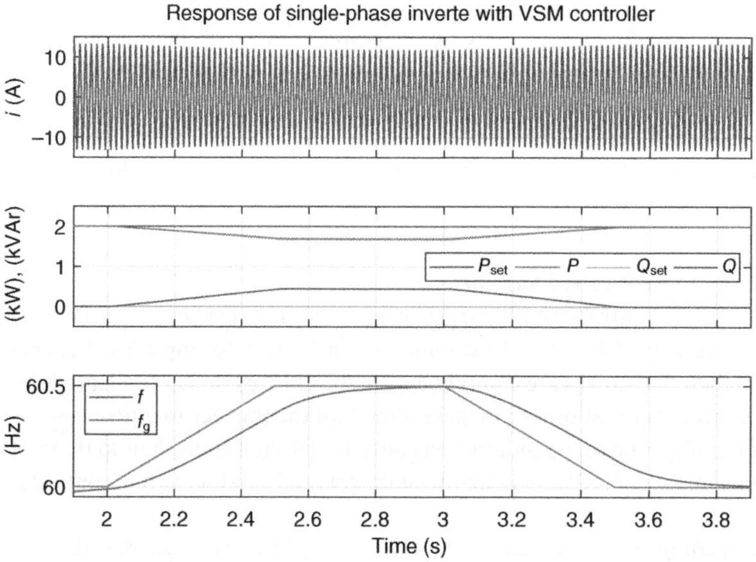

Figure 6.44 Responses of VSM controller: grid voltage frequency swing between $t = 2$ seconds and $t = 3.5$ seconds.

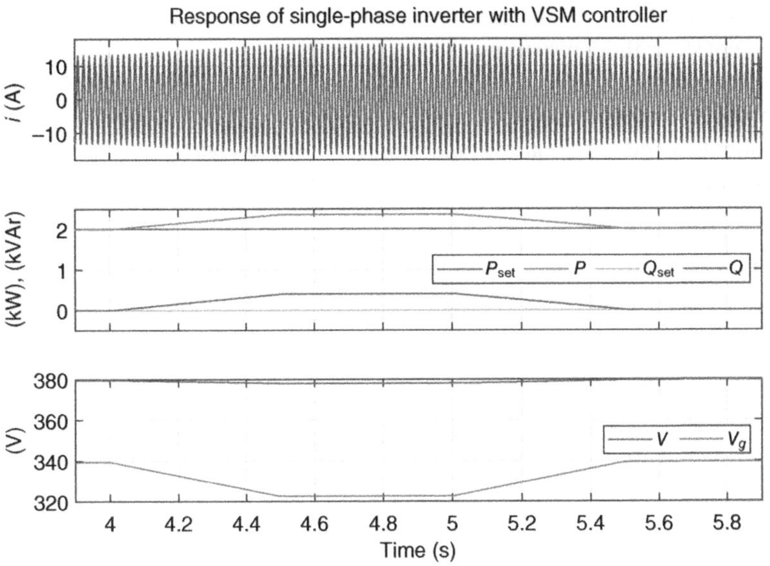

Figure 6.45 Responses of VSM controller: grid voltage magnitude swing between $t = 4$ seconds and $t = 5.5$ seconds.

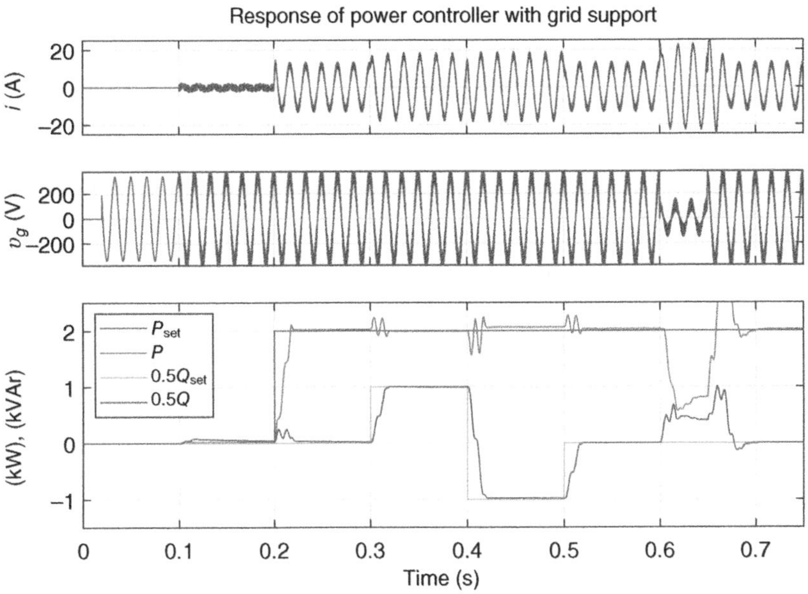

Figure 6.46 Responses of power controller *with grid support*: starting synchronization during $0 \leq t < 0.1$ seconds; inverter starts at $t = 0.1$ seconds; real and reactive power set-points at $t = 0.2, 0.3, 0.4,$ and 0.5 seconds; low-voltage grid fault at $t = 0.6$ seconds; and fault clears at $t = 0.65$ seconds.

inverter naturally reduces the real power and increases its reactive power. This is a correct fault ride-through operation. These responses of the controller is much similar to those of the VSM shown in Figure 6.42.

In this study, the power controller with grid support (PC-GS) was not able to stabilize the voltage when the grid got disconnected. We had two insert an extremely large capacitor in series with a large resistance across the local load to stabilize the voltage. This is not, however, a practical solution. Therefore, we do not show those responses.

Responses of the controller to a *grid frequency swing* is shown in Figure 6.47. The responses are relatively close to those of the VSM responses shown in Figure 6.44.

Responses of the controller to a *grid voltage magnitude swing* is shown in Figure 6.48. These are also pretty much similar to those of the VSM controller shown in Figure 6.45.

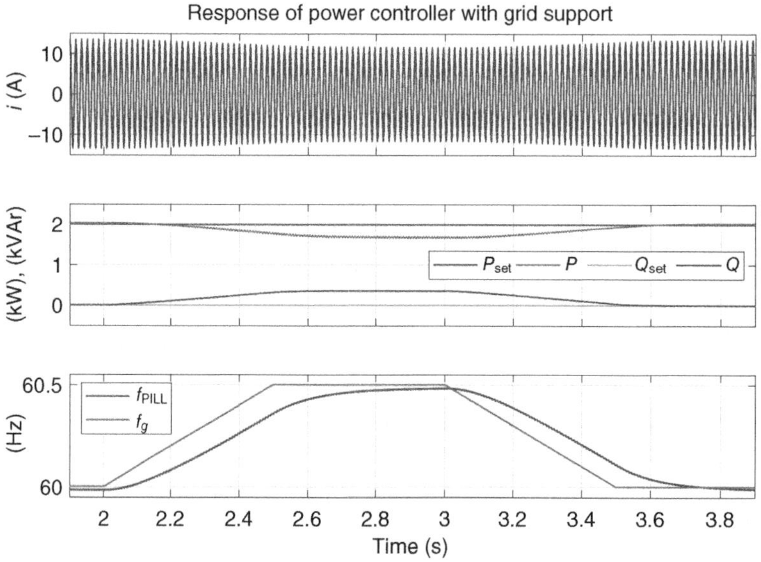

Figure 6.47 Responses of the power controller *with grid support*: grid voltage frequency swing between $t = 2$ seconds and $t = 3.5$ seconds.

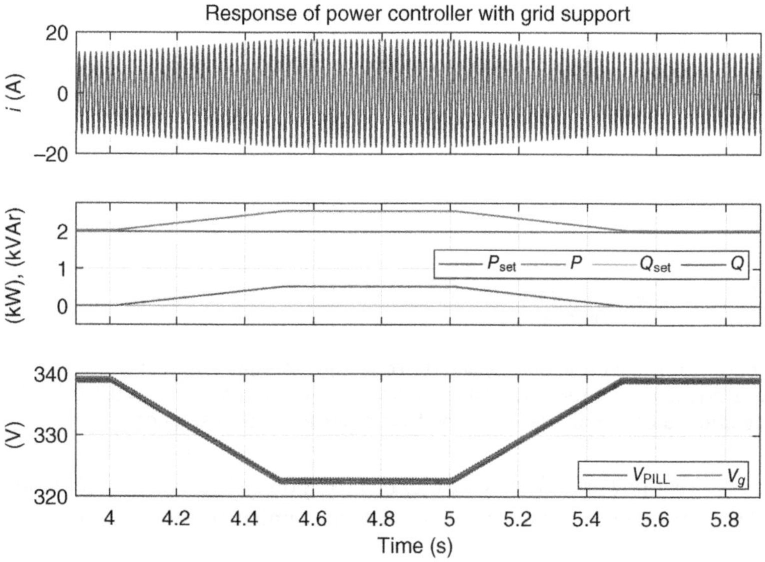

Figure 6.48 Responses of power controller *with grid support*: grid voltage magnitude swing between $t = 4$ seconds and $t = 5.5$ seconds.

Discussion on VSM and Power Controller with Grid Support (PC-GS) controllers: The VSM obviously beats the PC-GS in the transition from grid-connected to islanded condition. It can successfully "form a grid," hence the terminology *grid-forming controller* as opposed to just being a *grid-following controller*. The reason is probably due to the fact that it does not use a PLL. It rather emulates a PLL inside the VSM function. Therefore, it does not need to measure the grid voltage frequency, nor the grid voltage frequency and magnitude derivatives and build the voltage from them. Instead, it actually emulates an internal voltage, $v_1(t)$, independent from whether the grid voltage is available or not.[35] Overall, the VSM seems to be very stable in enabling different transitions. The PC-GS, however, is robust and fast during the grid-connected operation and it beats the VSM in such conditions. But its performance degrades for weak grid conditions while the VSM performs well during weak grid conditions as well. The VSM has a robust performance for weak grid conditions as well.

6.4 dc Voltage Control and Support

Consider the dc/ac converter shown in Figure 6.49. In Section 6.2, we discussed how to efficiently build and design a CFL for a grid-connected inverter. The control system receives real and reactive power set-points and executes while ensuring that the current is sinusoidal and within the pre-specified limits during abrupt/severe transients. In Section 6.3, we discussed how the controller can provide ac grid support functions. Two approaches of power controller with grid-support and VSM were introduced and studied. The CFL is still used in their cores.

These controllers address the operation of the converter as far as the ac grid side is concerned. These include the ac current quality, presence of grid non-ideality factors (such as grid voltage distortions, low-voltage transient faults, and weak grid conditions), and providing static and dynamic grid support.

The dc voltage support may also be provided by the controller. Similar to the ac situation, this support may be divided into static and dynamic supports. Both of these supports add an incremental power to the real power set-point, P_{set}, of the converter. Generally speaking, a rise in the dc voltage as an indication of excess of energy on the dc side, and vice versa.

Thus, the static support may be expressed by

$$\Delta P_{stat} = K_{stat}(v_{dc} - V_{dc}), \tag{6.42}$$

where $v_{dc}(t)$ is the actual dc side voltage and V_{dc} is its rated value. This means that the converter increases its power transfer from dc to ac when the dc voltage is above its nominal value, and decreases it otherwise. The positive constant K_{stat} determines how much the power is increased for a unit increase in the dc side voltage.

The dynamic (or inertia) support may be expressed by

$$\Delta P_{dyn} = K_{dyn}\dot{v}_{dc}(t). \tag{6.43}$$

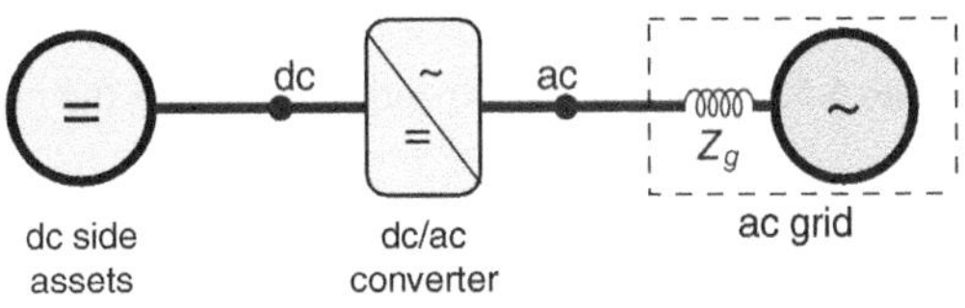

Figure 6.49 Converter interfacing a dc side to an ac side.

35 The problem of rigorously proving that the PC-GS is unstable (or has lower stability margins than the VSM) during grid-connected to islanded transition is open.

This means that the converter instantly increases its power transfer from dc to ac in proportion to the rate of change of dc side voltage. The positive constant K_{dyn} determines how much the power is increased for a unit increase in the dc side voltage rate of change.

In this section, we study in details the scenarios where the inverter is also responsible to control (and/or support) the dc side voltage.

6.4.1 System Modeling

The power system is shown in Figure 6.50 and the control block diagram of the system is shown in Figure 6.51 where C is the capacitance of the dc link and p_{in} is the input power to the dc link. Notice that for the dc-bus capacitor, the power balance indicates that

$$v_c(t)i_c(t) = Cv_c(t)\frac{d}{dt}v_c(t) = \frac{d}{dt}w_c(t) = p_{\text{in}} - p_{\text{out}}, \quad w_c(t) = \frac{1}{2}Cv_c(t)^2,$$

where $w_c(t)$ denotes the instantaneous stored energy in the capacitor C. The power p_{out} may be approximated with $v_g(t)i(t)$ by ignoring the inductor reactive power and the converter power losses.

The system model shown in Figure 6.51 is nonlinear. The nonlinearity corresponding to multiplication into the capacitor voltage is easily removed as shown in Figure 6.52. A feed-forward term $v_{\text{ff}}(t)$ may also be added that is provided by the PLL and compensates for the fundamental component of v_g. The resulted system is shown in Figure 6.53.

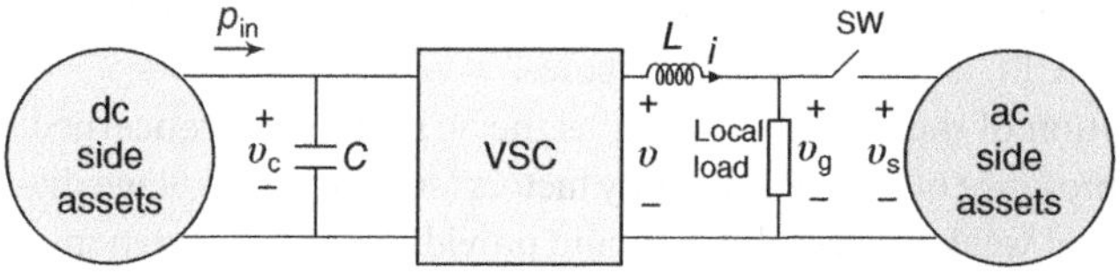

Figure 6.50 Circuit diagram of a converter with dc-side voltage to be controlled.

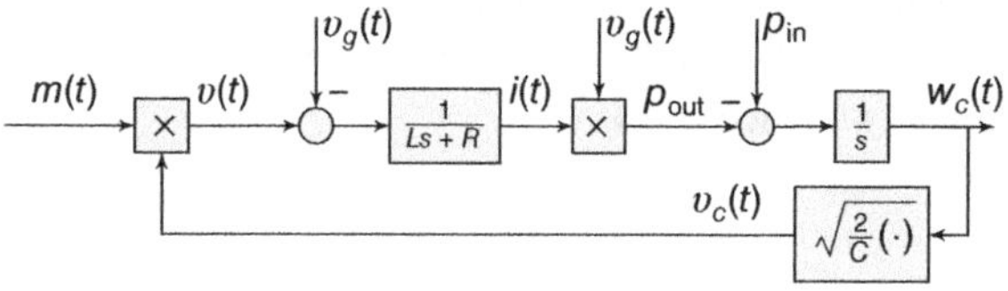

Figure 6.51 Block diagram corresponding to the dc side voltage control.

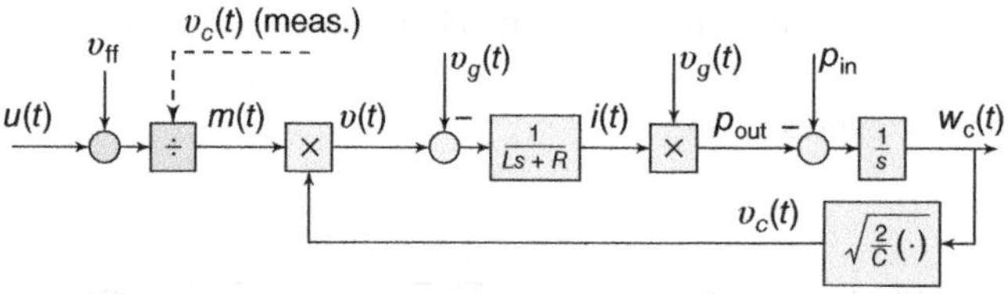

Figure 6.52 Feedback linearization to remove the nonlinearity at the input.

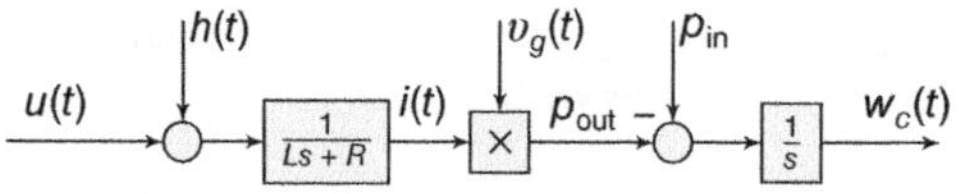

Figure 6.53 Linear equivalent model of Figure 6.52, $m(t) = \frac{u(t)+v_{\text{ff}}(t)}{v_c(t)}$

In a *two-stage converter*, the term p_{in} is controlled by the first-stage converter and is independent from the shown diagram. It may be treated as an external disturbance in this case. In a *single-stage PV converter*, p_{in} is nonlinearly (and uncertainly) dependent on v_c. In this case, the block diagram is shown in Figure 6.54. The PV nonlinearity/uncertainty makes this system nonlinear and its theoretical and rigorous analysis is challenging. In the rest of this section we treat p_{in} as a disturbance which is true in two-stage converters (or in general when the dc side is connected to multiple entities) and is also a valid assumption with some approximation in single-stage PV converter.

The energy is expressed by $w_c(t) = \frac{1}{2}Cv_c(t)^2$. Let us assume that the dc-bus voltage $v_c(t)$ is desired to be regulated at a given constant value V_c. The energy function may be linearized around this point using the Taylor's series concept as

$$w_c = \frac{1}{2}Cv_c^2 \approx \frac{1}{2}CV_c^2 + CV_c(v_c - V_c) = CV_cv_c - \frac{1}{2}CV_c^2 = CV_cv_c - W_c.$$

Using this equation, the model of Figure 6.55 can be derived.

Technically, the system of Figure 6.55 is linear but not time invariant due to the presence of block with multiplication into $v_g(t)$. The governing differential equations of this system are

$$\dot{w}_c = -v_g(t)i(t) + p_{\text{in}}$$
$$\dot{i}(t) = -\tfrac{R}{L}i(t) + \tfrac{1}{L}u(t) + \tfrac{1}{L}h(t).$$

In the common approach, the control system is broken down to two loops: internal current control loop and external dc voltage control loop as it will be explained in Section 6.4.2.

6.4.2 Control Structure and Design

We already designed a CFL for the internal subsystem of Figure 6.55 as shown in Figure 6.56. The current controller (CC) is fully studied in Section 6.2. The block denoted by "Figure 6.8" generates and limits the sinusoidal reference current $i_{\text{ref}}(t)$.

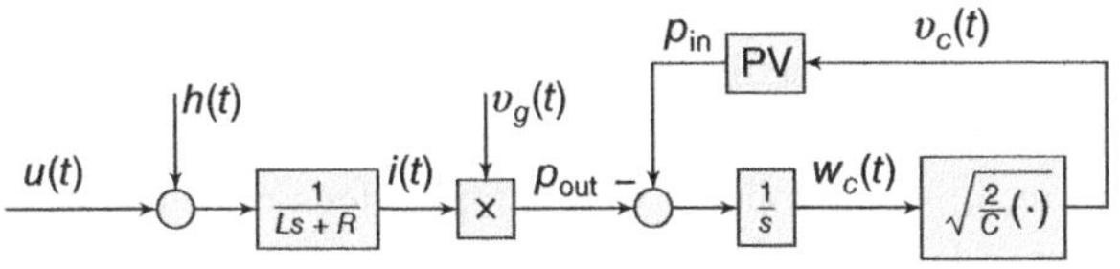

Figure 6.54 Control model when a single-stage PV converter is used.

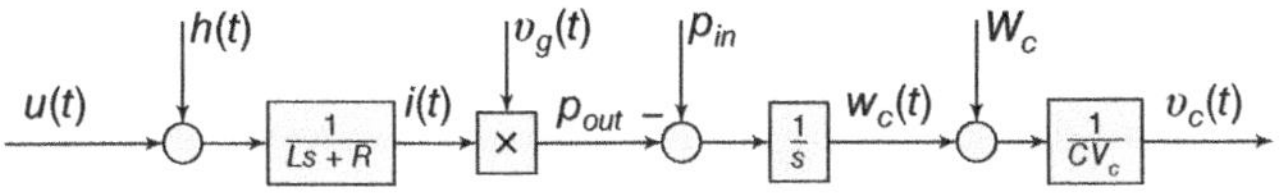

Figure 6.55 Equivalent model of Figure 6.53 in terms of capacitor voltage.

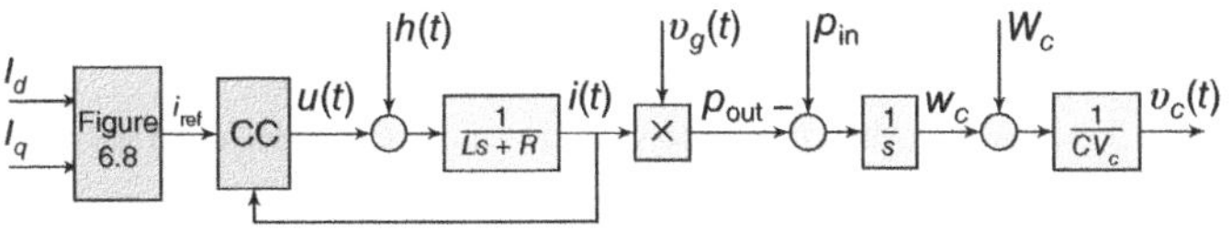

Figure 6.56 Current control is added to Figure 6.55.

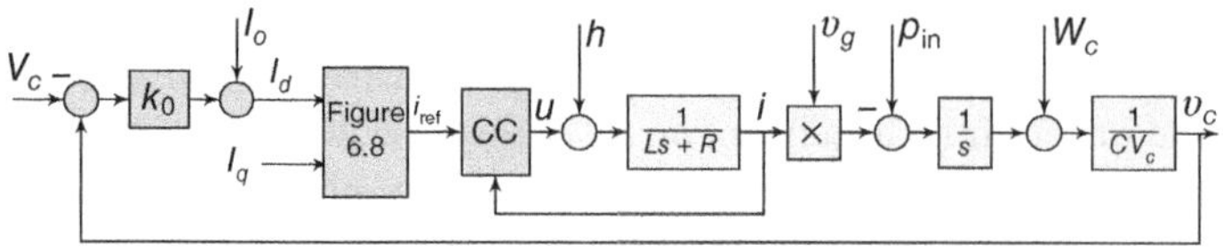

Figure 6.57 Capacitor voltage control is added to Figure 6.56.

The common two-loop control structure is shown in Figure 6.57. The capacitor voltage is regulated to V_c and this determines the amount of (real) power (or current) supplied by the inverter. The amount of reactive power (current), specified by I_q, is supplied to the loop externally.

In the dc voltage control loop of Figure 6.57, I_o denotes the rated current of the DER. Therefore, the controller establishes[36]

$$I_d = I_o + k_0(v_c - V_c) \tag{6.44}$$

which indicates that the capacitor voltage will, in steady state, converge to $v_c = V_c + \frac{i_d - I_o}{k_0}$ where i_d is the operating direct current. When the converter operates at its rated current of I_o, the dc voltage will also be at its set-point of V_c. If it operates at a lower/higher current, the dc voltage will also go down/up proportionally and linearly. Assume for instance that for the converter current i_d from zero to the rated value, we want that the capacitor voltage experiences a change of only $\alpha\%$ of V_c, then $\frac{I_o}{k_0} = 0.01\alpha V_c$ or

$$\boxed{k_0 = \frac{I_o}{0.01\alpha V_c}.} \tag{6.45}$$

For example, for a $V_c = 500$ V, $I_o = 11.8$ A (corresponding to 2 kW at 240 V rms grid), and $\alpha = 5\%$, it will be $k_0 = \frac{11.8}{25} = 0.47$.

Thus, the gain k_0 determines the range of variations of the dc voltage (in steady operation) when its current varies. In the existing literature, e.g. [5, 6], there is a common tendency to use a proportional-integrating (PI) function to accurately control the dc voltage and remove all the steady error from it. We believe, however, that there is no need to completely remove this error. This error, as long as it is within a reasonably small range, is not practically harmful and can even be useful in some applications. For instance, it can be used for dc bus signalling if multiple inverters share a common dc bus. This way, they will be informed of the operational status of other inverters through the dc voltage. Furthermore, this controller is very simple, does not have the problem of integrator windup, and has a faster and more robust performance compared to when a PI is used, as illustrated in [7].[37]

In order to perform a loop stability analysis and controller design, it is assumed that the CC is already designed. Moreover, it is assumed that the internal loop is sufficiently faster than the external loop. Therefore, from the standpoint of the external loop, the internal loop is assumed as a unity gain. Therefore, $i(t) = i_{ref}(t)$. The point denoted by p_{out} in Figure 6.57 is expressed by

$$\begin{aligned} p_{out} &= i(t)v_g(t) = [I_d\cos(\omega_o t) - I_q\sin(\omega_o t)]V_g\cos(\omega_o t) \\ &= \frac{V_g}{2}I_d + \frac{V_g}{2}I_d\cos(2\omega_o t) - \frac{V_g}{2}I_q\sin(2\omega_o t) = P + S\cos(2\omega_o t + \phi), \end{aligned}$$

36 In a bidirectional converter, I_o in Figure 6.57 may be set to zero. In this case, the capacitor voltage v_c will symmetrically vary around the rated value V_c: it is below V_c when power flows from ac side to dc side, and it will go above V_c when power flows from dc side to ac side.

37 If V_c is determined by another control loop, for instance a MPPT loop, the feedback of that loop will correct the error.

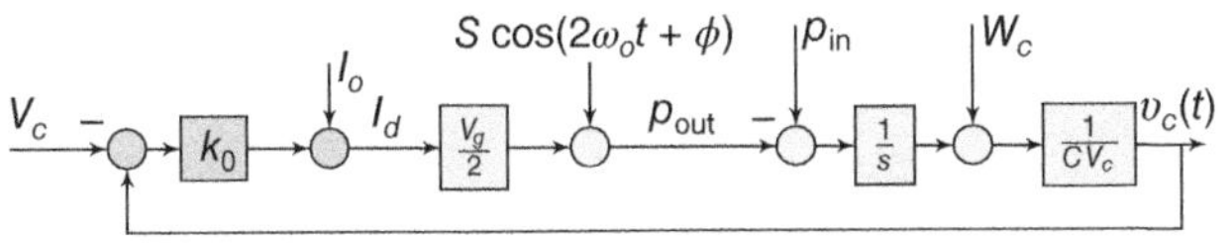

Figure 6.58 Simplified equivalent control model of Figure 6.57.

where $I = \sqrt{I_d{}^2 + I_q{}^2}$ and $\tan\phi = \frac{I_q}{I_d}$. Therefore, the control loop may be simplified and represented as Figure 6.58.

The control loop of Figure 6.58 is not LTI due to presence of different frequencies in the loop. We can study the loop from two standpoints: (i) double-frequency ripples and (ii) dc variables. These two domains are coupled but an approximate analysis that assumes independence of these domains is as follows.

From the standpoint of double-frequency ripples, a ripple with magnitude of S is injected to the loop at the summation point shown in Figure 6.58. This will cause a double-frequency oscillation with peak magnitude of

$$V_{c_2} = S \times \frac{1}{2\omega_o} \times \frac{1}{CV_c} = \frac{S}{2C\omega_o V_c} \tag{6.46}$$

in the bus voltage.[38] This value is obviously independent from the control system and is due to power balance nature of the system (assuming that p_{in} is dc resulting in the capacitor supplying the total double-frequency power). The peak double-frequency ripple in I_d is equal to $I_{d_2} = V_{c_2}k_0 = \frac{Sk_0}{2C\omega_o V_c}$. We also have $S = 0.5V_g I$. Therefore, the percentage ripple current is equal to

$$R_p = \frac{I_{d_2}}{I} \times 100 = \frac{V_g k_0}{4C\omega_o V_c} \times 100 \Rightarrow \boxed{R_p = \frac{25V_g k_0}{C\omega_o V_c}\%.} \tag{6.47}$$

This equation and the one in (6.45) indicate the trade-off that larger k_0 narrows the range of change of v_c (specified by α) but increases the level of 2-f ripples on the current I_d. Notice that the 2-f ripples on I_d translate into the third-order harmonic on the actual current, if not removed by the internal CFL.

The characteristics equation of the dc voltage loop is

$$1 + k_0 \frac{V_g}{2CV_c}\frac{1}{s} = 0$$

which means a single real pole at $s = -\beta$ where,

$$\boxed{\beta = \frac{k_0 V_g}{2CV_c}.} \tag{6.48}$$

Comparing this equation and (6.47), we can also write $\boxed{\beta = 0.02\omega_o R_p = 7.5R_p}$ for the 60 Hz system. This indicates the trade-off between the speed of the control loop and the level of ripples: the faster the loop (larger β) will cause larger ripples on the current. (Notice also that β should be, at maximum, still several times smaller than the poles of the internal CFL. But this is not normally a concern because the ripple constraint overrides it.) Based on this discussion, the following design algorithm may be presented.

38 This is the same as (6.3) but we derived it here in a different way.

Algorithm 6.1

Input:
- Grid voltage and frequency V_g (peak), ω_o
- Inverter real power rating P (its peak current: $I_o = \frac{P}{0.5V_g}$)
- Capacitor voltage V_c, Range of capacitor voltage change α%
- Allowable current ripple level R_p%

Output:
- Capacitance C
- Controller gain k_0

Step 1: Find k_0 from (6.45).

Step 2: Find C from (6.47).

Remark 6.1 The response time-constant is $\tau = \frac{1}{\beta}$ where β is given in (6.48) or simply $\beta = 7.5R_p$ (for a 60 Hz system). If this is too slow, we will need to increase R_p.

Remark 6.2 The ripple level R_p shown in (6.47) is the ripple at the reference current. The actual third-harmonic level at the output current may be lower than this if the current controller filters it.

Remark 6.3 The ripple on the dc voltage is given by (6.46) for the designed value of C. If it is beyond the acceptable range, C must be increased. This can be done by decreasing either the range of capacitor voltage change α or the allowable ripple level R_p (or both). The following example shows details pertaining to all these points.

Example 6.6 ***Capacitor Voltage Control***

Consider the same grid and inverter parameters given in Table 6.1. Assume that $\alpha = 5\%$ and $R_p = 5\%$, i.e. 5% (25 V of 500 V) deviation in the capacitor voltage, and 5% 2-f ripple. Then, following the algorithm, we get $k_0 = \frac{I_o}{0.01\alpha V_c} = \frac{2000/(0.5\times240\sqrt{2})}{0.01\times5\times500} = 0.47$ and $C = \frac{25V_g k_0}{\omega_o V_c R_p} = \frac{25\times240\sqrt{2}\times0.47}{377\times500\times5} = 4.2$ mF. For this design, we have $\beta = 7.5R_p = 37.7$ which corresponds to a time-constant of about 26.5 ms (about 1.5 ac cycle). The peak capacitor voltage ripple (when delivering S) is $V_{c_2} = \frac{2000\sqrt{2}}{2\times0.0042\times377\times500} = 1.8$ V which is certainly acceptable.

In this example, the same current controller of Example 6.1 is used. Figure 6.59 shows a sample of performance of this control system. The responses are as expected in terms of their dynamics (smooth and first-order) and transient times (of about 1.5 ac cycles the time constant). The THD of current is about 3% which is below the selected 5% ripple because the current controller further filters the third harmonic. There is 5% ripple on I_d (1.2 A peak-to-peak at 11.8 A). The capacitor voltage has a peak-to-peak ripple of 2.5 V (at this power transfer level). All objectives are achieved.

6.4.3 Removing 2-f Ripples from Control Loop*

The double-frequency component of the dc voltage penetrates into the current reference. The controller designed in Section 6.4.2 characterized the amount of such ripples on the current and offered a method of designing the capacitor to ensure that such ripples are under a desired level. As shown in Example 6.6, the 2-f ripples on the voltage may be quite small (1.25 V peak in that example, which is only 0.25% of the 500 V). Therefore, it seems that there is room to decrease the size of capacitor. However, in order to prevent these ripples to penetrate into the current reference, the dc voltage ripples must be either filtered or removed from the control loop.

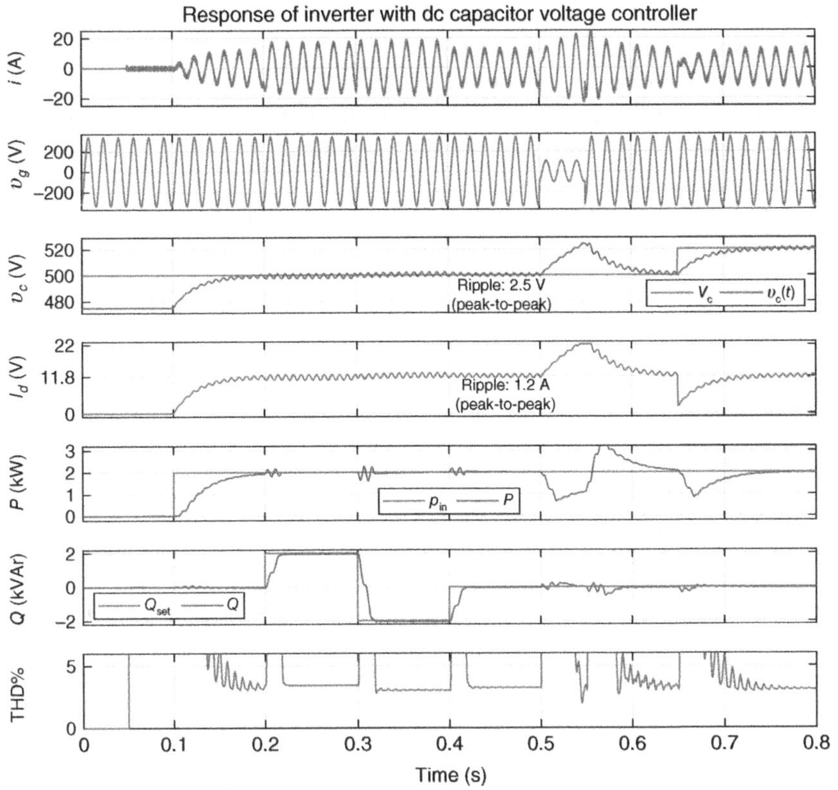

Figure 6.59 Responses of the inverter with dc voltage controller. PLL starts at $t = 0$. Converter starts at $t = 0.05$ seconds. Input power jumps from 0 to 2 kW at $t = 0.1$ seconds followed by reactive power jumps at ± 2 kVAr at $t = 0.2,\ 0.3$ and back to 0 VAr at $t = 0.4$ seconds. A deep low-voltage grid fault occurs at $t = 0.5$ seconds and clears at $t = 0.55$ seconds. At $t = 0.6$ seconds, the dc voltage reference V_c jumps to 520 V.

In this section, two methods of removing such ripples from the control loop are discussed. The first one is based an employing a notch filter centered at the double-frequency that can block the ripples. The second one is based on using the PLL information to construct and cancel the ripples adaptively. The advantage of both methods is that they facilitate further reduction of the size of the dc capacitance.

6.4.3.1 Notch Filtering Method

In this method, a second-order notch filter with transfer function $H_{\mathrm{NF}}(s) = \frac{s^2+4\omega_o^2}{s^2+4\zeta\omega_o s+4\omega_o^2}$ is used to block the double-frequency ripple as shown in Figure 6.60. Note that a notch filter is related to a band-pass filter through $H_{\mathrm{NF}}(s) = 1 - H_{\mathrm{BPF}}(s)$.

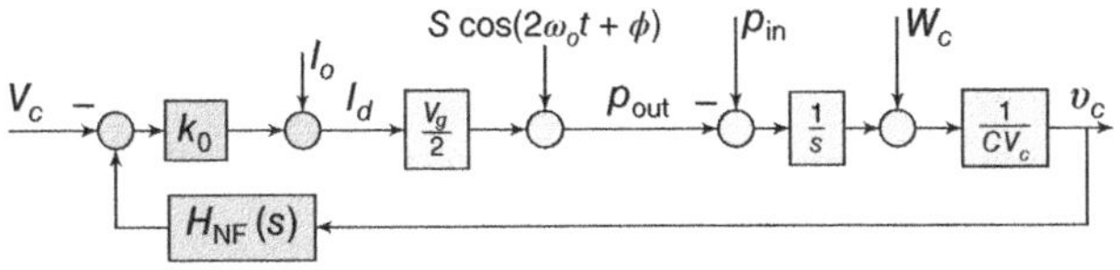

Figure 6.60 A notch filter added to the control loop of Figure 6.58.

Now, the current ripple constraint R_p is relaxed and the design criteria would consist of acceptable voltage fluctuations V_{c_2} and acceptable range of voltage offset α. And this will allow to decrease the capacitor value below the previous case (i.e. without a notch filter).

The design, however, should take into account that the NF increases the order of voltage control loop to 3. Therefore, the NF parameter ζ must be properly designed such that the dynamics of this loop is still desirable. This process is explained below.

We still assume that the CFL is fast and its coupling with the dc voltage loop is negligible. Thus, the dc voltage loop characteristic equation can be expressed as

$$1 + \lambda_1 \frac{1}{s} \frac{s^2 + 4\omega_o^2}{s^2 + 4\zeta\omega_o s + 4\omega_o^2} = 0, \tag{6.49}$$

where $\lambda_1 = \frac{k_0 V_g}{2CV_c}$. Rearranging the equation yields to

$$1 + \zeta \frac{4\omega_o s^2}{\left(s^2 + 4\omega_o^2\right)(s + \lambda_1)} = 0, \tag{6.50}$$

which is in the standard root-locus form versus the parameter ζ.

First we notice using a simple Routh-Hurwitz criterion that the roots of this equation are always stable for all positive values of ζ.[39] Then, we notice from the root-locus approach that there are two complex poles which approach the two zeros at origin and one real pole that moves to the left as ζ increases. The constraint is that these poles must be sufficiently slower than the inner CFL. Below is a design algorithm and a numerical example to further complete this discussion.

Algorithm 6.2

Input:
- Grid voltage and frequency V_g (peak), ω_o
- Inverter real power rating P (peak current: $I_o = \frac{P}{0.5V_g}$)
- Capacitor voltage V_c, Range of capacitor voltage change α%
- Allowable ripple level on the capacitor voltage V_{c_2}

Output:
- Capacitance C - Controller gain k_0 - NF parameter ζ

Step 1: Find k_0 from (6.45).

Step 2: Find C from (6.46).

Step 3: Find ζ from the root-locus of (6.38) such that the roots are sufficiently slower than the CC modes yet have sufficient damping and speed. If not possible, decrease V_{c_2} and repeat.

Example 6.7 ***Capacitor Voltage Control with Notch Filter***

Consider the same grid and inverter parameters given in Table 6.1. Assume that $\alpha = 5\%$, i.e. 5% (25 V of 500 V) deviation in the capacitor voltage, and $V_{c_2} = 5$ V, i.e. 1% double-frequency ripple in the capacitor voltage. Then, following the algorithm, we get $k_0 = \frac{I_o}{0.01\alpha V_c} = \frac{2000/(0.5\times240\sqrt{2})}{0.01\times5\times500} = 0.47$ and $C = \frac{P}{2\omega_o V_c V_{c_2}} = \frac{2000}{2\times377\times500\times5} = 1.1$ mF. Thus, the capacitor size is about 4 times smaller than

39 This is another advantage of using k_0 as opposed to using a PI which cannot guarantee the loop stability for all ζ [5].

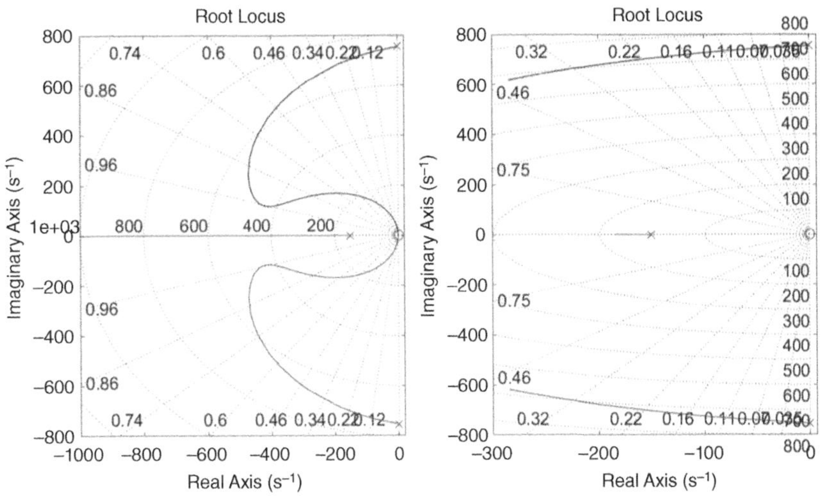

Figure 6.61 Root-locus of (6.38): complete picture (a) and up to $\zeta = 0.4$ (b).

Example 6.6. Next, we plot the root-locus of (6.38) versus the NF parameter ζ as shown in Figure 6.61. For a value of $\zeta = 0.4$, the roots are around $-185, -285 \pm j618$. These are still at least three times slower than the CFL where its poles are shown in Figure 6.11.

Figure 6.62 shows a sample of performance of this control system. The responses are as expected in terms of their dynamics and transient times. Expectedly, the responses are much faster than Figure 6.59. The THD of current is shown at about 1%. There is no ripple on I_d. The capacitor voltage has a peak ripple of 5 V as designed. All objectives are achieved.

6.4.3.2 Direct Ripple Cancellation Method

In this method, shown in Figure 6.63, the double-frequency ripple of the dc voltage is directly cancelled in the controller. The double-frequency ripple of the dc voltage is equal to $v_{c_2} = -\frac{S}{2\omega CV_c}\sin(2\omega t + \phi)$.[40] This equation may be expanded as

$$v_{c_2} = -\frac{S}{2\omega CV_c}\sin(2\omega t + \phi) = -\frac{P}{2\omega CV_c}\sin(2\omega t) - \frac{Q}{2\omega CV_c}\cos(2\omega t), \tag{6.51}$$

where P and Q are the real and reactive powers delivered by the inverter to the grid. These two powers can be approximated by $P = \frac{1}{2}I_d V_g$ and $Q = -\frac{1}{2}I_q V_g$ where I_d and I_q are real and reactive current references shown in Figure 6.57 and also redrawn in Figure 6.63. Notice that we have assumed the grid voltage in the form $v_g(t) = V_g\cos(\omega t)$. This means that ωt is the grid voltage phase angle and this piece of information is made available by the PLL. More precisely, $\sin(2\omega t) = 2s_d(t)s_q(t)$ and $\cos(2\omega t) = s_d^2(t) - s_q^2(t)$. This concludes that all the parameters and variables in (6.51) are available and thus, v_{c_2} can be estimated. Notice, however, that I_d and I_q must be slightly delayed or filtered to avoid formation of an algebraic loop in Figure 6.63.

40 We derived this equation before. However, it is easy to verify it by inspecting Figure 6.63. In Figure 6.63, p_{out} has a 2-f ripple of $S\cos(2\omega t + \phi)$. This passes through $\frac{1}{s}$ which means a magnitude of $\frac{1}{2\omega}$ and an angle of $-90°$ (converting cosine to sine). And finally passes through the block $\frac{1}{CV_c}$, and a negative sign, to generate the 2-f component of v_c.

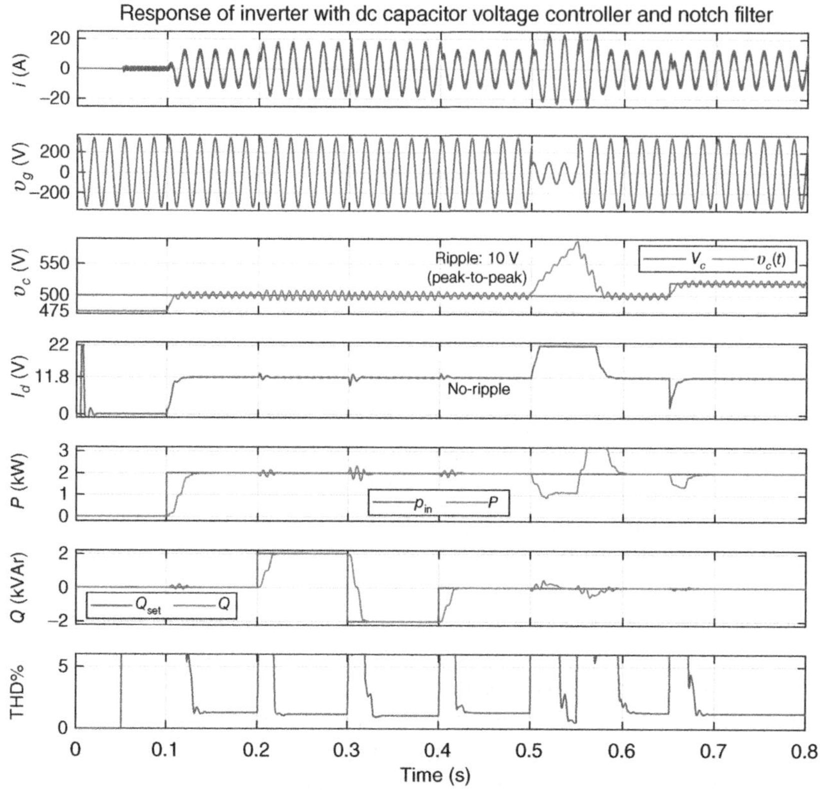

Figure 6.62 Responses of the inverter with dc voltage controller and notch filter. PLL starts at $t = 0$. Converter starts at $t = 0.05$ seconds. Input power jumps from 0 to 2 kW at $t = 0.1$ seconds followed by reactive power jumps at ±2 kVAr at $t = 0.2$, 0.3 and back to 0 VAr at $t = 0.4$ seconds. A deep low-voltage grid fault occurs at $t = 0.5$ seconds and clears at $t = 0.55$ seconds. At $t = 0.6$ seconds, the dc voltage reference V_c jumps to 520 V. Compare with Figure 6.59: faster responses, and lower THD, while a smaller capacitance.

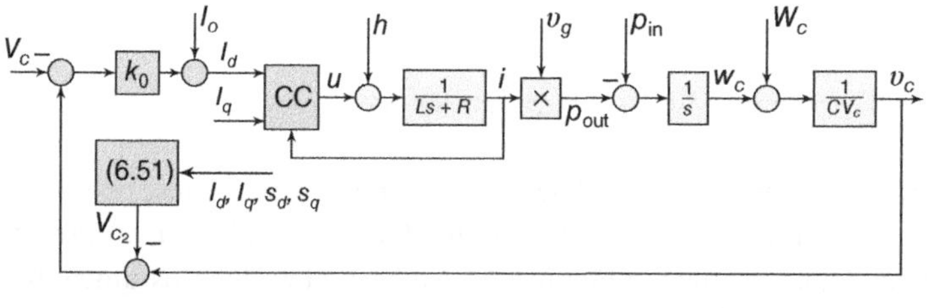

Figure 6.63 Direct cancellation of dc ripples using PLL information.

Example 6.8 ***Capacitor Voltage Control with Direct Ripple Cancellation***
Consider the same grid and inverter parameters given in Table 6.1. Assume that $\alpha = 5\%$ and $V_{c_2} = 5$ V, i.e. 5% (25 V of 500 V) deviation in the capacitor voltage, and 1% 2-f ripple in the capacitor voltage. Then, following the algorithm, we get $k_0 = \frac{I_o}{0.01\alpha V_c} = \frac{2000/(0.5\times240\sqrt{2})}{0.01\times5\times500} = 0.47$ and $C = \frac{P}{2\omega_o V_c V_{c2}} = \frac{2000}{2\times377\times500\times5} = 1.1$ mF. The pole of the dc voltage loop is at $\beta = \frac{k_0 V_g}{2CV_c} = 151$ from (6.48). This is still several times slower than the modes of the current control loop.

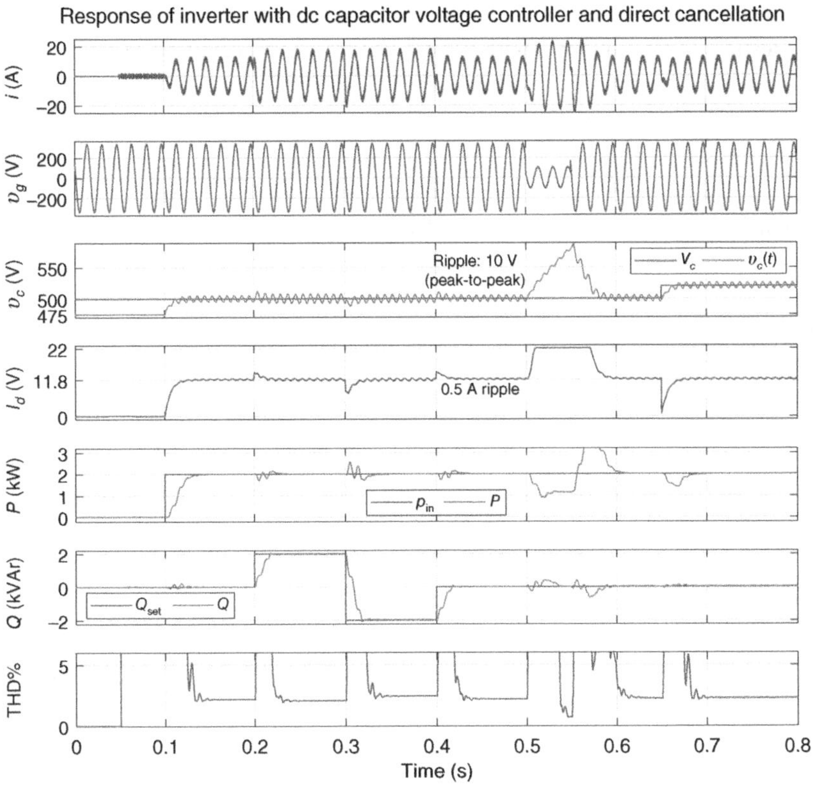

Figure 6.64 Responses of the inverter with dc voltage controller and direct ripple cancellation approach. PLL starts at $t = 0$. Converter starts at $t = 0.05$ seconds. Input power jumps from 0 to 2 kW at $t = 0.1$ seconds followed by reactive power jumps at ±2 kVAr at $t = 0.2,\ 0.3$ seconds and back to 0 VAr at $t = 0.4$ seconds. A deep low-voltage grid fault occurs at $t = 0.5$ seconds and clears at $t = 0.55$ seconds. At $t = 0.6$ seconds, the dc voltage reference V_c jumps to 520 V. Compare with Figure 6.62: slightly larger THD while no NF is used.

Figure 6.64 shows a sample of performance of this control system. The responses are as expected in terms of their dynamics and transient times. Expectedly, the responses are much faster than Figure 6.59. The THD of current is shown at about 2%. There is a small amount of ripple on I_d. The reason is because (6.51) calculates V_{c_2} based on the reference values of I_d and I_q not the actual values of i_d and i_q. The capacitor voltage has a peak ripple of 5 V as designed. Conclusion is that this method is simpler than the NF but some small level of ripples remain in the loop.

6.4.4 Obtaining Inertia from Capacitor*

The dc side capacitor C establishes the equation

$$Cv_c\dot{v}_c = p_{\text{in}} - p_{\text{out}}, \tag{6.52}$$

which means the output power is $p_{\text{out}} = p_{\text{in}} - Cv_c\dot{v}_c$. Assume that we can establish a relationship

$$v_c = V_c + \gamma(\omega - \omega_0), \tag{6.53}$$

where V_c is a nominal value for the capacitor voltage, ω_0 is the nominal value of the grid frequency, ω is the actual grid frequency or a variable that mimics it, and γ is a positive number. Then, the power supplied by he capacitor is $-Cv_c\dot{v}_c = -\gamma Cv_c\dot{\omega} \approx -\gamma CV_c\dot{\omega}$ which indicates an inertia response similar to the kinetic inertia of a rotating mass with the moment of inertia equal to[41]

$$\boxed{J = \gamma\frac{V_c}{\omega_0}C.} \tag{6.54}$$

This indicates that γ plays the role of "inertia amplification factor." By allowing the capacitor voltage to swing in a wider range tha the grid frequency, larger level of inertia is extracted from the capacitor. The condition is, however, that the capacitor voltage variations must mimic the grid frequency variations.

6.4.4.1 Non-VSM Approach

One approach to implement this concept is to measure the grid frequency ω_g (using a PLL for instance), and set the reference value for the capacitor voltage according to $V_c + \gamma(\omega_g - \omega_0)$ in the control system. The dc voltage controllers discussed in Section 6.4 can be used. The block diagram of this approach is shown in Figure 6.65 where $\Delta\omega = \omega_g - \omega_0$. The reference current generation and limitation of Figure 6.8 is used and the current controller (CC) includes the resonant controller and additional blocks as explained before.

6.4.4.2 VSM Approach

This approach uses the VSM concept where the grid frequency is tied to an internal (virtual) frequency (or virtual rotor speed). The internal frequency is directly generated using the dc voltage as shown in Figure 6.66 according to the equation

$$\omega = \omega_0 + \frac{1}{\gamma}(v_c - V_c), \tag{6.55}$$

where γ is a positive constant. This is indeed the same equation as (6.53) and thus, the same level of inertia is obtained for a given γ and V_c. The difference, however, is that in this approach, there is no need to measure the grid frequency. The capacitor voltage v_c (or more precisely, its transients) emulates the grid frequency (or its transients).

The magnitude loop in Figure 6.66 is the same as the VSM. It provides dynamic voltage support according to $M\dot{V} = Q_{set} - Q$. The static volt/var support can also be added by adding the term $K_v(V_0 - V_g)$ to Q_{set} where V_0 is the desired (rated) grid voltage and V_g is the measured grid voltage magnitude, as we discussed in details in Section 6.3.4.

The term D' shows a damping term similar to what is used in VSM, Figure 6.34. However, here its input has shifted to after the capacitor voltage integrator. Therefore, its output is also added directly to the angle. In order to have the same damping effect of Figure 6.34, it must satisfy

$$\boxed{D' = DCV_c,} \tag{6.56}$$

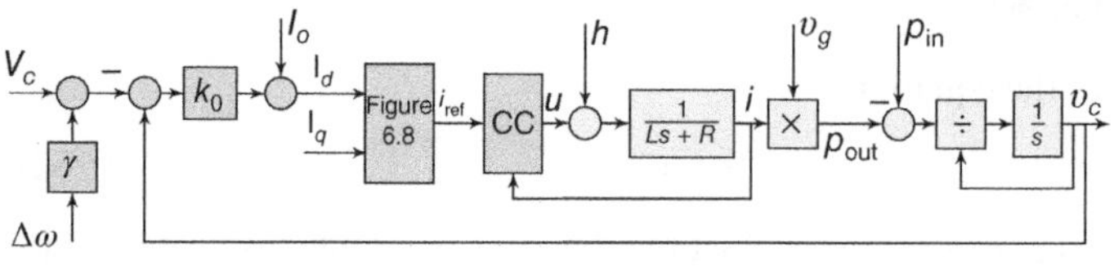

Figure 6.65 Obtaining inertia from capacitor: non-VSM approach.

41 The kinetic inertia of the rotor mass is $-J\omega\dot{\omega}$ which can well be approximated with $-J\omega_0\dot{\omega}$ since ω changes only up to a few percent around ω_0.

where D is calculated according to (6.31).

Finally, an extended version of the pre-synchronization shown in Figure 6.39 is used before starting the inverter. If the converter starts without pre-synchronization, although the current-limiting block ensures that the converter current remains within specified range, the capacitor voltage may grow (or drop) excessively. The extension is in the fact that the capacitor also needs to be emulated (virtually) during the pre-synchronization stage. Let us denote this virtual voltage as v_{c_v}, then it satisfies

$$Cv_{c_v}\dot{v}_{c_v} = p_{\text{in}} - p_{\text{out}} = 0 - 0.5Vi_{1_d}, \tag{6.57}$$

where i_{1_d} is shown in Figure 6.31. Figure 6.67 shows this equation.[42] Now, during pre-synchronization, i.e. $0 \le t \le t_{\text{start}}$, the virtual voltage v_{c_v} is used in the control loop instead of the actual voltage v_c. Once the voltage is synchronized, the inverter starts, and at the same time, the actual capacitor voltage is used in the control loop.

Example 6.9 ***Inertia Extraction from Capacitor: Non-VSM approach***

The approach of Figure 6.65 using the system and controller parameters of Example 6.6 is studied here. The inertia constant γ is selected at 25 meaning that every $\Delta\omega$ rad/s change in the frequency translates to $\Delta v_c = 25\Delta\omega$ V in the capacitor voltage. The reactive current reference I_q is determined according to $I_q = -\frac{Q_{\text{ref}}}{0.5V_g}$ where $Q_{\text{ref}} = Q_{\text{set}} + K_v(V_o - V_g)$ in which V_g is taken from the PLL. The constant K_v is chosen as $K_v = \frac{S}{0.2V_o} = 44.2$ Var/V and $Q_{\text{set}} = 0$.

Figure 6.68 shows the starting and also real/reactive power tracking performances of the converter. Smooth and accurate responses are observed.

Figure 6.69 shows its responses to a grid frequency swing of 0.5 Hz during the time $1 < t < 2.5$ seconds. Figure 6.69 shows that the converter provides an inertia power response to the frequency changes. Notice that a frequency offset (with no change, i.e. $1.5 < t < 2$ seconds) does not generate a power. The inertia power is about 350 W for a frequency ramp of 1 Hz/s. This complies with the theoretical calculation of $Cv_c\dot{v}_c \approx C\gamma V_c\dot{\omega} = 0.0042 \times 25 \times 500 \times 2\pi = 330$. The emulated moment of inertia is $J = \frac{\gamma CV_c}{\omega_o} = \frac{25\times 0.0042\times 500}{120\pi} = 0.14$ which corresponds to an inertia constant of $H = \frac{0.5J\omega_o^2}{S} = 3.3$ seconds (at $S = 3$ kVA). This indicates a rather large level of emulated inertia while a modest size of capacitor is used.

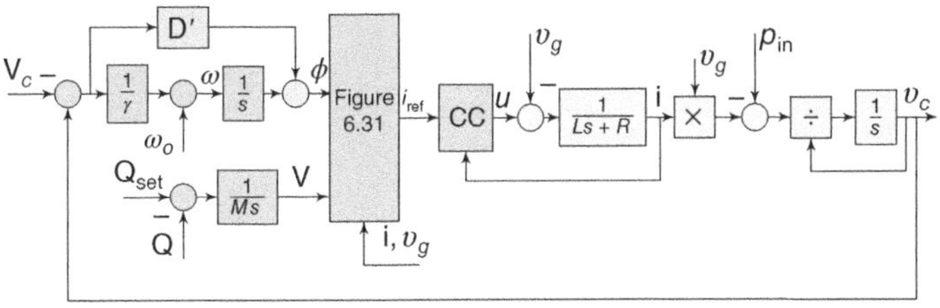

Figure 6.66 Obtaining inertia from capacitor: VSM approach.

Figure 6.67 Pre-synchronization mechanism for VSM approach.

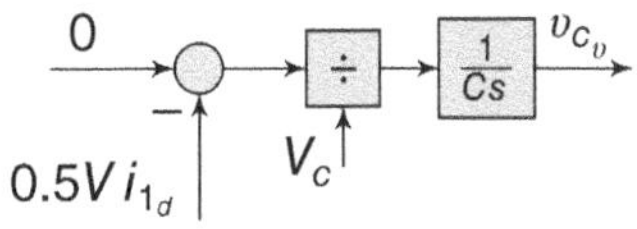

42 Notice that we have replaced the division by v_{c_v} with V_c (which is the nominal value of capacitor voltage) to obtain smoother transients.

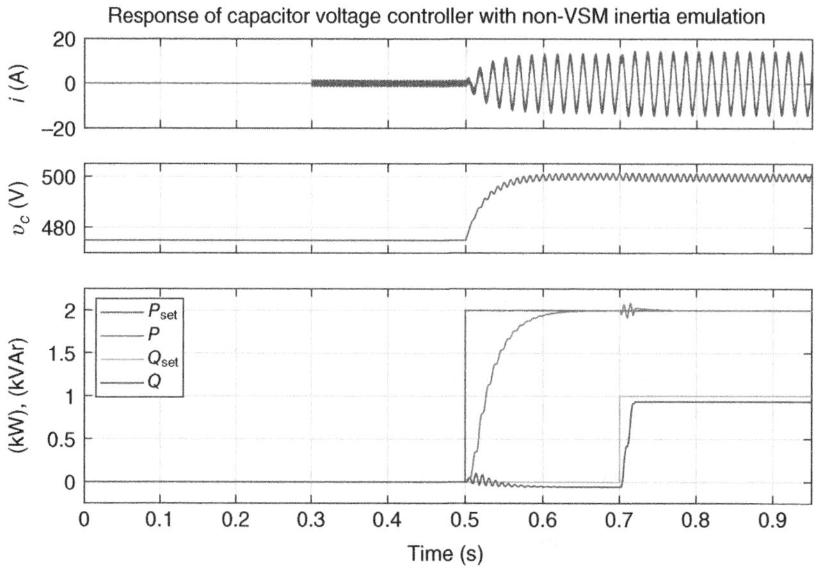

Figure 6.68 Responses of capacitor voltage controller with inertia: non-VSM approach. Starting synchronization during $0 \leq t < 0.3$ seconds; inverter starts at $t = 0.3$ seconds; real and reactive power set-points at $t = 0.5$ and 0.7 seconds.

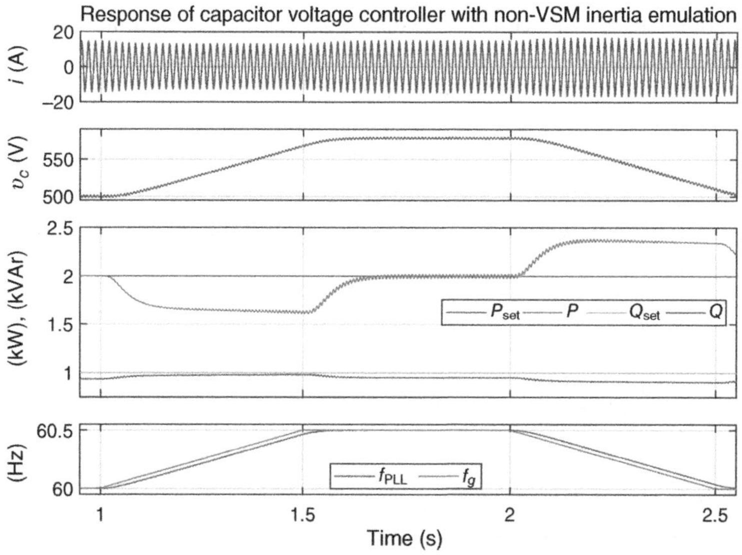

Figure 6.69 Responses of capacitor voltage controller with inertia: non-VSM approach. Grid frequency swing during $1 \leq t < 2.5$ seconds.

Figure 6.70 shows its responses to a grid voltage swing of 5% (17 V) during the time $3 < t < 4.5$ seconds. Figure 6.70 shows that the converter provides a volt-VAr (reactive power) support in response to this swing. A reactive power of about 750 Var is generated for a voltage offset of 17 V which complies with $K_v(V_o - V_g) = 44.2 \times 17$.

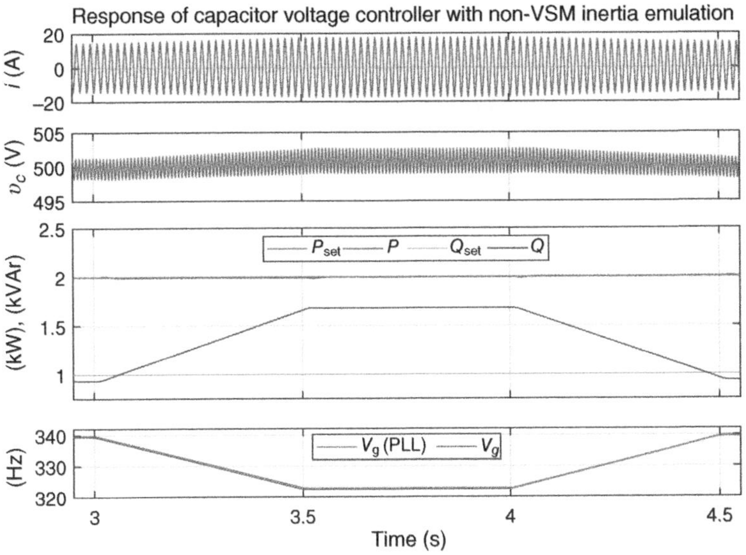

Figure 6.70 Responses of capacitor voltage controller with inertia: non-VSM approach. Grid voltage swing during $3 \leq t < 4.5$ seconds.

Example 6.10 ***Inertia Extraction from Capacitor: VSM Approach***

The approach of Figure 6.66 using the system and current controller parameters of Example 6.6 is studied here. The inertia constant γ is selected at 25 (to produce results comparable with Example 6.9). The controller gains are $D' = DCV_c = \frac{\mu CV_c}{V_o^2}$ and $M = \frac{\mu}{V_o}$ where $V_c = 500$ is the nominal capacitor voltage and $V_o = 240\sqrt{2}$ is the nominal peak grid voltage. The gain μ is designed using the root-locus approach and the corresponding equation (6.39) for $\theta = 0$ and $L_v = 5$ mH, $R_v = 1\ \Omega$. The value of μ will be 259.2 and the poles of system (as shown in Figure 6.36) are at -67 and $-67 \pm j360$. The reactive power reference is $Q_{\text{ref}} = Q_{\text{set}} + K_v(V_o - V_g)$ in which V_g is the grid voltage and the constant K_v is chosen as $K_v = \frac{S}{0.2V_o} = 44.2$ Var/V, and $Q_{\text{set}} = 0$.

Figure 6.71 shows the starting and also real/reactive power tracking performances of the converter. Smooth and accurate responses are observed.

Figure 6.72 shows its responses to a grid frequency swing of 0.5 Hz during the time $1 < t < 2.5$ seonds. Figure 6.72 shows that the converter provides an inertia power response to the frequency changes. Notice that a frequency offset (with no change, i.e. $1.5 < t < 2$ seconds) does not generate a power. The inertia power is about 350 W for a frequency ramp of 1 Hz/s. This complies with the theoretical calculation of $Cv_c\dot{v}_c \approx C\gamma V_c\dot{\omega} = 0.0042 \times 25 \times 500 \times 2\pi = 330$. The emulated moment of inertia is $J = \frac{\gamma CV_c}{\omega_o} = \frac{25\times0.0042\times500}{120\pi} = 0.14$ which corresponds to an inertia constant of $H = \frac{0.5J\omega_o^2}{S} = 3.3$ seconds (at $S = 3$ kVA). This indicates a rather large level of emulated inertia with a modest size of capacitor.

Figure 6.73 shows its responses to a grid voltage swing of 5% (17 V) during the time $3 < t < 4.5$ seconds. Figure 6.73 shows that the converter provides a volt-VAr (reactive power) support in response to this swing. A reactive power of about 750 Var is generated for a voltage offset of 17 V which complies with $K_v(V_o - V_g) = 44.2 \times 17$.[43]

43 The "inertia" component of the reactive power, i.e. the component $-M\dot{V}$, is negligible in this example. During the ramp of voltage with the slope of $\frac{17}{0.5} = 34$ V/s and $M = \frac{\mu}{V_o} = \frac{260}{340} = 0.77$, this component is only about $-0.77 \times 34 = -26$ VAr!

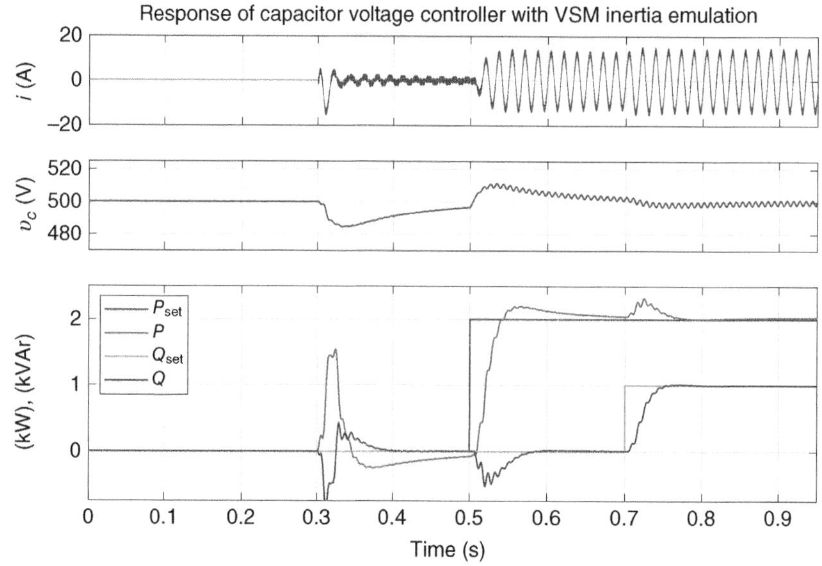

Figure 6.71 Responses of capacitor voltage controller with inertia: VSM Approach. Starting synchronization during $0 \leq t < 0.3$ seconds; inverter starts at $t = 0.3$ seconds; real and reactive power set-points at $t = 0.5$ and 0.7 seconds.

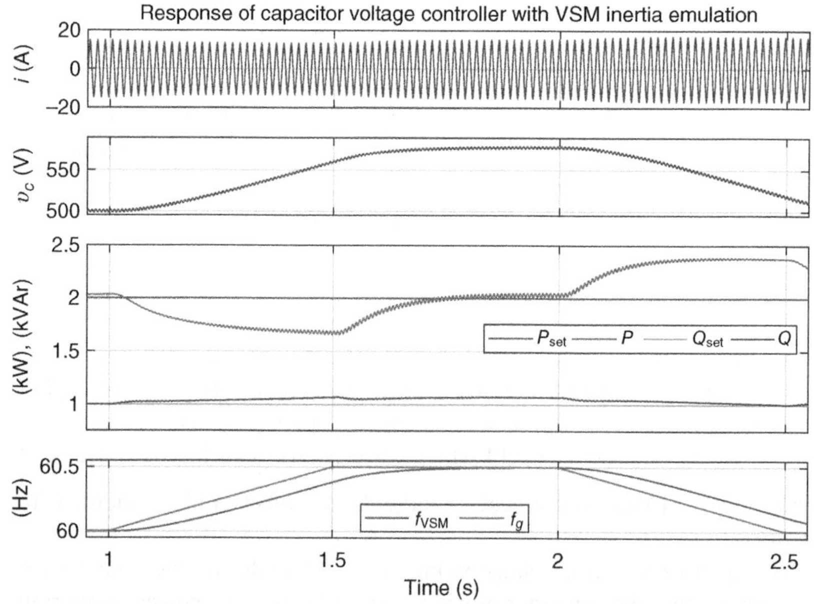

Figure 6.72 Responses of capacitor voltage controller with inertia: VSM approach. Grid frequency swing during $1 \leq t < 2.5$ seconds.

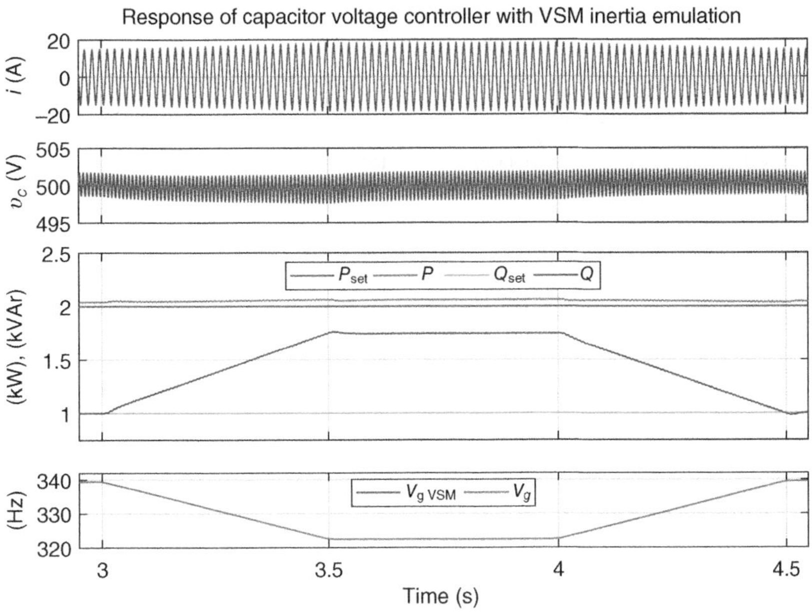

Figure 6.73 Responses of capacitor voltage controller with inertia: VSM approach. Grid voltage swing during $3 \leq t < 4.5$ seconds.

6.5 Load Voltage Control and Support

A single-phase VSC connected to an ac load is shown in Figure 6.74 where L and C constitute the interfacing output filter.[44] The dc side voltage is v_c. The voltage across the load is $v_g(t)$ and the converter current is $i(t)$.[45]

The control objectives can be stated as follows.

- Control the output voltage: A sinusoid with regulated magnitude and frequency at the desired values V_o and ω_o.
- Maintain this objective robustly despite system uncertainties including those in the filter (L, C), load, and dc side voltage.
- VSC should respond to load variations in a quick yet smooth manner.
- Prevent converter over-current.

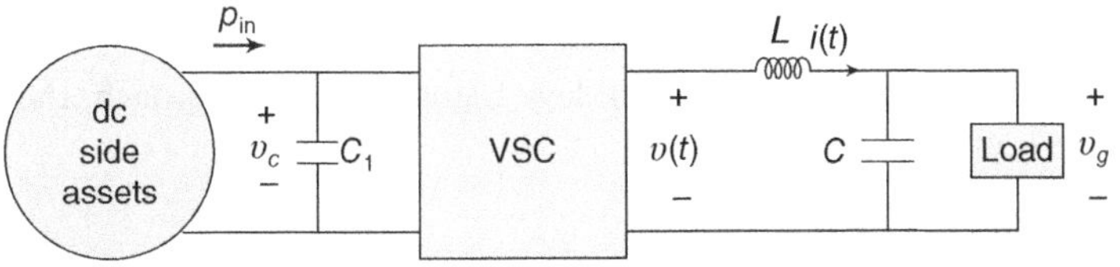

Figure 6.74 VSC connected to an ac load.

44 The parasitic resistances of the filter components do not impact the control system studies in this section and are not shown for simplicity.

45 We have used the same notation v_g for the ac side voltage that we used for grid-connected mode. This will simplify the notations and also better allows the subsequent discussions.

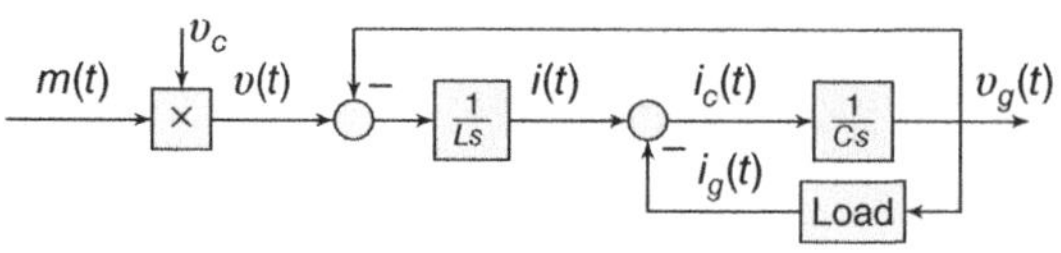

Figure 6.75 Control block diagram of the VSC connected to an ac load.

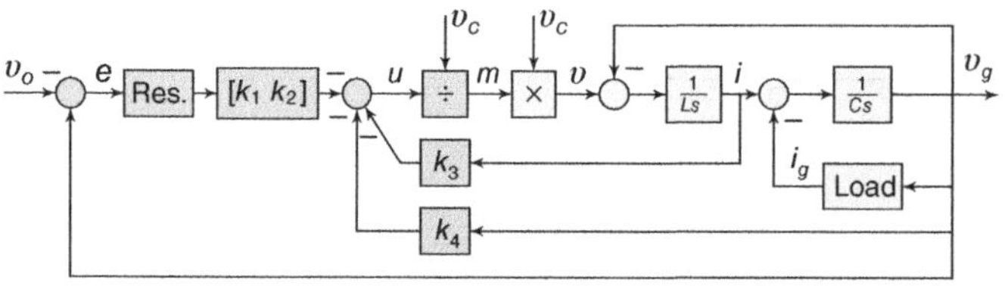

Figure 6.76 Direct voltage control of VSC connected to ac load.

The control block diagram of the system is shown in Figure 6.75 for a full-bridge converter. In a half-bridg converter, v_c changes to $0.5v_c$.

6.5.1 Direct Voltage Control Approach

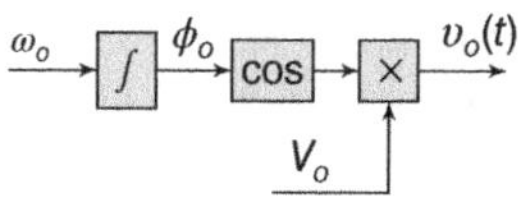

Figure 6.77 Voltage reference generation.

A direct voltage control (DVC) approach based on combined state and output feedback loop is shown in Figure 6.76. The controller is structured properly to allow application of the LQT method. The ac voltage reference is denoted by $v_o(t)$ which may be expressed as $v_o(t) = V_o \cos(\omega_o t)$, as shown in Figure 6.77. A resonant controller is used to achieve accurate tracking of the sinusoidal reference.

The load model may be considered in this design. Here, however, we consider the load current as a disturbance (even though it is technically not a disturbance because it is determined within the control loop) and design a robust controller (using the LQT approach). The extent of robustness of the controller (to work at different loads and different load types, e.g. resistive, or inductive load) should then be verified by simulations.

State space differential equations of this system may be expressed as

$$\dot{x}_1(t) = x_2(t) \qquad \dot{x}_2(t) = -\omega_o^2 x_1(t) + x_4(t) - v_o(t)$$
$$\dot{x}_3(t) = -\frac{1}{L}x_4(t) + \frac{1}{L}u(t) \quad \dot{x}_4(t) = \frac{1}{C}x_3(t) - \frac{1}{C}i_g(t),$$

where $x_4 = v_g$, $x_3 = i$, and x_2 and x_1 are the state variables of the resonant controller. The control input is

$$u(t) = -k_1x_1(t) - k_2x_2(t) - k_3x_3(t) - k_4x_4(t) = -Kx(t)$$

which is in the standard form of a linear full state feedback law. Using the LQT approach, the operator $\frac{d^2}{dt^2} + \omega_o^2$ is applied to arrive at[46]

$$\dot{z}_1(t) = z_2(t) \qquad \dot{z}_2(t) = -\omega_o^2 z_1(t) + z_4(t)$$
$$\dot{z}_3(t) = -\frac{1}{L}z_4(t) + \frac{1}{L}w(t) \quad \dot{z}_4(t) = \frac{1}{C}z_3(t),$$

46 Notice that, rigorously speaking, this transformation does not null the load current because it is not an external signal. Thus, the presented analysis is approximate as mentioned. Furthermore, the load current is distorted if the load is nonlinear. In this case, the control structure and the applied transformation can be enhanced (if need be) to ensure a clean output voltage. All this analysis and design procedure follows the same steps that were presented for grid-connected inverters with distorted grid and are not detailed here.

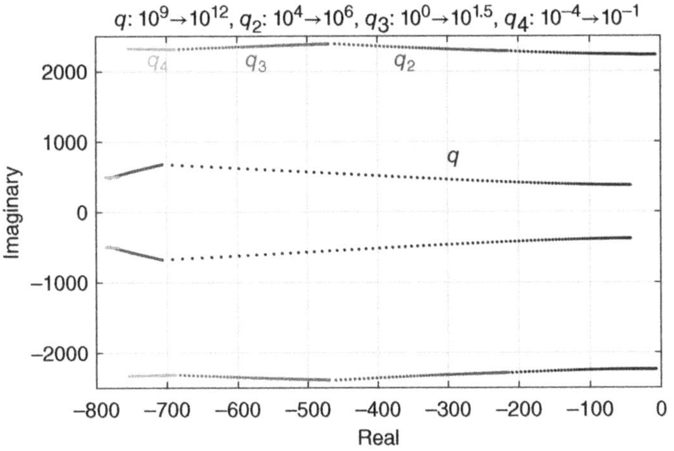

Figure 6.78 Evolution of closed-loop poles of DVC when elements of Q vary.

where

$$z_i(t) = \left(\frac{d^2}{dt^2} + \omega_o^2\right) x_i(t), \; i = 1, 2, 3, 4, \quad w(t) = \left(\frac{d^2}{dt^2} + \omega_o^2\right) u(t).$$

Notice particularly that

$$z_1(t) = e(t), \; z_2(t) = \dot{e}(t),$$

where e is the tracking error. In order to minimize the cost function

$$J = \int_0^{\infty} \left[qe(t)^2 + q_2\dot{e}(t)^2 + q_s z_3(t)^2 + q_4 z_4(t)^2 + w(t)^2\right] dt,$$

we choose

$$A = \begin{bmatrix} 0 & 1 & 0 & 0 \\ -\omega_o^2 & 0 & 0 & 1 \\ 0 & 0 & 0 & -\frac{1}{L} \\ 0 & 0 & \frac{1}{C} & 0 \end{bmatrix}, \; B = \begin{bmatrix} 0 \\ 0 \\ \frac{1}{L} \\ 0 \end{bmatrix}, \; Q = \begin{bmatrix} q & 0 & 0 & 0 \\ 0 & q_2 & 0 & 0 \\ 0 & 0 & q_3 & 0 \\ 0 & 0 & 0 & q_4 \end{bmatrix}$$

and solve in Matlab: ≫ `K = lqr(A, B, Q, 1)`.

Example 6.11 ***Direct Load Voltage Control***
An example of a design is shown in Figure 6.78. The system parameters are chosen as $L = 5$ mH, $C = 40$ μF, and $\omega_o = 120\pi$ rad/s. The Q matrix elements are consecutively adjusted as specified in Figure 6.78. At the final point, the eigenvalues are located at about $-754 \pm j2327, -768 \pm j505$. The controller gains are $K = [7.6e6 \;\; 2000 \;\; 15.2 \;\; 0.80]$.[47]

A sample of time-responses of the designed inverter is shown in Figure 6.79. The inverter starts at $t = 0.02$ seconds at no load condition. A parallel RL load (4 kW, 3 kVAr) is switched on at $t = 0.04$ seconds. The inverter succeeds to generate a sinusoidal voltage with 240 rms and 60 Hz frequency across the load terminals. At $t = 0.08$ seconds, the load terminals are short-circuited through a small impedance of 5 Ω. This causes the converter to increase its current to around 100 A

47 Notice that the LC resonance mode is actively damped by the controller. The pole location $-754 \pm j2327$ indicates a damping ratio of over 0.3.

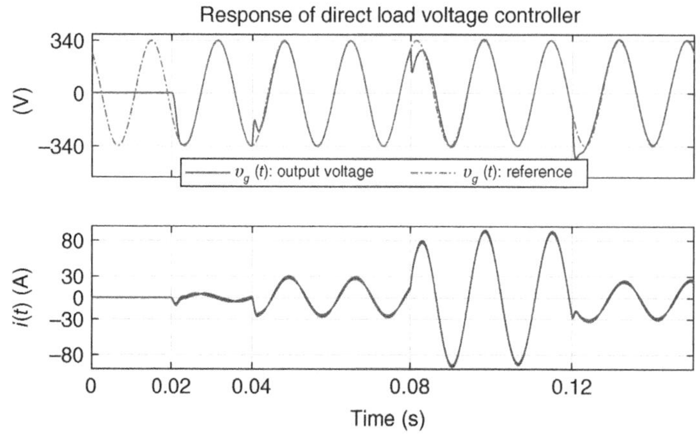

Figure 6.79 Responses of the DVC: Inverter starts at $t = 0.02$ seconds (no load), load is switched on at $t = 0.04$ seconds, a nonzero impedance short-circuit occurs at $t = 0.08$ seconds, clears at $t = 0.12$ seconds. Over-current happens during the short-circuit condition.

to be able to maintain the voltage at the given level of 240 V (rms). Thus, the converter will be subject to an over-current. At $t = 0.12$ seconds, the short-circuit clears and the current comes back to normal.

6.5.2 Voltage Control Loop with Current Limiting

In order to achieve converter current limiting, we use the CFL as an internal loop to an external voltage control loop as shown in Figure 6.80. Here, the reference current is generated by limiting an intermediary virtual current i_v. This virtual current satisfies[48]

$$v_o - v_g = R_v i_v, \tag{6.58}$$

where R_v is a virtual resistance between the reference voltage v_o and the actual voltage v_g. The reference current i_{ref} is achieved from i_v in two steps: first a dq transformation as shown in Figure 6.32 is used (where the angle ϕ is simply $\omega_o t$), and then the dq components are limited and the reference is generated as shown in Figure 6.8.

In Figure 6.80, the virtual current i_v is originally a sinusoidal function which is subsequently converted to dq domain. Alternatively, it is possible to directly generate the dq components of this current using $i_{v_d} = \frac{V_o - v_{gd}}{R_v}$ and $i_{v_q} = \frac{0 - v_{gq}}{R_v}$. To do this way, we will need to compute the dq components of V_g. This case, for the general case of $L_v s + R_v$, is shown in Figure 6.31. Both methods appear to be equivalent.

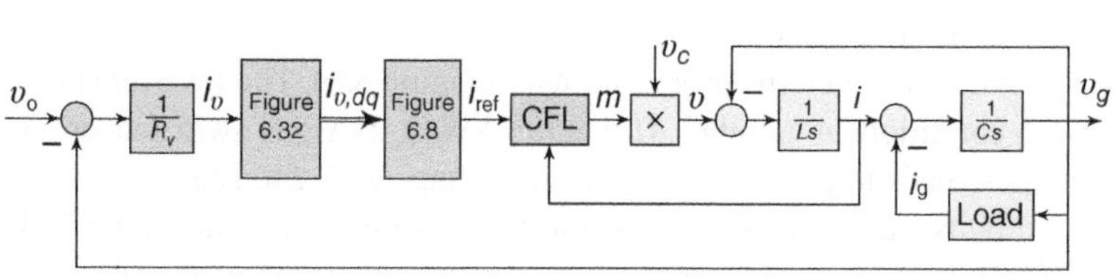

Figure 6.80 Voltage control of VSC connected to ac load with current limiting.

48 In a more general case, an RL impedance may be used which replaces R_v with $L_v s + R_v$.

Regarding the design of R_v, or both L_v and R_v, it must be noted that this represents a virtual impedance between v_o and v_g where the current i flows in it. Therefore, it must be small enough to ensure that v_g remains close to v_o for the range of i from zero to full load. Meanwhile, it should be large enough to allow the necessary dynamics for the control process. We suggest around 10% voltage drop across this impedance for the full range of load variations.

Example 6.12 ***Load Voltage Control with Current Limiting***

For Example 6.11, it is desired to limit the converter current under 40 A. Thus, the controller of Figure 6.80 is used. The same CFL gains used in Example 6.1, i.e. Example 6.1, are used. The system parameters are chosen as $L = 5$ mH, $C = 20$ μF, $V_o = 240\sqrt{2}$ V, and $\omega_o = 120\pi$ rad/s. A resistance of $R = 10$ Ω is added in series with the filter capacitance for passive damping of resonance mode.[49] The virtual resistance R_v is selected at 1. Assuming that the converter power rating is 5 kVA, its current is about 20 A (rms). This current will cause a voltage drop of $20R_v$, in this case 20 V (rms), across the virtual resistor which is reasonable.

A sample of time-responses of the designed inverter is shown in Figure 6.81. The inverter starts at $t = 0.02$ seconds at no load condition. A parallel RL load (4 kW, 3 kVAr) is switched on at $t = 0.05$ seconds. The inverter succeeds to generate a sinusoidal voltage with 240 rms and 60 Hz frequency across the load terminals. At $t = 0.1$ seconds, the load terminals are short-circuited through a small impedance of 5 Ω. This causes the converter to increase its current but is successfully limited to 40 A by the controller. As a result, of course, the load voltage drops to about 150 V (peak). At $t = 0.15$ seconds, the short-circuit clears and the current comes back to normal.

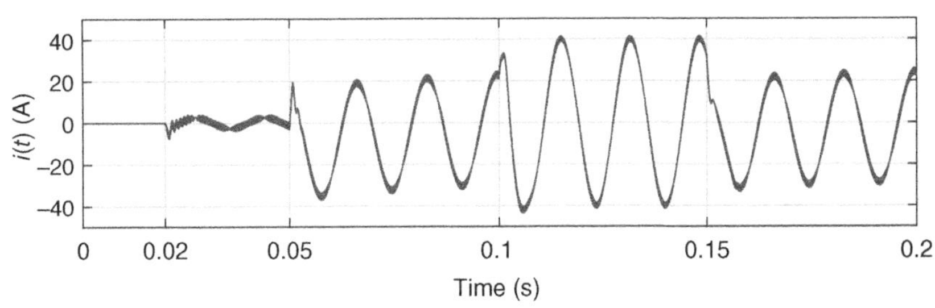

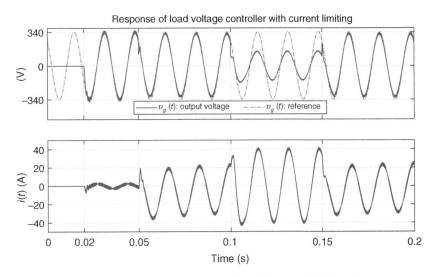

Figure 6.81 Responses of the voltage controller with current limiting: Inverter starts at $t = 0.02$ seconds (no load), load is switched on at $t = 0.05$ seconds, a nonzero impedance short-circuit occurs at $t = 0.1$ seconds, clears at $t = 0.15$ seconds. Current is succesfully limited at 40 A during the short-circuit condition.

49 This way, the LC filter characteristics will be $LCs^2 + RCs + 1 = 0$ which indicates a damping ratio of around 0.3. This resistance introduces a power loss of about 1% at full power of 4 kW. Of course, the current controller may also be upgraded to have active damping if removing this losses is necessary. Refer to Section 7.7.2.1 for an active damping approach.

6.5.3 Deriving Grid-Forming Controllers

The DVC of Figure 6.76 tightly regulates the output voltage v_g to the reference voltage v_o. The voltage control with current limiting approach shown in Figure 6.80 allows the output voltage v_g to deviate from v_o through the equation $v_g = v_o - (L_v s + R_v)i$. Now, assume a situation that v_g is not exclusively controlled by this converter. Other entities (possibly other converters, loads, and even a grid connected at some distance) may be present. In this scenario, the converter should "support" the output voltage rather than controlling it. To this end, the reference voltage v_o must be further relaxed to allow a cooperative impact of all entities on v_g. A couple of strategies to determine the output voltage reference v_o is discussed here and shown that the voltage control method will converge to the the grid forming controllers such as the VSM.

6.5.3.1 Power Droops Strategy

The previous way of generating the voltage reference $v_o(t)$ is shown in Figure 6.77 where its frequency and magnitude are fixed at ω_o and V_o. The common power droop laws manipulate these two quantities according to a linear droop of real and reactive powers expressed by

$$\omega = \omega_o + K_p(P_{\text{set}} - P), \quad V = V_o + K_q(Q_{\text{set}} - Q) \tag{6.59}$$

as shown in Figure 6.82. Here, K_p and K_q are positive, P_{set} and Q_{set} signify some real and reactive power set-points, and P and Q are actual real and reactive powers of the converter.

The droop principles are inspired from the operation of synchronous generators. The virtual frequency ω will be analogous to the rotor speed. It will copy the frequency of the output voltage which may be under the influence of other entities. An increase in this frequency indicates an excess of power in the system, according to the principles of operation of synchronous generators. This will, thus, linearly decrease the converter output power. A similar argument may also be done for the reactive power and its relationship to the voltage magnitude.

As we discussed in Section 6.3, in an inverter-dominated power system, these relationships need not be held necessarily. For instance, it will be plausible and even advantageous to have a "reverse" droop law between the frequency and reactive power, and also the voltage magnitude and real power. However, if SGs are still present and constitute a sizeable share in the system generation, the conventional droop terms may be combined with the "reverse" ones through a controlling angle as discussed in Section 6.3.

The real and reactive powers P and Q in Figure 6.82 are calculated according to $P = 0.5V_1 i_{v_d}$ and $Q = -0.5V_1 i_{v_q}$ where $i_{v_{dq}}$ are shown in Figure 6.80. It is common to use a simple first-order low-pass filter (LPF) before applying these powers to the droop laws, as shown in Figure 6.83.

6.5.3.2 Swing Equation Strategy (VSM Approach)

Another approach to generate the reference voltage is based on directly emulating the swing equation of a synchronous machine. This is $P_m - P_e = J\omega\dot{\omega}$ where P_m is the input mechanical power, P_e is the output electrical power, ω is the rotor speed, and J is its moment of inertia. This approach is show in Figure 6.84. Notice that the division by ω is replaced by ω_o since this frequency

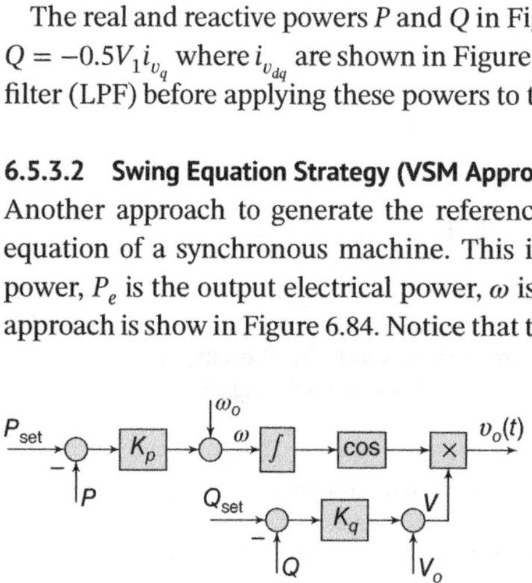

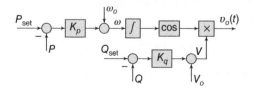

Figure 6.82 Voltage reference generation using the power droops.

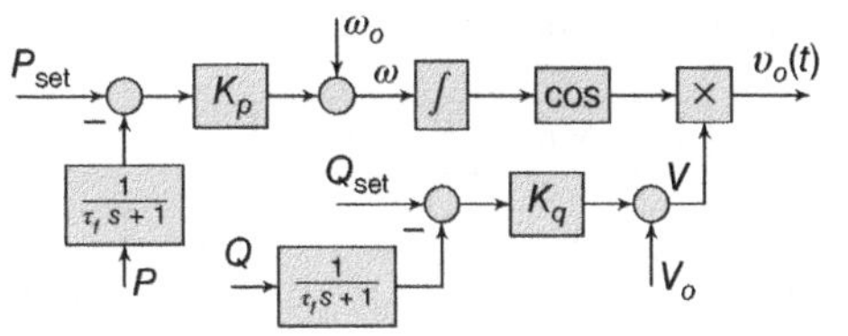

Figure 6.83 Voltage reference generation using the filtered droops.

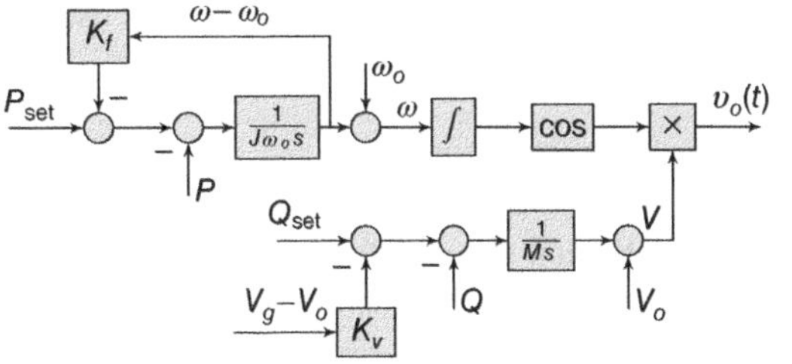

Figure 6.84 Voltage reference generation using the swing equation (VSM).

is not supposed to change much anyway. Moreover, inspired from the governor function of the synchronous generator, a droop term $K_f(\omega_o - \omega)$ is added to the power set-point P_{set} to form the total input power (analogous to P_m).

In Figure 6.84, the magnitude dynamics is formed, again inspired from the automatic voltage regulator (AVR) function in a synchronous generator, by adding a droop term $K_v(V_g - V)$ to the reactive power set-point. Here, V_g is the measured output voltage magnitude. Figure 6.84 shows the basic structure of what is called the virtual synchronous machine (VSM) and is also called the synchronverter in some literature. In Section 6.5.3.3, we establish the close equivalence between this structure and the previous one shown in Figure 6.83.

6.5.3.3 Analogy Between the Two Approaches

In the method based on filtered power droops shown in Figure 6.83, the equation describing the dynamics of internal frequency ω is

$$\omega = \omega_o + K_p\left(P_{set} - \frac{P}{\tau_f s + 1}\right) \Rightarrow \tau_f\dot{\omega} + \omega = \omega_o + K_p(P_{set} - P). \tag{6.60}$$

Notice that ω_o and P_{set} are constant and their derivatives are zero. For the method based on swing equation shown in Figure 6.84, dynamics of ω are represented by

$$J\omega_o\dot{\omega} = P_{set} + K_f(\omega_o - \omega) - P \Rightarrow J\omega_o K_f^{-1}\dot{\omega} + \omega = \omega_o + K_f^{-1}(P_{set} - P). \tag{6.61}$$

It is apparent that the two equations are identical for

$$\boxed{\tau_f = J\omega_o K_f^{-1},\ K_p = K_f^{-1}.} \tag{6.62}$$

In the method based on filtered power droops shown in Figure 6.83, the equation describing the dynamics of internal magnitude V is

$$V = V_o + K_q\left(Q_{set} - \frac{Q}{\tau_f s + 1}\right) \Rightarrow \tau_f\dot{V} + V = V_o + K_q(Q_{set} - Q). \tag{6.63}$$

Notice that V_o and Q_{set} are constant and their derivatives are zero. For the method based on swing equation shown in Figure 6.84, dynamics of V are represented by

$$M\dot{V} = Q_{set} + K_v(V_o - V_g) - Q \Rightarrow MK_v^{-1}\dot{V} + V_g = V_o + K_v^{-1}(Q_{set} - Q). \tag{6.64}$$

It is apparent that the two equations are "almost" identical for

$$\boxed{\tau_f = MK_v^{-1},\ K_q = K_v^{-1}.} \tag{6.65}$$

The only difference is that V on the left side of (6.63) is replaced by V_g on the left side of (6.64).

It is concluded from this discussion that the filter time-constant τ_f is directly related to the "virtual" moment of inertia. When the filter is excluded, it corresponds to zero-inertia dynamics. In general, presence of some level of inertia is useful to allow more flexible and robust dynamics.

6.5.3.4 Damping Strategies

The VSM shown in Figure 6.84 suffers from lack of adequate damping when the damping provided by the K_f branch is not sufficiently large. There are multiple approaches to introduce more damping into this system [8–11]. One approach that we prefer is shown in Figure 6.85. As discussed in Section 6.3.4, this term establishes an analogy between the VSM and ePLL, facilitates stability analysis, and we used this to design K_d.

An alternative damping strategy is shown in Figure 6.86. This is equivalent with the one in Figure 6.85 if the coefficient K_d is properly adjusted to count for the gain $J\omega_o$. This approach does nor have the algebraic loop issue. We already used this method in the method for extracting inertia from capacitor in Section 6.4.4, the VSM Approach, shown in Figure 6.66.

There are several other damping strategies that have been introduced in the literature. In [9], the derivative of real power P is added to ω. It is quite easy to understand why this performs a dampping: the real power P can be written as $P_{\max} \sin \delta$. Thus, $\dot{P}$ will be proportional to $\dot{\delta}$ which is a measure of $\omega - \omega_g$, the difference between rotor speed and grid frequency. In [10], the derivative of $v_g(t) \sin \phi$ is added to the magnitude V. Again, it is readily observed that $v_g \sin \phi$ contains the term $\sin(\phi - \phi_g)$ and thus, its derivative will supply the term $\omega - \omega_g$. More discussion on the VSM damping strategies are presented in Chapter 7.

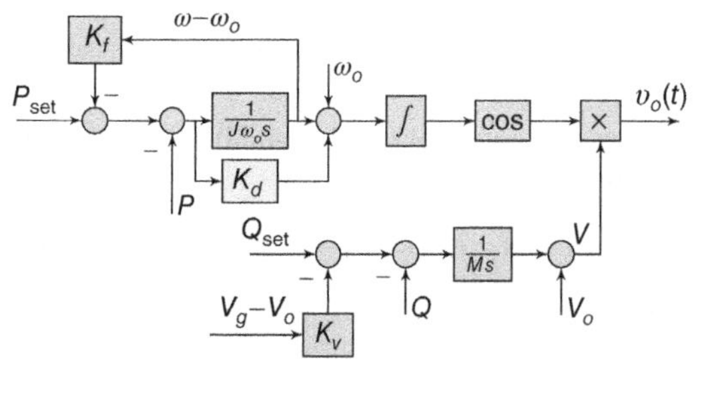

Figure 6.85 Additional damping term K_d.

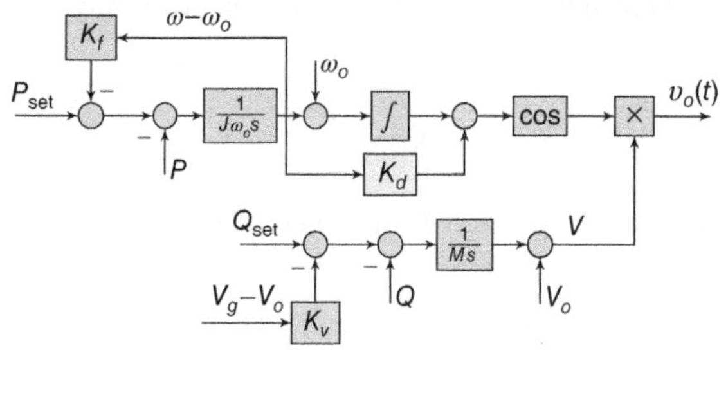

Figure 6.86 Alternative additional damping term K_d.

6.5.4 Discussion

This section showed that how the load voltage controller for "standalone" operation, when added with a CFL and its reference voltage is relaxed, converged to the grid forming controller that we derived for "grid-connected" operation in Section 6.3.4. This should of course not be a surprise because a grid-forming controller must be able to operate the converter in both modes and may be derived from either one of them as the starting point.

6.6 DERs in a Hybrid ac/dc Network

DERs may exist in many forms such as energy storage (battery), solar, and wind.[50] They may be used to build an ac micro-grid, a dc micro-grid or a hybrid micro-grid with both ac and ac subsystems [12, 13]. As far as the interconnection of inverters are concerned, two different topologies may be considered: common dc bus or common ac bus as shown in Figures 6.87 and 6.88 respectively. Each topology offers some advantages and some disadvantages. Generally speaking (and assuming effective controllers), a shared dc bus means a more stable and reliable dc bus and a shared ac bus

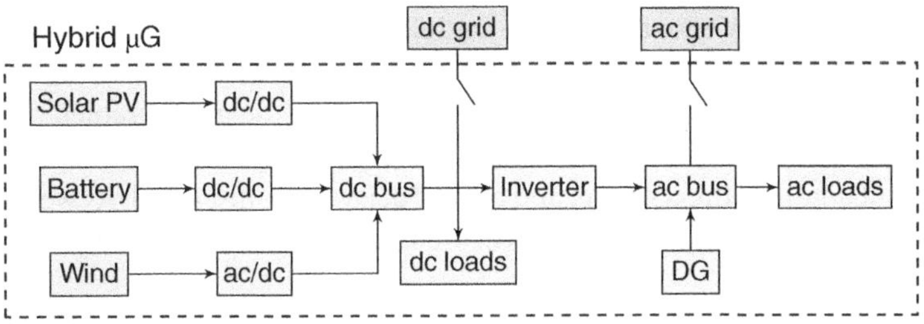

Figure 6.87 Shared dc bus approach.

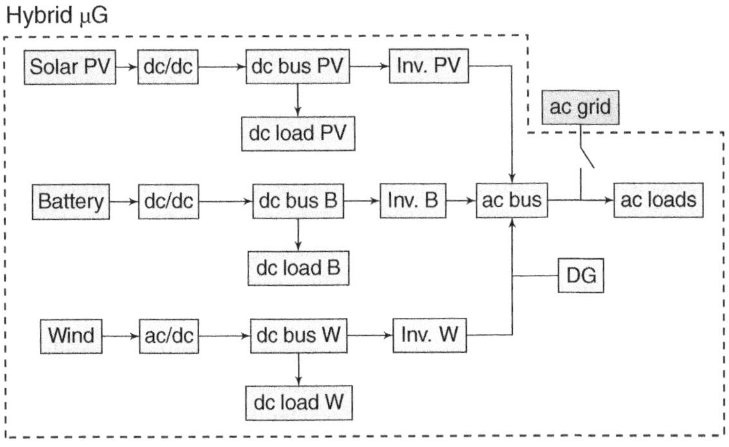

Figure 6.88 Shared ac bus approach.

50 These are the sources that use a converter. Directly-coupled sources (without converter) such as backup diesel generators (DGs), or conventional synchronous generators, may also be part of a hybrid system. The converter-based loads (dc or ac) are also considered as DERs.

means a more stable and reliable ac bus. An external ac grid and an external dc grid may also be present. The microgrid (μG) can be either connected to or disconnected from either or both of these grids.

Power management and control of hybrid systems with many DERs is a delicate research and engineering work. In the common dc bus topology, one option could be to let the Inverter control (or support) the ac bus and the battery control (or support) the dc bus. The renewable sources operate at MPPT by default and should only shift away from this mode when the dc bus voltage tends to rise above its acceptable limits.

In the common ac bus, each DER unit is responsible for controlling its own dc bus. As far as the ac bus control is concerned, one option is to let the battery control it. The renewable units operate at MPPT by default and should only shift away from this mode if the power at the ac bus tends to rise above its acceptable limits; which will manifest as either a frequency rise or voltage rise or both. Meanwhile, they can also provide support to the ac bus voltage.

6.7 Summary and Conclusion

This chapter discussed the modeling and control aspects of single-phase ac converters. First, the concept of power control was discussed. A CFL or current controller (CC) is used as an internal loop to make it possible to limit the converter current. The ePLL was reviewed as an efficient means of synchronizing the converter with the grid, and also of estimating the grid parameters. Multiple aspects such such as how to reject the offset and harmonics from the converter current, higher-order interface filters, and how to address system and control delays were discussed. As well, enhancing the controller by including the PLL state variables in order to strengthen its performance in weak grid conditions were discussed.

The power control approach, as presented above, is a grid-following (GFL) or grid-feeding approach. This means that the converter assumes the presence of a grid and follows it, or feeds into it. It cannot operate when the grid is not available. This can be done by grid-forming (GFM) converters which, not only they can operate when a grid is available, they can also form (or establish) a grid (or a stable voltage) on their own when the external grid is not available. Thus, we introduced grid-support functions and derived the GFM controllers, e.g. the VSM approach, using a grid-support perspective. Stability analyses and design of VSM controller was discussed in details.

Toward the end, the chapter dealt with the problem of load voltage control in a standalone (grid-isolated) application. It was shown that, interestingly enough but not surprisingly, the GFM controllers are also emerged from the load voltage controllers when the reference voltage is relaxed and properly calculated; as they emerge from power controllers when grid-support functions are considered.

In all the discussions and proposed controllers, we have made a particular attention to making sure that an internal CFL is present. This seems to be a requirement for the practically important aspect of keeping the converter current limited during short-term faults and similar incidents. Moreover, we have developed a systematic way of designing the CFL control gains using a convenient and efficient optimal control approach, called the linear quadratic tracker (LQT). The LQT ensures robust performance, and convenient design, without engaging in the concerns about stability during the design process. This approach has been consistently used across the many examples in the chapter and its desired features are demonstrated.

Problems

6.1 Prove (6.1).

6.2 Consider the resonant controller shown in Figure 6.6 and assume that the controller is totally replaced by a simple proportional (P) gain, i.e. $u(t) = ke(t)$. This will not be able to completely remove the steady-state error.

(a) Write down the expression of $i_{ref}(t)$ (corresponding to the power set-points P_{set} and Q_{set}), i.e. (6.5) where assume $s_d = \cos(\omega t)$ and $s_q = \sin(\omega t)$. Derive an expression for $i(t)$ in the steady state in terms of the system parameters L, R, k, and power set-points. Derive expressions for P and Q in the steady state.

(b) Show that as k becomes larger, the errors (deviations in the powers from their set-points) become smaller. Remember that k is limited by the bandwidth of the loop which must be several times smaller than the switching frequency.

(c) For the numerical values of $L = 5$ mH, $R = 30$ mΩ, $P_{set} = 2$ kW, $Q_{set} = 2$ kVAr, $V_g =$ 240 V (rms), 60 Hz, $f_{sw} = 4$ kHz (unipolar PWM), find the value of k which places the pole of the closed-loop at −2000. (This is perhaps the largest value for this pole!)

(d) For the numerical values of the previous part and the computed k, find the actual values of P and Q supplied by the inverter to the grid. Compute the percentage error in both powers.

6.3 With reference to the analysis presented in Section 6.2.4.1, first notice that the magnitude equation of the ePLL is $\dot{A} = \mu e \cos\phi = \mu(v_g - A\cos\phi)\cos\phi = \mu(V_g\cos\phi_g - A\cos\phi)\cos\phi$ and for e to become zero there could be two solutions $(A, \phi) = (V_g, \phi_g)$ or $(A, \phi) = (-V_g, \phi_g + \pi)$ with the angle remaining in $[0, 2\pi)$.

(a) show that nonlinear equations of the ePLL without considering the high-frequency terms may be expressed as

$$\begin{aligned}\dot{A} &= -\tfrac{1}{2}\mu[A - V_g\cos(\Delta\phi)]\\ \dot{\Delta\omega} &= -\tfrac{\mu_3 V_g}{2A}\sin(\Delta\phi)\\ \dot{\Delta\phi} &= \Delta\omega - \tfrac{\mu V_g}{2A}\sin(\Delta\phi),\end{aligned} \tag{6.66}$$

where $\Delta\omega = \omega - \omega_g$, and $\Delta\phi = \phi - \phi_g$.

(b) Show that their linearization around the point $(A, \Delta\omega, \Delta\phi) = (V_g, 0, 0)$ is given by

$$\begin{aligned}\dot{\Delta A} &= -\tfrac{1}{2}\mu\Delta A\\ \dot{\Delta\omega} &= -\tfrac{\mu_3}{2}\Delta\phi\\ \dot{\Delta\phi} &= \Delta\omega - \tfrac{\mu}{2}\Delta\phi,\end{aligned} \tag{6.67}$$

where $\Delta A = A - V_g$. Prove that this is a stable system.

(c) Show that their linearization around the point $(A, \Delta\omega, \Delta\phi) = (-V_g, 0, \pi)$ is given by

$$\begin{aligned}\dot{\Delta A} &= -\tfrac{1}{2}\mu\Delta A\\ \dot{\Delta\omega} &= -\tfrac{\mu_3}{2}\Delta\phi_1\\ \dot{\Delta\phi_1} &= \Delta\omega - \tfrac{\mu}{2}\Delta\phi_1,\end{aligned} \tag{6.68}$$

where $\Delta A = A - (-V_g) = A + V_g$, and $\Delta\phi_1 = \Delta\phi - \pi$. Thus, this also is a stable system.

(d) Follow a similar procedure to prove that the second equilibrium point, i.e. $(A, \Delta\omega, \Delta\phi) = (-V_g, 0, \pi)$, becomes unstable if the division by A is replaced by a division by $|A|$ in the ePLL equations. Also prove that this does not change the stability of the first equilibrium point, i.e. $(A, \Delta\omega, \Delta\phi) = (V_g, 0, 0)$.

6.4 Use the controller gain vector K designed in Example 6.3 to show that the eigenvalues of $A - BK$ (A and B are defined right before the example) do not experience a significant change when P_{set} varies from zero to 2 kW; or when Q_{set} varies from −2 to 2 kVAr.

6.5 Use the two values of the controller gain vector K designed in Examples 6.1 and 6.3 to show that the eigenvalues of $A - BK$ (A and B are defined right before Example 6.3) experience a smaller range of change when L_s varies from 0 to 10 mH for the K of Example 6.3 compared with the K of Examples 6.1.

6.6 Using $x_1 = V\cos\phi$ and $x_2 = V\sin\phi$, prove Eqs. (6.32) and (6.33).

6.7 In Eq. (6.34), assume that P_{set} and Q_{set} are given. Find the VSM voltage magnitude V that corresponds to this amount of power transfer. This value can be substituted in the term $\frac{1}{V^2}$ in (6.34) in order to include the effect of power set-points on the system eigenvalues.

6.8 Prove the validity of the expression for $G(s)$ given in (6.35) and the values of z_1 and z_2 given in (6.36).

6.9 Prove that the three roots of (6.38) are aligned vertically at the real value of $-\frac{R_v}{3L_v}$ for the value of μ given in (6.39).
Hint: Write the characteristic equation in the form of $(s+\alpha)(s^2+2\alpha s+\beta^2)$ and find α, β, and μ.

References

1 IEEE standard for interconnection and interoperability of distributed energy resources with associated electric power systems interfaces. *IEEE Std 1547-2018 (Revision of IEEE Std 1547-2003)*, pages 1–138, 2018.

2 Masoud Karimi-Ghartemani. *Enhanced phase-locked loop structures for power and energy applications*. John Wiley & Sons, 2014.

3 Thomas S Basso and Richard DeBlasio. IEEE 1547 series of standards: interconnection issues. *IEEE Transactions on Power Electronics*, 19(5):1159–1162, 2004.

4 Thaer Qunais and Masoud Karimi-Ghartemani. Systematic modeling of a class of microgrids and its application to impact analysis of cross-coupling droop terms. *IEEE Transactions on Energy Conversion*, 34(3):1632–1643, 2019.

5 Masoud Karimi-Ghartemani, Sayed Ali Khajehoddin, Praveen Jain, and Alireza Bakhshai. A systematic approach to DC-bus control design in single-phase grid-connected renewable converters. *IEEE Transactions on Power Electronics*, 28(7):3158–3166, 2013.

6 F Blaabjerg, R Teodorescu, M Liserre, and A V Timbus. Overview of control and grid synchronization for distributed power generation systems. *IEEE Transactions on Industrial Electronics*, 53(5):1398–1409, 2006.

7 Ali Zakerian, Roshan Sharma, and Masoud Karimi-Ghartemani. Improving stability and power sharing of bidirectional power converters by relaxing DC capacitor voltage. *IEEE Transactions on Power Electronics*.

8 Prasanna Piya and Masoud Karimi-Ghartemani. A stability analysis and efficiency improvement of synchronverter. In *Applied Power Electronics Conference and Exposition (APEC)*, pages 3165–3171. IEEE, 2016.

9 Shuan Dong and Yu Christine Chen. Adjusting synchronverter dynamic response speed via damping correction loop. *IEEE Transactions on Energy Conversion*, 32(2):608–619, 2017.

10 Mohammad Ebrahimi, S Ali Khajehoddin, and Masoud Karimi-Ghartemani. An improved damping method for virtual synchronous machines. *IEEE Transactions on Sustainable Energy*, 10(3):1491–1500, 2019.

11 Masoud Karimi-Ghartemani. Universal integrated synchronization and control for single-phase DC/AC converters. *IEEE Transactions on Power Electronics*, 30(3):1544–1557, 2014.

12 Peng Wang, Lalit Goel, Xiong Liu, and Fook Hoong Choo. Harmonizing AC and DC: a hybrid ac/dc future grid solution. *IEEE Power and Energy Magazine*, 11(3):76–83, 2013.

13 Peng Wang, Jianfang Xiao, Chi Jin, Xiaoqing Han, and Wenping Qin. *Hybrid AC/DC micro-grids: solution for high efficient future power systems*. Springer, 2017.

7

Three-Phase DERs

Modeling of three-phase voltage source converter (VSC) was discussed in Section 2.8. It was shown that in the absence of a zero-sequence path for the current, the system can be modeled and controlled in the two-phase stationary ($\alpha\beta$) or rotating (dq) domains rather than the original three-phase *abc* domain. The dq domain has conventionally been more popular due the property that the signals are of dc nature and simple integrating controllers can be used (instead of resonant controllers) to remove the steady-state errors (for the balanced grid situations).

This chapter delves into details of various control approaches for a distributed energy resource (DER) employing a three-phase VSC. The flow of the chapter is much similar to Chapters 5 and 6. We start with the power control approach using a current feedback loop (CFL), also called the grid-following (GFL) control. This is called the vector current control (VCC) in three-phase systems and is the most common approach due to its ability to limit the current. Both stationary, $\alpha\beta$, and synchronous, dq, domain controllers are discussed.

We then discuss the grid support aspects and derive the virtual synchronous machine (VSM) topology which is a grid-forming (GFM) control from a grid-support perspective (rather than emulating the synchronous machines). The dc voltage control/support approaches are subsequently discussed. Particularly, extracting inertia from dc capacitor is studied.

Finally, the vector voltage control (VVC) for controlling the output voltage in the standalone mode is discussed and it is shown how, by adding an inner CFL to allow current limiting, the GFM controllers (including the VSM concept) are derived.

7.1 Introduction

This section presents an introduction on some background material needed for better understanding of three-phase systems.

7.1.1 Symmetrical Components

A brief explanation of symmetrical sequence components in a three-phase system is given first. A *positive-sequence* three-phase set of signals is defined as three sinusoids at a given frequency with

Modeling and Control of Modern Electrical Energy Systems, First Edition. Masoud Karimi-Ghartemani.

identical magnitudes where their phase-angles are respectively delayed 120°. Mathematically, we can express it as[1]

$$x_p(t) = \begin{bmatrix} x_{pa}(t) \\ x_{pb}(t) \\ x_{pc}(t) \end{bmatrix} = \begin{bmatrix} X_p \cos(\theta_p) \\ X_p \cos(\theta_p - \frac{2\pi}{3}) \\ X_p \cos(\theta_p - \frac{4\pi}{3}) \end{bmatrix} = X_p C_p(\theta_p), \tag{7.1}$$

where $\theta_p(t) = \omega t + \delta_p$[2] and $C_p(\theta) = [\cos(\theta) \quad \cos(\theta - \frac{2\pi}{3}) \quad \cos(\theta - \frac{4\pi}{3})]^T$.[3]

A *negative-sequence* three-phase set of signals is defined as three sinusoids at a given frequency with identical magnitudes where their phase-angles are respectively advanced 120°. Mathematically, we can express it as

$$x_n(t) = \begin{bmatrix} x_{na}(t) \\ x_{nb}(t) \\ x_{nc}(t) \end{bmatrix} = \begin{bmatrix} X_n \cos(\theta_n) \\ X_n \cos(\theta_n + \frac{2\pi}{3}) \\ X_n \cos(\theta_n + \frac{4\pi}{3}) \end{bmatrix} = X_n C_n(\theta_n), \tag{7.2}$$

where $\theta_n(t) = \omega t + \delta_n$ and $C_n(\theta) = [\cos(\theta) \quad \cos(\theta + \frac{2\pi}{3}) \quad \cos(\theta + \frac{4\pi}{3})]^T$. Since $\cos(x) = \cos(-x)$, it is readily observed that $C_n(\theta) = C_p(-\theta)$. Thus, a negative-sequence signal may be considered as a positive-sequence signal rotating at reverse angle direction, or we can say at a negative frequency.

A *zero-sequence* three-phase set of signals is defined as three sinusoids at a given frequency with identical magnitudes and phase angles. Mathematically,

$$x_z(t) = \begin{bmatrix} x_{za}(t) \\ x_{zb}(t) \\ x_{zc}(t) \end{bmatrix} = \begin{bmatrix} X_z \cos(\theta_z) \\ X_z \cos(\theta_z) \\ X_z \cos(\theta_z) \end{bmatrix} = X_z C_z(\theta_z), \tag{7.3}$$

where $\theta_z(t) = \omega t + \delta_z$ and $C_z(\theta) = [\cos(\theta) \quad \cos(\theta) \quad \cos(\theta)]^T$.

It can be proved that every arbitrary set of sinusoidal three-phase signals at a given frequency can be expressed as sum of these three components [1].[4] In other words, an arbitrary sinusoidal three-phase function $x(t)$ can be written as

$$x(t) = \begin{bmatrix} x_a(t) \\ x_b(t) \\ x_c(t) \end{bmatrix} = \begin{bmatrix} X_a \cos(\theta_a) \\ X_b \cos(\theta_b) \\ X_c \cos(\theta_c) \end{bmatrix} = x_p(t) + x_n(t) + x_z(t). \tag{7.4}$$

These three sequences are called the *symmetrical components* of the signal $x(t)$.

It is particularly observed that $x_{pa}(t) + x_{pb}(t) + x_{pc}(t) = x_{na}(t) + x_{nb}(t) + x_{nc}(t) = 0$ for all times.[5] This immediately entails that the zero-sequence component is given by

$$x_{za}(t) = \frac{1}{3}[x_a(t) + x_b(t) + x_c(t)]. \tag{7.5}$$

Ideally, voltages and currents are balanced positive-sequence in a power system. However, due to imbalance of practical three-phase loads (and generators and transmission lines to a lower degree) and single-phase loads and generators (such as small PV generators that are interfaced with the

1 Choice of cosine function versus the sine is arbitrary. This leads to more convenient notations.

2 Mathematically, and to be more accurate, the term ωt may be written as $\int^t \omega(\tau)d\tau$.

3 Note that $\cos(\theta - \frac{4\pi}{3})$ may be written as $\cos(\theta + \frac{2\pi}{3})$. As well, we continue to use the superscript $(\cdot)^T$ to denote vector and matrix transpose.

4 The proof is indeed quite simple. See Problem 7.1.

5 This is easily shown by simple trigonometric relationships, and/or using phasors. See Problem 7.2.

three-phase power system), the whole system has some small level of imbalance. The level of admissible imbalance is regulated by standards. The unbalance factor (UF) is defined as the ratio of the negative-sequence magnitude to the positive-sequence magnitude, that is

$$\mathrm{UF} = \frac{X_n}{X_p} \times 100, \tag{7.6}$$

and is often used to quantify the level of unbalance.[6]

7.1.2 Powers in a Three-Phase System

Consider a three-phase circuit component with the voltage $v(t)$ and the current $i(t)$ flowing into it. In a three-phase system without a neutral connection, the KCL implies that $i_a(t) + i_b(t) + i_c(t) = 0$ for all times. This implies, from (7.5), that no zero-sequence current can flow. Therefore, we can consider a fictitious connection between the neutral point of the supply to the neutral point of this component where zero current flows in it. Zero current also means $i_a + i_b + i_c$ because this is zero! This means that the three-phase system can be broken down into three single-phase systems. This is shown in Figure 7.1. Therefore, the instantaneous power will be the sum of three single-phase powers as

$$p(t) = p_a(t) + p_b(t) + p_c(t) = v_a(t)i_a(t) + v_b(t)i_b(t) + v_c(t)i_c(t) = v(t)^T i(t). \tag{7.7}$$

The *complex power* for a single-phase element is defined as $\vec{S} = \frac{1}{2}\vec{V}\vec{I}^* = P + jQ$ where $\vec{V}$ and $\vec{I}$ are the voltage and current phasors,[7] and P and Q are the real and reactive powers of that element.[8] Therefore, the complex power of the three-phase element will be

$$\vec{S} = \vec{S}_a + \vec{S}_b + \vec{S}_c = \frac{1}{2}\vec{V}_a\vec{I}_a^* * + \frac{1}{2}\vec{V}_b\vec{I}_b^* * + \frac{1}{2}\vec{V}_c\vec{I}_c^* *= \vec{S}_a + \vec{S}_b + \vec{S}_c. \tag{7.8}$$

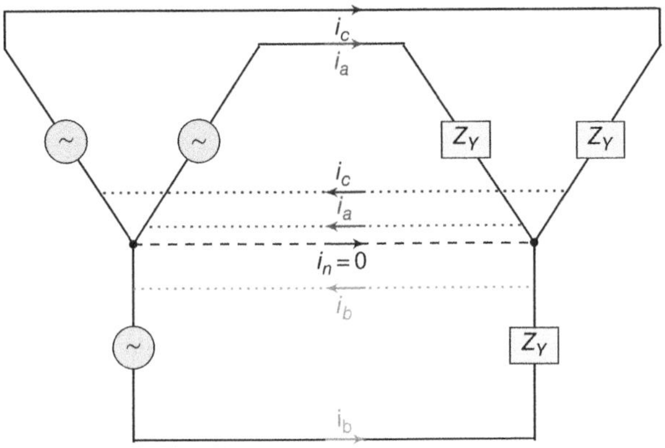

Figure 7.1 A three phase source-load connection. Three fictitious lines are also drawn to show how the three-phase circuit is equivalent to three single-phase subsystems.

6 It is typically under 5% in distribution and below 2% in transmission systems.
7 Here, the phasor's magnitude is considered equal to the peak of the waveform not the rms value.
8 Note that in the single-phase case, $v(t) = V\cos(\omega t)$ and $i(t) = I\cos(\omega t + \phi)$,

$$p(t) = v(t)i(t) = \frac{1}{2}VI\cos(\phi)[1 + \cos(2\omega t)] - \frac{1}{2}VI\sin(\phi)\sin(2\omega t)$$

and $P = \frac{1}{2}VI\cos(\phi)$ and $Q = -\frac{1}{2}VI\sin(\phi)$.

7.1.2.1 Balanced Situation

Assume that the voltage and current are balanced (positive-sequence), i.e. $v(t) = VC_p(\theta)$ and $i(t) = IC_p(\theta + \phi)$. Then, (7.7) and (7.8) imply that (i) $p(t)$ is constant, that is no double-frequency ripples in instantaneous power,[9] and (ii) the real and reactive powers of the three-phase system will be

$$P = \frac{3}{2}VI\cos\phi, \qquad Q = -\frac{3}{2}VI\sin(\phi), \tag{7.9}$$

where V and I are the peak voltage and current values, and ϕ is the angle of current with reference to the voltage.[10]

7.1.2.2 Unbalanced Situation

When the voltage and current are unbalanced, they may be written as[11]

$$\begin{aligned} v(t) &= V_pC_p(\theta_p) + V_nC_n(\theta_n) + V_zC_z(\theta_z), \\ i(t) &= I_pC_p(\theta_p + \phi_p) + I_nC_n(\theta_n + \phi_n), \end{aligned}$$

and the real power will be[12]

$$\begin{aligned} p(t) = v(t)^T i(t) = {} & \overbrace{\frac{3}{2}V_pI_p\cos(\phi_p) + \frac{3}{2}V_nI_n\cos(\phi_n)}^{\text{constant}} \\ & \underbrace{+\frac{3}{2}V_pI_n\cos(\theta_p + \theta_n + \phi_n) + \frac{3}{2}V_nI_p\cos(\theta_p + \theta_n + \phi_p)}_{\text{double-frequency}} \end{aligned}. \tag{7.10}$$

This indicates that (i) the unbalance causes a change in the average of power, i.e. $\frac{3}{2}V_nI_n\cos(\phi_n)$ and (ii) it also creates a double-frequency ripple component in it. Both factors intensify as the unbalance factor in voltage and/or current goes up.

7.1.3 Space Phasor Concept and Notation

An alternative, yet efficient, approach to analyze the three-phase systems and simplify the power expressions is to use the $\alpha\beta$ domain which is equivalent with the space phasor concept. As shown in Chapter 2, by defining

$$\vec{x} = \frac{2}{3}\left(x_a + x_b e^{j\frac{2\pi}{3}} + x_c e^{-j\frac{2\pi}{3}}\right) = x_\alpha + jx_\beta, \tag{7.11}$$

called the *space phasor*, the three *abc* phases of the system are reduced to two phases denoted by x_α and x_β. In terms of real-valued vector notations, we use $x_{\alpha\beta} = [x_\alpha \;\; x_\beta]^T$. Obviously, there is a one-to-one correspondence (and equivalence) between complex phasor notation and real-valued vector notation. It is only that sometimes the complex notation is more convenient and sometimes the real-valued vector notation.

---

9 This is a great advantage of three-phase systems and allows construction of large electromechanical machines without the mechanical stress on their rotating shafts caused by double-frequency ripples.
10 See Problem 7.3.
11 No zero-sequence current flows in a system without neutral connection.
12 See Problem 7.4.

7.1.3.1 Space Phasor of a Positive-Sequence Signal

Consider the three-phase balanced (positive-sequence) signal

$$x(t) = \left[X\cos(\theta)\ \ X\cos\left(\theta - \frac{2\pi}{3}\right)\ \ X\cos\left(\theta + \frac{2\pi}{3}\right)\right]^T.$$

It can readily be shown from (7.11) that

$$\vec{x} = X\cos(\theta) + jX\sin(\theta) = Xe^{j\theta} = X\vec{s}_{d_p}(\theta),\ \ \vec{s}_{d_p}(\theta) = e^{j\theta}, \tag{7.12}$$

which indicates, on the complex plane, a vector with magnitude X rotating *counter-clockwise* at speed $\omega = \dot{\theta}$. In terms of real-valued vector notation,[13]

$$x = \begin{bmatrix} X\cos(\theta) \\ X\sin(\theta) \end{bmatrix} = Xs_{d_p}(\theta),\ \ s_{d_p}(\theta) = \begin{bmatrix} \cos(\theta) \\ \sin(\theta) \end{bmatrix}.$$

In a more general case where an initial angle δ is also present, i.e.

$$x(t) = \left[X\cos(\theta+\delta)\ \ X\cos\left(\theta + \delta - \frac{2\pi}{3}\right)\ \ X\cos\left(\theta + \delta + \frac{2\pi}{3}\right)\right]^T,$$

again, it is easy to show that[14]

$$\vec{x} = X\cos(\delta)e^{j\theta} - X\sin(\delta)e^{j(\theta-\frac{\pi}{2})} = X_d\vec{s}_{d_p}(\theta) - X_q\vec{s}_{q_p}(\theta), \tag{7.13}$$

where

$$X_d = X\cos(\delta),\ \ X_q = X\sin(\delta),\ \ \vec{s}_{q_p}(\theta) = e^{j(\theta-\frac{\pi}{2})}.$$

Notice that $\vec{s}_{q_p}$ is 90° delayed version of $\vec{s}_{d_p}$. In terms of real-valued vector notation,

$$x = X_d s_{d_p}(\theta) - X_q s_{q_p}(\theta),\ \ s_{d_p}(\theta) = \begin{bmatrix} \cos(\theta) \\ \sin(\theta) \end{bmatrix},\ \ s_{q_p}(\theta) = \begin{bmatrix} \sin(\theta) \\ -\cos(\theta) \end{bmatrix}.$$

7.1.3.2 Space Phasor of a Negative-Sequence Waveform

Consider the three-phase negative-sequence signal

$$x(t) = \left[X\cos(\theta)\ \ X\cos\left(\theta + \frac{2\pi}{3}\right)\ \ X\cos\left(\theta - \frac{2\pi}{3}\right)\right]^T.$$

It can readily be shown from (7.11) that

$$\vec{x} = X\cos(\theta) - jX\sin(\theta) = Xe^{-j\theta} = X\vec{s}_{d_n}(\theta),\ \ \vec{s}_{d_n}(\theta) = e^{-j\theta}, \tag{7.14}$$

which indicates, on the complex plane, a vector with magnitude X rotating *clockwise* at speed ω. In terms of real-valued vector notation,

$$x = \begin{bmatrix} X\cos(\theta) \\ -X\sin(\theta) \end{bmatrix} = Xs_{d_n}(\theta),\ \ s_{d_n}(\theta) = \begin{bmatrix} \cos(\theta) \\ -\sin(\theta) \end{bmatrix}.$$

In a more general case where an initial angle δ is also present, i.e.

$$x(t) = \left[X\cos(\theta+\delta)\ \ X\cos\left(\theta + \delta + \frac{2\pi}{3}\right)\ \ X\cos\left(\theta + \delta - \frac{2\pi}{3}\right)\right]^T,$$

again, it is easy to show that

$$\vec{x} = X\cos(\delta)e^{-j\theta} - X\sin(\delta)e^{-j(\theta-\frac{\pi}{2})} = X_d\vec{s}_{d_n}(\theta) - X_q\vec{s}_{q_n}(\theta), \tag{7.15}$$

---

13 We drop the index $\alpha\beta$ to simplify the notations.
14 See Problem 7.6.

where

$$X_d = X\cos(\delta),\quad X_q = X\sin(\delta),\quad \vec{s}_{q_n}(\theta) = e^{-j(\theta-\frac{\pi}{2})}.$$

Note that $\vec{s}_{q_n}$ is 90° delayed version of $\vec{s}_{d_n}$. In terms of real-valued vector notation,

$$x = X_d s_{d_n}(\theta) - X_q s_{q_n}(\theta),\quad s_{d_n}(\theta) = \begin{bmatrix} \cos(\theta) \\ -\sin(\theta) \end{bmatrix},\quad s_{q_n}(\theta) = \begin{bmatrix} \sin(\theta) \\ \cos(\theta) \end{bmatrix}.$$

To summarize, for both the positive-sequence and negative-sequence signals, we defined two unit vectors: direct and quadrature. These are

$$\boxed{\vec{s}_{d_p}(\theta_p) = e^{j\theta_p},\ \vec{s}_{q_p}(\theta_p) = e^{j(\theta_p-\frac{\pi}{2})},\ \vec{s}_{d_n}(\theta_n) = e^{-j\theta_n},\ \vec{s}_{q_n}(\theta_n) = e^{-j(\theta_n-\frac{\pi}{2})},} \tag{7.16}$$

in terms of complex vectors, or

$$\boxed{s_{d_p} = \begin{bmatrix} \cos(\theta_p) \\ \sin(\theta_p) \end{bmatrix},\ s_{q_p} = \begin{bmatrix} \sin(\theta_p) \\ -\cos(\theta_p) \end{bmatrix},\ s_{d_n} = \begin{bmatrix} \cos(\theta_n) \\ -\sin(\theta_n) \end{bmatrix},\ s_{q_n} = \begin{bmatrix} \sin(\theta_n) \\ \cos(\theta_n) \end{bmatrix},} \tag{7.17}$$

in terms of real-valued vectors. And then we represented each sequence in terms of these basis functions.

7.1.3.3 Power Definitions and Expressions Using Space Phasor

The two-phase $\alpha\beta$ presentation (or the space phasor concept) is accurately applicable as long as no zero-sequence current flows. Assume that $\vec{v}$ and $\vec{i}$ represent the voltage and current space phasors across a circuit element. The *complex power* $\vec{S}$ is

$$\vec{S} = \frac{3}{2}\vec{v}\,\vec{i}^{\,*}, \tag{7.18}$$

where $\vec{i}^{\,*}$ is the complex conjugate of $\vec{i}$. The real and imaginary parts of $\vec{S}$ give the real and reactive powers, i.e.

$$\vec{S} = P + jQ. \tag{7.19}$$

Writing $\vec{v} = v_\alpha + jv_\beta$ and $\vec{i} = i_\alpha + ji_\beta$, we get

$$\vec{S} = \frac{3}{2}(v_\alpha + jv_\beta)(i_\alpha - ji_\beta) = P + jQ, \tag{7.20}$$

which leads to

$$\boxed{P = \frac{3}{2}(v_\alpha i_\alpha + v_\beta i_\beta),\quad Q = \frac{3}{2}(v_\beta i_\alpha - v_\alpha i_\beta).} \tag{7.21}$$

Balanced Situation In this case, i.e. when both the voltage and current are positive-sequence, we get $\vec{v} = V\vec{s}_{d_p}$ and $\vec{i} = I_d\vec{s}_{d_p} - I_q\vec{s}_{q_p}$, where $I_d = I\cos(\phi)$, $I_q = I\sin(\phi)$ and ϕ is the angle of current with reference to the voltage.

Therefore, substituting in the complex power definition,[15]

$$\vec{S} = \frac{3}{2}V\vec{s}_{d_p}(I_d\vec{s}_{d_p} - I_q\vec{s}_{q_p})^* = \frac{3}{2}VI_d - j\frac{3}{2}VI_q, \tag{7.22}$$

which implies $P = \frac{3}{2}VI_d = \frac{3}{2}VI\cos(\phi)$ and $Q = -\frac{3}{2}VI_q = -\frac{3}{2}VI\sin(\phi)$.

15 Note that $\vec{s}_{d_p}\vec{s}_{d_p}^{\,*} = 1$ and $\vec{s}_{d_p}\vec{s}_{q_p}^{\,*} = j$.

Unbalanced Situation In this case, let us write the phase-a voltage and current as

$$v_a(t) = V_p\cos(\theta_p) + V_n\cos(\theta_n), \quad i_a(t) = I_p\cos(\theta_p + \phi_p) + I_n\cos(\theta_n + \phi_n).$$

Then, the space phasors are expressed as

$$\vec{v} = V_p\vec{s}_{d_p} + V_n\vec{s}_{d_n}, \quad \vec{i} = I_{d_p}\vec{s}_{d_p} - I_{q_p}\vec{s}_{q_p} + I_{d_n}\vec{s}_{d_n} - I_{q_n}\vec{s}_{q_n},$$

where

$$I_{d_p} = I_p\cos(\phi_p),\ I_{q_p} = I_p\sin(\phi_p),\ I_{d_n} = I_n\cos(\phi_n),\ I_{q_n} = I_n\sin(\phi_n).$$

Now, substituting in the complex power equation,

$$\begin{aligned}
\vec{S} &= \tfrac{3}{2}\left(V_p\vec{s}_{d_p} + V_n\vec{s}_{d_n}\right)\left(I_{d_p}\vec{s}_{d_p} - I_{q_p}\vec{s}_{q_p} + I_{d_n}\vec{s}_{d_n} - I_{q_n}\vec{s}_{q_n}\right)^* \\
&= \tfrac{3}{2}\left(V_p\vec{s}_{d_p} + V_n\vec{s}_{d_n}\right)\left(I_{d_p}\vec{s}_{d_p}^{\,*} - I_{q_p}\vec{s}_{q_p}^{\,*} + I_{d_n}\vec{s}_{d_n}^{\,*} - I_{q_n}\vec{s}_{q_n}^{\,*}\right) \\
&= \tfrac{3}{2}\left(V_pe^{j\theta_p} + V_ne^{-j\theta_n}\right)\left(I_{d_p}e^{-j\theta_p} - I_{q_p}e^{-j\left(\theta_p-\frac{\pi}{2}\right)} + I_{d_n}e^{j\theta_n} - I_{q_n}e^{j\left(\theta_n-\frac{\pi}{2}\right)}\right) \\
&= \tfrac{3}{2}\left(V_pI_{d_p} - jV_pI_{q_p} + V_pI_{d_n}e^{j(\theta_p+\theta_n)} + jV_pI_{q_n}e^{j(\theta_p+\theta_n)}\right. \\
&\quad \left. + V_nI_{d_p}e^{-j(\theta_p+\theta_n)} - jV_nI_{q_p}e^{-j(\theta_p+\theta_n)} + V_nI_{d_n} + jV_nI_{q_n}\right).
\end{aligned}$$

The real part of this expression gives the real power P,

$$\begin{aligned}
P &= \tfrac{3}{2}[V_pI_{d_p} + V_nI_{d_n} + V_pI_{d_n}\cos(\theta_p + \theta_n) - V_pI_{q_n}\sin(\theta_p + \theta_n) \\
&\quad + V_nI_{d_p}\cos(\theta_p + \theta_n) - V_nI_{q_p}\sin(\theta_p + \theta_n)] \\
&= \overbrace{\frac{3}{2}[V_pI_p\cos(\phi_p) + V_nI_n\cos(\phi_n)]}^{\text{constant}} \\
&\quad \underbrace{+\frac{3}{2}[V_pI_n\cos(\theta_p + \theta_n + \phi_n) + V_nI_p\cos(\theta_p + \theta_n + \phi_p)]}_{\text{double-frequency}},
\end{aligned} \tag{7.23}$$

which confirms (7.10). The imaginary part of $\vec{S}$ is equal to the reactive power Q,

$$\begin{aligned}
Q &= \tfrac{3}{2}[-V_pI_{q_p} + V_nI_{q_n} + V_pI_{d_n}\sin(\theta_p + \theta_n) + V_pI_{q_n}\cos(\theta_p + \theta_n) \\
&\quad - V_nI_{d_p}\sin(\theta_p + \theta_n) - V_nI_{q_p}\cos(\theta_p + \theta_n)] \\
&= \overbrace{\frac{3}{2}[-V_pI_p\sin(\phi_p) + V_nI_n\sin(\phi_n)]}^{\text{constant}} \\
&\quad \underbrace{+\frac{3}{2}[V_pI_n\cos(\theta_p + \theta_n - \phi_n) - V_nI_p\cos(\theta_p + \theta_n - \phi_p)]}_{\text{double-frequency}}.
\end{aligned} \tag{7.24}$$

The synchronizing signals s_{d_p}, s_{q_p}, s_{d_n}, s_{q_n}, and often the grid voltage magnitudes V_{g_p} and V_{g_n}, in addition to other possible variables are normally provided by a three-phase phase-locked loop (PLL). Therefore, Section 7.2 is devoted to an overview of the three-phase PLL.

7.2 Three-Phase PLL

The representation of a three-phase signal as its three sequence components is very convenient and simplifies the calculations, analysis, and understanding of three-phase power systems.

The common three-phase PLLs are thus framed to operate according to these components, rather than using individual single-phase PLLs. This chapter provides a brief yet effective overview of the most common PLL and its extensions. Their principles of operation, modeling, and design aspects are studied.

7.2.1 SRF-PLL

This has been the most common PLL for three-phase applications. It directly provides the synchronizing signals associated with the positive-sequence of a given input signal. Figure 7.2 shows the block diagram structure of the SRF (synchronous reference frame)-PLL. Here, $u_{abc}(t)$ denotes the three-phase signal, e.g. the measured grid voltage. The variables ω and θ denote the estimated frequency and phase angle by the PLL, while ω_o is the rated system frequency. The two gains h_0 and h_1 are the proportional (P) and the integrating (I) gains. The direct signal u_d is passed through a simple low-pass filter (LPF) with the transfer function $\frac{h_2}{s+h_2}$ and is used to normalize the quadrature signal u_q. This will make the system dynamics (at the linear level) independent from the input signal magnitude as shown below.

7.2.1.1 Principles of Operation

The *abc* to *dq* transformation is defined as $u_{dqo} = T^{dq}_{abc} u_{abc}$ where

$$T^{dq}_{abc} = \frac{2}{3} \begin{bmatrix} \cos(\theta) & \cos(\theta - \frac{2\pi}{3}) & \cos(\theta + \frac{2\pi}{3}) \\ -\sin(\theta) & -\sin(\theta - \frac{2\pi}{3}) & -\sin(\theta + \frac{2\pi}{3}) \end{bmatrix}. \tag{7.25}$$

For a balanced input signal $u(t) = UC_p(\phi)$, it is very easy to show that[16]

$$u_d = U\cos(\phi - \theta), \quad u_q = U\sin(\phi - \theta). \tag{7.26}$$

The PLL control loop regulates u_q to zero, thanks to the presence of integrating term, hence pushing θ to become equal to (or lock to) ϕ.[17] The magnitude satisfies $U = \sqrt{u_d^2 + u_q^2}$ for all operating conditions but it can be approximated by u_d alone since the PLL pushes u_q to zero anyway.

7.2.1.2 Approximate Linear Analysis and Design

The equations of this PLL are summarized as

$$\dot{U}_d + h_2 U_d = h_2 u_d, \quad \dot{\omega} = h_1 w_q, \quad \dot{\theta} = \omega + h_0 w_q, \quad w_q = \frac{u_q}{U_d}, \tag{7.27}$$

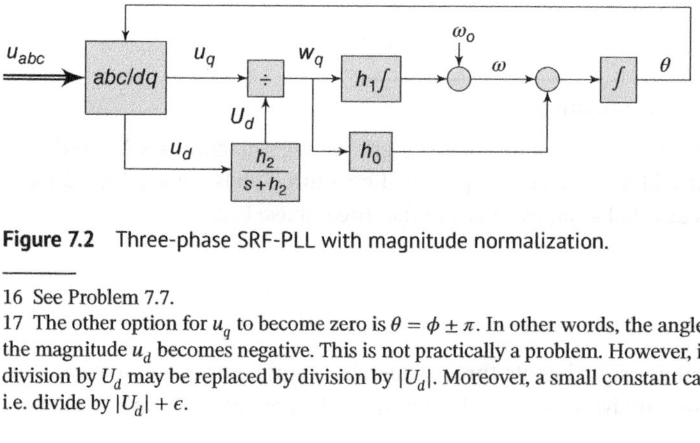

Figure 7.2 Three-phase SRF-PLL with magnitude normalization.

16 See Problem 7.7.
17 The other option for u_q to become zero is $\theta = \phi \pm \pi$. In other words, the angle may have 180° shift in which case the magnitude u_d becomes negative. This is not practically a problem. However, if avoiding this is desired, the division by U_d may be replaced by division by $|U_d|$. Moreover, a small constant can be added to avoid divide-by zero, i.e. divide by $|U_d| + \epsilon$.

where u_d and u_q are given in (7.26). The three variables U_d, ω, and θ tend toward the magnitude, frequency, and angle of the input, respectively. Therefore, these nonlinear equations may be linearized around the equilibrium point. As a result, the signal w_q may be approximated by $(\phi - \theta)$ in a simple linear analysis. Therefore, the loop is characterized by the equation

$$1 + \left(h_0 + \frac{h_1}{s}\right)\frac{1}{s} = 0 \Rightarrow s^2 + h_0 s + h_1 = 0. \tag{7.28}$$

Denote $h_0 = 2\zeta\omega_n$ and $h_1 = \omega_n^2$ where ζ and ω_n represent the well-known damping ratio and natural frequency (or bandwidth) concepts. We suggest to have a damping ratio in the range $\boxed{1 \le \zeta \le 1.5}$ and a bandwidth in the range $\boxed{50 \le \omega_n \le 150}$ for the normal power system applications.[18] For instance for $\zeta = 1.25$ and $\omega_n = 100$, we get the values of $\boxed{h_0 = 250,\ \ h_1 = 10\,000}$. The poles of the loop will be at -50 and -200. The slower mode represents the frequency and the faster one represents the phase angle dynamics. The respective time-constants of about 20 and 5 ms are quite reasonable for these two variables.

7.2.1.3 Alternative Presentations

An alternative representation of the SRF-PLL of Figure 7.2 is shown in Figure 7.3. Here, simply the abc/dq transformation is shifted around and the vectors $C_p(\theta) = [\cos(\theta)\cos(\theta - \frac{2\pi}{3})\cos(\theta - \frac{4\pi}{3})]^T$ and $S_p(\theta) = [\sin(\theta)\sin(\theta - \frac{2\pi}{3})\sin(\theta - \frac{4\pi}{3})]^T$ are used. The block $u_{abc} \cdot C_p(\theta)$ performs the vector dot product $u_{abc}^T C_p(\theta)$. The equivalence of Figures 7.2 and 7.3 is clear.

Figure 7.4 shows the same concept in the $\alpha\beta$ domain. The vectors s_{d_p} and s_{q_p} are defined as before, i.e. $s_{d_p}(\theta) = [\cos(\theta) \quad \sin(\theta)]^T$ and $s_{q_p}(\theta) = [\sin(\theta) \quad -\cos(\theta)]^T$. This representation tends to be more modular for subsequent extensions.

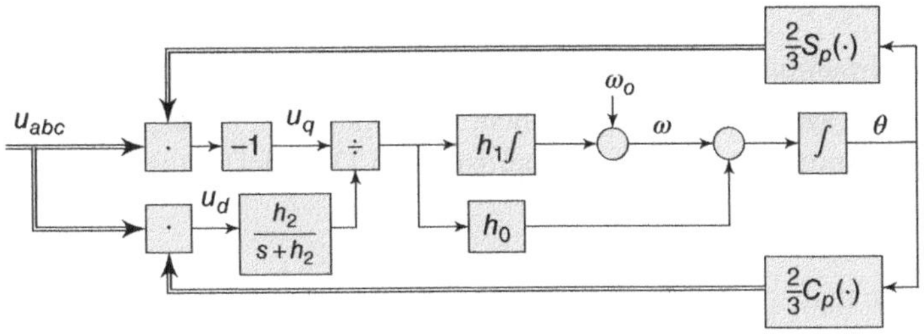

Figure 7.3 Alternative presentation of SRF-PLL of Figure 7.2.

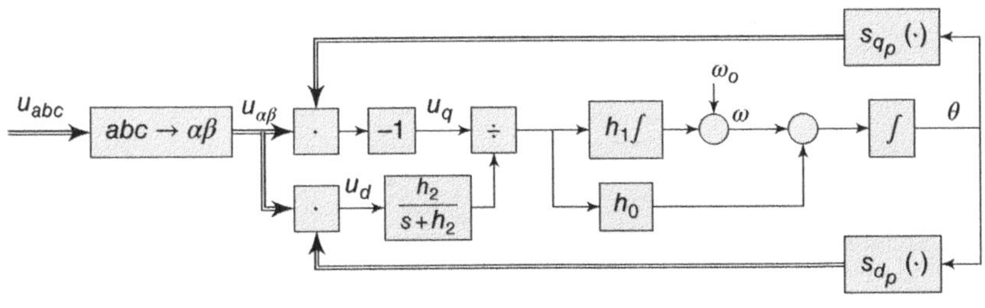

Figure 7.4 Alternative presentation of SRF-PLL of Figures 7.2 and 7.3.

18 Selecting too large of a bandwidth makes the frequency estimation fast but makes the whole system sensitive to disturbances and practical conditions such as weak grid conditions.

7.2.2 Three-Phase Enhanced PLL (ePLL)

The three-phase enhanced phase-locked loop (ePLL) is introduced in a few different versions. Here, we introduce its basic structure first and study its properties. This will be shown to be equivalent to the SRF-PLL while having a different structure. Then, building on this ePLL structure, two more complete versions of the ePLL are presented to address dc offset and also the unbalance conditions.

7.2.2.1 Basic ePLL Structure

The basic three-phase ePLL is derived from the SRF-PLL of Figure 7.4 and is shown in Figure 7.5. The main difference is in adding a subtraction block at the input where $e_{\alpha\beta} = u_{\alpha\beta} - U_d s_{d_p}(\theta)$, called the error, is performed. The LPF block is also replaced by an integrator with the gain h_2.

7.2.2.2 Analysis of Basic ePLL

It is readily observed that u_q in the ePLL is identical with u_q in the SRF-PLL. The subtraction unit introduces an additional term proportional to $s_{d_p}(\phi)^T s_{q_p}(\phi)$. But this term is zero for all ϕ. Meanwhile, it can be proved that U_d in ePLL is also identical with the output of LPF in SRF-PLL.[19] Therefore, the ePLL is indeed equivalent with the SRF-PLL. But, as we will show, the ePLL allows further extensions in a convenient and modular way.

An interesting fact is that it is possible to define a suitable change of variables to derive a linear time-invariant (LTI) representation for the ePLL (or SRF-PLL). This is stated in the following theorem.

Theorem 7.1 *Assume that $h_2 = h_0$, and $h_1 = 0$. The ePLL reduces to a completely LTI system in terms of the new state variables $x = U_d s_{d_p}(\theta)$.*

Before stating the proof, it is worthwhile noting that the condition $h_1 = 0$ indicates that the frequency is not updated. In the power system applications where frequency does not experience a large change, this is an acceptable assumption to derive an approximate LTI model for the purposes of system analysis and design.

Proof: To prove this, we first notice that

$$\dot{\theta} = \omega - \frac{h_0}{U_d} s_{q_p}(\theta)^T e_{\alpha\beta} = \omega - \frac{h_0}{U_d}\sin(\theta)e_\alpha + \frac{h_0}{U_d}\cos(\theta)e_\beta,$$
$$\dot{U}_d = h_0 s_{d_p}(\theta)^T e_{\alpha\beta} = h_0\cos(\theta)e_\alpha + h_0\sin(\theta)e_\beta.$$

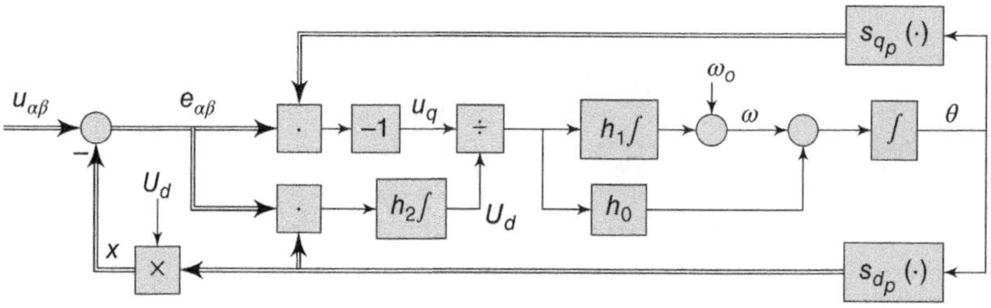

Figure 7.5 Basic three-phase ePLL.

19 See Problem 7.8.

Define $x = U_d s_{d_p}(\theta)$ which means $x_1 = U_d \cos(\theta)$ and $x_2 = U_d \sin(\theta)$. Then, taking derivatives and using simple trigonometric relationships lead to[20]

$$\begin{aligned} \dot{x}_1 &= \dot{U}_d \cos(\theta) - U_d\dot{\theta}\sin(\theta) = -\omega x_2 + h_0 e_\alpha \\ \dot{x}_2 &= \dot{U}_d \sin(\theta) + U_d\dot{\theta}\cos(\theta) = \omega x_1 + h_0 e_\beta. \end{aligned} \tag{7.29}$$

Since $e_{\alpha\beta} = u_{\alpha\beta} - x$, we can get

$$\begin{aligned} \dot{x}_1 &= -h_0 x_1 - \omega x_2 + h_0 u_\alpha \\ \dot{x}_2 &= \omega x_1 - h_0 x_2 + h_0 u_\beta. \end{aligned} \tag{7.30}$$

In transfer function representation,[21]

$$X(s) = T(s)U_{\alpha\beta}(s), \quad T(s) = \frac{h_0}{(s+h_0)^2 + \omega^2}\begin{bmatrix} s+h_0 & -\omega \\ \omega & s+h_0 \end{bmatrix}. \tag{7.31}$$

■

This analysis further shows that the poles of the ePLL (as a closed-loop system without considering the frequency dynamics) are located at $-h_0 \pm j\omega_o$ which is $-250 \pm j377$ for $h_0 = 250$ and $\omega_o = 377$. This indicates a time-constant of around 4 ms and a damping ratio of about 0.55 which are sufficient.

The above model is in $\alpha\beta$ or stationary domain. It shows that the ePLL can be represented by a fully LTI model (without approximation or Jacobian linearization) if the frequency dynamics are ignored.[22] The following theorem derives a full nonlinear model of the ePLL in dq or synchronous frame.[23]

Theorem 7.2 *Using the variables $z = [U_d \;\; \Delta\omega \;\; \delta]^T$, where $\Delta\omega = \omega - \omega_o$ and $\delta = \theta - \omega_o t$, the ePLL is represented by*

$$\dot{z}_1 = -h_2 z_1 + h_2 u_d, \quad \dot{z}_2 = h_1\frac{u_q}{z_1}, \quad \dot{z}_3 = z_2 + h_0\frac{u_q}{z_1}. \tag{7.32}$$

Proof: This can be readily proved by inspecting the ePLL blocks.[24] ■

The first equation in (7.32) is linear but the other two are nonlinear. The nonlinear ones may be linearized around the equilibrium point, for example, if a stability analysis using eigenvalues is to be done. As a matter of fact, since the only nonlinear term is $\frac{u_q}{U_d}$ and since u_q will tend to zero anyway, this nonlinear term may be simply linearized by $\frac{1}{U_d^*}u_q$ where U_d^* is the value of U_d at the equilibrium point. In other words, the linear version of (7.32) may be expressed as

$$\dot{z}_1 = -h_2 z_1 + h_2 u_d, \quad \dot{z}_2 = \frac{h_1}{Z_1}u_q, \quad \dot{z}_3 = z_2 + \frac{h_0}{Z_1}u_q, \tag{7.33}$$

where Z_1 is the equilibrium (operating) value of z_1.

20 See Problem 7.9.
21 See Problem 7.10.
22 We use this LTI model for the design of ePLL parameters. It is also used in a control approach to improve the converter performance in weak grid conditions.
23 Apart from theoretical value, and usefulness for performing system stability studies, this model can then be linearized and used to improve the converter controller for weak grid conditions.
24 See Problem 7.11.

7.2.3 ePLL with Negative-Sequence Estimation

Figure 7.6 shows the ePLL that is upgraded to estimate the negative-sequence and reject its impacts on the accuracy of estimated variables. This is an important feature because negative-sequence is often present in practical systems at a small degree and can also be present at large degree during transient fault conditions. The estimated negative-sequence is shown by y. The unit vectors $s_{d_n}(\cdot)$ and $s_{q_n}(\cdot)$ are defined as before, i.e. $s_{d_n}(\theta) = [\cos(\theta) \;\; -\sin(\theta)]^T$ and $s_{q_n}(\theta) = [\sin(\theta) \;\; \cos(\theta)]^T$.

The state-space equations of the ePLL with negative-sequence estimation in terms of new state variables x and y are summarized as[25]

$$\begin{aligned} \dot{x}_1 &= -\omega x_2 + h_0 e_\alpha, \quad \dot{x}_2 = \omega x_1 + h_0 e_\beta \\ \dot{y}_1 &= \omega y_2 + h_0 e_\alpha, \quad \dot{y}_2 = -\omega y_1 + h_0 e_\beta. \end{aligned} \tag{7.34}$$

Using $e_{\alpha\beta} = u_{\alpha\beta} - x - y$, we get

$$\begin{aligned} \dot{x}_1 &= -h_0 x_1 - \omega x_2 - h_0 y_1 + h_0 u_\alpha, \quad \dot{x}_2 = \omega x_1 - h_0 x_2 - h_0 y_2 + h_0 u_\beta \\ \dot{y}_1 &= -h_0 x_1 - h_0 y_1 + \omega y_2 + h_0 u_\alpha, \quad \dot{y}_2 = -h_0 x_2 - \omega y_1 - h_0 y_2 + h_0 u_\beta. \end{aligned} \tag{7.35}$$

Thus, the characteristic matrix of this ePLL is given by

$$A = \begin{bmatrix} -h_0 & -\omega & -h_0 & 0 \\ \omega & -h_0 & 0 & -h_0 \\ -h_0 & 0 & -h_0 & \omega \\ 0 & -h_0 & -\omega & -h_0 \end{bmatrix}. \tag{7.36}$$

The characteristic equation of this matrix is given by[26]

$$\det(sI - A) = (s^2 + 2h_0 s + \omega^2)^2. \tag{7.37}$$

This means that the eigenvalues are repeated at $-h_0 \pm j\sqrt{\omega_o^2 - h_0^2}$. For the values of $h_0 = 250$ and $\omega_o = 377$, it will be $-250 \pm j282$ which indicates a damping ratio of around 0.66. Interestingly, the

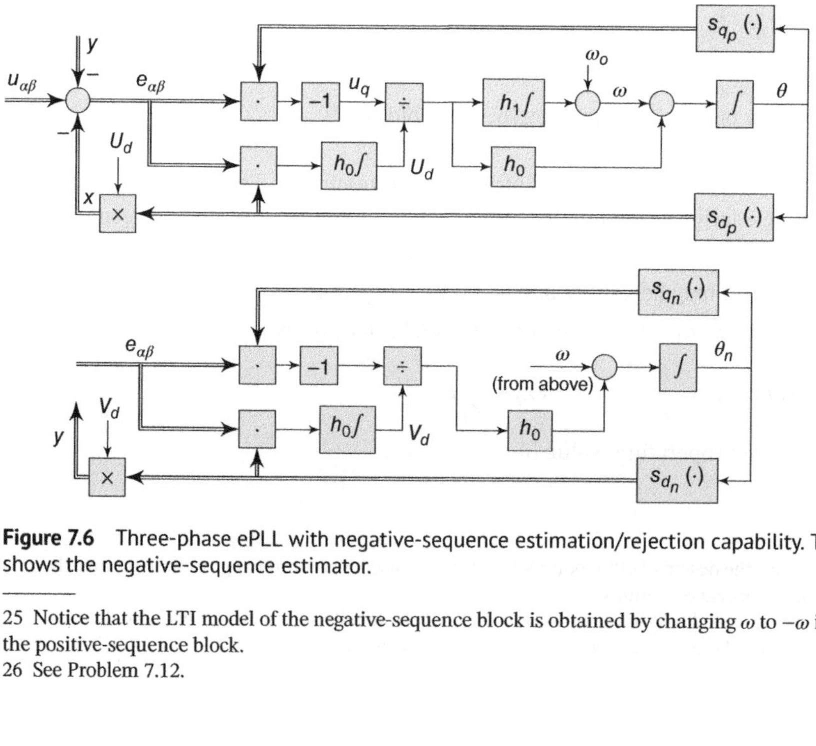

Figure 7.6 Three-phase ePLL with negative-sequence estimation/rejection capability. The lower block shows the negative-sequence estimator.

---

25 Notice that the LTI model of the negative-sequence block is obtained by changing ω to $-\omega$ in the LTI model of the positive-sequence block.
26 See Problem 7.12.

damping ratio has increased from the original value of around 0.55 for the basic ePLL (i.e. the one without negative-sequence unit). This means that this ePLL will have smoother responses despite the fact that it has dynamics of higher order.

7.2.4 ePLL with Negative-seq and dc Estimation

Figure 7.7 shows a more complete version comprising both the dc and negative-sequence extraction units. The estimated dc vector is shown by z and the negative-sequence by y.

By inspecting the two previous ePLLs, the equations of this ePLL are derived as

$$\begin{aligned} \dot{x}_1 &= -\omega x_2 + h_0 e_\alpha, & \dot{x}_2 &= \omega x_1 + h_0 e_\beta \\ \dot{y}_1 &= \omega y_2 + h_0 e_\alpha, & \dot{y}_2 &= -\omega y_1 + h_0 e_\beta \\ \dot{z}_1 &= h_3 e_\alpha, & \dot{z}_2 &= h_3 e_\beta. \end{aligned} \tag{7.38}$$

And using $e_{\alpha\beta} = u_{\alpha\beta} - x - y - z$, we get

$$\begin{aligned} \dot{x}_1 &= -h_0 x_1 - \omega x_2 - h_0 y_1 - h_0 z_1 + h_0 u_\alpha, \\ \dot{x}_2 &= \omega x_1 - h_0 x_2 - h_0 y_2 - h_0 z_2 + h_0 u_\beta, \\ \dot{y}_1 &= -h_0 x_1 - h_0 z_1 - h_0 y_1 + \omega y_2 + h_0 u_\alpha, \\ \dot{y}_2 &= -h_0 x_2 - h_0 z_2 - \omega y_1 - h_0 y_2 + h_0 u_\beta, \\ \dot{z}_1 &= -h_3 x_1 - h_3 z_1 - h_3 y_1 + h_3 u_\alpha, \\ \dot{z}_2 &= -h_3 x_2 - h_3 z_2 - h_3 y_2 + h_3 u_\beta. \end{aligned} \tag{7.39}$$

The characteristic matrix is

$$A = \begin{bmatrix} -h_0 & -\omega & -h_0 & 0 & -h_0 & 0 \\ \omega & -h_0 & 0 & -h_0 & 0 & -h_0 \\ -h_0 & 0 & -h_0 & \omega & -h_0 & 0 \\ 0 & -h_0 & -\omega & -h_0 & 0 & -h_0 \\ -h_3 & 0 & -h_3 & 0 & -h_3 & 0 \\ 0 & -h_3 & 0 & -h_3 & 0 & -h_3 \end{bmatrix}. \tag{7.40}$$

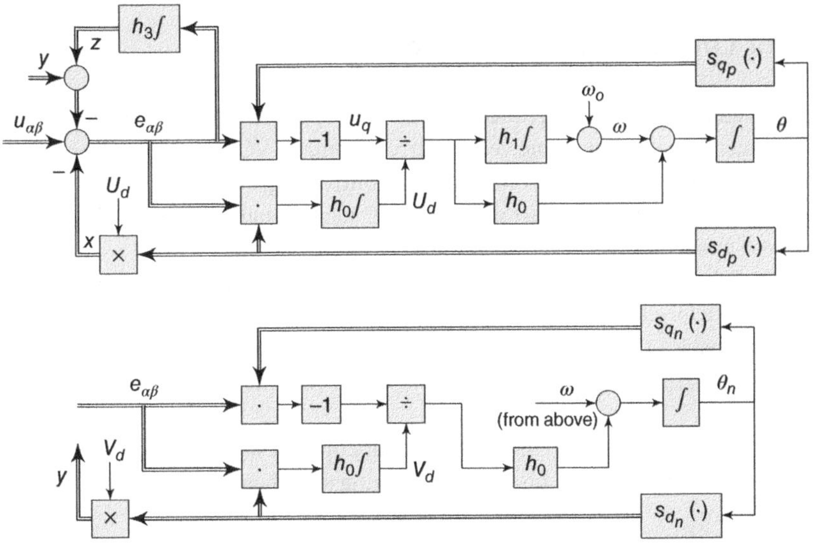

Figure 7.7 Three-phase ePLL with negative-sequence and dc component estimation/rejection.

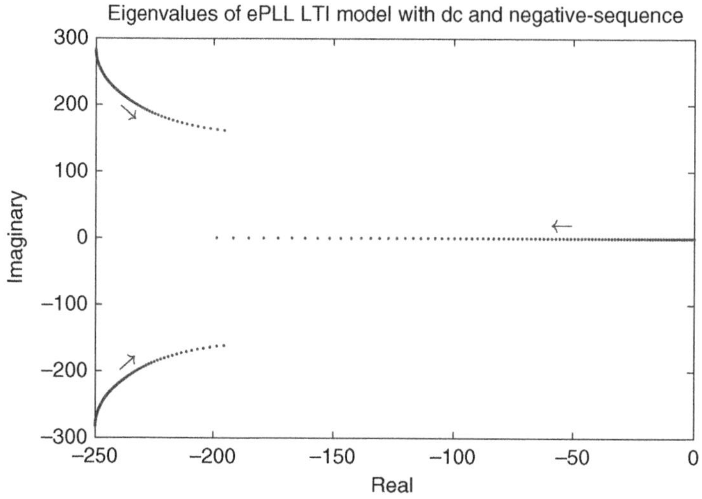

Figure 7.8 Poles of the ePLL with dc and negative-sequence rejection loops, i.e. eigenvalues of (7.40), when h_3 varies from 0 to 90 ($h_0 = 250$).

Figure 7.8 shows an example. Here, $\omega = 377$ rad/s, $h_0 = 250$, and h_3 varies from 0 to 90. Interestingly, the poles are repeated.[27] It is observed that at $h_3 = 90$, there are two set of poles at $-196 \pm j162$ and -199. This seems to be a reasonable design: the complex poles are well damped (with a damping factor around 0.77), and sufficiently distant from the imaginary axis. Thus, this parameters set of

$$\boxed{h_0 = 250, \quad h_1 = 10\,000, \quad h_3 = 90,} \tag{7.41}$$

is used in the examples in this chapter.

7.3 Vector Current Control in Stationary Domain

Power control via a CFL is commonly used in grid-connected inverters. The CFL may be formed in stationary or in synchronous domain. The name vector current control (VCC) is also used for this method. The SRF control is more popular as it offers simple control loops using integrating compensators, not resonant controllers. We will start with the VCC in stationary frame which greatly resembles the single-phase case and then we move to the VCC in SRF domain.

The three-phase VSC using a simple inductor interface filter L is shown in Figure 7.9. More elaborated filters such as inductor-capacitor-inductor (LCL) topologies are also used. The inverter interfaces a dc side with voltage V_{dc} to a three-phase ac side shown by the voltage $v_g(t)$. The resistor R models the cable's resistance, the inductor's parasitic resistance, and the conducting resistance of switches.

The sinusoidal PWM technique to generate the switching functions is shown in Figure 7.10 for leg-a of the converter. The variable m_a is the phase-a modulating signal and $c(t)$ is the triangular (or saw-tooth) carrier signal.

---

27 We will leave it as a problem to prove analytically that the eigenvalues of matrix A in (7.40) are repeated.

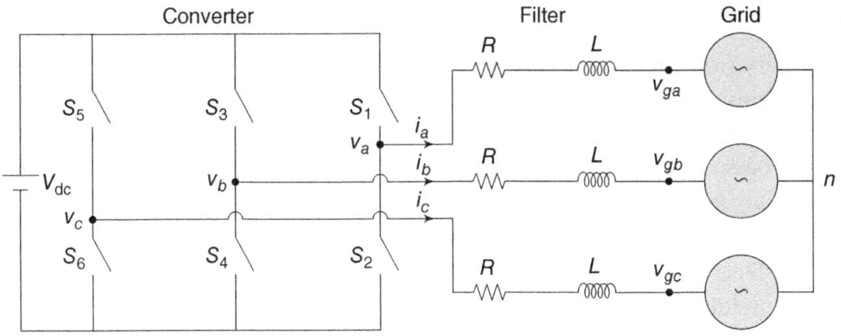

Figure 7.9 Grid-connected three-phase voltage source converter.

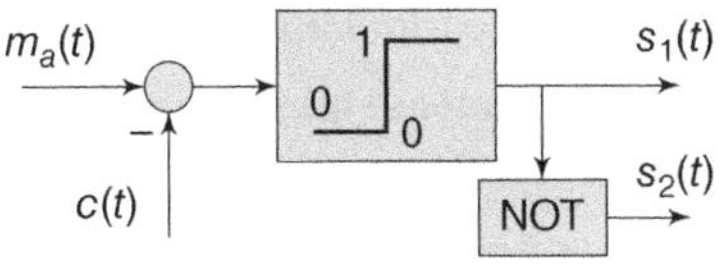

Figure 7.10 PWM for phase-a of the three-phase VSC.

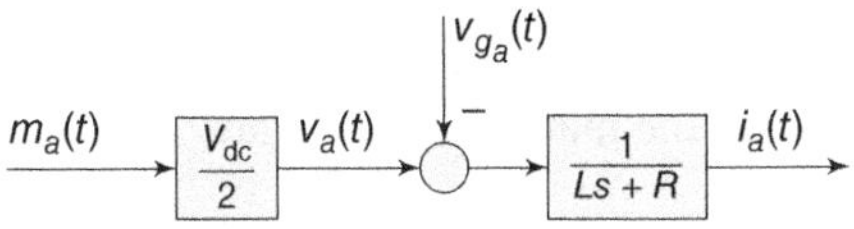

Figure 7.11 Per-phase average or control model of the VSC (phase-a is shown).

It was shown in Chapter 2 that the three-phase VSC is equivalent to three single-phase half-bridge VSCs. Figure 7.11 shows the control block diagram for phase-a. The voltages are measured with respect to a common reference point.

The circuit equations are

$$L\frac{d}{dt}i_x(t) = v_x(t) - Ri_x(t) - v_{g_x}(t), \quad x = a, b, c. \tag{7.42}$$

By transferring them to $\alpha\beta$ domain, we get[28]

$$L\frac{d}{dt}i_{\alpha\beta}(t) = v_{\alpha\beta}(t) - Ri_{\alpha\beta}(t) - v_{g_{\alpha\beta}}(t). \tag{7.43}$$

The equivalent model in $\alpha\beta$ domain is shown in Figure 7.12 for α signal. The model for the β axis is similar and decoupled (independent) from the α axis. Thus, the system is modeled by two decoupled single-phase systems.

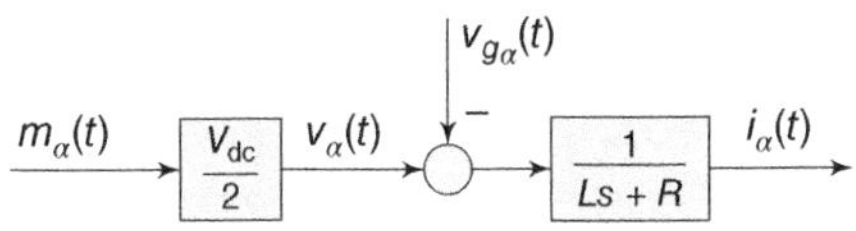

Figure 7.12 Per-phase average or control model in $\alpha\beta$ (or stationary) domain.

28 Equation (7.43) may be considered as one single complex equation or two real equations.

7.3.1 Controller Structure

The CFL in the stationary ($\alpha\beta$) frame may then be formed exactly similar to the single-phase VSC. This is shown in Figure 7.13. Here the output current is measured, converted to $\alpha\beta$ domain, the reference currents are also generated (in $\alpha\beta$ domain) and once subtracted to generate the tracking errors, they are passed through resonant controllers to close the two feedback loops. The feed-forward terms $v_{\mathrm{ff}_{\alpha,\beta}}$ are generated by the PLL and are used to achieve initial synchronization for a soft start.

The outputs of controllers, i.e. m_α and m_β, will then be transformed from $\alpha\beta$ domain to abc domain to generate the modulating signals in the abc domain to be forwarded to the PWM block. The controller gains are designed optimally using the linear quadratic tracker (LQT) method discussed for single-phase VSC in Chapter 6. Figure 7.14 shows the complete control loop diagram in stationary domain. Notice that the connecting lines denote two or three signals (depending whether it is $\alpha\beta$ or abc, respectively). The PLL generates two sets of signals: (i) the feed-forward terms $v_{\mathrm{ff}_{\alpha,\beta}}$ for initial synchronization and (ii) the synchronizing references s_d and s_q that are used to generate and limit the current reference as explained below.

The process of designing the single-phase CFL feedback gains k_1, k_2, and k_3 using LQT is described in Section 6.2.5 which is reviewed here for easy reference. Define the state variables of resonant controller as $X_1(s) = \frac{1}{s^2+\omega_o^2}E(s)$ and $X_2(s) = \frac{s}{s^2+\omega_o^2}E(s)$ where $E(s)$ is the tracking error $e(t)$. This also means $(s^2 + \omega_0^2)X_1(s) = E(s)$, $(s^2 + \omega_0^2)X_2(s) = sE(s)$.

The entire state equations of the system (including the resonant controller and the converter model) is summarized as $\dot{x} = Ax + Bu + B_{\mathrm{ref}}i_{\mathrm{ref}}$ where $x_3 = i$ and the matrices/vectors are

$$A = \begin{bmatrix} 0 & 1 & 0 \\ -\omega_0^2 & 0 & 1 \\ 0 & 0 & -\frac{R}{L} \end{bmatrix}, \quad B = \begin{bmatrix} 0 \\ 0 \\ \frac{1}{L} \end{bmatrix}, \quad B_{\mathrm{ref}} = \begin{bmatrix} 0 \\ -1 \\ 0 \end{bmatrix}. \tag{7.44}$$

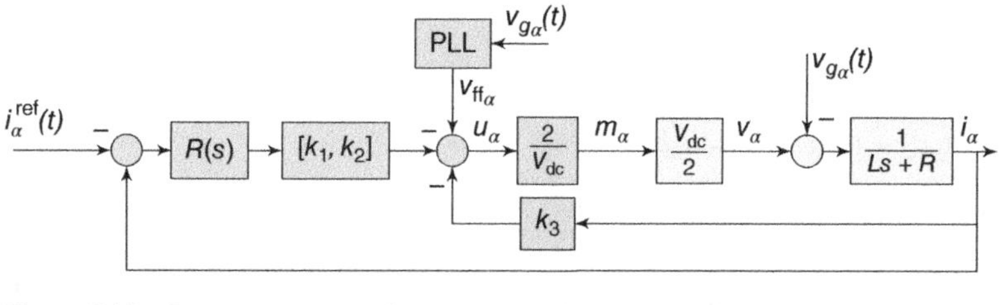

Figure 7.13 Per-phase current feedback loop in $\alpha\beta$ (or stationary) domain.

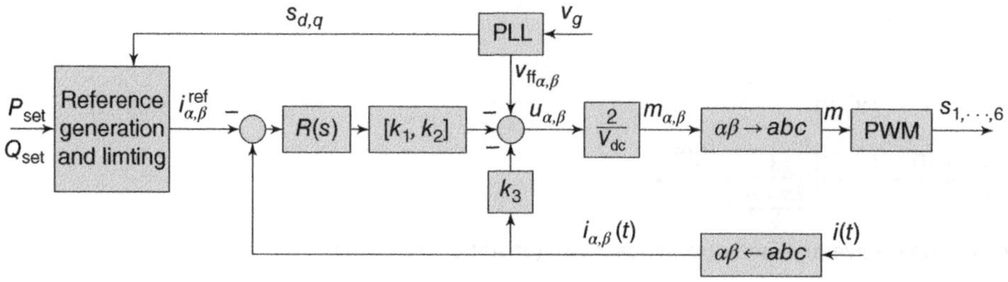

Figure 7.14 Complete diagram of power control approach using CFLs in $\alpha\beta$ domain.

The control law is $u(t) = -k_1x_1(t) - k_2x_2(t) - k_3x_3(t) = -Kx(t)$ which is a full linear state feedback. In order to proceed with the LQT method, apply the operator $\frac{d^2}{dt^2} + \omega_o^2$ to both sides of state equations to obtain $\dot{z} = Az + Bw$ where $z_i(t) = \ddot{x}_i(t) + \omega_o^2x_i(t)$ for $i = 1, 2, 3$ and $w(t) = \ddot{u}(t) + \omega_o^2u(t)$. Notice that $\frac{d^2}{dt^2}i_{\text{ref}}(t) + \omega_o^2 i_{\text{ref}}(t) = 0$ because the current reference is a pure sinusoidal at ω_o. Notice also that $w(t) = -Kz(t)$, and that $z_1(t) = e(t)$, $z_2(t) = \dot{e}(t)$ which are the tracking error and its derivative.

Therefore, using the LQT technique, the cost function $J = \int_0^\infty [qe(t)^2 + q_2\dot{e}(t)^2 + q_3z_3(t)^2 + w(t)^2]dt$ may be minimized using Matlab: $\gg$ `K = lqr(A, B, Q, 1)` where $Q = \text{diag}([q \;\; q_2 \;\; q_3])$.

7.3.2 Current Reference Generation and Limiting

In the three-phase applications, different current reference scenarios may be used depending on the desired specifications and system conditions. For instance, one scenario is to generate a balanced current regardless of whether the grid voltage is balanced or not. Another scenario is to let the current become unbalanced, for example, in order to remove the power ripples. Some of these scenarios are discussed in this section.

When the grid voltage is balanced, it is normally required for the converter to exchange balanced current with the grid, unless there are some unbalanced local load where the converter is desired to compensate for its current. In cases where the grid voltage is slightly unbalanced, without any grid voltage faults happening, again the converter may exchange balanced current, or slightly unbalanced current. Finally, during the grid voltage faults (which are mostly unsymmetrical faults), the converter may be required to ride through the fault and exchange balanced or unbalanced current to satisfy certain objectives [2–4]. In all these situations, the real and reactive power set-points must be executed by the converter.

7.3.2.1 Balanced Current

In this case, a balanced current is generated *regardless* of whether the voltage is balanced or unbalanced. The current reference generation and limitation mechanism is shown in Figure 7.15. Its operation is the same as the one for single-phase case shown in Figure 6.8 except that s_{d_p} and s_{q_p} are vectors here. The reference current i_{ref} is also a vector representing the current in $\alpha\beta$ domain. More explanation follows.

Consider the space-phasor grid voltage expression $\vec{v}_g = V_{g_p}\vec{s}_{d_p} + V_{g_n}\vec{s}_{d_n}$ where V_{g_p} and V_{g_n} denote its positive- and negative-sequence magnitudes, and $\vec{s}_{d_p} = e^{j\theta_p}$ and $\vec{s}_{d_n} = e^{-j\theta_n}$ in which θ_p and θ_n are total phase angles of positive- and negative-sequence components. A positive-sequence current

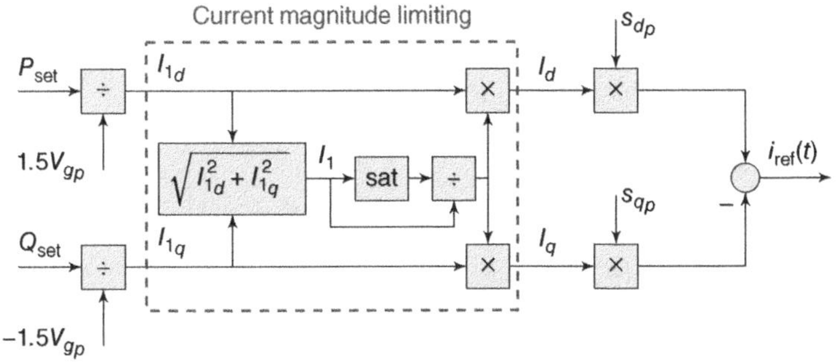

Figure 7.15 Generating and limiting the balanced reference current in stationary frame.

can also be expressed as $\vec{i} = I_d \vec{s}_{d_p} - I_q \vec{s}_{q_p}$ where $I_d = I\cos(\phi)$, $I_q = I\sin(\phi)$, I is the magnitude, and ϕ is the phase angle of the current with reference to the voltage. Note that in real-valued vector notations, we have

$$s_{d_p} = \begin{bmatrix} \cos(\theta_p) \\ \sin(\theta_p) \end{bmatrix}, \; s_{q_p} = \begin{bmatrix} \sin(\theta_p) \\ -\cos(\theta_p) \end{bmatrix}, \; s_{d_n} = \begin{bmatrix} \cos(\theta_n) \\ -\sin(\theta_n) \end{bmatrix}, \; s_{q_n} = \begin{bmatrix} \sin(\theta_n) \\ \cos(\theta_n) \end{bmatrix}. \tag{7.45}$$

The complex power is equal to

$$\begin{aligned}
\vec{S} &= \tfrac{3}{2}\vec{v}\,\vec{i}^{\,*} = \tfrac{3}{2}\left(V_{g_p}\vec{s}_{d_p} + V_{g_n}\vec{s}_{d_n}\right)\left(I_d\vec{s}_{d_p} - I_q\vec{s}_{q_p}\right)^* \\
&= \tfrac{3}{2}\left(V_{g_p}e^{j\theta_p} + V_{g_n}e^{-j\theta_n}\right)\left(I_d e^{-j\theta_p} - I_q e^{-j\left(\theta_p - \frac{\pi}{2}\right)}\right) \\
&= \tfrac{3}{2}V_{g_p}I_d - j\tfrac{3}{2}V_{g_p}I_q + \tfrac{3}{2}V_{g_n}I_d e^{-j(\theta_p+\theta_n)} - j\tfrac{3}{2}V_{g_n}I_q e^{-j(\theta_p+\theta_n)} \\
&= \underbrace{\frac{3}{2}\left[V_{g_p}I_d + V_{g_n}I\cos\left(\theta_p + \theta_n + \phi\right)\right]}_{P} + \underbrace{j\frac{3}{2}\left[-V_{g_p}I_q - V_{g_n}I\sin\left(\theta_p + \theta_n + \phi\right)\right]}_{Q}.
\end{aligned}$$

This derivation shows that both the real and reactive powers will have double-frequency pulsations if the grid voltage is unbalanced. Ignoring the 2-f ripples – or assuming that the grid voltage is balanced – the current reference (in the $\alpha\beta$ domain) is processed based on the real and reactive power set-points P_{set} and Q_{set} according to

$$\boxed{i_{ref}(t) = I_d s_{d_p} - I_q s_{q_p}, \quad I_d = \frac{P_{set}}{1.5V_{g_p}}, \quad I_q = -\frac{Q_{set}}{1.5V_{g_p}},} \tag{7.46}$$

where V_{g_p} is the magnitude of the positive-sequence component of the grid voltage.

7.3.2.2 Unbalanced Current

In this case, we assume that the current can be unbalanced. Its full expression is therefore,

$$\vec{i} = I_{d_p}\vec{s}_{d_p} - I_{q_p}\vec{s}_{q_p} + I_{d_n}\vec{s}_{d_n} - I_{q_n}\vec{s}_{q_n} \tag{7.47}$$

where $I_{d_p} = I_p\cos(\phi_p)$, $I_{q_p} = I_p\sin(\phi_p)$, $I_{d_n} = I_n\cos(\phi_n)$, $I_{q_n} = I_n\sin(\phi_n)$, I_p and I_n are the magnitudes of its positive- and negative-sequence components, and ϕ_p and ϕ_n are their phase angles with reference to the corresponding component of the voltage.

The complex power is equal to

$$\begin{aligned}
\vec{S} &= \tfrac{3}{2}\vec{v}\,\vec{i}^{\,*} = \tfrac{3}{2}\left(V_{g_p}\vec{s}_{d_p} + V_{g_n}\vec{s}_{d_n}\right)\left(I_{d_p}\vec{s}_{d_p} - I_{q_p}\vec{s}_{q_p} + I_{d_n}\vec{s}_{d_n} - I_{q_n}\vec{s}_{q_n}\right)^* \\
&= \tfrac{3}{2}\left(V_{g_p}e^{j\theta_p} + V_{g_n}e^{-j\theta_n}\right)\left(I_{d_p}e^{-j\theta_p} - I_{q_p}e^{-j\left(\theta_p - \frac{\pi}{2}\right)} + I_{d_n}e^{j\theta_n} - I_{q_n}e^{j\left(\theta_n - \frac{\pi}{2}\right)}\right) \\
&= \tfrac{3}{2}V_{g_p}I_{d_p} - j\tfrac{3}{2}V_{g_p}I_{q_p} + \tfrac{3}{2}V_{g_p}I_{d_n}e^{j(\theta_p+\theta_n)} + j\tfrac{3}{2}V_{g_p}I_{q_n}e^{j(\theta_p+\theta_n)} \\
&\quad + \tfrac{3}{2}V_{g_n}I_{d_p}e^{-j(\theta_p+\theta_n)} - j\tfrac{3}{2}V_{g_n}I_{q_p}e^{-j(\theta_p+\theta_n)} + \tfrac{3}{2}V_{g_n}I_{d_n} + j\tfrac{3}{2}V_{g_n}I_{q_n} \\
&= \tfrac{3}{2}\left[V_{g_p}I_{d_p} + V_{g_n}I_{d_n} + V_{g_p}I_n\cos\left(\theta_p + \theta_n + \phi_n\right) + V_{g_n}I_p\cos\left(\theta_p + \theta_n + \phi_p\right)\right] \\
&\quad + j\tfrac{3}{2}\left[-V_{g_p}I_{q_p} + V_{g_n}I_{q_n} + V_{g_p}I_n\sin\left(\theta_p + \theta_n + \phi_n\right) - V_{g_n}I_p\sin\left(\theta_p + \theta_n + \phi_p\right)\right].
\end{aligned}$$

This equation opens up multiple options to generate the current reference when the grid voltage is unbalanced. Notice that in addition to normal small grid voltage unbalance conditions, this study covers the cases that the grid voltage is highly unbalanced during a short interval of time, for example when a transient grid fault occurs. The converter may be required to have fault ride-through or low-voltage ride-through (LVRT) property. Some of the possible scenarios for reference current generation are discussed below.

Balanced Current That is $I_n = I_{d_n} = I_{q_n} = 0$. This reduces to the same scenario discussed in Section 7.3.2.1. The reference current is generated from (7.46).

Unbalanced Current with Given Angles That is ϕ_p and ϕ_n are given.[29] Then, the magnitudes of current positive- and negative-sequences, I_p and I_n, are calculated from

$$\begin{aligned} &\tfrac{3}{2}V_{g_p}I_p\cos(\phi_p) + \tfrac{3}{2}V_{g_n}I_n\cos(\phi_n) = P_{\text{set}} \\ &\tfrac{3}{2}V_{g_p}I_p\sin(\phi_p) - \tfrac{3}{2}V_{g_n}I_n\sin(\phi_n) = -Q_{\text{set}}, \end{aligned} \tag{7.48}$$

and the reference current is computed from (7.47).

Any other situations where two of the four unknowns $(I_p, I_n, \phi_p, \phi_n)$ are given can be solved for the other two unknowns from Eq. (7.48).

Remove 2-f Ripples from Real Power One scenario for unbalanced current generation is to remove the 2-f real power oscillations while the real and reactive power set-points P_{set} and Q_{set} are followed. For this to happen, it is readily seen that the following conditions must be satisfied:

$$\begin{aligned} &\tfrac{3}{2}V_{g_p}I_p\cos(\phi_p) + \tfrac{3}{2}V_{g_n}I_n\cos(\phi_n) = P_{\text{set}} \\ &\tfrac{3}{2}V_{g_p}I_p\sin(\phi_p) - \tfrac{3}{2}V_{g_n}I_n\sin(\phi_n) = -Q_{\text{set}} \\ &\phi_n = \phi_p + \pi \\ &V_{g_p}I_n = V_{g_n}I_p. \end{aligned} \tag{7.49}$$

The last equation implies that $I_n = \frac{V_{gn}}{V_{gp}}I_p$. Define the voltage unbalance factor $U_v = \frac{V_{gn}}{V_{gp}}$, then, by combining the equations, we get

$$I_{d_p} = \frac{P_{\text{set}}}{\tfrac{3}{2}(V_{g_p} - U_v V_{g_n})},\ I_{q_p} = \frac{Q_{\text{set}}}{-\tfrac{3}{2}(V_{g_p} + U_v V_{g_n})},\ I_{d_n} = -U_v I_{d_p},\ I_{q_n} = -U_v I_{q_p}.$$

In terms of the real-valued $\alpha\beta$ domain, the current reference is

$$i_{\text{ref}}(t) = I_{d_p}s_{d_p} - I_{q_p}s_{q_p} + I_{d_n}s_{d_n} - I_{q_n}s_{q_n}. \tag{7.50}$$

Finally, the block diagram representation of this current generation mechanism is shown in Figure 7.16.

Note that the "average magnitude" of the current reference (before limiting) is

$$\sqrt{I_{d_p}^2 + I_{q_p}^2 + I_{d_n}^2 + I_{q_n}^2} = \sqrt{(1 + U_v^2)\left(I_{d_p}^2 + I_{q_p}^2\right)}.$$

This is used to properly adjust the limiting mechanism.[30]

This approach generates an unbalanced current which transfers the desired real and reactive powers to the grid while ensuring that the real power has no 2-f pulsations. Similar derivations may be done for removing the double-frequency pulsations from the reactive power.

29 For instance, for $\phi_p = \phi_n = 0$, the current is "parallel" with the voltage. In other words, both positive- and negative-sequences of the current and voltage have the same angle [3].

30 Although this helps limiting the "average magnitude current" during extreme unbalance (or grid fault) situations, it does not guarantee that individual *abc* currents remain limited within the prespecified range for those extreme conditions.

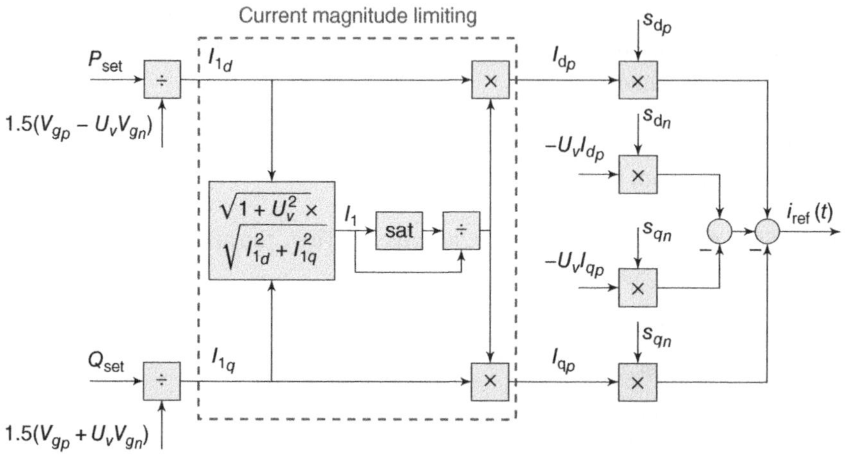

Figure 7.16 Generating and limiting the unbalanced reference current in stationary frame.

Example 7.1 *(Power Controller in Stationary Frame)*

The system parameters used in this example, and subsequent ones in this chapter, are given in Table 7.1. According to $S = \frac{3}{2}VI$, the maximum peak current is $\frac{2S}{3V} = \frac{2\times10\ 000\sqrt{2}}{3\times120\sqrt{2}} = 55.6$ A. To limit the current within ±70 A (25% above the peak rated current), we set the current limits in the "sat" block to ±70 A.

For this set of parameters, a view of the closed-loop poles of the CFL[31] is shown in Figure 7.17 when q varies from 10^{10} to $10^{13.25}$ (blue); q_2 varies from 10^4 to $10^{7.25}$ (red); and q_3 varies from 10^{-1} to $10^{0.75}$ (magenta). At final point, the poles are located at -1050, $-1000 \pm j1000$. These poles correspond to a rather fast response with about 1 ms time constant to allow tight current control and limiting.[32] The two complex poles have a damping ratio of about 0.7. At the final point, the controller gains are equal to $K = [k_1\ \ k_2\ \ k_3] = [3.36 \times 10^6\ \ 8 \times 10^3\ \ 6]$.[33]

Table 7.1 Converter system parameters used for the examples in this chapter.

Parameter	Symbol	Value	Unit
dc side voltage	V_{dc}	500	V
Grid side voltage (rms) (L-G)	V_g	120	V
Grid side frequency	f_o (ω_o)	60 (120π)	Hz (rad/s)
Inductance	L	2	mH
Parasitic resistance	R	20	mΩ
Switching frequency	f_{sw}	4	kHz
Power rating	P_{rate}	10	kW
Reactive power rating	Q_{rate}	±10	kVAr
Current limits	$I_{max,\ min}$	±70	A

31 That is eigenvalues of $A - BK$ where $K = \text{lqr}(A, B, Q, 1)$ and (A, B) is defined in (7.44).
32 This time-constant of 1 ms for the control system is one-fourth of the switching cycle and is reasonable.
33 In Example 6.1, we designed the LQT controller for $L = 5$ mH. Now, it was easily possible to adjust that design and use it here: since 2 mH is 40% of 5 mH, the controller gains of Example 6.1 will be multiplied into 0.4 to get the

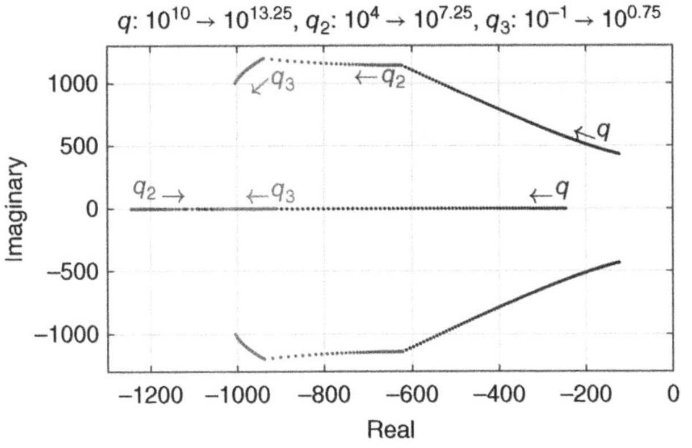

Figure 7.17 Poles of the closed-loop system using LQT approach.

Three system responses are examined: (i) balanced grid voltage and balanced current generation of Figure 7.15; (ii) unbalanced grid voltage and balanced current generation of Figure 7.15; and (iii) unbalanced current generation of Figure 7.16.

At $t = 0.05$ seconds, the gating commands are applied to the switches. Prior to this time, the ePLL is functioning to prepare a pre-synchronization reference for soft-start.[34] The inverter starts at $t = 0.05$ seconds. At $t = 0.1$ seconds, the real power reference of 10 kW (and zero reactive power, i.e. unity power factor operation) is applied. At $t = 0.15$ seconds, the reactive power set-point of 10 kVAr is applied (i.e. the power factor of 0.7 lagging). The change of reactive power to 0 and subsequently to −10 kVAr (i.e. the power factor of 0.7 leading) at $t = 0.2$ seconds and $t = 0.25$ seconds are also applied. The reactive power is restored to 0 at $t = 0.3$ seconds. At $t = 0.35$ seconds, the grid voltage experiences a deep drop of 50% (down to about 85 V peak) and restores to normal at $t = 0.4$ seconds.

Figure 7.18 shows the inverter responses when the grid is balanced and the inverter also generates a balanced current according to Figure 7.15. The responses show that the converter generates balanced sinusoidal currents. The controller succeeds to (i) yield a soft-start, (ii) execute real and reactive power commands quickly and smoothly, (iii) generate sinusoidal and balanced current, and (iv) limit the converter current.

Figure 7.19 shows the inverter responses when the grid is unbalanced and the inverter generates a balanced current according to Figure 7.15. The responses show that the converter generates balanced sinusoidal currents but the powers pulsate at double the frequency. The controller succeeds to (i) yield a soft-start, (ii) execute real and reactive power commands quickly and smoothly, (iii) generate sinusoidal and balanced current, and (iv) limit the converter current. Figure 7.20 repeats the same simulation but with the difference that a transient deep single-phase (line-to-ground: L-G) grid fault happens during 0.35–0.4 seconds. The converter continues to feed the same desired average real and reactive powers to the grid and maintains the balanced current during the fault. The current is within the limits. Figure 7.21 repeats the same simulation for a transient deep

---

controller gains for 2 mH. Note that the plant transfer function is $\frac{1}{Ls}$ and thus, the controller gains are proportional to the value of L.

34 The PLL used in this example and subsequent examples is the one with negative-sequence and dc offset estimation capability shown in Figure 7.7. The parameters of the ePLL are set at $h_0 = 250$, $h_1 = 10\,000$, $h_3 = 90$ as discussed. The PLL signal $V_g s_d(t)$ is used for the feed-forward term v_{ff} in the controller.

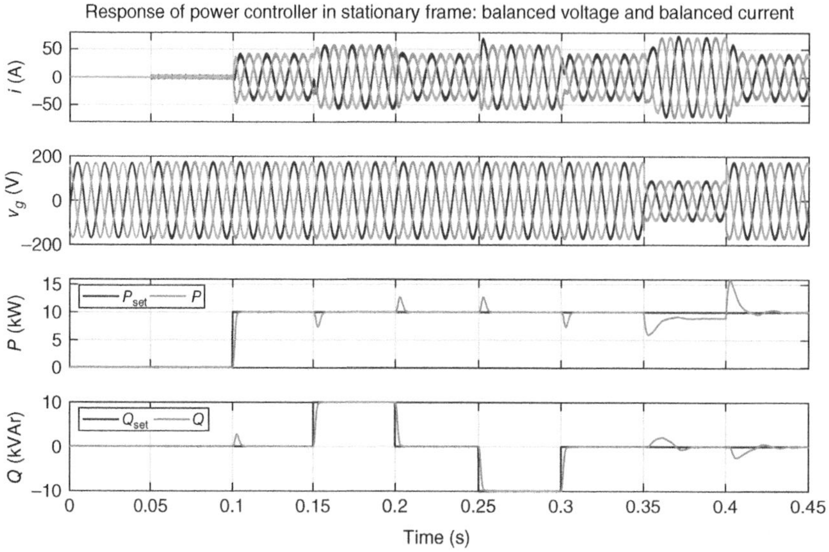

Figure 7.18 Responses of power controller in stationary domain: generating balanced current while the grid is also balanced. $t = 0$: ePLL starts, $t = 0.05$ seconds: converter starts, $t = 0.1,\ 0.15,\ 0.20,\ 0.25,$ 0.3 seconds: real and reactive power commands applied, $t = 0.35$ seconds: grid fault (voltage sag down to 30%), $t = 0.4$ seconds: fault is cleared.

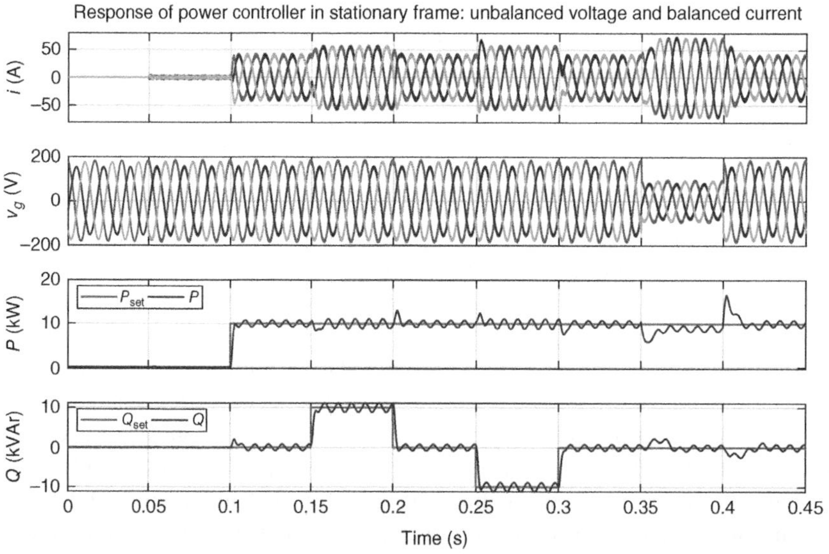

Figure 7.19 Responses of power controller in stationary domain: generating balanced current while the grid is unbalanced; causing power pulsations.

double-phase (line-line-to-ground: L-L-G) grid fault happening during 0.35–0.4 seconds. The converter continues to feed balanced current but has to reduce its real power to ensure that the current is limited at 70 A peak. In general, to implement a LVRT strategy, the set-points of real and reactive powers must be adjusted, during the fault interval, according to that given LVRT specifications.

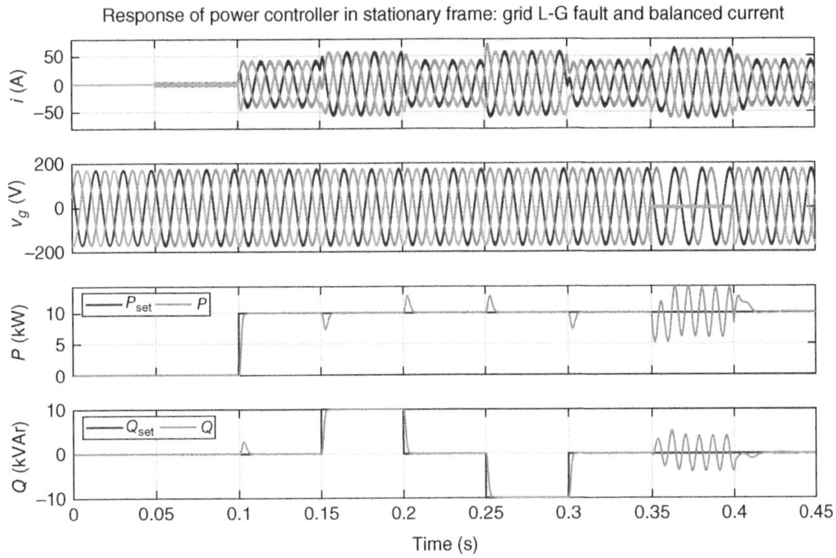

Figure 7.20 Responses of power controller in stationary domain: generating balanced current while the grid is unbalanced (under a transient L-G fault).

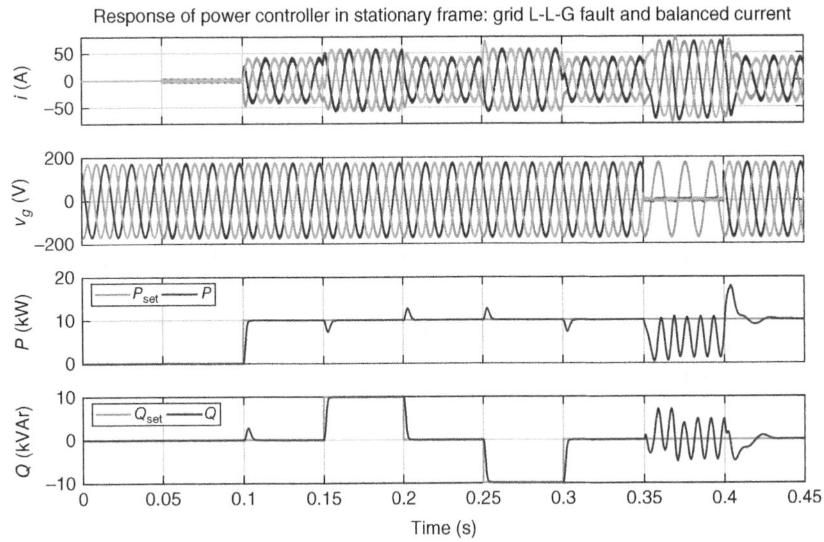

Figure 7.21 Responses of power controller in stationary domain: generating balanced current while the grid is unbalanced (under a transient L-L-G fault).

Figure 7.22 shows the inverter responses when the grid is unbalanced and the inverter generates an unbalanced current to remove the real power pulsations according to Figure 7.16. The controller succeeds to (i) yield a soft-start, (ii) execute real and reactive power commands quickly and smoothly, (iii) generate sinusoidal and unbalanced current to remove to real power pulsations, and (iv) limit the converter current. Two points are observed from this study: (i) the level of pulsations on the reactive power has, however, increased compared with the previous case where

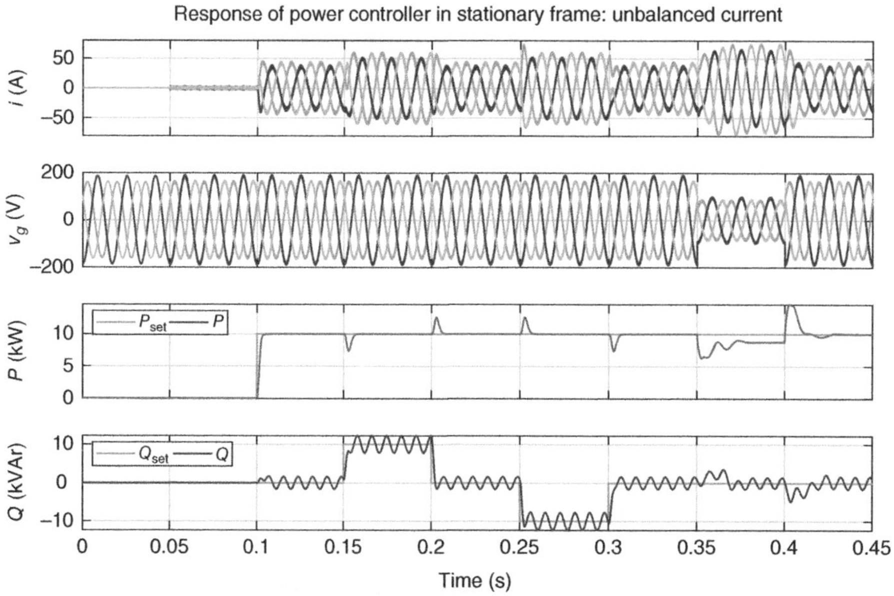

Figure 7.22 Responses of power controller in stationary domain: generating unbalanced current while the grid is unbalanced; removing real power pulsations.

the current was balanced,[35] (ii) the unbalanced current with larger magnitude has crossed the limit of 70 A slightly because the controller limits the "average magnitude current" in *dq* domain and this does not ensure that the currents in *abc* domain are accurately limited within the same range.

Responses of the controller to an extreme condition is shown in Figure 7.23. Here, a grid impedance of 2 mH is added to weaken the voltage at the point of grid coupling, and the inverter output filter inductance is increased from 2 to 3 mH (modeling an unknown uncertainty). Figure 7.23 shows that, compared to the based case shown in Figure 7.18, the responses exhibit some oscillatory modes. These oscillations can lead to instability for larger grid impedances.

7.3.3 Harmonics, Higher-Order Filters, System Delays

Due to the complete structural similarity of the VCC in stationary domain with its single-phase counterpart discussed in Chapter 6, all these aspects (i.e. upgrading the controller to reject the dc and harmonics from the current, upgrading the controller to work for higher-order filters such as LCL filters, and upgrading the controller to address system and control delays) can be addressed in the same way discussed there. For the weak grid condition, and including the PLL in the controller design, however, the three-phase scenario is different because the type of PLL used couples the two (α and β) phases. Therefore, the modification must be done in a way that it considers the coupling between these two phases through the PLL. This topic is discussed in Section 7.3.4.

7.3.4 Weak Grid Conditions and Including PLL in Controller*

Weak grid condition occurs when the voltage at the point of connection of the converter to the grid is weak. Thus, this voltage, i.e. v_g, is a function of the converter current and can experience

35 It can be mathematically proved that removing the ripples on the real power causes the ripples on the reactive power to go up. See Problem 7.13.

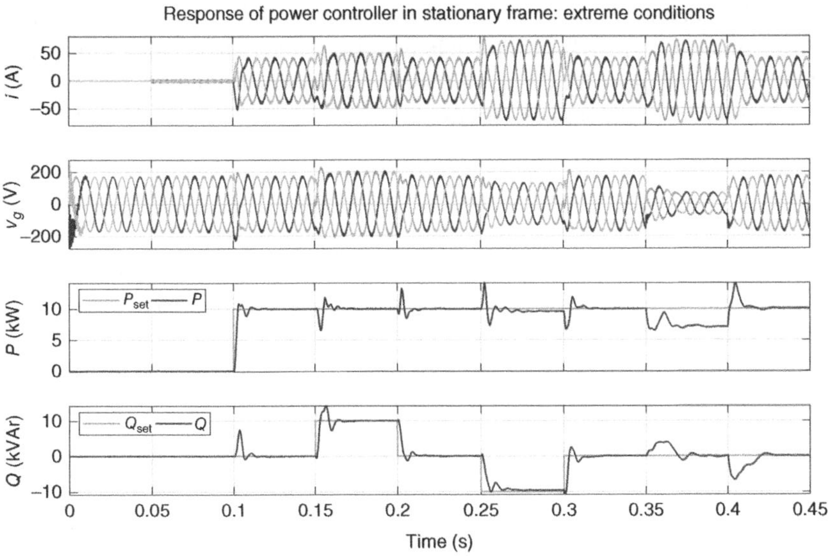

Figure 7.23 Responses of power controller in stationary domain: extreme conditions.

fluctuations as a result of interactions between the grid and the converter. Such fluctuations can lead to increased oscillations and instabilities. A converter working in weak grid conditions, such as one operating distant from main grid lines (for example in offshore wind turbines), must be mindful of exciting or amplifying such fluctuations. In this section, the power controller (in stationary domain) is modified to improve its performance in weak grid conditions.

The grid voltage v_g is directly used by the PLL to generate synchronizing signals for the controller. Thus, the PLL critically serves as the interface between the grid and the converter controller. The approach adopted in this section is to use additional feedback terms from the PLL in order to more effectively present the grid voltage fluctuations to the controller. As a result, the entire controller (including the PLL and resonant controllers) will have higher ability to respond to grid voltage dynamics.

The LTI model of the three-phase ePLL introduced in Theorem 7.1, shown in (7.30), indicates a two-input two-output coupled system. Therefore, in order to incorporate these equations in the controller, we propose to use a coupled (multi-input, multi-output: MIMO) topology as shown in Figure 7.24. Here, the state variables of the resonant controllers, the PLL, and the converter (currents) are all merged and used in a multivariable format to account for all couplings, as detailed below.

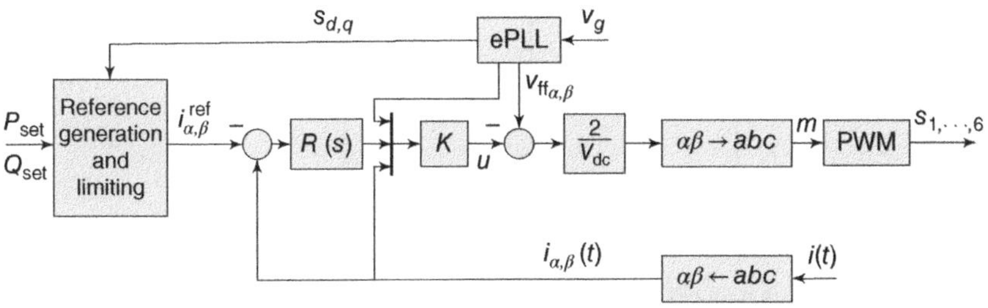

Figure 7.24 Power control approach in stationary domain with PLL modeling.

The common weak grid modeling is to assume that the weak grid voltage v_g represents a stiff grid voltage v_s behind an impedance $sL_s + R_s$ (as shown in Figure 6.20 for a single-phase situation). Therefore, it will satisfy $v_g = v_s + L_s\frac{d}{dt}i + R_s i$ where i is the current flowing to the grid. Meanwhile, let us rewrite the ePLL LTI model equations of (7.30) as

$$\begin{aligned}\dot{x}_5 &= -h_0 x_5 - \omega x_6 + h_0 v_{g_\alpha},\\ \dot{x}_6 &= \omega x_5 - h_0 x_6 + h_0 v_{g_\beta},\end{aligned} \quad \Rightarrow \quad \dot{x}_{\text{PLL}} = \begin{bmatrix} -h_0 & -\omega \\ \omega & -h_0 \end{bmatrix} x_{\text{PLL}} + h_0 v_g. \tag{7.51}$$

where $x_{\text{PLL}} = [x_5 \ x_6]^T = V_g s_d(\theta)$ which means $x_5 = V_g\cos(\theta)$ and $x_6 = V_g\sin(\theta)$.[36] Substituting $L\frac{d}{dt}i + Ri = u + x_{\text{PLL}} - v_g$ into $v_g = v_s + L_s\frac{d}{dt}i + R_s i$ will lead to an expression for v_g as[37,38]

$$v_g = \frac{LR_s - RL_s}{L + L_s}i + \frac{L_s}{L + L_s}(u + x_{\text{PLL}}) + \frac{L}{L + L_s}v_s. \tag{7.52}$$

Substituting (7.52) into (7.51) results in the final state equations of the PLL as

$$\dot{x}_{\text{PLL}} = \begin{bmatrix} \frac{-h_0 L}{L+L_s} & -\omega \\ \omega & \frac{-h_0 L}{L+L_s} \end{bmatrix} x_{\text{PLL}} + h_0\frac{LR_s - RL_s}{L + L_s}i + \frac{h_0 L_s}{L + L_s}u + \frac{h_0 L}{L + L_s}v_s. \tag{7.53}$$

The differential equation for the current is

$$\frac{d}{dt}i = -\frac{R + R_s}{L + L_s}i + \frac{1}{L + L_s}(u + x_{\text{PLL}} - v_s). \tag{7.54}$$

The state equations of 2I2O resonant controller may be expressed as

$$\dot{x}_r = \begin{bmatrix} 0 & 0 & 1 & 0 \\ 0 & 0 & 0 & 1 \\ -\omega^2 & 0 & 0 & 0 \\ 0 & -\omega^2 & 0 & 0 \end{bmatrix} x_r + \begin{bmatrix} 0 & 0 \\ 0 & 0 \\ 1 & 0 \\ 0 & 1 \end{bmatrix} e \tag{7.55}$$

where $e = i - i_{\text{ref}}$. With this state-space description for the resonant controller, the definitions of four state variables of the resonant controller are

$$X_r(s) = \frac{1}{s^2 + \omega^2}[E_1(s) \ E_2(s) \ sE_1(s) \ sE_2(s)]^T.$$

The entire system has eight state variables: x_r (four variables), x_{PLL} (two variables), and i (two variables). Define $x = [x_r \ x_{\text{PLL}} \ i]^T$ and write down the state space equations of the entire system as $\dot{x} = Ax + Bu + B_{\text{ref}}i_{\text{ref}} + B_s v_s$ with

$$A = \begin{bmatrix} 0_2 & I_2 & 0_2 & 0_2 \\ -\omega^2 I_2 & 0_2 & 0_2 & I_2 \\ 0_2 & 0_2 & A_{\text{PLL}} & h_0' I_2 \\ 0_2 & 0_2 & \frac{1}{L'}I_2 & -\frac{R'}{L'}I_2 \end{bmatrix}, \quad B = \begin{bmatrix} 0_2 \\ 0_2 \\ \frac{h_0 L_s}{L'}I_2 \\ \frac{1}{L'}I_2 \end{bmatrix}, \quad B_{\text{ref}} = \begin{bmatrix} 0_2 \\ I_2 \\ 0_2 \\ 0_2 \end{bmatrix}, \tag{7.56}$$

36 To simplify the discussion, we assume that an ePLL without negative-sequence and dc-offset loops is used.
37 Note that the feed-forward term $v_{\text{ffa},\beta}$ is equal to x_{PLL} in this case.
38 Some intermediary steps are skipped here due to similarity with the single-phase equations. See Problem 7.14.

where A_{PLL} is the 2×2 matrix in (7.51), 0_2 is the 2×2 zero matrix, I_2 is the 2×2 identity matrix, $h_0' = h_0 \frac{LR_s - RL_s}{L+L_s}$, $R' = R + R_s$, and $L' = L + L_s$. The current reference i_{ref} may be written as[39]

$$i_{\text{ref}} = \frac{P_{\text{set}}}{1.5V_g} s_d + \frac{Q_{\text{set}}}{1.5V_g} s_q = \frac{1}{1.5V_g^2} \begin{bmatrix} P_{\text{set}} & Q_{\text{set}} \\ -Q_{\text{set}} & P_{\text{set}} \end{bmatrix} x_{\text{PLL}} \approx A_{\text{ref}} x_{\text{PLL}} \tag{7.57}$$

which is a nonlinear equation because $V_g^2 = x_5^2 + x_6^2$. However, this nonlinearity may be approximated by using the nominal value of the grid voltage magnitude. Subsequently, this matrix, i.e. A_{ref}, can be inserted properly in (7.56) to place the blue-colored 0_2. In the rest of the discussion in this section, we ignore this term for simplicity.[40] It can be shown (as is done in Chapter 6 for single-phase power controller) that this term does not have noticeable impact on the results.

In order to use the LQT approach, we apply the linear operator $\frac{d^2}{dt^2} + \omega^2$ to the both sides of the state equation. Notice that since v_s is a assumed to be stiff at the frequency of ω, then $\left(\frac{d^2}{dt^2} + \omega^2\right) v_s = 0$. Defining $z = \left(\frac{d^2}{dt^2} + \omega^2\right) x$ and $w = \left(\frac{d^2}{dt^2} + \omega^2\right) u$, the transformed state equation will be $\dot{z} = Az + Bw$. Notice that $u = -Kx$, thus, $w = -Kz$, and also that

$$z_1 = e_1, \quad z_2 = e_2, \quad z_3 = \dot{e}_1, \quad z_4 = \dot{e}_2.$$

Thus, by defining the cost function

$$J = \int_0^{\infty} \left(q_1 e_1^2 + q_2 e_2^2 + q_3 \dot{e}_1^2 + q_4 \dot{e}_2^2 + \sum_{i=5}^{8} q_i z_i^2 + w_1^2 + w_2^2 \right) dt$$

and solving in MATLAB: ≫ `K=lqr(A,B,Q,eye(2))` with $Q = \text{diag}([q_1, \ldots, q_8])$, the optimal controller gains K are obtained.

Example 7.2 ***(Power Controller in Stationary Frame with PLL Feedback)***
The method discussed above is used in this example to design the feedback gains for the system parameters given in Table 7.1. As for the grid impedance, the nominal values of $L_s = 2$ mH and $R_s = 0.020\ \Omega$ are used.

For this set of parameters, a view of the closed-loop root-locus is shown in Figure 7.25 when the cost function of $J = \int_0^{\infty} \left[q\left(e_1^2 + e_2^2\right) + q_2\left(\dot{e}_1^2 + \dot{e}_2^2\right) + q_3\left(z_5^2 + z_6^2\right) + q_4\left(z_7^2 + z_8^2\right)\right] dt$ is considered and q varies from 10^{10} to $10^{13.75}$ (blue); q_2 varies from 10^4 to $10^{7.75}$ (red); q_3 varies from 10^{-1} to $10^{1.75}$ (magenta), and q_4 varies from 10^{-1} to 10^1 (cyan). At final point, the poles are located at $-912 \pm j994$, $-1016 \pm j930$, $-1005 \pm j62$, and $-225 \pm j379$. The last two correspond to the ePLL modes. The others show a rather fast response with about 1 ms time constant to allow tight current control and limiting.[41] At the final point, the controller gains matrix K is equal to

$$\begin{bmatrix} 5.84 \times 10^6 & -14.9 \times 10^4 & 14\,500 & 370 & 1 & 0.34 & 11.7 & -0.15 \\ 14.9 \times 10^4 & 5.84 \times 10^6 & 370 & 14\,500 & -0.34 & 1 & 0.15 & 11.7 \end{bmatrix}.$$

Here, a grid impedance of 2.5 mH is added to weaken the voltage at the point of grid coupling, and the inverter output filter inductance is increased from 2 to 3 mH (modeling an unknown uncertainty). Figures 7.26 and 7.27 show, respectively, the responses of the power controller without and with the PLL feedback signals. At $t = 0.05$ seconds, the gating commands are applied to the switches. Prior to this time, the ePLL is functioning to prepare a pre-synchronization reference for soft-start. The inverter starts at $t = 0.05$ seconds. At $t = 0.1$ seconds, the real power

39 See Problem 7.15.
40 Which is to say we design for zero power set-points.
41 Similar to those in Example 7.1.

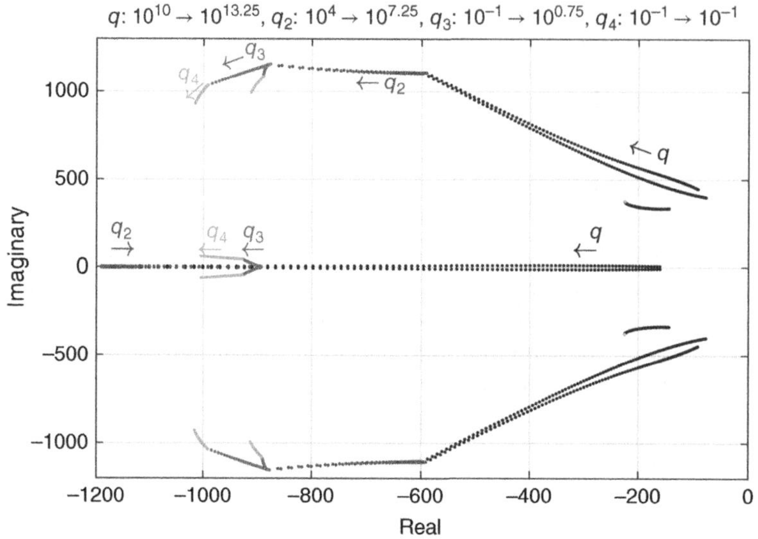

Figure 7.25 Poles of the power controller with PLL feedback terms (in stationary domain) using LQT approach.

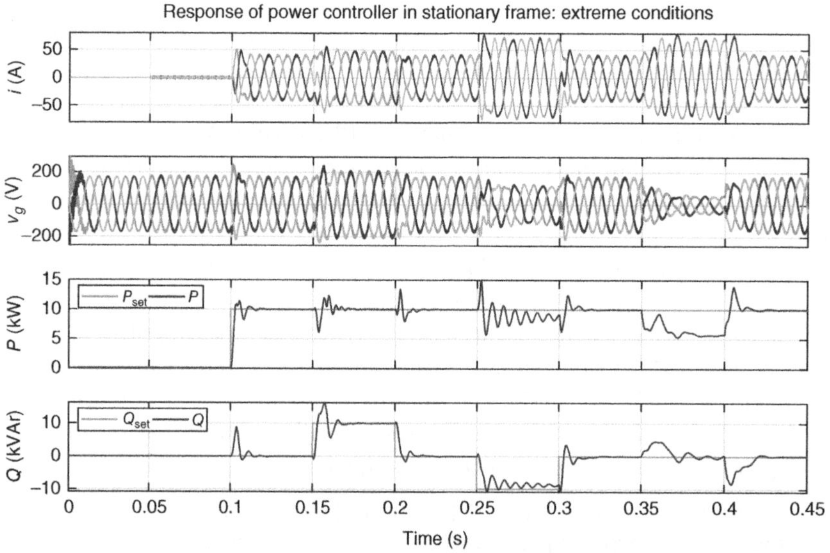

Figure 7.26 Responses of power controller in stationary domain: extreme conditions.

reference of 10 kW (and zero reactive power, i.e. unity power factor operation) is applied. At $t = 0.15$ seconds, the reactive power set-point of 10 kVAr is applied (i.e. the power factor of 0.7 lagging). The change of reactive power to 0 and subsequently to −10 kVAr (i.e. the power factor of 0.7 leading) at $t = 0.2$ seconds and $t = 0.25$ seconds are also applied. The reactive power is restored to 0 at $t = 0.3$ seconds. At $t = 0.35$ seconds, the grid voltage v_s experiences a deep drop of 50% (down to about 85 V peak) and restores to normal at $t = 0.4$ seconds. The controller with the PLL feedback terms clearly shows more robust responses (with a lower level of oscillations).

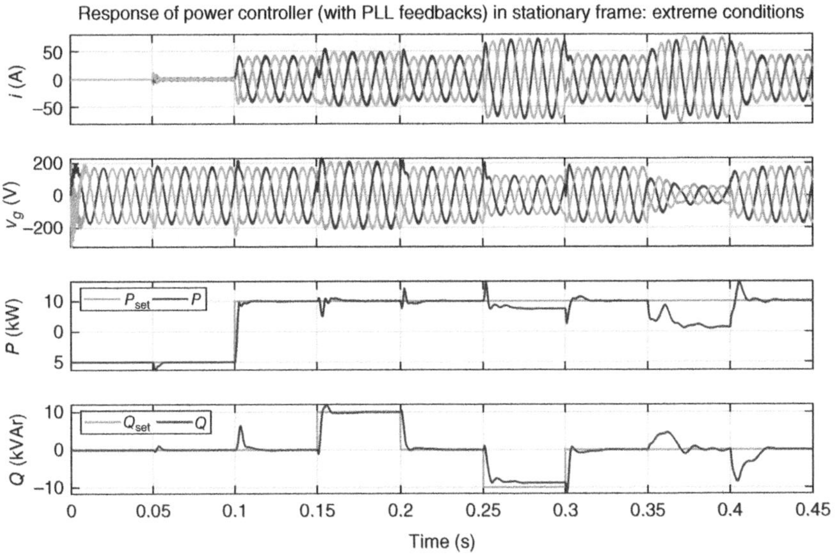

Figure 7.27 Responses of power controller (with PLL feedback terms) in stationary domain: extreme conditions.

7.4 Vector Current Control in Synchronous Reference Frame

The VCC in rotating frame is the most common approach for three-phase grid-connected converters. An angle, commonly the grid voltage angle obtained from a PLL, is used to transform the $\alpha\beta$ variables into the dq domain. In this domain, the variables are slowly varying. Therefore, proportional-integrating (PI) type compensators (as opposed to resonant type compensators of stationary domain) are used. This section studies this topic in details.

The SRF is defined by the transformation

$$x_{dq} = e^{-j\theta}x_{\alpha\beta} \quad \text{or} \quad \begin{bmatrix} x_d \\ x_q \end{bmatrix} = \begin{bmatrix} \cos(\theta) & \sin(\theta) \\ -\sin(\theta) & \cos(\theta) \end{bmatrix} \begin{bmatrix} x_\alpha \\ x_\beta \end{bmatrix}, \tag{7.58}$$

over the $\alpha\beta$ frame where the angle θ should rotate with the same frequency of the system, i.e. $\theta = \int \omega dt$. Commonly the grid voltage angle is used. This transformation implies that

$$\dot{x}_{dq} = -j\omega x_{dq} + e^{-j\theta}\dot{x}_{\alpha\beta}. \tag{7.59}$$

Multiplying into L and substituting from (7.43) results in

$$\begin{aligned} L\frac{d}{dt}i_{dq} &= -j\omega L i_{dq} + e^{-j\theta}L\frac{d}{dt}i_{\alpha\beta} \\ &= -j\omega L i_{dq} + e^{-j\theta}\left[v_{\alpha\beta}(t) - Ri_{\alpha\beta}(t) - v_{s_{\alpha\beta}}(t)\right] \\ &= -j\omega L i_{dq} + v_{dq} - Ri_{dq} - v_{g_{dq}} \end{aligned} \tag{7.60}$$

which simplifies to the following two real equations

$$\begin{aligned} L\frac{d}{dt}i_d &= -Ri_d + \omega L i_q + v_d - v_{g_d} \\ L\frac{d}{dt}i_q &= -Ri_q - \omega L i_d + v_q - v_{g_q}. \end{aligned} \tag{7.61}$$

These two equations summarize the dq-frame or synchronous frame control model of the three-phase VSC which is shown in Figure 7.28. Notice that the dq-frame model represents a

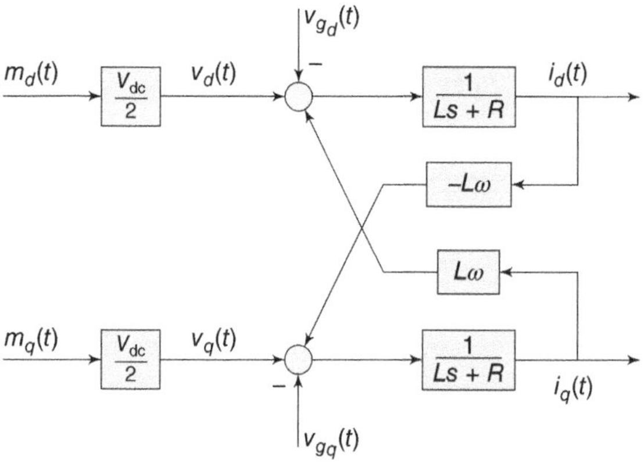

Figure 7.28 Control model of the three-phase VSC in *dq* (or synchronous) domain.

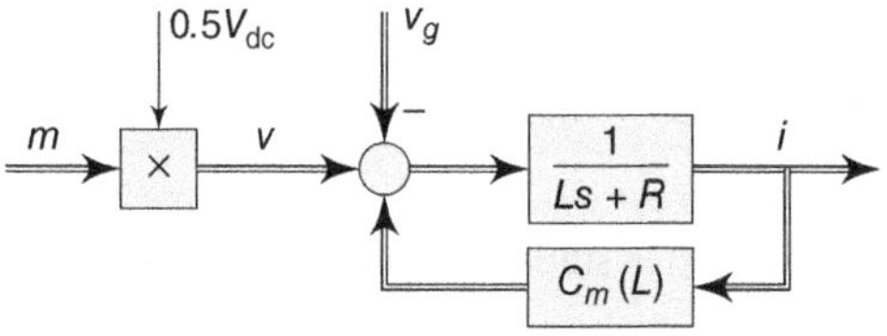

Figure 7.29 Control block diagram of VSC: valid for both stationary and rotating frame.

two-input two-output model similar to $\alpha\beta$ frame. However, the dq-frame model is coupled due to those $L\omega$ terms while the $\alpha\beta$-frame model is decoupled.

The model of Figure 7.28 may be shown in a double-line connection as depicted in Figure 7.29. This diagram is valid for both $\alpha\beta$ and dq frames with the note that the coupling matrix C_m is zero in $\alpha\beta$ and is equal to $C_m(L) = \begin{bmatrix} 0 & L\omega \\ -L\omega & 0 \end{bmatrix}$ in dq domain.

7.4.1 Control Structure

The control structure is shown in Figure 7.30. The dq transformation angle θ is the angle of the grid voltage positive-sequence component and is obtained by using a three-phase PLL. This PLL will also generate v_{g_d} and v_{g_q}. This angle is used to calculate the dq components of the current, i.e. i_d and i_q. When the controller calculates m_d and m_q, the same angle is used to generate m_a, m_b, and m_c and forward them to the PWM block.

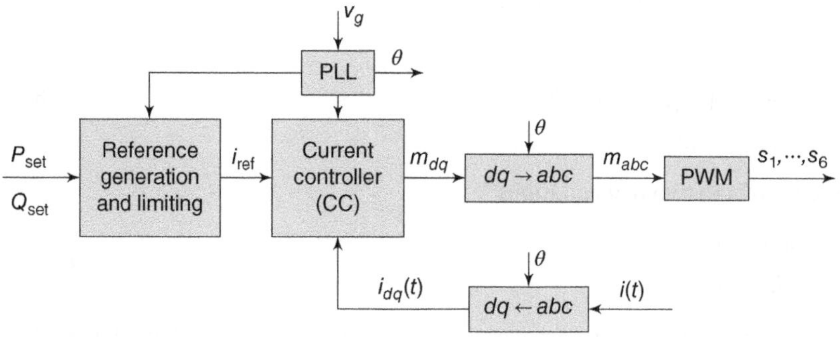

Figure 7.30 General power control structure based on VCC in synchronous domain.

Apart from the two *abc*/*dq* and *dq*/*abc* transformations, the controller has three main blocks: (i) reference generation and limiting, (ii) current controller (CC), and (iii) PLL. The PLL described earlier with negative-sequence and dc offset estimation/rejection capability is the preferred choice. There are multiple approaches for building the CC block. The conventional CFL using integrating and resonant functions, and also an advanced approach based on multivariable optimal control theory will be discussed in this chapter. Two different outlooks to reference generation and limiting will also be explained.

7.4.2 Current Reference Generation and Limiting

Different current reference generation scenarios may be considered. Similar to the stationary domain, we discuss two cases here: balanced current; unbalanced current to remove power ripples. As an introduction, we derive expressions for the powers in *dq* domain first. Again, no flow of zero-sequence current (no neutral connection) is assumed. Then, for a circuit element with the voltage v and current i, the complex power is

$$\vec{S} = \frac{3}{2}\vec{v}\,\vec{i}^{\,*} *= \frac{3}{2}(v_d + jv_q)e^{j\theta}[(i_d + ji_q)e^{j\theta}]^* *= \frac{3}{2}(v_d + jv_q)(i_d - ji_q), \tag{7.62}$$

which indicates

$$\boxed{P = \tfrac{3}{2}(v_d i_d + v_q i_q), \quad Q = \tfrac{3}{2}(v_q i_d - v_d i_q).} \tag{7.63}$$

Now, assuming that the voltage v comprises a positive and a negative-sequence component, its space-phasor can be written as $\vec{v} = V_p\vec{s}_{d_p} + V_n\vec{s}_{d_n}$. Similarly, the current can be written as $\vec{i} = I_{d_p}\vec{s}_{d_p} - I_{q_p}\vec{s}_{q_p} + I_{d_n}\vec{s}_{d_n} - I_{q_n}\vec{s}_{q_n}$. Also, assume that the transformation angle θ is equal to the angle of the positive-sequence component of the voltage. Then, the voltage and current vectors in *dq* domain are expressed as

$$\begin{aligned}
\vec{v}e^{-j\theta_p} &= V_p + V_n e^{-j(\theta_p+\theta_n)} = v_d + jv_q \\
\vec{i}e^{-j\theta_p} &= I_{d_p} + jI_{q_p} + I_{d_n}e^{-j(\theta_p+\theta_n)} - jI_{q_n}e^{-j(\theta_p+\theta_n)} = i_d + ji_q.
\end{aligned}$$

This equation implies that

$$\boxed{v_d = V_p + V_n\cos(\theta_p + \theta_n), \quad v_q = -V_n\sin(\theta_p + \theta_n),} \tag{7.64}$$

which indicates double-frequency components with equal magnitude on both v_d and v_q when the voltage is unbalanced. When the voltage is balanced, $v_d = V_p$ and $v_q = 0$.

7.4.2.1 Balanced Current

If the current is balanced (positive-sequence), $I_{d_n} = I_{q_n} = 0$, the complex power will be

$$S = \frac{3}{2}\vec{v}\,\vec{i}^{\,*} = \frac{3}{2}\left(V_p + V_n e^{-j(\theta_p+\theta_n)}\right)(I_{d_p} - jI_{q_p})$$

which leads to the previously derived equations for the real and reactive powers

$$\begin{aligned}
P &= \tfrac{3}{2}[V_p I_{d_p} + V_n I_{d_p}\cos(\theta_p + \theta_n) - V_n I_{q_p}\sin(\theta_p + \theta_n)] \\
&= \tfrac{3}{2}[V_p I_p\cos\phi + V_n I_p\cos(\theta_p + \theta_n + \phi)] \\
Q &= -\tfrac{3}{2}[V_p I_{q_p} + V_n I_{q_p}\cos(\theta_p + \theta_n) + V_n I_{d_p}\sin(\theta_p + \theta_n)] \\
&= -\tfrac{3}{2}[V_p I_p\sin(\phi) + V_n I_p\sin(\theta_p + \theta_n + \phi)],
\end{aligned}$$

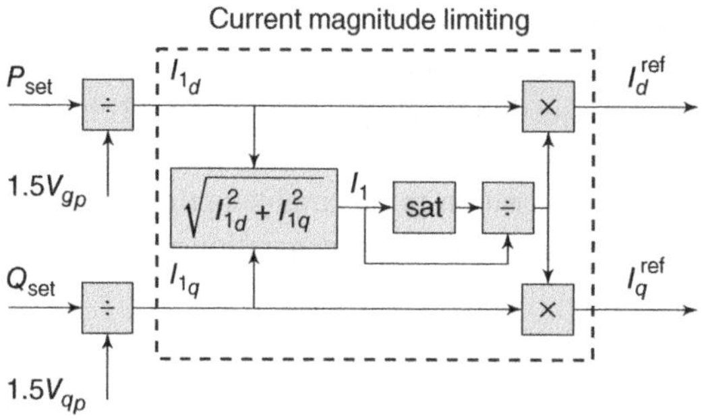

Figure 7.31 Generating and limiting the (balanced) reference current in synchronous frame.

which again confirms that they pulsate at the double frequency if the voltage is unbalance, i.e. if $V_n \neq 0$. If the voltage is balanced, $V_n = 0$, then the powers will become constant and equal to

$$P = \frac{3}{2}V_p I_{d_p} = \frac{3}{2}V_p I_p \cos(\phi), \quad Q = -\frac{3}{2}V_p I_{q_p} = -\frac{3}{2}V_p I_p \sin(\phi).$$

This scenario is summarized in Figure 7.31 which shows how to generate the reference currents in the dq domain from the desired set-points of real and reactive power.[42] The "sat" block limits the current magnitude between the specified limits. Notice that the reference currents are constant as long as the power set-points are constant.

7.4.2.2 Unbalanced Current

For unbalanced current, the complex power will be

$$S = \frac{3}{2}\left(V_p + V_n e^{-j(\theta_p+\theta_n)}\right)\left(I_{d_p} - jI_{q_p} + I_{d_n}e^{j(\theta_p+\theta_n)} - jI_{q_n}e^{j(\theta_p+\theta_n)}\right)$$

which leads to the previously derived equations for the real and reactive powers

$$P = \tfrac{3}{2}[V_p I_{d_p} + V_n I_{d_n} + V_p I_n \cos(\theta_p + \theta_n + \phi_n) + V_n I_p \cos(\theta_p + \theta_n + \phi_p)]$$

$$Q = \tfrac{3}{2}[-V_p I_{q_p} + V_n I_{q_n} + V_p I_n \sin(\theta_p + \theta_n + \phi_n) - V_n I_p \sin(\theta_p + \theta_n + \phi_p)]$$

where ϕ_p and ϕ_n are the relative angles of current positive-sequence and negative-sequence components with respect to the corresponding voltage component.

Same scenarios discussed in Section 7.3.2.2 may also be considered here. For example, the above equation confirms that it is possible to choose the negative-sequence of the current properly so as to remove the double-frequency component of the real power (or the reactive power). Thus, for removing the ripples on the real power, $V_p I_n = V_n I_p$ and $\phi_n = \pi + \phi_p$ are necessary. Defining the voltage unbalance factor $U_v = \frac{V_n}{V_p}$, the conditions for removing the real power pulsations lead to

$$I_{d_p} = \frac{P_{set}}{\frac{3}{2}(V_p - U_v V_n)}, \quad I_{q_p} = \frac{Q_{set}}{-\frac{3}{2}(V_p + U_v V_n)}, \quad I_{d_n} = -U_v I_{d_p}, \quad I_{q_n} = -U_v I_{q_p}.$$

The resulting current reference generation and limiting is shown in Figure 7.32[43] where $\vec{s}_{d_n}$ and $\vec{s}_{q_n}$ are functions of $\theta_p + \theta_n$ as

$$\vec{s}_{d_n}(\theta_p + \theta_n) = \begin{bmatrix} \cos(\theta_p + \theta_n) \\ -\sin(\theta_p + \theta_n) \end{bmatrix}, \quad \vec{s}_{q_n}(\theta_p + \theta_n) = \begin{bmatrix} \sin(\theta_p + \theta_n) \\ \cos(\theta_p + \theta_n) \end{bmatrix}.$$

42 In Figure 7.31, we have used V_{g_p} instead of V_p to emphasize the grid voltage. Moreover, in Figure 7.31, we have dropped the unnecessary index p for the positive-sequence in the currents, for simplicity.
43 Compare with Figure 7.16.

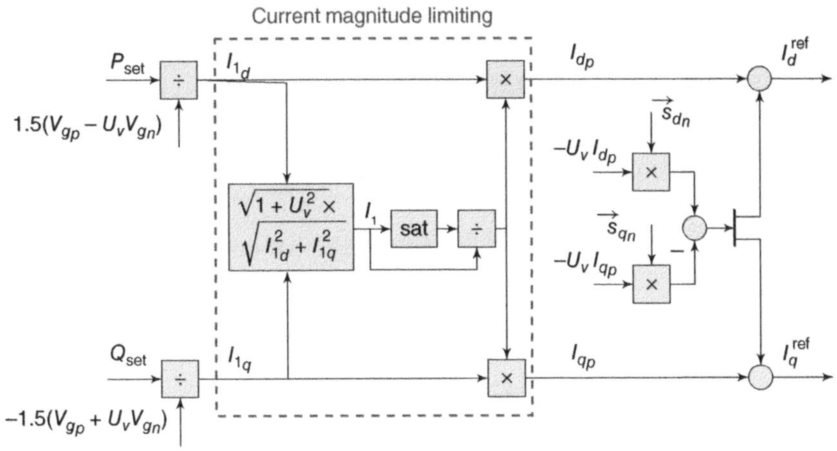

Figure 7.32 Generating and limiting the (unbalanced) reference in synchronous frame.

It must be noted that the reference currents generated in Figure 7.32 are not constant, and they pulsate at double the fundamental frequency, despite the fact that the power set-points are constant. Therefore, as we will see shortly, simple integrating (PI) compensators cannot track them accurately. Instead, resonant (PR) type compensators (at double frequency) are necessary to completely remove the steady state error. This may be considered as a minor drawback of SRF compared with stationary frame. In the stationary domain, regardless of whether the current is balanced or unbalanced, PR compensators (at the fundamental frequency) are adequate. Moreover, as explained in Section 7.3.2.2, the current-limiting mechanism used in Figure 7.32 only limits the "average magnitude" which means it may not be able to limit the individual *abc* currents when the level of unbalance is large, for example during transient grid voltage faults.

7.4.3 SISO and MIMO Control Approaches

One common control methodology, shown in Figure 7.33, treats the system as two separate single-input single-output (SISO) control systems. It has two control loops and it comprises the

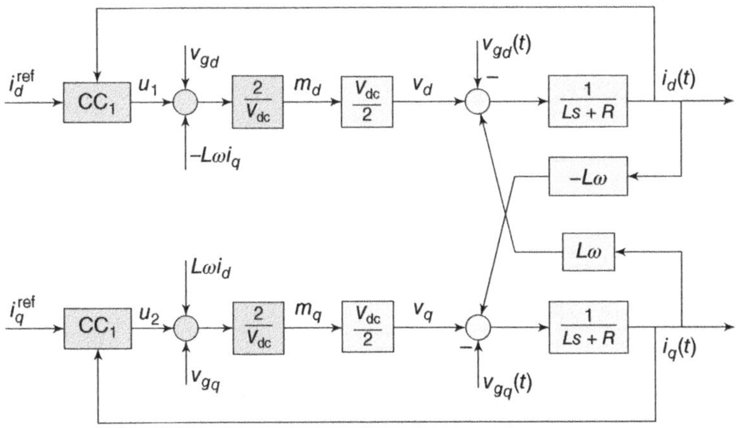

Figure 7.33 SISO VCC approach in synchronous reference frame.

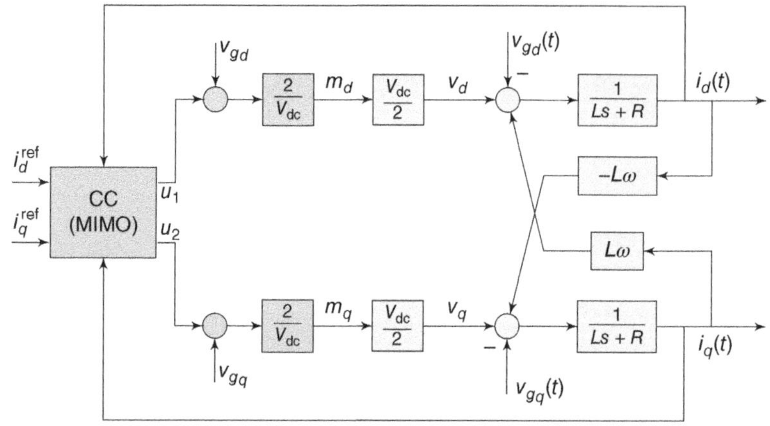

Figure 7.34 MIMO VCC approach in synchronous reference frame.

two decoupling terms ($-L\omega i_q$ and $L\omega i_d$), the two soft-start feed-forward terms (v_{g_d} and v_{g_q}), the inverter gain adjustment factor $\left(\frac{2}{V_{dc}}\right)$ and the CC_1 which is the core CC, often in the form of a PI controller (k_2 and $\frac{k_1}{s}$).[44]

In the MIMO control approach, Figure 7.34, the decoupling terms are not employed. Rather, the controller allows coupling between the channels. The decoupling terms have been reported to be source of multiple problems, e.g. sensitivity to filter inductance and also instabilities in weak grid conditions [5]. The MIMO controller avoids them and leads to improved responses [6].

7.4.4 Optimal Design of PI Controller Gains

This section presents the optimal design of the CC for both SISO and MIMO approaches in the case where they are integrating (PI) functions. The integrating compensator ensures that the steady-state error to *constant commands* is zero. Therefore, they are good for *balanced current* situations where the reference currents are constant. In the unbalanced current injection scenarios, resonant functions are better option to completely remove the steady state errors.

7.4.4.1 SISO Design

Here the controller output is formulated according to

$$u_1 = -k_1 \int \left(i_d - i_d^{ref}\right) dt - k_2 i_d, \quad u_2 = -k_1 \int \left(i_q - i_q^{ref}\right) dt - k_2 i_q,$$

or equivalently

$$\begin{bmatrix} u_1 \\ u_2 \end{bmatrix} = -k_1 \begin{bmatrix} \int \left(i_d - i_d^{ref}\right) dt \\ \int \left(i_q - i_q^{ref}\right) dt \end{bmatrix} - k_2 \begin{bmatrix} i_d \\ i_q \end{bmatrix}. \tag{7.65}$$

The two gains k_1 and k_2 may be designed optimally using the LQT approach. In fact, this design is the same as the dc case we studied in detail in Chapter 5. Here, we give a quick review.

44 Same controller gains are used for both loops.

For each loop, by defining x_1 as the output of integrator, x_2 as the corresponding current component (i_d or i_q), and v as the corresponding control input (v_d or v_q), the state equations for this loop will be

$$\dot{x}_1(t) = e(t) = x_2(t) - i_{\text{ref}}, \quad \dot{x}_2(t) = -\tfrac{R}{L}x_2(t) + \tfrac{1}{L}v(t),$$

where e is the tracking error of the corresponding loop. Apply the operator $\frac{d}{dt}$ to both sides of the two above equations to obtain

$$\dot{z}_1(t) = z_2(t), \quad \dot{z}_2(t) = -\tfrac{R}{L}z_2(t) + \tfrac{1}{L}w(t),$$

where $z_1(t) = \dot{x}_1(t) = e(t)$, $z_2(t) = \dot{x}_2(t)$, and $w(t) = \dot{v}(t)$. Notice that $\frac{d}{dt}i_{\text{ref}} = 0$. Also notice that

$$w(t) = \dot{v}(t) = -k_1\dot{x}_1(t) - k_2\dot{x}_2(t) = -k_1z_1(t) - k_2z_2(t) = -Kz(t),$$

where $K = [k_1 \ k_2]$ and $z(t) = [z_1(t) \ \ z_2(t)]^T$. Therefore, the state space representation in terms of z variables is

$$\dot{z}(t) = Az(t) + Bw(t), \text{ where } A = \begin{bmatrix} 0 & 1 \\ 0 & -\frac{R}{L} \end{bmatrix}, \ B = \begin{bmatrix} 0 \\ \frac{1}{L} \end{bmatrix}.$$

To minimize the cost function $J = \int_0^\infty \left[qz_1^2(t) + q_2z_2^2(t) + v^2(t)\right] dt$ for some $q > 0$, $q_2 \geq 0$, the matrix Q should be chosen as $Q = \text{diag}\{[q, q_2]\}$, and the LQR problem is solved in MATLAB: ≫ `K = lqr(A, B, Q, 1)`.

7.4.4.2 MIMO Design

The MIMO controller considers all possible coupling terms between the two channels. Mathematically, it extends (7.65) to

$$\begin{bmatrix} u_1 \\ u_2 \end{bmatrix} = -\underbrace{\begin{bmatrix} k_{11} & k_{12} \\ k_{13} & k_{14} \end{bmatrix}}_{k_1}\begin{bmatrix} \int \left(i_d - i_d^{\text{ref}}\right) dt \\ \int \left(i_q - i_q^{\text{ref}}\right) dt \end{bmatrix} - \underbrace{\begin{bmatrix} k_{21} & k_{22} \\ k_{23} & k_{24} \end{bmatrix}}_{k_2}\begin{bmatrix} i_d \\ i_q \end{bmatrix}. \tag{7.66}$$

Define the two outputs of integrators as x_1 and x_2. Then,

$$\dot{x}_1 = e_1 = i_d - i_d^{\text{ref}} = x_3 - i_d^{\text{ref}}, \quad \dot{x}_2 = e_2 = i_q - i_q^{\text{ref}} = x_4 - i_q^{\text{ref}},$$

where $x_3 = i_d$ and $x_4 = i_q$. The equations for x_3 and x_4 are

$$\dot{x}_3 = -\frac{R}{L}x_3 + \omega x_4 + \frac{1}{L}v_d, \quad \dot{x}_4 = -\frac{R}{L}x_4 - \omega x_3 + \frac{1}{L}v_q.$$

Augmenting these four equations results in

$$\dot{x} = Ax + Bv + Ei_{\text{ref}},$$

where

$$A = \begin{bmatrix} 0 & 0 & 1 & 0 \\ 0 & 0 & 0 & 1 \\ 0 & 0 & -\frac{R}{L} & \omega \\ 0 & 0 & -\omega & -\frac{R}{L} \end{bmatrix}, \quad B = \begin{bmatrix} 0 & 0 \\ 0 & 0 \\ \frac{1}{L} & 0 \\ 0 & \frac{1}{L} \end{bmatrix}, \quad E = \begin{bmatrix} -1 & 0 \\ 0 & -1 \\ 0 & 0 \\ 0 & 0 \end{bmatrix}.$$

The control law is

$$v = \begin{bmatrix} v_d \\ v_q \end{bmatrix} = -\left[\underbrace{\begin{matrix} k_{11} & k_{12} \\ k_{13} & k_{14} \end{matrix}}_{k_1} \middle| \underbrace{\begin{matrix} k_{21} & k_{22} \\ k_{23} & k_{24} \end{matrix}}_{k_2}\right] x = -Kx.$$

Now, we apply the operator $\frac{d}{dt}$ to the above state and control equations to obtain

$$\dot{z} = Az + Bw, \quad w = -Kz$$

where $z = \dot{x}$ and $w = \dot{v}$. Notice that the derivative of the reference current is zero. Moreover, notice that $z_1 = \dot{x}_1 = e_1$, $z_2 = \dot{x}_2 = e_2$, and z_3 and z_4 are the rate of the change of the currents. This latter equation can be optimally solved for K using the LQR technique. The cost function of

$$J = \int_0^\infty \left(q_1 e_1^2 + q_2 e_2^2 + q_3 z_3^2 + q_4 z_4^2 + w_1^2 + w_2^2\right) dt$$

is considered where $q_1 > 0, q_2 > 0, q_3 \geq 0, q_4 \geq 0$. Then, choose $Q = \text{diag}\{[q_1, \dots, q_4]\}$, solve for K in MATLAB: ≫ `K = lqr(A, B, Q, eye(2))`.

Example 7.3 ***SISO and MIMO PI Designs and Performances***
For the set of system parameters shown in Table 7.1, the closed-loop poles of the SISO control system when q and q_2 vary, respectively, between 10^3 to $10^{6.5}$ (blue) and 10^{-2} to $10^{0.5}$ (red) are shown in Figure 7.35. At final point, the poles are located at $-801 \pm j497$, and the controller gains are equal to $K = [k_1 \;\; k_2] = [1778 \;\; 3.2]$. Note that the same design is applied to both control loops.

For the same set of system parameters shown in Table 7.1, the closed-loop poles of the MIMO control system when $q_1 = q_2$ and $q_3 = q_4$ vary, respectively, between 10^3 to $10^{6.5}$ (blue) and 10^{-2} to $10^{0.5}$ (red) are shown in Figure 7.36. At final point, the poles are located at $-883 \pm j716, -705 \pm j339$, and the controller gains are equal to

$$K = \left[\begin{array}{cc|cc} 1730 & -411 & 3.2 & 0 \\ 411 & 1730 & 0 & 3.2 \end{array}\right].$$

It is observed that the optimal controller employs the cross-coupling term $411x_2 = 411\int e_2 dt$ from loop 2 to loop 1 and the term $-411x_1 = -411\int e_1 dt$ from loop 1 to loop 2. Interestingly though, it does not use the cross-coupling terms from the currents.

Figure 7.37 shows the inverter responses when the grid is balanced and the inverter also generates a balanced current according to Figure 7.31. The responses show that the converter generates balanced sinusoidal currents. The controller succeeds to (i) yield a soft-start, (ii) execute real and

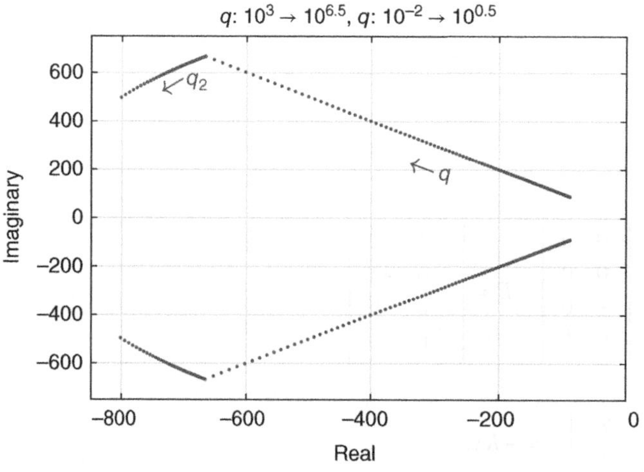

Figure 7.35 Poles of the closed-loop system using SISO LQT approach.

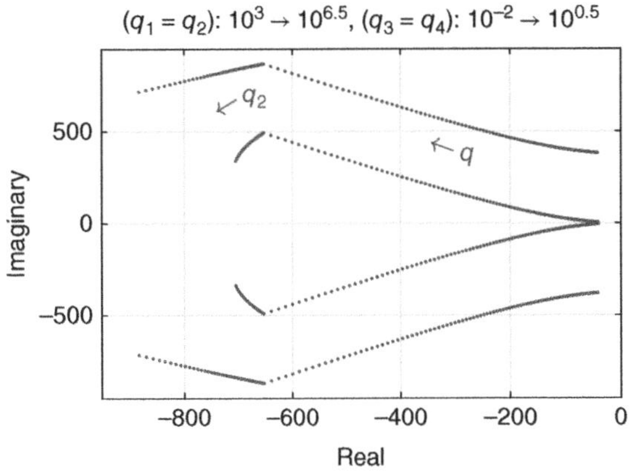

Figure 7.36 Poles of the closed-loop system using MIMO LQT approach.

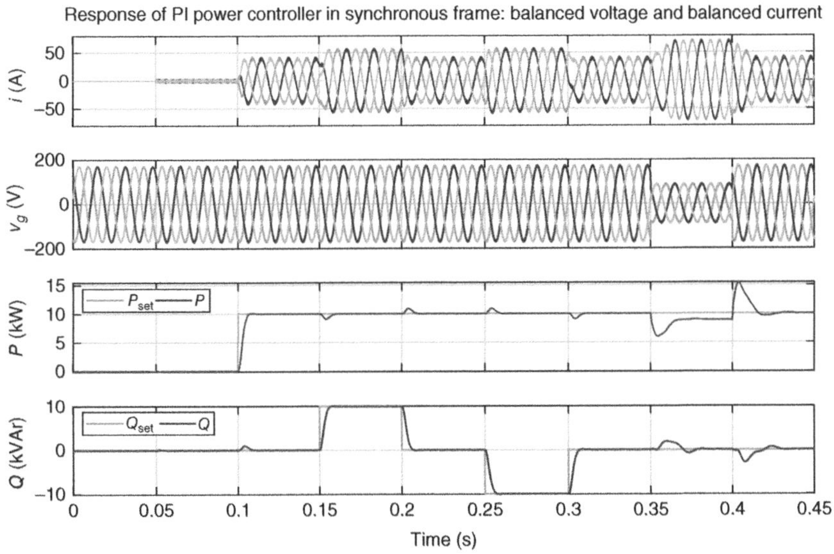

Figure 7.37 Responses of the PI power controller in synchronous frame generating balanced current while the grid is also balanced. $t = 0$: ePLL starts, $t = 0.05$ seconds: converter starts, $t = 0.1,\ 0.15,\ 0.20,\ 0.25,\ 0.3$ seconds: real and reactive power commands applied, $t = 0.35$ seconds: grid fault (voltage sag down to 30%), $t = 0.4$ seconds: fault is cleared.

reactive power commands quickly and smoothly, (iii) generate sinusoidal and balanced current, and (iv) limit the converter current.

Figure 7.38 shows the inverter responses when the grid is unbalanced and the inverter generates a balanced current according to Figure 7.31. The responses show that the converter generates balanced sinusoidal currents but the powers pulsate at double the frequency. The controller succeeds to (i) yield a soft-start, (ii) execute real and reactive power commands quickly and smoothly, (iii) generate sinusoidal and balanced current, and (iv) limit the converter current. Figure 7.39 repeats the same simulation but with the difference that a transient deep single-phase (line-to-ground: L-G) grid fault happens during 0.35–0.4 seconds. The converter continues to feed the same desired average real and reactive powers to the grid and maintains the balanced current during the fault.

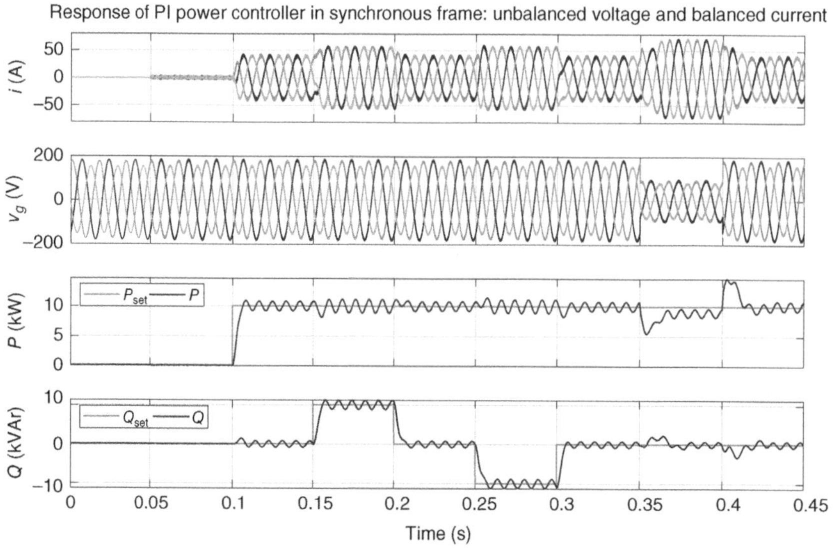

Figure 7.38 Responses of the PI power controller in synchronous frame generating balanced current while the grid is unbalanced; causing power pulsations.

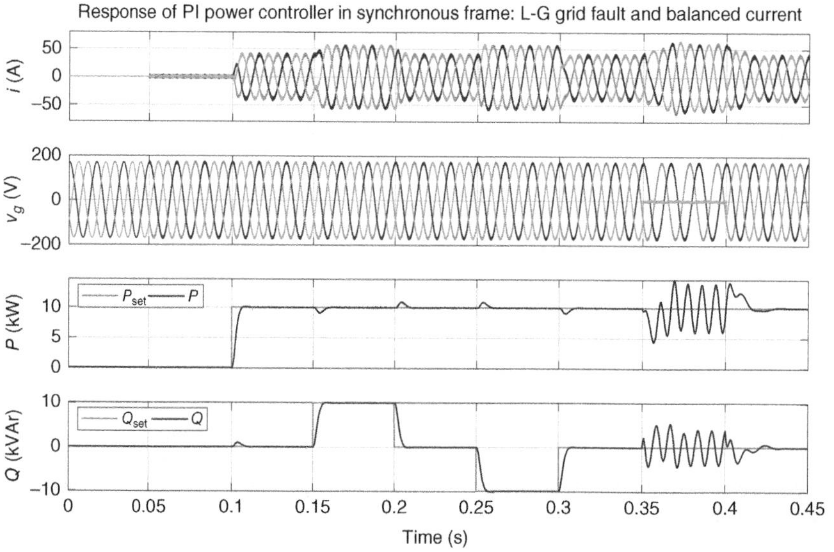

Figure 7.39 Responses of power controller in synchronous frame: generating balanced current while the grid is unbalanced (under a transient L-G fault).

The current is within the limits. Figure 7.40 repeats the same simulation for a transient deep double-phase (line-line-to-ground: L-L-G) grid fault happening during 0.35–0.4 seconds. The converter continues to feed balanced current but has to reduce its real power to ensure that the current is limited at 70 A peak. In general, to implement a LVRT strategy, the set-points of real and reactive powers must be adjusted, during the fault interval, according to that given LVRT specifications.

Figure 7.41 shows the inverter responses when the grid is unbalanced and the inverter generates an unbalanced current to remove the real power pulsations according to Figure 7.32. The controller succeeds to (i) yield a soft-start, (ii) execute real and reactive power commands quickly and

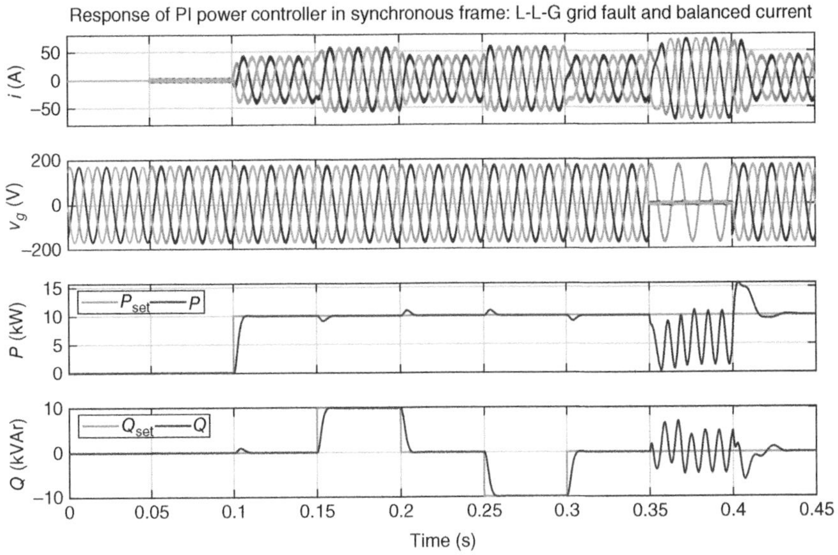

Figure 7.40 Responses of power controller in synchronous frame: generating balanced current while the grid is unbalanced (under a transient L-L-G fault).

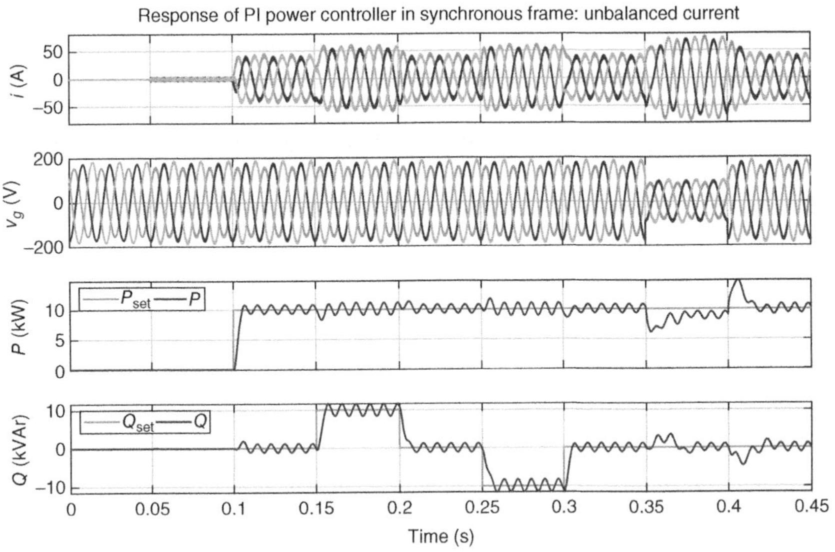

Figure 7.41 Responses of the PI power controller in synchronous frame generating unbalanced current while the grid is unbalanced; failing to remove real power pulsations.

smoothly, (iii) generate sinusoidal current, and (iv) limit the converter current. It, however, fails to remove the real power pulsations. The reason is that the PI compensators cannot accurately track/reject double-frequency commands/disturbances. The proportional-integrating-resonant (PIR) compensators (discussed later) will solve this problem. However, it is possible to "partially" address the problem by modifying the feed-forward terms $v_{g_{dq}}$ from using (7.64), to simply use $v_{g_d} = V_{g_p}$ and $v_{g_q} = 0$. This will remove the double-frequency from these two voltages and reduce the pulsations on the real power as shown in Figure 7.42.

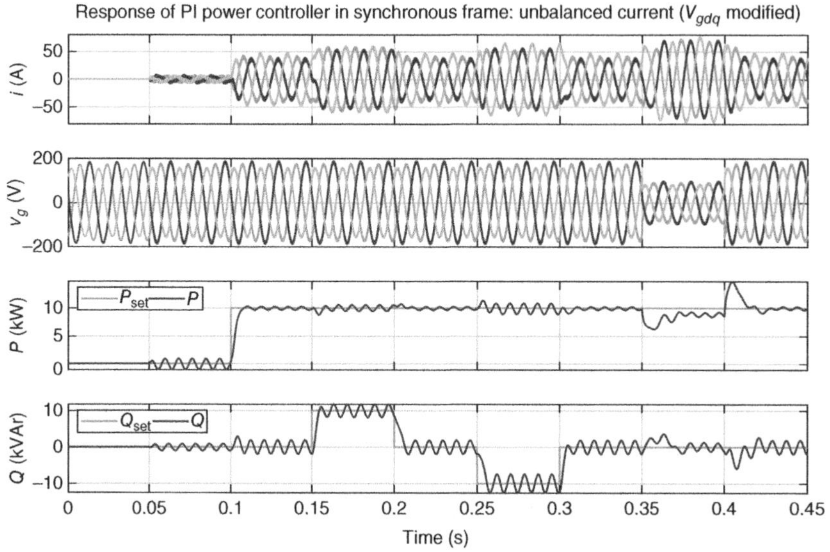

Figure 7.42 Responses of the PI power controller in synchronous frame generating unbalanced current while the grid is unbalanced; mitigating the real power pulsations (by using different feed-forward terms: $v_{g_d} = V_{g_p}$ and $v_{g_q} = 0$ instead of (7.64)).

The SISO and MIMO controllers perform much similarly during normal conditions. However, the MIMO controller performs better during extreme operating conditions, as detailed in [6]. An example is shown here where the extreme condition is defined as follows. A grid impedance of 2 mH is added to weaken the voltage at the point of grid coupling, and the inverter output filter inductance is increased from 2 to 3 mH (modeling an unknown uncertainty). Figures 7.43 and 7.44 show the responses of the SISO and MIMO controllers to different power commands and a grid voltage sag. As observed, the MIMO controller exhibits slightly more robust and stable responses with lower level of oscillations in both the inverter current and the grid voltage.

7.4.5 Optimal Design of PIR Controller Gains

This section presents the optimal design of the CC for both SISO and MIMO approaches in the case where they are PIR functions. This compensator ensures that the steady-state error to *constant* and *sinusoidal* commands is zero. Therefore, they are good for *both* balanced and unbalanced current situations.

7.4.5.1 SISO Design

Here the controller output, in Laplace domain, may be formulated according to[45]

$$u_1 = -\frac{k_3 s^2 + k_2 s + k_1}{s(s^2 + 4\omega^2)} e_1 - k_4 i_d, \quad e_1 = i_d - i_d^{\text{ref}}.$$

45 We formulate the solution for the d-axis loop. The same design is valid for the q-axis controller.

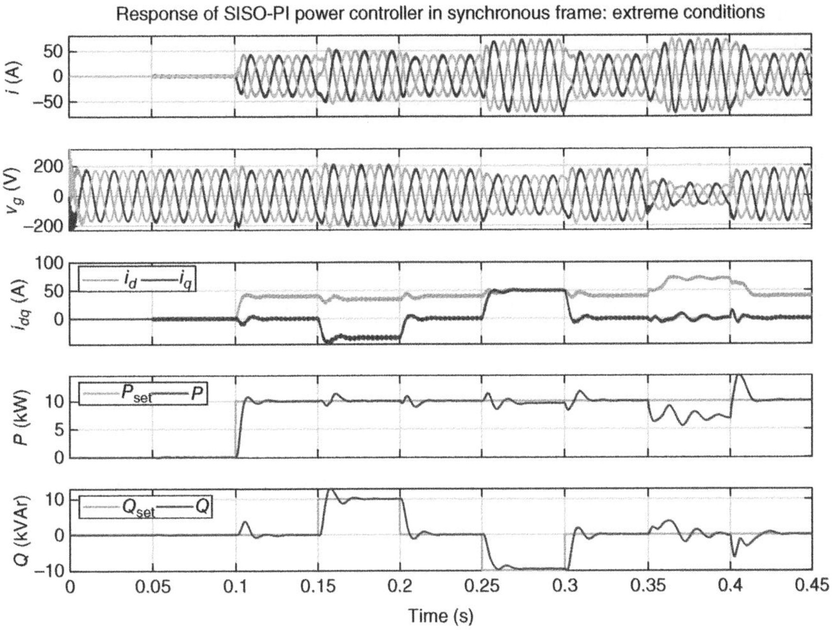

Figure 7.43 Responses of the SISO-PI power controller in synchronous frame for the extreme conditions (weak grid, and filter inductance uncertainty).

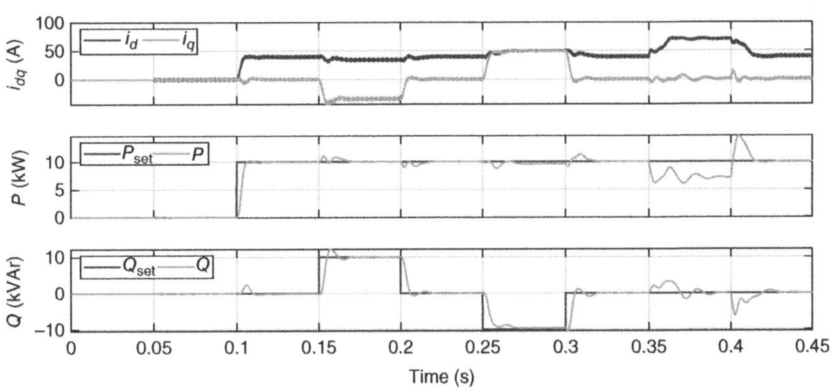

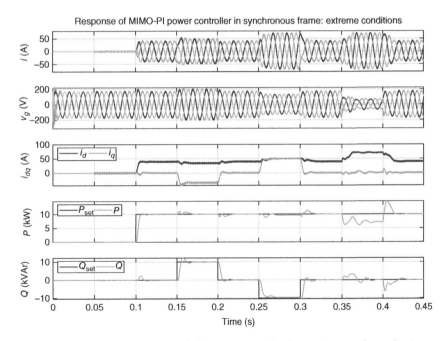

Figure 7.44 Responses of the MIMO-PI power controller in synchronous frame for the extreme conditions (weak grid, and filter inductance uncertainty).

Note that $s(s^2 + 4\omega^2)$ indicates that the compensator has an integrating and a resonant term.[46] Define the state variables

$$x_1 = \frac{1}{s(s^2 + 4\omega^2)} e_1, \quad x_2 = \frac{s}{s(s^2 + 4\omega^2)} e_1, \quad x_3 = \frac{s^2}{s(s^2 + 4\omega^2)} e_1, \quad x_4 = i_d,$$

to derive the state space and control equations

$$\dot{x}_1 = x_2, \quad \dot{x}_2 = x_3, \quad \dot{x}_3 = e_1 - 4\omega^2 x_2 = -4\omega^2 x_2 + x_4 - i_{\text{ref}},$$
$$\dot{x}_4 = -\frac{R}{L} x_4 + \frac{1}{L} v, \quad v = -Kx,$$

where $v = v_d = -[k_1 \ k_2 \ k_3 \ k_4]x = -Kx$ is the control input. Apply the operator $\frac{d}{dt}(\frac{d^2}{dt^2} + 4\omega^2)$ to obtain

$$\dot{z}_1 = z_2, \quad \dot{z}_2 = z_3, \quad \dot{z}_3 = -4\omega^2 z_2 + z_4, \quad \dot{z}_4 = -\frac{R}{L} z_4 + \frac{1}{L} w, \quad w = -Kz,$$

where $z_1(t) = e_1(t)$, $z_2(t) = \dot{e}_1(t)$, $z_3(t) = \ddot{e}_1(t)$, and $w(t) = \frac{d}{dt}(\frac{d^2}{dt^2} + 4\omega^2)v(t)$. Notice that $\frac{d}{dt}(\frac{d^2}{dt^2} + 4\omega^2)i_{\text{ref}} = 0$. Therefore, the state space representation in terms of z variables is $\dot{z}(t) = Az(t) + Bw(t)$ where

$$A = \begin{bmatrix} 0 & 1 & 0 & 0 \\ 0 & 0 & 1 & 0 \\ 0 & -4\omega^2 & 0 & 1 \\ 0 & 0 & 0 & -\frac{R}{L} \end{bmatrix} \quad B = \begin{bmatrix} 0 \\ 0 \\ 0 \\ \frac{1}{L} \end{bmatrix}.$$

To minimize the cost function $J = \int_0^\infty \left[qe_1^2 + q_2\dot{e}_1^2 + q_3\ddot{e}_1^2 + q_4 z_4^2 + w^2(t)\right] dt$ for some $q > 0$, $q_i \geq 0$ $(i = 1, 2, 3)$, the matrix Q should be chosen as $Q = \text{diag}\{[q, q_2, q_3, q_4]\}$, and the LQR problem is solved in MATLAB: ≫ `K = lqr(A, B, Q, 1)`. It is worthwhile noting that $z_4 = \frac{d}{dt}\left(\frac{d^2}{dt^2} + 4\omega^2\right) i_d$ and will converge to zero in steady state condition. Therefore, the cost function is mathematically well defined.

7.4.5.2 MIMO Design

The control input in the MIMO controller may be written as $v = -Kx$ where the control input v is a two-dimensional vector (for the d and q loops), the state vector x is eight-dimensional (including six state variables of the controller and two currents), and K will be a 2×8 matrix. More specifically, the state variables may be considered as

$$x_1 = \frac{1}{s(s^2+4\omega^2)} e_1, \quad x_2 = \frac{1}{s(s^2+4\omega^2)} e_2, \quad x_3 = \frac{s}{s(s^2+4\omega^2)} e_1, \quad x_4 = \frac{s}{s(s^2+4\omega^2)} e_2,$$
$$x_5 = \frac{s^2}{s(s^2+4\omega^2)} e_1, \quad x_6 = \frac{s^2}{s(s^2+4\omega^2)} e_2, \quad x_7 = i_d, \quad x_8 = i_d,$$

to derive the state space and control equations

$$\begin{aligned} &\dot{x}_1 = x_3, \quad \dot{x}_2 = x_4, \quad \dot{x}_3 = x_5, \quad \dot{x}_4 = x_6, \\ &\dot{x}_5 = -4\omega^2 x_3 + e_1 = -4\omega^2 x_3 + x_7 - i_d^{\text{ref}}, \\ &\dot{x}_6 = -4\omega^2 x_4 + e_2 = -4\omega^2 x_4 + x_8 - i_q^{\text{ref}}, \\ &\dot{x}_7 = -\frac{R}{L} x_7 + \omega x_8 + \frac{1}{L} v_1, \quad \dot{x}_8 = -\frac{R}{L} x_8 - \omega x_7 + \frac{1}{L} v_2. \end{aligned} \tag{7.67}$$

46 More specifically, this transfer function may be written as

$$\frac{k_3 s^2 + k_2 s + k_1}{s(s^2 + 4\omega^2)} = \frac{\frac{k_1}{4\omega^2}}{s} + \frac{\left(k_3 - \frac{k_1}{4\omega^2}\right) s + k_2}{s^2 + 4\omega^2}.$$

Note that $e_1 = i_d - i_d^{\text{ref}}$ and $e_2 = i_q - i_q^{\text{ref}}$. Augmenting these eight equations results in $\dot{x} = Ax + Bv + Ei_{\text{ref}}$, where

$$A = \begin{bmatrix} 0 & 0 & 1 & 0 & 0 & 0 & 0 & 0 \\ 0 & 0 & 0 & 1 & 0 & 0 & 0 & 0 \\ 0 & 0 & 0 & 0 & 1 & 0 & 0 & 0 \\ 0 & 0 & 0 & 0 & 0 & 1 & 0 & 0 \\ 0 & 0 & -4\omega^2 & 0 & 0 & 0 & 1 & 0 \\ 0 & 0 & 0 & -4\omega^2 & 0 & 0 & 0 & 1 \\ 0 & 0 & 0 & 0 & 0 & 0 & -\frac{R}{L} & \omega \\ 0 & 0 & 0 & 0 & 0 & 0 & -\omega & -\frac{R}{L} \end{bmatrix}, \quad B = \begin{bmatrix} 0 & 0 \\ 0 & 0 \\ 0 & 0 \\ 0 & 0 \\ 0 & 0 \\ 0 & 0 \\ \frac{1}{L} & 0 \\ 0 & \frac{1}{L} \end{bmatrix}, \quad E = \begin{bmatrix} 0 & 0 \\ 0 & 0 \\ 0 & 0 \\ 0 & 0 \\ -1 & 0 \\ 0 & -1 \\ 0 & 0 \\ 0 & 0 \end{bmatrix}.$$

Now, we apply the operator $\frac{d}{dt}\left(\frac{d^2}{dt^2} + 4\omega^2\right)$ to the above state and control equations to obtain $\dot{z} = Az + Bw, \quad w = -Kz$ where $z = \frac{d}{dt}\left(\frac{d^2}{dt^2} + 4\omega^2\right)x$ and $w = \frac{d}{dt}\left(\frac{d^2}{dt^2} + 4\omega^2\right)v$. Notice that $z_1 = e_1$, $z_2 = e_2$, $z_3 = \dot{e}_1$, $z_4 = \dot{e}_2$, $z_5 = \ddot{e}_1$, and $z_6 = \ddot{e}_2$. This can be optimally solved for K using the LQR technique. The cost function of

$$J = \int_0^\infty (q_1 e_1^2 + q_2 e_2^2 + q_3 \dot{e}_1^2 + q_4 \dot{e}_2^2 + q_5 \ddot{e}_1^2 + q_6 \ddot{e}_2^2 + q_7 z_7^2 + q_8 z_8^2 + w_1^2 + w_2^2)\, dt$$

is considered where $q_1 > 0$, $q_2 > 0$, $q_i \geq 0\ (i = 3, \ldots, 8)$. Then, choose $Q = \text{diag}\{[q_1, \ldots, q_8]\}$, solve for K in MATLAB: ≫ `K = lqr(A, B, Q, eye(2))`.

Example 7.4 ***SISO and MIMO PIR Designs***

For the set of system parameters shown in Table 7.1, the closed-loop poles of the SISO PIR control system when q, q_2, and q_3 vary, respectively, between 10^{14} to 10^{19} (blue), 10^9 to 10^{13} (red), and 10^3 to 10^7 (magenta) are shown in Figure 7.45. At final point, the poles are located at $-787 \pm j402$ and $-706 \pm j1235$, and the controller gains are equal to $K = [k_1\ k_2\ k_3\ k_4] = [3.16e9\ \ 5.2e6\ \ 8920\ \ 5.95]$. Note that the same design is applied to both control loops.

For the same set of system parameters shown in Table 7.1, the closed-loop poles of the MIMO control system when $q_1 = q_2$, $q_3 = q_4$, and $q_5 = q_6$ vary, respectively, between 10^{14} to 10^{19} (blue), 10^9 to 10^{13} (red), and 10^3 to 10^7 (magenta) are shown in Figure 7.46. At final point, the poles

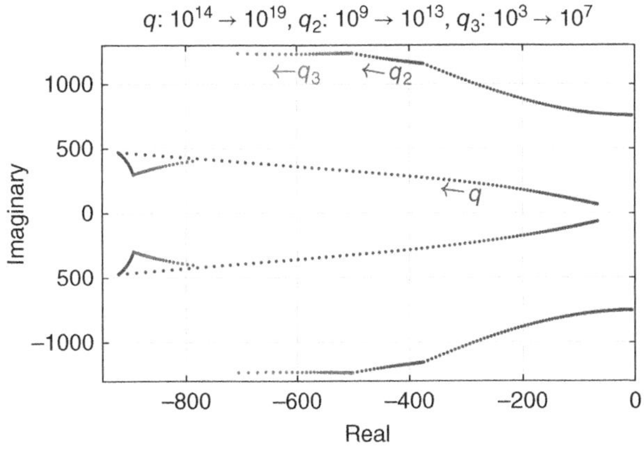

Figure 7.45 Poles of the closed-loop system using LQT approach for SISO PIR controller.

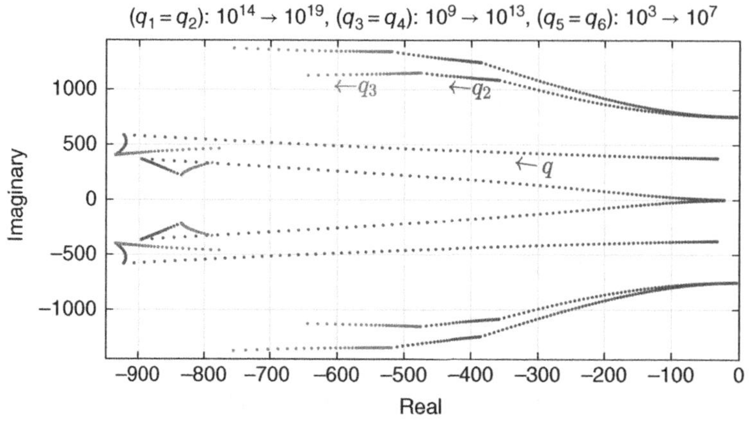

Figure 7.46 Poles of the closed-loop system using LQT approach for MIMO-PIR controller.

are located at $-778 \pm j462$, $-789 \pm j332$, $-644 \pm j1129$, $-756 \pm j1276$, and the controller gains are equal to

$$K = \left[\begin{array}{cccccc|cc} 3.13e9 & -4.8e8 & 5.1e6 & -8.8e5 & 8802 & -1414 & 5.91 & 0 \\ 4.8e8 & 3.13e9 & 08.8e5 & 5.1e6 & 1414 & 8802 & 0 & 5.91 \end{array}\right].$$

It is observed that the optimal controller employs multiple cross-coupling terms between the two control channels. Interestingly, the currents are not cross-coupled in the controller.

Figure 7.47 shows the inverter responses when the grid is unbalanced and the inverter generates an unbalanced current to remove the real power pulsations according to Figure 7.32. The controller succeeds to (i) yield a soft-start, (ii) execute real and reactive power commands quickly and smoothly, (iii) generate sinusoidal current to remove the real power pulsations, and (iv) limit the converter current.

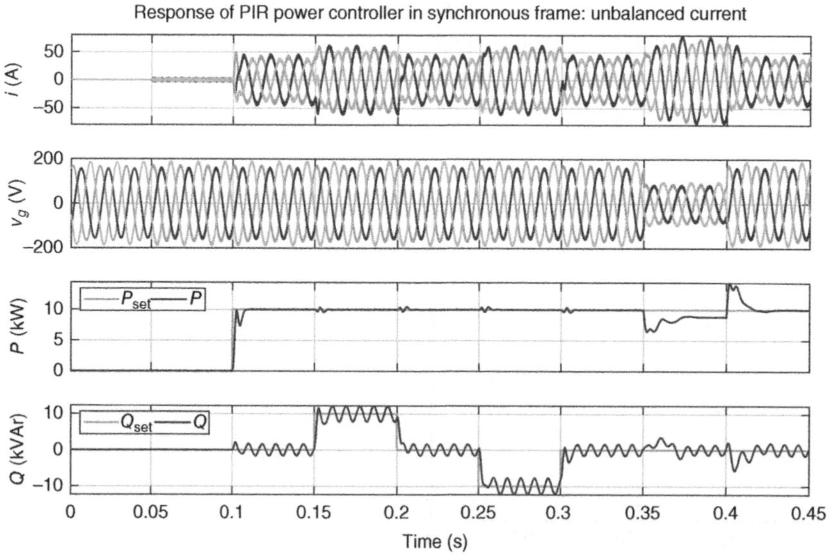

Figure 7.47 Responses of the PIR power controller in synchronous frame generating unbalanced current while the grid is unbalanced; removing the real power pulsations.

The SISO and MIMO controllers perform much similarly during normal conditions, such as strong grid, balanced grid, absence of system uncertainties, etc. However, the MIMO controller performs better during extreme operating conditions. An example is shown here where the extreme condition is defined as follows. A grid impedance of 1.5 mH is added to weaken the voltage at the point of grid coupling, and the inverter output filter inductance is increased from 2 to 2.5 mH (modeling an unknown uncertainty). Figures 7.48 and 7.49 show the responses of the

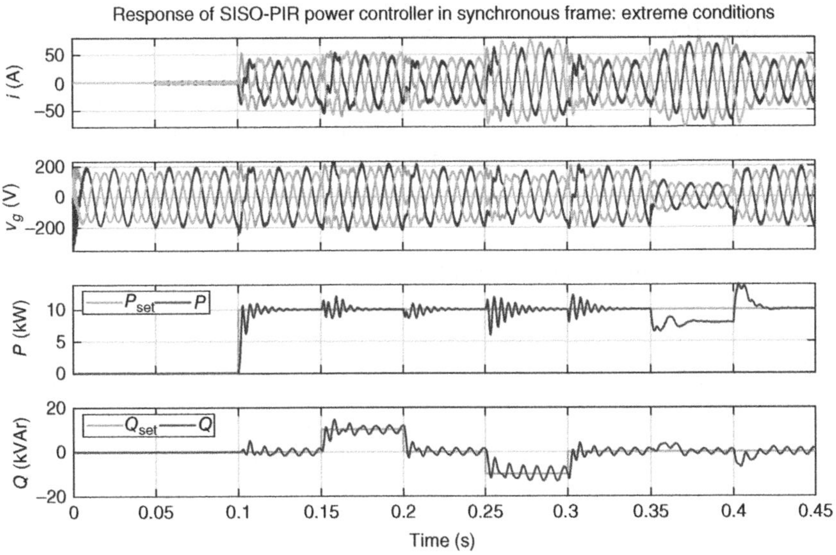

Figure 7.48 Responses of the SISO-PIR power controller in synchronous frame for the extreme conditions (weak grid, and filter inductance uncertainty).

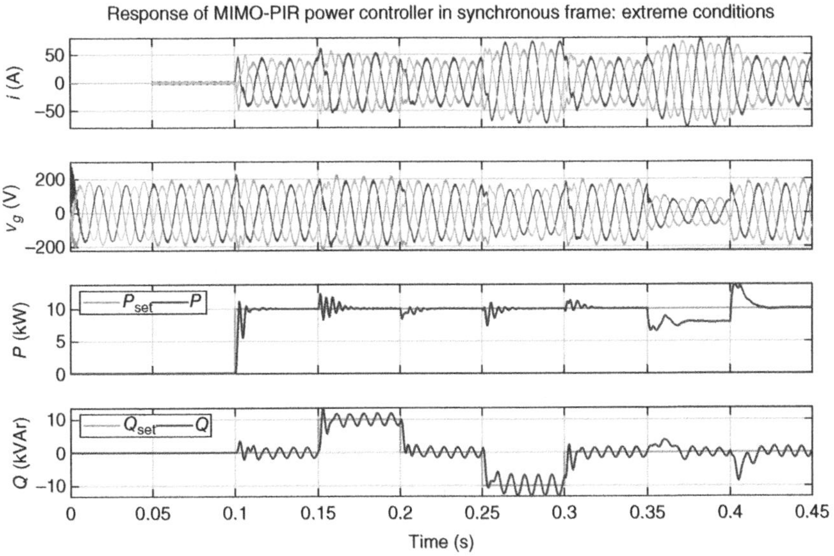

Figure 7.49 Responses of the MIMO-PIR power controller in synchronous frame for the extreme conditions (weak grid, and filter inductance uncertainty).

SISO and MIMO PIR controllers to different power commands and a grid voltage sag. As observed, the MIMO controller exhibits more robust and stable responses with lower level of oscillations in both the inverter current and the grid voltage.

7.4.6 Harmonics, Higher-Order Filters, Control Delays*

Given the detailed control models and approaches discussed above for the VCC in synchronous domain, all the aspects of (i) rejection of dc and harmonics components from the current waveform, (ii) higher-order filters, and (iii) system and control delays, can be addressed by similar strategies discussed for the VCC in stationary domain. Some of the aspects appear to be more convenient to address in the stationary domain, for example the dc offset rejection which is achieved by adding an integrator, while in the synchronous domain, we need to add resonant controllers at the fundamental frequency. However, mathematically speaking, the approaches for both domains are from the same nature. Therefore, these aspects are not further examined here.

7.4.7 Weak Grid Conditions and Including PLL in Controller*

The three-phase PLL linear model derived and presented in (7.33) may be used to include the ePLL state variables in the VCC in synchronous frame. To this end, one approach is to express the grid voltage v_g as a stiff voltage v_s behind an impedance $sL_s + R_s$ (as shown in Figure 6.20 for a single-phase situation). Therefore, it will satisfy $v_g = v_s + L_s\frac{d}{dt}i + R_s i$ where i is the current flowing to the grid. Transforming this to dq domain using the PLL angle θ leads to

$$v_{g_d} = v_{s_d} + L_s\frac{d}{dt}i_d - L_s\omega i_q + R_s i_d, \quad v_{g_q} = v_{s_q} + L_s\frac{d}{dt}i_q + L_s\omega i_d + R_s i_q. \tag{7.68}$$

Expressing $v_{s_\alpha} = V_s\cos(\theta_s)$ and $v_{s_\beta} = V_s\sin(\theta_s)$ results in $v_{s_d} = V_s\cos(\delta)$ and $v_{s_q} = -V_s\sin(\delta)$ where $\delta = \theta - \theta_s$. Moreover, we note that $x_7 = i_d$ and $x_8 = i_q$ satisfy the last two equations in (7.67). Therefore, by substituting the equations, and linearizing $\cos(\delta)$ and $\sin(\delta)$ terms as 1 and δ (around $\delta = 0$), respectively, results in

$$v_{g_d} = V_s + \left(R_s - L_s\frac{R}{L}\right)x_7 + \frac{L_s}{L}v_1, \quad v_{g_q} = -V_s\delta + \left(R_s - L_s\frac{R}{L}\right)x_8 + \frac{L_s}{L}v_2, \tag{7.69}$$

where v_1 and v_2 are the control inputs in dq domain.

The PLL linearized equations (7.33) are rewritten as

$$\dot{x}_9 = -h_2x_9 + h_2v_{g_d}, \quad \dot{x}_{10} = \frac{h_1}{X_9}v_{qq}, \quad \dot{x}_{11} = x_{10} + \frac{h_0}{X_9}v_{g_q}, \tag{7.70}$$

where x_9 is U_d in the PLL, x_{10} is $\Delta\omega$, $x_{11} = \delta = \theta - \theta_s$, and X_9 is the equilibrium (operating) value of x_9.[47] By substituting v_{g_d} and v_{g_q} from (7.69) into (7.70), the final form of PLL equations are derived as

$$\begin{aligned}
\dot{x}_9 &= -h_2x_9 + h_2V_s + h_2\left(R_s - L_s\tfrac{R}{L}\right)x_7 + h_2\tfrac{L_s}{L}v_1,\\
\dot{x}_{10} &= -h_1x_{11} + \tfrac{h_1}{V_s}\left(R_s - L_s\tfrac{R}{L}\right)x_8 + \tfrac{h_1}{V_s}\tfrac{L_s}{L}v_2,\\
\dot{x}_{11} &= x_{10} - h_0x_{11} + \tfrac{h_0}{V_s}\left(R_s - L_s\tfrac{R}{L}\right)x_8 + \tfrac{h_0}{V_s}\tfrac{L_s}{L}v_2.
\end{aligned} \tag{7.71}$$

This equation together with (7.67) represent an 11th order LTI system. Now, we apply the operator $\frac{d}{dt}\left(\frac{d^2}{dt^2} + 4\omega^2\right)$ to the 11th-order state and control equations to obtain $\dot{z} = Az + Bw, \ \ w = -Kz$

47 For simplicity, we do this analysis for zero power transfer which means the equilibrium point of PLL will be $X_9 = V_s, X_{10} = 0$, and $X_{11} = 0$. A more general analysis is presented in [7].

where $z = \frac{d}{dt}\left(\frac{d^2}{dt^2} + 4\omega^2\right)x$ and $w = \frac{d}{dt}\left(\frac{d^2}{dt^2} + 4\omega^2\right)v$. Notice that $z_1 = e_1$, $z_2 = e_2$, $z_3 = \dot{e}_1$, $z_4 = \dot{e}_2$, $z_5 = \ddot{e}_1$, and $z_6 = \ddot{e}_2$. This can be optimally solved for K using the LQR technique. The cost function of

$$J = \int_0^\infty \left(q_1 e_1^2 + q_2 e_2^2 + q_3 \dot{e}_1^2 + q_4 \dot{e}_2^2 + q_5 \ddot{e}_1^2 + q_6 \ddot{e}_2^2 + \sum_{i=7}^{11} q_i z_i^2 + w_1^2 + w_2^2 \right) dt$$

is considered where $q_1 > 0, q_2 > 0, q_i \geq 0\,(i = 3, \ldots, 11)$. The last three state variables represent the PLL. Then, choose $Q = \text{diag}\{[q_1, \ldots, q_{11}]\}$, solve for K in MATLAB: ≫ `K = lqr(A, B, Q, eye(2))`.

Example 7.5 ***MIMO-PIR Controller with PLL Modeling***

For the set of system parameters shown in Table 7.1, and the nominal grid impedance of $L_s = 2$ mH and $R_s = 20$ mΩ, the closed-loop poles of the MIMO-PIR control system with PLL modeling when q, q_2, and q_3 vary, respectively, between 10^{14} to 10^{19} (blue), 10^9 to 10^{13} (red), and 10^3 to 10^7 (magenta) (same as Example 7.4), and subsequently q_9, q_{10}, and q_{11} vary, respectively, between 10^0 to $10^{1.5}$ (black), 10^0 to $10^{2.5}$ (green), and 10^3 to $10^{5.5}$ (cyan) are shown in Figure 7.50. At final point, the poles are located at $-1241 \pm j135$, $-798 \pm j267$, $-557 \pm j940$, $-723 \pm j1198$, -50, -197, and -246. Notice that the last three are the PLL poles slightly shifted from the original values ($-50, -200, -250$) due to feedback. The 2×11 controller gain matrix K is equal to

$$\left[\begin{array}{cccccc|cc|ccc} 3.1e9 & -5.6e8 & 5.0e6 & -9.6e5 & 9103 & -1671 & 6.51 & -0.13 & 0.26 & 0.045 & 42.7 \\ 5.6e8 & 3.1e9 & 9.6e5 & 5.0e6 & 1674 & 9076 & 0.14 & 6.47 & -0.27 & -0.123 & 42.6 \end{array}\right].$$

The controller with PLL gains feedback performs better during extreme operating conditions. An example is shown here where the extreme condition is defined as follows. A grid impedance of 2 mH is added to weaken the voltage at the point of grid coupling, and the inverter output filter inductance is increased from 2 to 2.5 mH (modeling an unknown uncertainty). Figures 7.51 and 7.52 show the responses of the MIMO-PIR controllers without and with the PLL modeling, respectively.[48] The

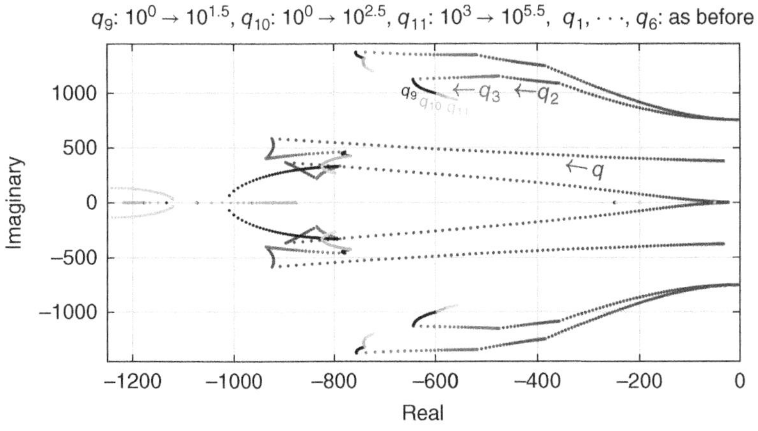

Figure 7.50 Poles of the closed-loop system using LQT approach for MIMO-PIR controller with PLL modeling and feedback.

48 These results are for unbalanced current generation (to remove real power pulsations). The balanced current situation leads to same conclusions as well.

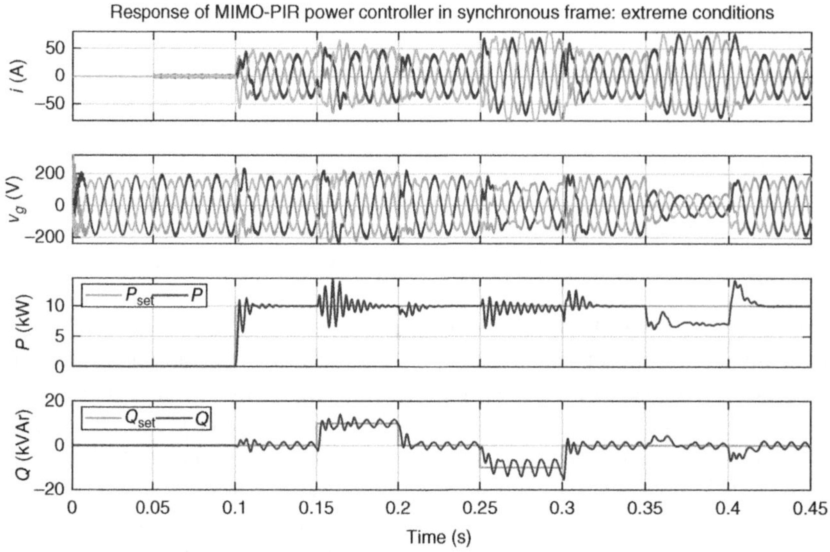

Figure 7.51 Responses of the MIMO-PIR controller (without PLL modeling) in synchronous frame for the extreme conditions (weak grid, and filter inductance uncertainty).

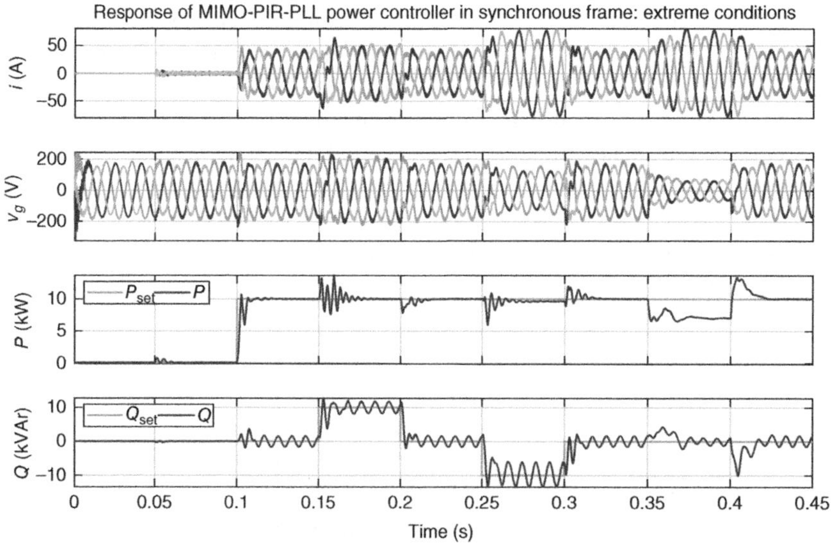

Figure 7.52 Responses of the MIMO-PIR power controller (with PLL modeled and used in feedback controller) in synchronous frame for the extreme conditions (weak grid, and filter inductance uncertainty).

simulation scenario is the same as defined in Example 7.4. As observed, the controller with PLL feedback exhibits more robust and stable responses with lower level of oscillations in both the inverter current and the grid voltage. As a result, it can tolerate larger level of grid weaknesses. Or at a given grid weakness, it can tolerate more severe disturbances at a higher level of power transfer capacity.

7.4.8 Remarks on VCC in $\alpha\beta$ and dq Frames

The VCC controller with PI compensators has been widely used. This has been due to its simple analysis and design stage. It performs satisfactorily in normal conditions. However, in more dynamic conditions where it is desired to upgrade the controller to address more requirements (such as unbalanced grid, polluted with distortions grid, weak grid, etc.), it appears that the VCC controller in stationary domain may be more advantageous in terms of the level of complexity of control system as well as the transparency in the design. The LQT design approach can be used to design efficient controllers in both domains. The MIMO approach is particularly interesting in developing strong controllers. Some more specific remarks are stated below.

- We showed that if the grid voltage is unbalanced, and the controller is desired to feed unbalanced current (to remove power ripples for instance), the PIR compensators (the resonant at the double-frequency) must be used in the VCC approach in synchronous frame. The VCC controller in stationary domain, however, only using a PR (with the resonant at the fundamental frequency) can do the same job.
- The dc offset rejection is done in stationary domain by adding an integrating compensator. This requires a resonant at the fundamental frequency in the synchronous domain.
- The current reference generation and limiting is done almost with very similar blocks in both approaches.

7.5 Grid-Supportive Controls

With reference to the discussion presented in Section 6.3 (for single-phase system), similar methodologies for deriving grid-supportive controls for three-phase converters may be followed. Specifically, static (or steady-state) support as well as dynamic (or inertia) support may be combined and used. As far as the controller implementation, there are two major approaches: enhancing the power controller to provide grid support, and VSM approach. These two types of controllers are discussed in this section.

7.5.1 Power Controller with Grid Support

In this approach, the real and reactive power references are generated from their set-points and the grid-support increments from (6.22). These dynamic power references are now used instead of the power set-points in the power controller (with VCC either in stationary or in SRF). In this approach, the grid voltage magnitude and frequency, and their rate of change (or derivative), are measured and used to compute the power references.

The power controller must be fast enough to follow the power references and respond to fast dynamic commands. To prevent overloading of the converter (mainly during the steady-state or large time scales), the power reference vector of (6.22) must be limited such that its magnitude remains under the apparent power rating of the converter. This method is not further discussed here. More details, in a single-phase context, are provided in Section 6.3.

7.5.2 Virtual Synchronous Machine (VSM)

In this approach, the controller emulates the grid frequency based on the swing equation of a synchronous machine. There are various ways to derive the VSM controller and there are in fact various

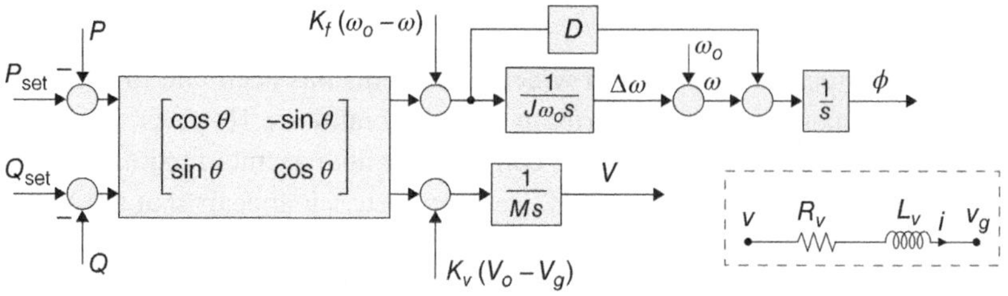

Figure 7.53 Core of VSM controller.

versions of VSM controller. One approach of deriving the VSM was discussed in Section 6.3 for single-phase systems.[49] Developing the same concept for a three-phase system is indeed easier than the single-phase because the quadrature signals are made available easily. The derived structure is shown in Figure 7.53.

The intermediary voltage $v(t) = \left[V\cos(\phi)\ \ V\cos\left(\phi - \frac{2\pi}{3}\right)\ \ V\cos\left(\phi + \frac{2\pi}{3}\right)\right]^T$ (in $\alpha\beta$ frame: $v(t) = [V\cos(\phi)\ \ V\sin(\phi)]^T$, in dq frame: $v(t) = [V\ \ 0]^T$) together with the virtual impedance $L_v s + R_v$ are used to improve the stability and also to facilitate current limiting. The moment of inertia is J, the magnitude loop gain is M, and the damping gain is D. There are different ways of realizing the damping effect as discussed in Section 7.7.3.4. However, we prefer the one used here, as shown in Figure 7.53, due to its suitability to facilitate stability analysis and design as discussed in Section 7.5.3. These five, i.e. (L_v, R_v, J, M, D), are design parameters.

The current limiting mechanism is shown in Figure 7.54, where

$$s_d(\phi) = \begin{bmatrix} \cos(\phi) \\ \sin(\phi) \end{bmatrix}, \quad s_q(\phi) = \begin{bmatrix} \sin(\phi) \\ -\cos(\phi) \end{bmatrix}, \tag{7.72}$$

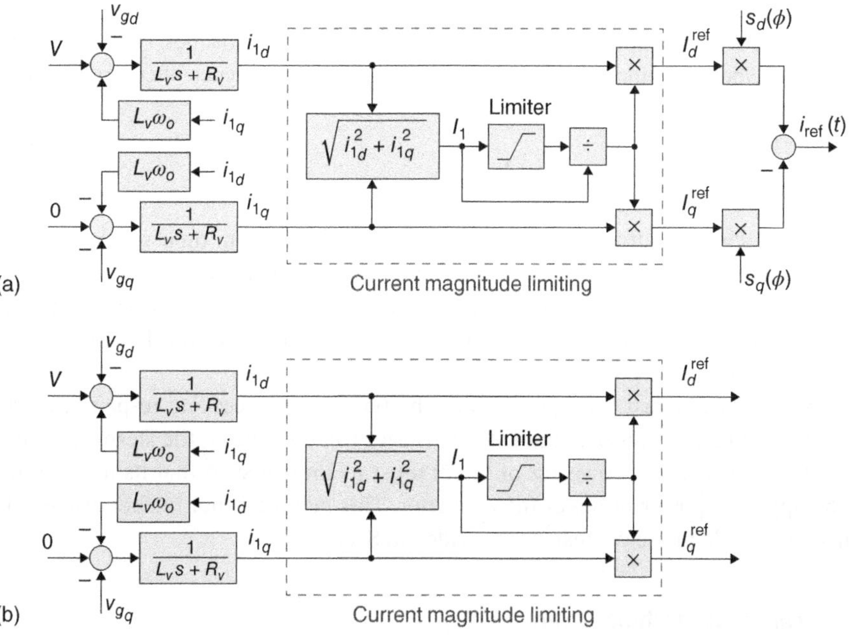

Figure 7.54 Current reference generation and limiting mechanism in stationary frame (a) and in rotating frame (b).

49 Other approaches include copying (with some simplifying assumptions) the equations of a synchronous generator, e.g. [8–10], or using optimization, e.g. [11, 12].

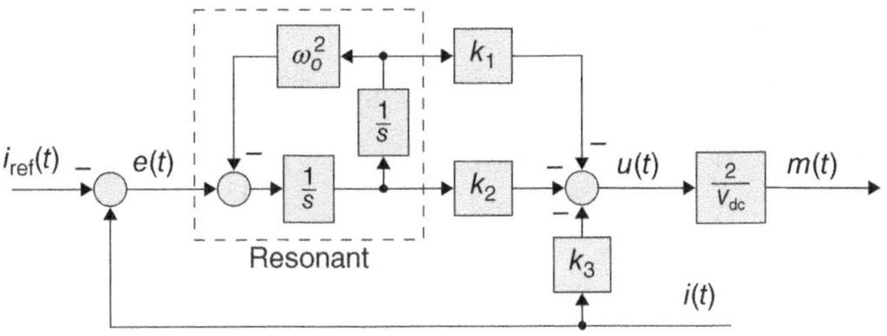

Figure 7.55 Single-line diagram of the current feedback loop (CFL) in stationary frame.

and the grid voltage dq components are calculated using the internal angle ϕ (without using a PLL) according to

$$\begin{bmatrix} v_{g_d} \\ v_{g_q} \end{bmatrix} = \begin{bmatrix} \cos(\phi) & \sin(\phi) \\ -\sin(\phi) & \cos(\phi) \end{bmatrix} \begin{bmatrix} v_{g_\alpha} \\ v_{g_\beta} \end{bmatrix}. \tag{7.73}$$

The current reference i_{ref} is then forwarded to the CFL as shown in Figure 7.55. Note that $i_{ref}(t)$ and $i(t)$ are in stationary frame and each have α and β components. Therefore, Figure 7.55 is used for both channels. Also notice that no feed-forward control term from the grid voltages are used in the CFL of the VSM in order to allow seamless transition between grid-connected and standalone modes of operation.

If the CFL is devised in synchronous frame, I_d^{ref} and I_q^{ref} are used as the current reference components, and the resonant controllers are replaced with integrating functions. In this case, the currents i_d and i_q (for use in CFL) are obtained using the internal angle ϕ on i_α and i_β. Here, either the SISO-PI approach or the MIMO-PI approach (as discussed earlier in this chapter) may be used.[50]

The grid voltage magnitude V_g used in the droop term $K_v(V_o - V_g)$ can be calculated using

$$\boxed{V_g = \sqrt{v_{g_\alpha}^2 + v_{g_\beta}^2}.} \tag{7.74}$$

The real and reactive powers P and Q used in the VSM loop can be calculated using

$$\boxed{P = 1.5Vi_{1_d}, \quad Q = -1.5Vi_{1_q}.} \tag{7.75}$$

Remark Equation (7.75) indicates that the powers are calculated at the virtual terminal associated with $v(t)$. The virtual impedance $L_v s + R_v$ stands between this point and the actual terminal associated with the grid voltage v_g. Therefore, if the power set-points P_{set} and Q_{set} are meant to be the powers at the grid terminals, they can be properly adjusted to count for the "virtual" real and reactive powers consumed in this impedance. Thus, the following terms must be added to them, respectively.

$$P_{v,set} = 1.5R_v I_{set}^2, \quad Q_{v,set} = 1.5L_v\omega_o I_{set}^2, \quad I_{set} = \frac{\sqrt{P_{set}^2 + Q_{set}^2}}{1.5V_o}. \tag{7.76}$$

50 Note that the voltage $v(t)$ as defined here is balanced. Therefore, when the grid voltage is unbalanced, the current will be unbalanced. The level of unbalance of the current depends on the level of unbalance of the grid voltage and the size of virtual impedance. If it is desired to have a direct control over the negative-sequence of current, the definition of voltage $v(t)$ must be extended to accommodate a negative-sequence component in it as well. This will be in the form $v(t) = [V\cos(\phi) \;\; V\sin(\phi)]^T + [V'\cos(\phi') \;\; -V'\sin(\phi')]^T$. The core VSM must then be extended to generate V' and ϕ' as well. A similar problem has been treated in [13].

7.5.3 Stability Analysis and Design of VSM

The angle and magnitude differential equations may be written from Figure 7.53 as

$$\begin{bmatrix} \dot{\phi} \\ \dot{V} \end{bmatrix} = \begin{bmatrix} \omega + DK_f(\omega_o - \omega) \\ \frac{K_v}{M}(V_o - V_g) \end{bmatrix} + \begin{bmatrix} D & 0 \\ 0 & \frac{1}{M} \end{bmatrix} \begin{bmatrix} \cos\theta & -\sin\theta \\ \sin\theta & \cos\theta \end{bmatrix} \begin{bmatrix} P_{set} - P \\ Q_{set} - Q \end{bmatrix}. \tag{7.77}$$

Define $x = [x_1 \ x_2]^T = [V\cos\phi \ V\sin\phi]^T$, then

$$\dot{x} = \dot{V}\begin{bmatrix} \cos\phi \\ \sin\phi \end{bmatrix} + V\dot{\phi}\begin{bmatrix} -\sin\phi \\ \cos\phi \end{bmatrix} = \begin{bmatrix} -x_2 & x_1 \\ x_1 & x_2 \end{bmatrix} \begin{bmatrix} \dot{\phi} \\ \frac{\dot{V}}{V} \end{bmatrix}. \tag{7.78}$$

For this analysis, we assume that the droop terms are neglected (which is a justified simplifying assumption for grid-connected operation). Then

$$\dot{x} = \begin{bmatrix} -\omega x_2 \\ \omega x_1 \end{bmatrix} + \begin{bmatrix} -x_2 & x_1 \\ x_1 & x_2 \end{bmatrix} \begin{bmatrix} D & 0 \\ 0 & \frac{1}{MV} \end{bmatrix} \begin{bmatrix} \cos\theta & -\sin\theta \\ \sin\theta & \cos\theta \end{bmatrix} \begin{bmatrix} P_{set} - P \\ Q_{set} - Q \end{bmatrix}. \tag{7.79}$$

Let's assume

$$\boxed{D = \frac{1}{MV} = \frac{\mu}{V^2}} \tag{7.80}$$

where $\mu > 0$. Notice also that the powers are expressed as

$$\begin{bmatrix} P \\ Q \end{bmatrix} = \begin{bmatrix} 1.5Vi_{1_d} \\ -1.5Vi_{1_q} \end{bmatrix} = 1.5V \begin{bmatrix} \cos\phi & \sin\phi \\ \sin\phi & -\cos\phi \end{bmatrix} i_{ref} \tag{7.81}$$

where i_{ref} is a two-dimensional vector comprising the α and β components. Substituting from (7.81) and doing some manipulations, (7.79) reduces to[51]

$$\dot{x} = \left(\begin{bmatrix} 0 & -\omega \\ \omega & 0 \end{bmatrix} + \frac{\mu}{V^2} \begin{bmatrix} \sin\theta & \cos\theta \\ \cos\theta & -\sin\theta \end{bmatrix} \begin{bmatrix} P_{set} & Q_{set} \\ Q_{set} & -P_{set} \end{bmatrix} \right) \begin{bmatrix} x_1 \\ x_2 \end{bmatrix} - \frac{3\mu}{2} \begin{bmatrix} \sin\theta & -\cos\theta \\ \cos\theta & \sin\theta \end{bmatrix} i_{ref}. \tag{7.82}$$

Assuming zero power set-points, and constant frequency condition $\omega = \omega_o$, this reduces to an LTI system represented by the state-space equations

$$\dot{x} = \begin{bmatrix} 0 & -\omega \\ \omega & 0 \end{bmatrix} \begin{bmatrix} x_1 \\ x_2 \end{bmatrix} - 1.5\mu \begin{bmatrix} \sin\theta & -\cos\theta \\ \cos\theta & \sin\theta \end{bmatrix} i_{ref}. \tag{7.83}$$

Therefore, the control block diagram of the system may be represented as in Figure 7.56, where I_2 is 2×2 identity matrix. Notice that $v(t) = x(t)$. This is a two-input two-output coupled control system.

The equations pertaining to the virtual impedance are

$$\frac{d}{dt} i_{ref} = -\frac{R_v}{L_v} i_{ref} + \frac{1}{L_v} v(t) - \frac{1}{L_v} v_g(t). \tag{7.84}$$

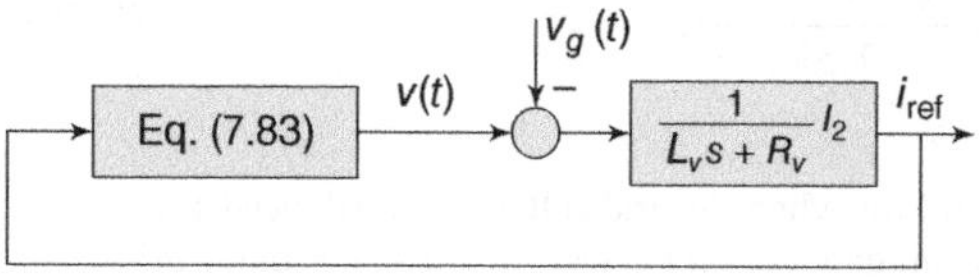

Figure 7.56 Approximate LTI model of the grid-connected VSM for stability analysis and design.

---

51 See Problem 7.16.

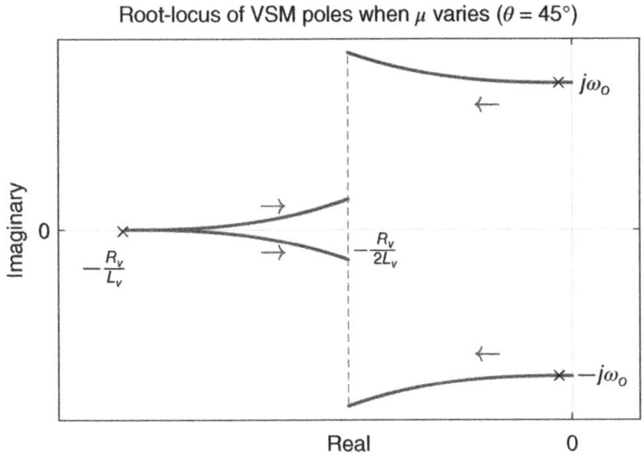

Figure 7.57 Eigenvalues of the 4 × 4 matrix in (7.85) when μ increases.

Augmenting (7.83) and (7.84) leads to

$$\frac{d}{dt}\begin{bmatrix} x \\ i_{\text{ref}} \end{bmatrix} = \begin{bmatrix} 0 & -\omega & -1.5\mu\sin\,\theta & 1.5\mu\cos\,\theta \\ \omega & 0 & -1.5\mu\cos\,\theta & -1.5\mu\sin\,\theta \\ \frac{1}{L_v} & 0 & -\frac{R_v}{L_v} & 0 \\ 0 & \frac{1}{L_v} & 0 & -\frac{R_v}{L_v} \end{bmatrix}\begin{bmatrix} x \\ i_{\text{ref}} \end{bmatrix} - \begin{bmatrix} 0_2 \\ I_2 \end{bmatrix} v_g. \tag{7.85}$$

The eigenvalues of the main 4 × 4 matrix in (7.85) determine the stability of the closed-loop system. Figure 7.57 shows variations of eigenvalues of this matrix when μ increases from 0 to the suggested value of

$$\boxed{\mu = \frac{R_v\omega}{3\cos\,\theta}.} \tag{7.86}$$

It is observed that originally, for $\mu = 0$, the poles are at $-\frac{R_v}{L_v}$ (two poles), and $\pm j\omega$. As μ increases, there is a critical value where all four poles are vertically aligned at $-\frac{R_v}{L_v}$. This seems to us to be a good location for poles. This critical, suggested value for μ is given in (7.86).[52] For a value of $R_v = 2$ and $L_v = 0.005$, i.e. $\frac{R_v}{L_v} = 400$, this critical value of μ is 355 at $\theta = 45°$.

For the design of R_v and L_v, following is the suggestion. (i) Choose the magnitude of this impedance, i.e. $Z_v = \sqrt{R_v^2 + L_v^2\omega_o^2}$, equal to k_1 pu.[53] (ii) Choose $\frac{R_v}{L_v}$ at $k_2\omega_o$.[54] This will lead to

$$\boxed{L_v = \frac{k_1 Z_b}{\omega_o\sqrt{1+k_2^2}}, \quad R_v = k_2\omega_o L_v,} \tag{7.87}$$

where $Z_b = \frac{V_b^2}{S_b}$ and S_b and V_b are the base values of the (total VA) power and (line-to-line rms) voltage of the inverter.

As for the value of θ, as it was discussed in single-phase VSM in Chapter 6, we note that conventional large synchronous generators interfaced with inductive lines use $\theta = 0$. For a DER interfaced

52 See Problem 7.17.
53 In our opinion, k_1 can be any number between 0.05 and 0.5 pu for the DER applications.
54 Suggested value for k_2 is between 0.5 and 3. For a 60 Hz system, $k_2 = 2$ means that $\frac{R_v}{2L_v}$ will be 377 which shows a quite reasonable distance of poles from the imaginary axis for the DER applications.

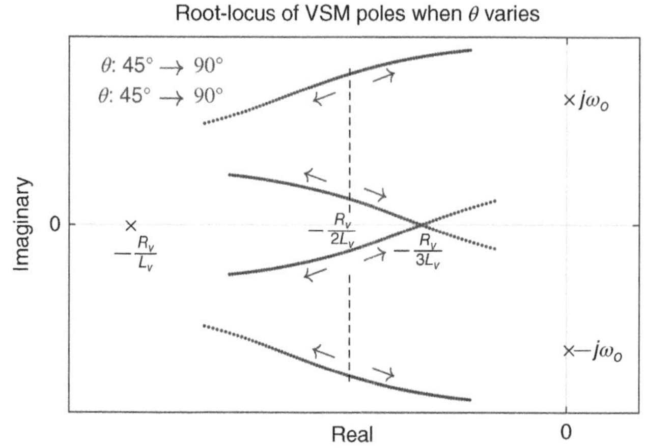

Figure 7.58 Eigenvalues of the 4 × 4 matrix in (7.85) when θ varies while the design of μ is done using (7.86) at $\theta = 45°$, and R_v and L_v are selected from (7.87).

with a distribution system, however, larger values of θ seem more advantageous [14]. Figure 7.58 shows the locus of the closed-loop poles when θ varies, while μ is fixed at the value of (7.86) for $\theta = 45°$, and R_v and L_v are fixed at the values of (7.87). For both cases where θ approaches zero (corresponding to an SG scenario), and when θ goes above 45°, the damping of the poles go down. However, this graph and also the simulation results shown later confirm that this design (i.e. the design for $\theta = 45°$) is good enough for the entire range of θ. For a specific θ, the design of μ, R_v, and L_v can be further adjusted and tuned.

7.5.3.1 Start-up Synchronization

This is to synchronize the VSM output voltage with the grid voltage prior to the start of applying the gating signals, to ensure that there will be a soft starting of the converter. For this, we set P_{set} and Q_{set} initially to zero. Then, during the pre-synchronization time, we virtually close the feedback loop (without operating the converter) by emulating the filter and the grid as shown in Figure 7.59.

During this time, the converter is not switching, and the actual current i is zero. By regulating the virtual current i_v to zero (setting $i_r = 0$), the converter output voltage v (or u) will be synchronized with the grid voltage v_g. When i_v is sufficiently small (indicating that the synchronization is complete), the feedback current i_x is replaced by the actual current i, the reference point i_r is replaced with the i_{ref} (generated by the VSM controller), and the converter starts switching. This process is expressed by

$$
\begin{aligned}
i_x &= \begin{cases} i_v, & t < t_{start}; \\ i(t) \text{ [measured current]}, & t \geq t_{start}. \end{cases} \\
i_r &= \begin{cases} 0, & t < t_{start}; \\ i_{ref}(t) \text{ [calculated by VSM controller]}, & t \geq t_{start}. \end{cases}
\end{aligned}
\tag{7.88}
$$

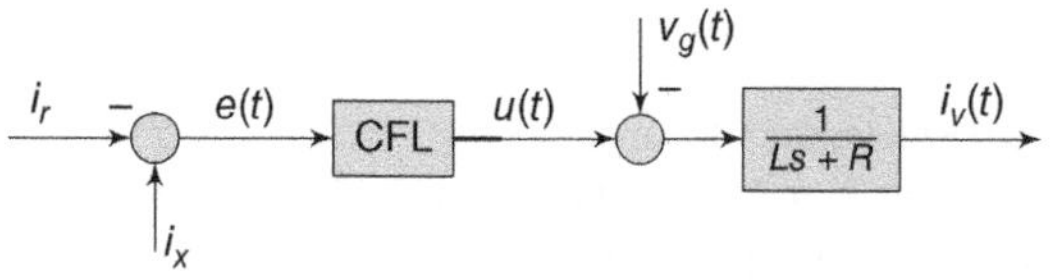

Figure 7.59 Mechanism for achieving synchronization prior to starting the inverter.

and shown in Figure 7.59. Note that the modulating signal is obtained from $m(t) = \frac{u(t)}{0.5v_c}$ where v_c is the dc side voltage.[55]

7.5.3.2 Grid-Connection Synchronization

The VSM is a GFM controller and can operate the inverter in both standalone and grid-connected modes. To seamlessly transition from standalone (isolated) operation to grid-connected mode, prior to closing the switch SW in Figure 7.60, the converter side voltage v_g must be sufficiently synchronized with the grid side voltage v_s to avoid possible voltage disturbances for the load and current transients for the converter.

A synchronization mechanism is shown in Figure 7.61 where the difference between the converter-side voltage v_g and the grid-side voltage v_s, projected on the two axes d and q, is used to manipulate the converter powers in a feedback loop. The positive constants K_{sd} and K_{sq} determine how strongly the synchronization pushes the voltage v_g close to v_s. A reasonable selection for these parameters is $K_{sd} = K_{sq} = K_v$, where K_v is the droop coefficient of the voltage versus reactive power loop. Notice that by increasing the real power, the converter increases v_{g_d} and brings it closer to v_{s_d}. However, it will need to decrease its reactive power to increase v_{g_q}. The equations $P = 1.5Vi_d$ and $Q = -1.5Vi_q$ explain this: increasing real power directly increases V (leading to an increase in v_{g_d}) but decreasing Q increases i_q (leading to an increase in v_{g_q}). A limiter block, denoted by "sat" in Figure 7.61, is used to limit the synchronization power within $\pm S$ to prevent possible transients subsequent to enabling the synchronization block.

7.5.3.3 Comment on Adaptive Parameters

Refer to explanation given in Section 6.3.4 pertaining to single-phase VSM.

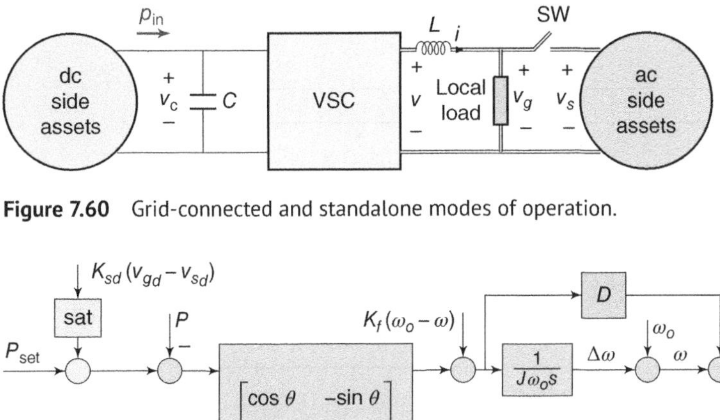

Figure 7.60 Grid-connected and standalone modes of operation.

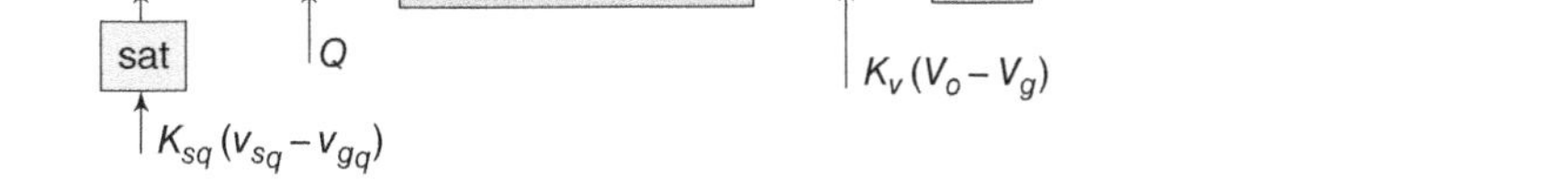

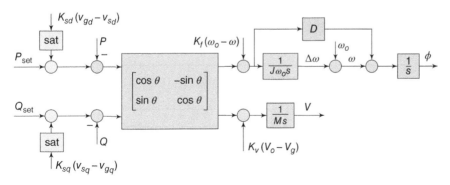

Figure 7.61 Mechanism to achieve synchronization prior to transition from standalone to grid-connected operation.

55 If the CFL is implemented in rotating frame, the transformation to stationary frame must also be done to get the actual modulation signal.

Example 7.6 ***Three-Phase VSM***

Consider the inverter and system with the same parameters given in Table 7.1. A three-phase local load of 7.2 kVA (half the rated inverter power) at 0.8 PF lagging (in the form of a series RL connection) is present and connected to the inverter terminals. A capacitor of 5 μF (per phase) is connected across the load terminals to filter the switching noises during standalone operation. A resistor of 1 Ω is connected in series with this capacitor to passively damp the LC resonance mode.[56] The switching frequency is doubled to 8 kHz for better voltage quality in standalone mode. In this example, we design a complete VSM-based controller for this inverter and study its performance.

The controller parameters are selected as follows. (i) The virtual impedance parameters are selected according to (7.87) at $L_v = 1.4$ mH, $R_v = 1.1$ Ω for the selection of $k_1 = 0.4$ and $k_2 = 2$. (ii) The controller parameter $\mu = 194$ is selected according to (7.86), assuming $\theta = 45°$, to place all three poles on a vertical line at $-\frac{R_v}{2L_v} = -377$. (iii) The parameter θ is arbitrarily chosen at 45°. (iv) The moment of inertia J is selected at $J = 0.1$. This will correspond to an inertia constant of about $H = 0.5$ seconds. (v) The droop coefficients are selected at $K_v = \frac{S}{0.2V_o} = 417$ Var/V, $K_f = \frac{S}{0.05\omega_o} = 750$ W s/rad. (vi) The grid-connection synchronization gain is selected at $K_s = K_v$. (vii) The same LQT-based CC (CFL) of Example 7.1 is used.

Figure 7.62 shows the *starting, power set-points tracking, and grid low-voltage ride through responses* of the controller. During $0 \leq t < 0.1$ seconds, the controller synchronizes to the grid

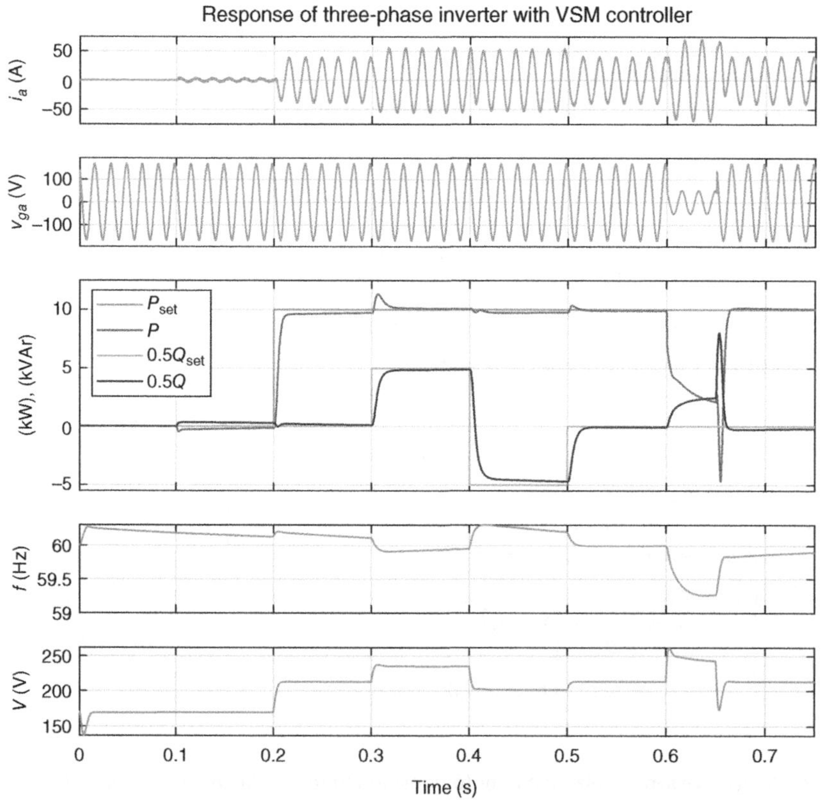

Figure 7.62 Responses of three-phase VSM controller: starting synchronization during $0 \leq t < 0.1$ seconds; inverter starts at $t = 0.1$ seconds; real and reactive power set-points at $t = 0.2, 0.3, 0.4$, and 0.5 seconds; low-voltage grid fault at $t = 0.6$ seconds; and fault clears at $t = 0.65$ seconds.

56 See Sections 5.9.2 and 7.7.2.1 for some active damping approaches.

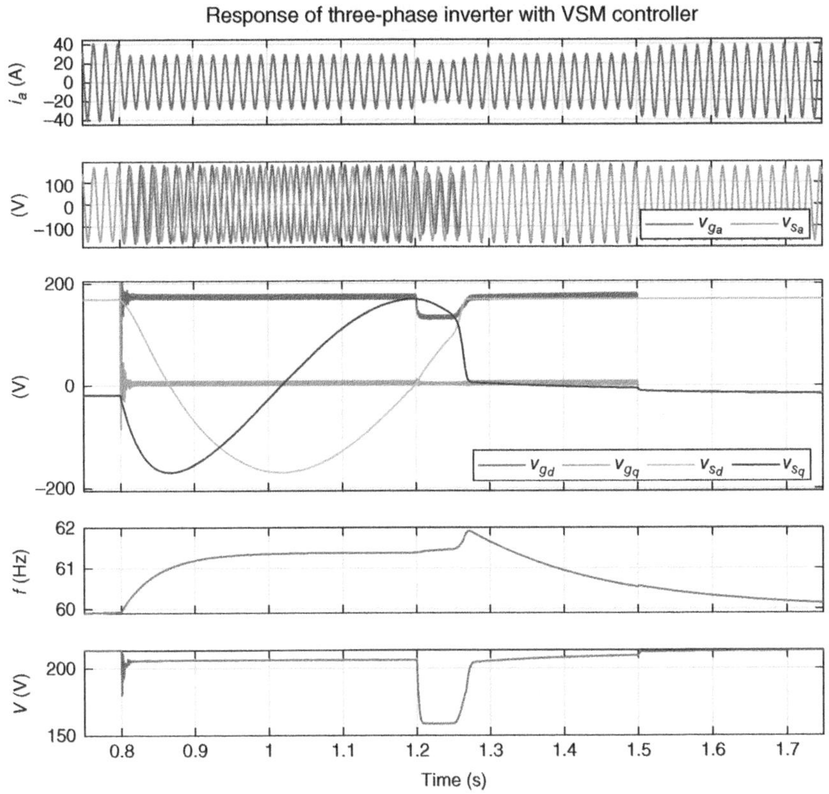

Figure 7.63 Responses of three-phase VSM controller: transition from grid-connected to standalone at $t = 0.8$ seconds; enabling the synchronization mechanism at $t = 1.2$ seconds; and reconnecting to the grid at $t = 1.5$ seconds.

voltage according to Figure 7.59. At $t = 0.1$ seconds, the inverter starts operating. The real and reactive power set-points are faithfully tracked. The current is limited during the grid voltage fault. During the fault, the inverter naturally reduces the real power and increases its reactive power. This is a correct fault ride-through operation. The internal VSM frequency and voltage variables (i.e. $f = \frac{\omega}{2\pi}$ and V) are also shown in Figure 7.62.

Figure 7.63 shows the *transition from grid-connected to standalone* (at $t = 0.8$ seconds), activation of *synchronization mechanism* of Figure 7.61 at $t = 1.2$ seconds, and *reconnection to grid* at $t = 1.5$ seconds. It is observed that all transitions are made seamlessly. The synchronization is achieved smoothly and quickly. Once it is enabled, it pushes v_{g_d} and v_{g_q} to the vicinity of v_{s_d} and v_{s_q}.

Responses of the controller to a *grid frequency swing* is shown in Figure 7.64. As the grid frequency goes up, the inverter decreases its real power. Note that it also increases the reactive powers because we have chosen $\theta = 45°$. If $\theta = 0°$ is chosen, the converter only responds to the frequency swing by changing its real power (similar to existing synchronous generators). It should also be noted that the power swings include both the dynamic (inertia) and the static (droop) components. However, since the inertia component is relatively small, only the droop component is visible in Figure 7.64.[57]

57 The real power, for example, is equal to

$$P = P_{set} + \cos(\theta)[-J\omega_o\dot{\omega} + K_f(\omega_o - \omega_g)] + \sin(\theta)[-MV\dot{V} + K_v(V_o - V_g)].$$

All these components can be evaluated. For instance, during the 1 Hz/s frequency ramp-up region, the inertia power is about $-J\omega\dot{\omega} = 0.1 \times 377 \times 2\pi = 237$ W. The droop power (for 0.5 Hz frequency deviation) reaches up to $K_f\Delta\omega = -750 \times \pi = -2360$ W.

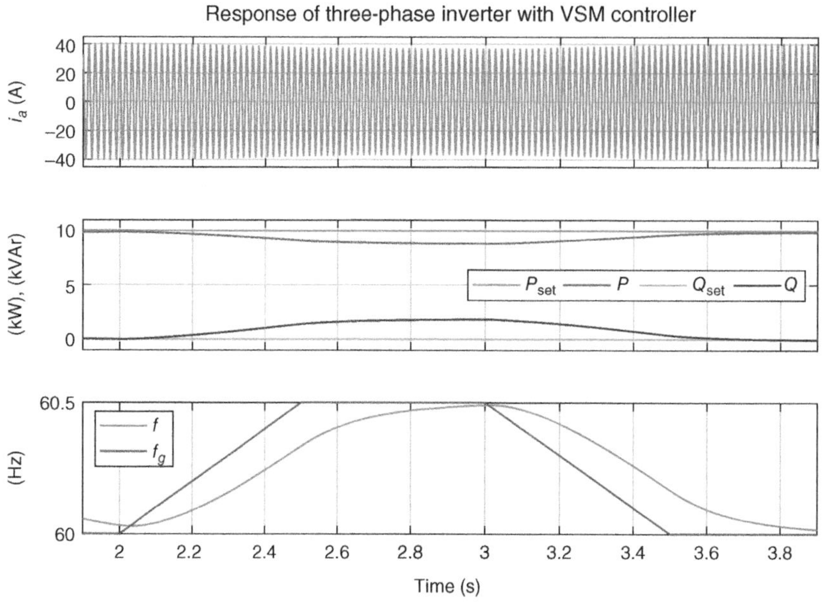

Figure 7.64 Responses of three-phase VSM controller: grid voltage frequency swing between $t = 2$ seconds and $t = 3.5$ seconds.

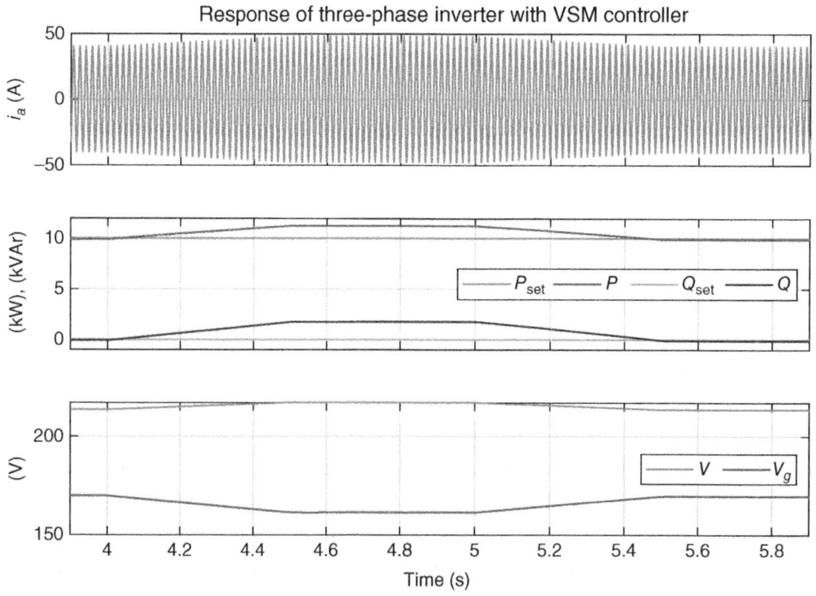

Figure 7.65 Responses of three-phase VSM controller: grid voltage magnitude swing between $t = 4$ seconds and $t = 5.5$ seconds.

Responses of the controller to a *grid voltage magnitude swing* is shown in Figure 7.65. As the grid voltage magnitude goes down, the inverter increases its power. It increases both the real and reactive powers because we have chosen $\theta = 45°$. If $\theta = 0°$ is chosen, the converter will only respond to the voltage magnitude swing by changing its reactive power (similar to existing synchronous generators).

7.6 dc Side Voltage Control and Support

Consider the dc/ac converter shown in Figure 7.66. In Section 7.4, we discussed how to efficiently build and design a CFL for this converter. The control system receives real and reactive power set-points and executes them while ensuring that the current is sinusoidal and within the prespecified limits during abrupt/severe transients. Then, we discussed how the controller can provide ac grid support functions. Two approaches of power controller with grid-support and VSM were introduced and studied. The CFL is still used in their cores.

The dc voltage support may also be provided by the controller. Similar to the ac situation, this support may be divided into static and dynamic supports. Both of these supports add an incremental power to the real power set-point, P_{set}, of the converter, flowing from dc to ac side.

Generally speaking, a dc voltage rise as an indication of the excess of power and its fall indicates shortage of power. Therefore, the static support may be defined and expressed by

$$\Delta P_{\text{stat}} = K_{\text{stat}}(v_c - V_c), \tag{7.89}$$

where $v_c(t)$ is the actual dc side voltage and V_c is its rated value. This means that when the dc voltage is above/below its rated value (indicating an excess/shortage of power on dc side), the converter increases/decreases its power transfer from dc to ac, helping to support the dc side. The positive constant K_{stat} determines how much the power is increased for a unit increase in the dc side voltage.

The dynamic (or inertia) support may be expressed by

$$\Delta P_{\text{dyn}} = K_{\text{dyn}}\dot{v}_c(t). \tag{7.90}$$

This means that the converter instantly increases its power transfer from dc to ac in proportion to the rate of change of dc side voltage. A sudden change of dc side voltage (producing large $\dot{v}_c$) causes the converter to respond quickly to revert it. The positive constant K_{dyn} determines how much the power is increased for a unit increase in the dc side voltage rate of change.

In this section, we study in details the scenarios where the converter is also responsible to control (and/or support) the dc side voltage.

7.6.1 System Modeling

The power system is shown in Figure 7.67 where C is the capacitance of the dc link and p_{in} is the input power to the dc link. The power balance in dc-bus capacitor indicates that

$$v_c(t)i_c(t) = Cv_c(t)\frac{d}{dt}v_c(t) = \frac{d}{dt}w_c(t) = p_{\text{in}} - p_{\text{out}}, \quad w_c(t) = \frac{1}{2}Cv_c(t)^2,$$

where $w_c(t)$ denotes the instantaneous stored energy in the capacitor C. The power p_{out} may be approximated with

$$p_{\text{out}} = 1.5v_g(t)\cdot i(t) = 1.5(v_{g_\alpha}i_\alpha + v_{g_\beta}i_\beta) = 1.5(v_{g_d}i_d + v_{g_q}i_q)$$

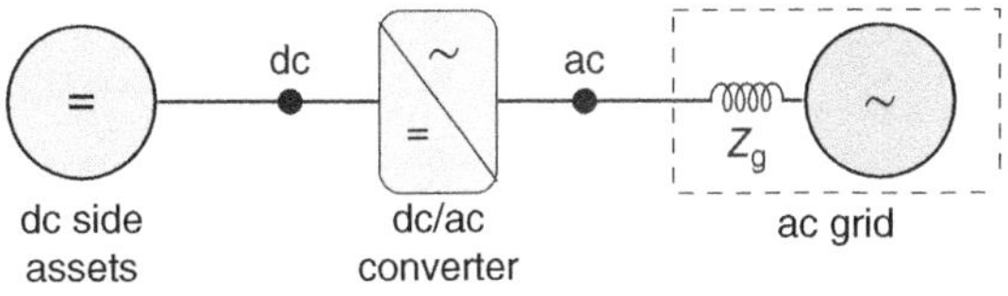

Figure 7.66 Converter interfacing a dc side to an ac side.

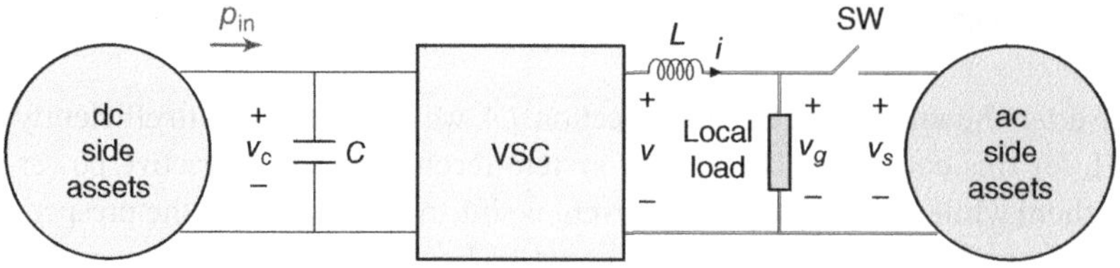

Figure 7.67 Circuit diagram of a three-phase converter with dc-side voltage to be controlled.

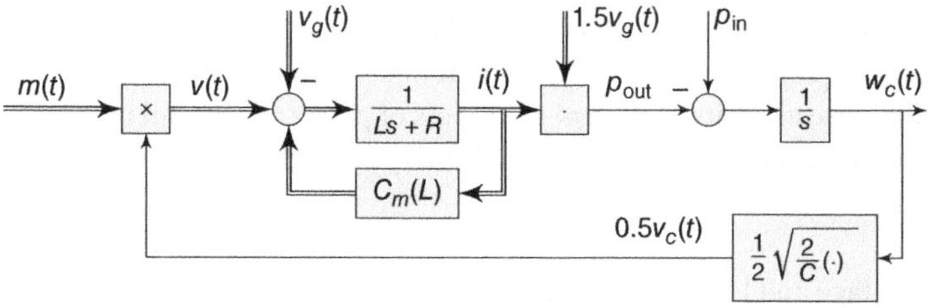

Figure 7.68 Block diagram corresponding to the dc-side voltage control in three-phase converter.

by ignoring the instantaneous power of the interfacing filter and the converter power losses. Notice that the last equation simplifies to $p_{\text{out}} = 1.5v_{g_d}i_d$ if the dq transformation is done using the angle of v_g.

The control block diagram of the system is shown in Figure 7.68 which is valid for both $\alpha\beta$ and dq frames with the note that the coupling matrix C_m is zero in $\alpha\beta$ and is equal to $C_m(L) = \begin{bmatrix} 0 & L\omega \\ -L\omega & 0 \end{bmatrix}$ in dq domain.

The system model shown in Figure 7.68 is not directly suitable for controller design and analysis. It has a nonlinearity corresponding to multiplication into the capacitor voltage at the input. This can be easily removed using a division by $0.5v_c$. A feed-forward term $v_{\text{ff}}(t)$ may also be added that is provided by the PLL and compensates for the fundamental component of v_g, normally in dq frame. The coupling term $C_m(L)i$ may also be removed at the input. The resulted system is shown in Figure 7.69 where the new control signal $u(t)$ is defined as

$$u(t) = 0.5v_c(t)m(t) - v_{\text{ff}}(t) + C_m(L)i. \tag{7.91}$$

Let us assume that the dc-bus voltage $v_c(t)$ is desired to be regulated at a given constant value V_c. The energy is expressed by $w_c(t) = \frac{1}{2}Cv_c(t)^2$. The energy function may be linearized around this point using the Taylor's series concept as

$$w_c = \frac{1}{2}Cv_c^2 \approx \frac{1}{2}CV_c^2 + CV_c(v_c - V_c) = CV_cv_c - \frac{1}{2}CV_c^2 = CV_cv_c - W_c.$$

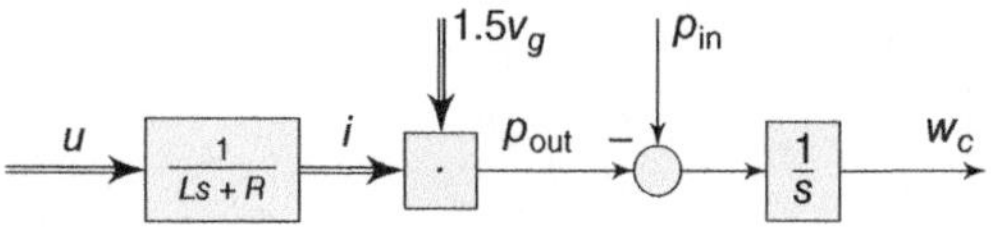

Figure 7.69 Simplifying the model of Figure 7.68 using $m(t) = \frac{u(t)+v_{\text{ff}}(t)-C_m(L)i}{0.5v_c(t)}$.

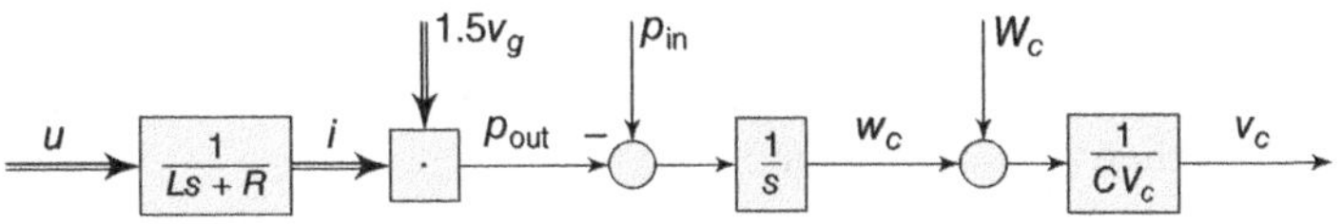

Figure 7.70 Linearizing w_c to v_c in the model of Figure 7.69.

Using this equation, the model of Figure 7.70 is derived.[58]

7.6.2 Control Structure and Design

The common approach considers two control loops: internal current control loop and external dc voltage control loop. This will allow converter current limiting. We already designed a CFL for the internal subsystem of Figure 7.70 as shown in Figure 7.71. The CC may be implemented in stationary frame (using PR compensators) or in rotation frame (using PI or PIR compensators, SISO or MIMO) as fully studied in Sections 7.3 and 7.4.

The two-loop control structure is shown in Figure 7.72. The capacitor voltage is controlled to stay around the reference value of V_c. Its output determines the amount of (real) power (or current, I_d) supplied by the inverter. The amount of reactive power (current, specified by I_q) is supplied to the loop externally. The block denoted by CML stands for current magnitude limiting and is shown in Figure 7.15.

In the dc voltage control loop of Figure 7.72, assume that I_o is the rated current of the DER. Therefore, the controller establishes

$$I_d = I_o + k_0(v_c - V_c) \tag{7.92}$$

which indicates that the capacitor voltage will, in steady state, converge to $v_c = V_c + \frac{i_d - I_o}{k_0}$ where i_d is the operating direct current.[59] When the converter operates at its rated current of I_o, the dc voltage will also be at its set-point of V_c. If it operates at a lower/higher current, the dc voltage will

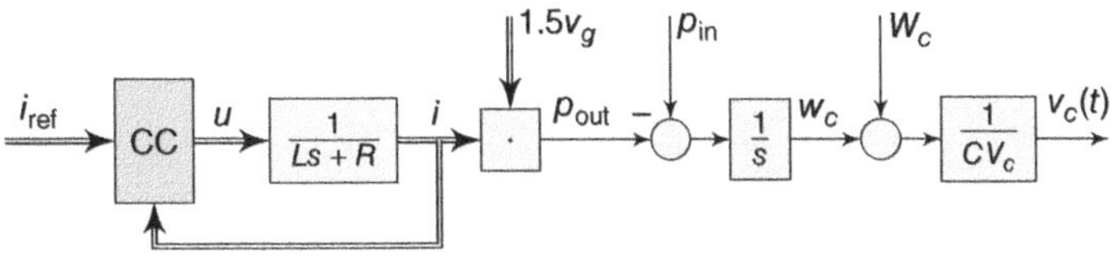

Figure 7.71 Current control is added to Figure 7.70.

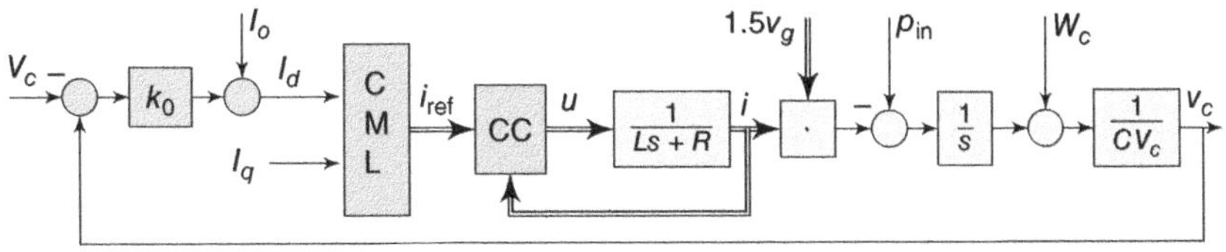

Figure 7.72 Capacitor voltage control is added to Figure 7.71.

58 It is also possible to close the feedback loop on w_c or v_c^2 (instead of v_c) to naturally avoid this nonlinearity. However, the voltage feedback loop is more common and we adopt this here. This nonlinearity has very minimal, if any, effect of the obtained results because the magnitude of changes of v_c are not sizeable compared with the rated value V_c. This is similar to the common simplifying assumption of constant grid frequency for design purposes.
59 Note that I_d is the reference value and i_d is the actual value.

also be lower/higher than the rated voltage, proportionally and linearly.[60] Assume for instance that the converter current i_d can change in the range of $[0 \;\; I_o]$ and we want that the capacitor voltage experiences a change of only $\alpha\%$ of V_c, then $\frac{I_o}{k_0} = 0.01\alpha V_c$ or

$$\boxed{k_0 = \frac{I_o}{0.01\alpha V_c}.} \tag{7.93}$$

For example, for a $V_c = 500$ V, $I_o = 39.3$ A (corresponding to 10 kW at 120 V rms grid), and $\alpha = 5\%$, it will be $k_0 = \frac{39.3}{25} = 1.57$.

Thus, the gain k_0 determines the range of variations of the dc voltage (in steady operation) when the converter current varies. In the existing literature, e.g. [15, 16], there is a common tendency to use a PI function to accurately control the dc voltage and remove all the steady error from it. We believe, however, that there is no need to completely remove this error. This error, as long as it is within a reasonably small range, is not practically harmful and can even be useful in some applications. For instance, it can be used for dc bus signalling if multiple converters share a common dc bus. This way, they will be informed of the operational status of other converters through the dc voltage. Furthermore, this controller is very simple, avoids design complications, does not have the problem of integrator windup, and has a faster and more robust performance compared to when a PI is used [17].[61]

In order to perform a loop stability analysis and controller design, it is assumed that the CC is already designed. Moreover, it is assumed that the internal loop is sufficiently faster than the external loop. Therefore, from the standpoint of the external loop, the internal loop is assumed as a unity gain. Therefore, $i(t) = i_{\text{ref}}(t)$, and the output power p_{out} may be expressed and simplified by

$$p_{\text{out}} = 1.5i(t) \cdot v_g(t) = 1.5(v_{g_d} I_d + v_{g_q} I_q) = 1.5V_g I_d, \tag{7.94}$$

where V_g is the rated value of the grid voltage magnitude (line-to-neutral, peak).[62] Therefore, the control loop may be simplified and represented as Figure 7.73.

The characteristics equation of the dc voltage loop is

$$1 + k_0 \frac{3V_g}{2CV_c} \frac{1}{s} = 0$$

which means a single real pole at $s = -\beta$, where

$$\boxed{\beta = \frac{3k_0 V_g}{2CV_c}.} \tag{7.95}$$

Based on this discussion, the following design algorithm may be presented.

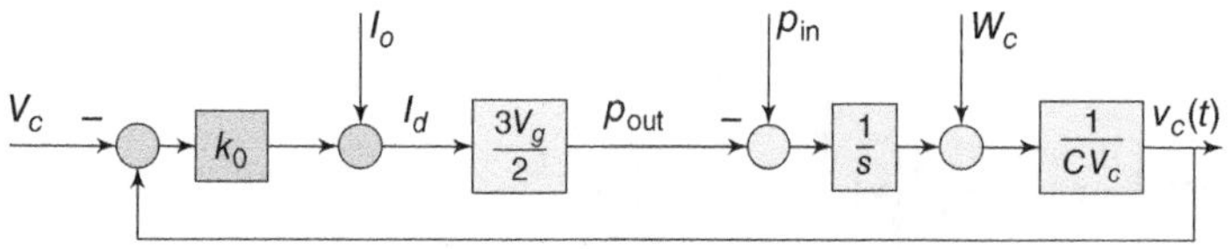

Figure 7.73 Simplified equivalent control model of Figure 7.72.

60 If the converter is bidirectional, I_o may be set to zero. This way, the capacitor voltage is at its rated value of V_c when it supplies no real power. This voltage will go up/down linearly with the real power of the converter.
61 Moreover, in applications where V_c is determined by another control loop, for instance an MPPT loop in a single-stage PV inverter, the feedback of that loop will correct the dc voltage offsets.
62 In (7.94), we have assumed the grid voltage to be balanced and have a magnitude of V_g. As well, the power across the converter and its filter are neglected. These are simplifying and acceptable assumptions for the sake of basic analysis and design purposes. They may be included for a more precise analysis [17].

Algorithm 7.1

```
Input:
- Grid voltage and frequency: Vg (peak, line-to-neutral), ω0
- Inverter real power rating: P (peak current: Io = P/(1.5Vg))
- Capacitor voltage: Vc, Its relative change: α%
Output:
- Capacitance C                    - Controller gain k0
Step 1: Find k0 from (7.93).       Step 2: Find C from (7.95) such that 1/β is
at least a few times larger than the time-constant of CC loop.
```

Remark 7.1 The value of capacitor designed above satisfies the control conditions. Normally, this capacitor has to satisfy some other constraints in practice. For example, when a transient grid voltage fault occurs, the capacitor voltage should stay within acceptable limits. Or when an unbalanced grid voltage occurs, the double-frequency ripple of the capacitor voltage must be within the acceptable range. All these constraints can easily be addressed by choosing the capacitor sufficiently large.

Example 7.7 ***Capacitor Voltage Control***
Consider the same grid and inverter parameters given in Table 7.1. Assume that $\alpha = 5\%$, i.e. 5% (25 V of 500 V) deviation in the capacitor voltage. Then, following the algorithm, we get $I_o = \frac{P}{1.5V_g} = \frac{10\ 000}{1.5\times120\sqrt{2}} = 39.3$ A and $k_0 = \frac{I_o}{0.01\alpha V_c} = \frac{39.3}{0.01\times5\times500} = 1.57$. Equation (7.95) indicates that $\beta = \frac{3k_0 V_g}{2CV_c} = \frac{0.8}{C}$. For a capacitance of $C = 10$ mF, for instance, $\beta = 80$, which corresponds to a time-constant which is surely several times larger than the time-constant of CC loop. The poles of CC loop are shown in Figure 7.17.

In this example, the same CC of Example 7.1 is used. Figure 7.74 shows a sample of performance of this control system. The responses are as expected in terms of their dynamics (smooth and first-order) and their overall time-constant (of about $\frac{1}{\beta} = 12.5$ ms). The capacitor voltage changes between the minimum value of 475 V (at no power) and 500 V (at full power). The converter current is successfully limited during the grid voltage fault, i.e. $0.3 \leq t \leq 0.35$ seconds. During this time, the capacitor absorbs the input power and its voltage rises to about 550 V. It quickly goes back to 500 V when the fault clears at $t = 0.35$ seconds. All the intended objectives are achieved.

7.6.3 Obtaining Inertia from dc Side Capacitor*

The dc side capacitor C establishes the equation

$$Cv_c\dot{v}_c = p_{\text{in}} - p_{\text{out}}, \tag{7.96}$$

which means the output power is $p_{\text{out}} = p_{\text{in}} - Cv_c\dot{v}_c$. Assume that we can establish a relationship

$$v_c = V_c + \gamma(\omega - \omega_0), \tag{7.97}$$

where V_c is a nominal value for the capacitor voltage, ω_0 is the nominal value of the grid frequency, ω is the actual grid frequency or a variable that mimics it, and γ is a positive number. Then, the power supplied by the capacitor is $-Cv_c\dot{v}_c = -\gamma Cv_c\dot{\omega} \approx -\gamma CV_c\dot{\omega}$.[63] This indicates an

63 We assume $\gamma(\omega - \omega_0)$ is much smaller than V_c.

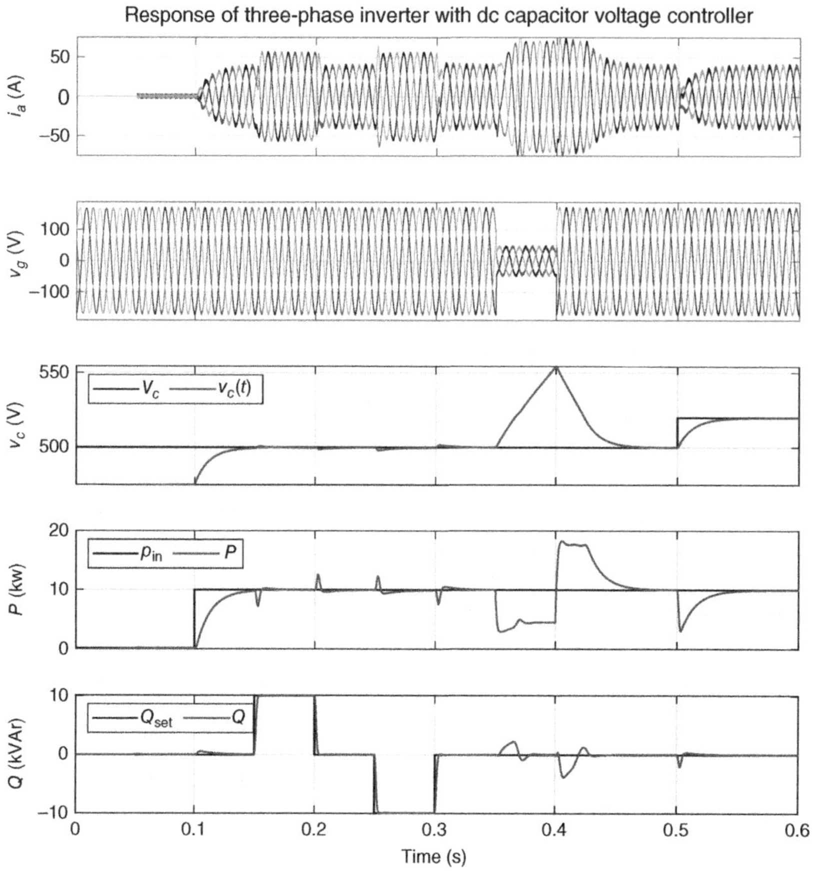

Figure 7.74 Responses of the three-phase inverter with dc voltage controller. PLL starts at $t = 0$ seconds. Converter starts at $t = 0.05$ seconds. Input power jumps from 0 to 10 kW at $t = 0.1$ seconds followed by reactive power jumps of 10 , 0, −10, 0 kVAr at $t = 0.15$, 0.2, 0.25, 0.3 seconds, respectively. A deep (75%) low-voltage grid fault occurs at $t = 0.35$ seconds and clears at $t = 0.4$ seconds. At $t = 0.5$ seconds, the dc voltage reference V_c jumps to 520 V.

inertia response similar to the kinetic inertia of a rotating mass with the moment of inertia equal to[64]

$$\boxed{J = \gamma \frac{V_c}{\omega_o} C.} \tag{7.98}$$

This means that γ plays the role of "inertia amplification factor." By allowing the capacitor voltage to swing in a wider range than the grid frequency, larger level of inertia is extracted from the capacitor. The condition is, however, that the capacitor voltage variations must mimic the grid frequency variations.

7.6.3.1 Non-VSM Approach

One approach to implement this concept is to measure the grid frequency ω_g (using a PLL, for instance), and set the reference value for the capacitor voltage according to $V_c + \gamma(\omega_g - \omega_o)$ in the control system. The capacitor voltage controller discussed in Section 7.6.2 can be used. The block diagram of this approach is shown in Figure 7.75 where $\Delta\omega = \omega_g - \omega_o$.

64 The kinetic inertia of the rotor mass is $-J\omega\dot{\omega}$ which can well be approximated with $-J\omega_o\dot{\omega}$ since ω is around ω_o.

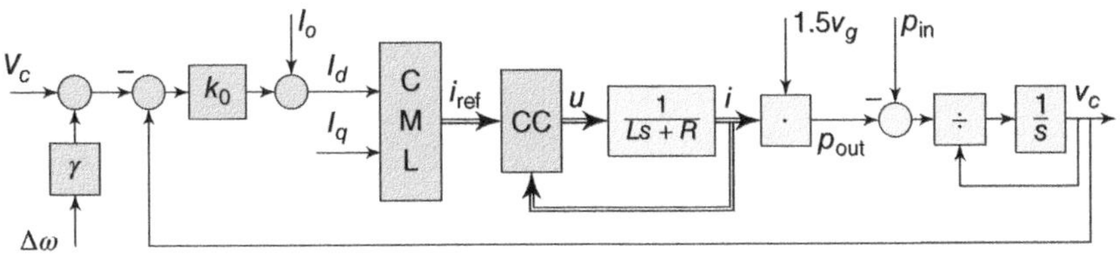

Figure 7.75 Obtaining inertia from capacitor: non-VSM approach.

7.6.3.2 VSM Approach

This approach uses the VSM concept where the grid frequency is tied to an internal (virtual) frequency (or virtual rotor speed). The internal frequency is directly generated using the dc voltage as shown in Figure 7.76 according to the equation

$$\omega = \omega_o + \frac{1}{\gamma}(v_c - V_c), \tag{7.99}$$

where γ is a positive constant. This is indeed the same equation as (7.97) and thus, the same level of inertia is obtained for a given γ and V_c. The difference, however, is that in this approach, there is no need to measure the grid frequency. The capacitor voltage v_c (or more precisely, its transients $v_c - V_c$) emulates the grid frequency (or its transients).

The magnitude loop in Figure 7.76 is the same as that in the VSM. It provides dynamic voltage support according to $M\dot{V} = Q_{set} - Q$. The static volt/var support can also be added by adding the term $K_v(V_o - V_g)$ to Q_{set} where V_o is the desired (rated) grid voltage and V_g is the measured grid voltage magnitude, as we discussed in details in Section 7.5.2.

The term D' shows a damping term similar to what is used in VSM, Figure 7.53. However, here its input has shifted to after the capacitor voltage integrator. Therefore, its output is also added directly to the angle. In order to have the same damping effect of Figure 7.53, it must satisfy

$$\boxed{D' = DCV_c,} \tag{7.100}$$

where D is calculated according to (7.80).

Finally, an extended version of the pre-synchronization shown in Figure 7.59 is used before starting the inverter. If the converter starts without pre-synchronization, although the current-limiting block ensures that the converter current remains within specified range, the capacitor voltage may grow (or drop) excessively. The extension is in the fact that the capacitor also needs to be emulated (virtually) during the pre-synchronization stage. Let us denote this virtual voltage as v_{c_v}, then it satisfies

$$Cv_{c_v}\dot{v}_{c_v} = p_{in} - p_{out} = 0 - 1.5Vi_{1_d} \tag{7.101}$$

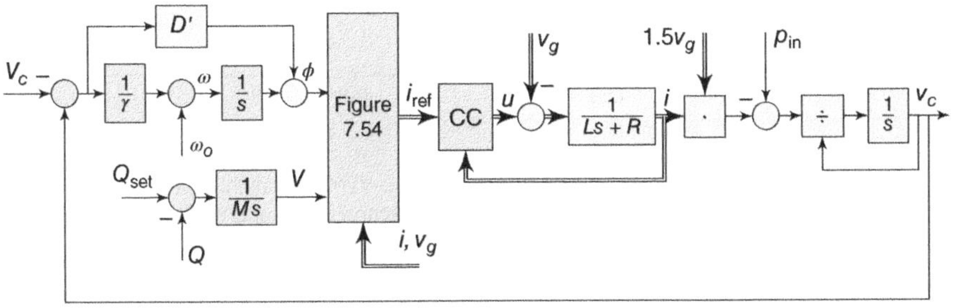

Figure 7.76 Obtaining inertia from capacitor: VSM approach.

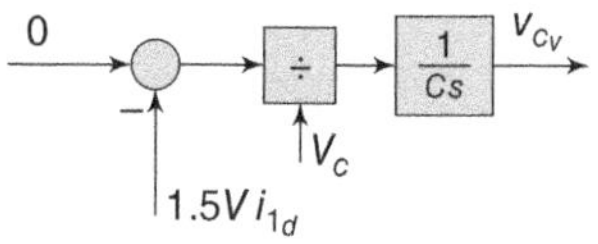

Figure 7.77 Pre-synchronization mechanism for VSM approach.

where i_{1_d} is shown in Figure 7.54. Figure 7.77 shows this equation.[65] Now, during pre-synchronization, i.e. $0 \leq t \leq t_{\text{start}}$, the virtual voltage v_{c_v} is used in the control loop instead of the actual voltage v_c. Once the voltage is synchronized, the inverter starts, and at the same time, the actual capacitor voltage is used in the control loop.

Example 7.8 ***Inertia Extraction from Capacitor: Non-VSM Approach***
The approach of Figure 7.75 using the system and controller parameters of Example 7.7 is studied here. The inertia constant γ is selected at 25 meaning that every $\Delta\omega$ rad/s change in the frequency translates to $\Delta v_c = 25\Delta\omega$ V change in the capacitor voltage. The reactive current reference I_q is determined according to $I_q = -\frac{Q_{\text{ref}}}{1.5V_{g_p}}$ where $Q_{\text{ref}} = Q_{\text{set}} + K_v(V_o - V_{g_p})$ in which V_{g_p} is the grid voltage magnitude (positive-sequence) taken from the PLL. The constant K_v is chosen as $K_v = \frac{S}{0.2V_o} = 417$ VAr/V and Q_{set} is a set-point for reactive power.

Figure 7.78 shows the starting and also real/reactive power tracking performances of the converter. Smooth and accurate responses are observed.

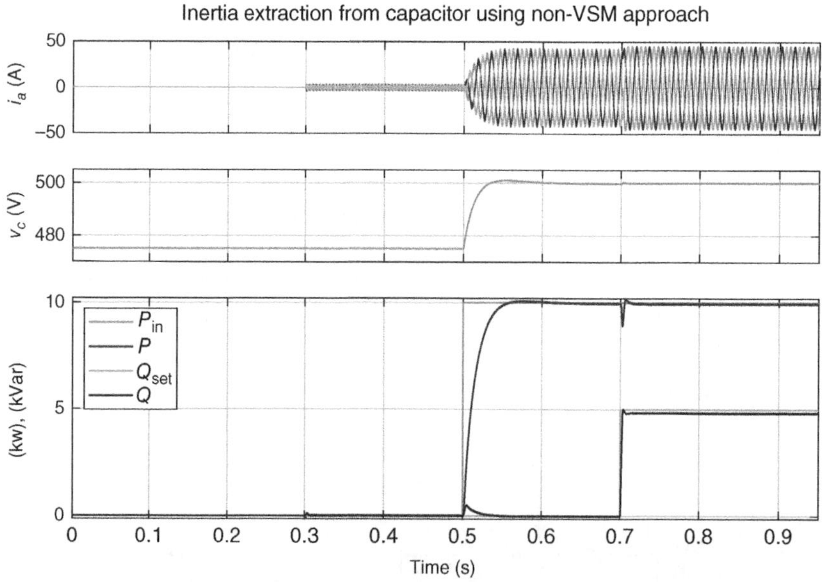

Figure 7.78 Responses of capacitor voltage controller with inertia: non-VSM approach. Starting synchronization during $0 \leq t < 0.3$ seconds; inverter starts at $t = 0.3$ seconds; real and reactive power set-points at $t = 0.5$ and 0.7 seconds.

65 Notice that we have replaced the division by v_{c_v} with V_c (which is the nominal value of capacitor voltage) to obtain smoother transients.

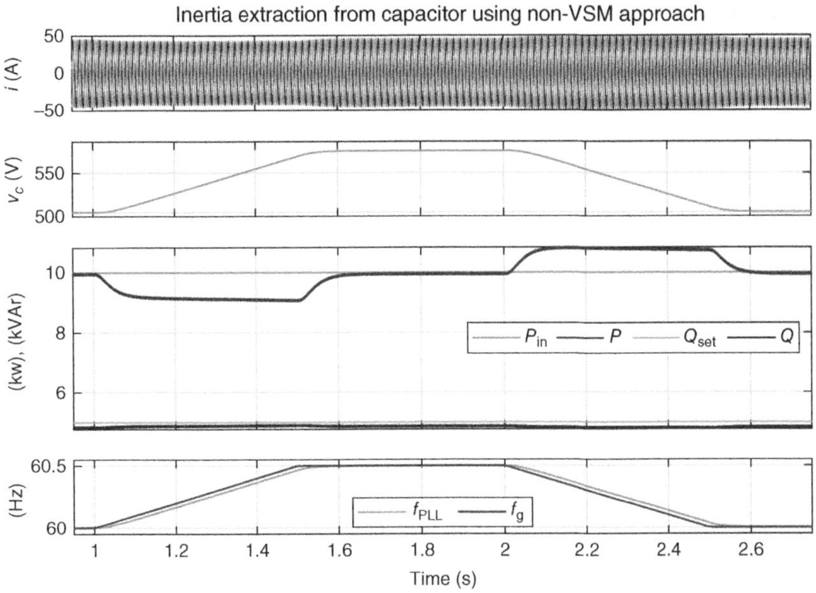

Figure 7.79 Responses of capacitor voltage controller with inertia: non-VSM approach. Grid frequency swing during $1 \leq t < 2.5$ seconds.

Figure 7.79 shows its responses to a grid frequency swing of 0.5 Hz during the time $1 < t < 2.5$ seconds. Figure 7.79 shows that the converter provides an inertia power response to the frequency changes. Notice that a frequency offset (with no change, i.e. during $1.5 < t < 2$ seconds) does not generate a power. The inertia power is about 900 W for a frequency ramp of 1 Hz/s from Figure 7.79. This complies well with the theoretical calculation of $Cv_c\dot{v}_c \approx C\gamma V_c\dot{\omega} = 0.01 \times 25 \times 500 \times 2\pi = 785$. The emulated moment of inertia is $J = \frac{\gamma C V_c}{\omega_o} = \frac{25\times0.01\times500}{120\pi} = 0.33$ which corresponds to an inertia constant of $H = \frac{0.5J\omega_o^2}{S} = 1.7$ seconds (at $S = 10\sqrt{2}$ kVA). This indicates a rather large level of emulated inertia while a modest size of capacitor is used.

Figure 7.80 shows its responses to a grid voltage swing of 5% (8.5 V) during the time $3 < t < 4.5$ seconds. Figure 7.80 shows that the converter provides a volt-VAr (reactive power) support in response to this swing. According to Figure 7.80, a reactive power of about 3.5 kVAr is generated for a voltage offset of 8.5 V (i.e. during $3.5 \leq t \leq 4$ seconds) which complies well with $K_v(V_o - V_g) = 417 \times 8.5$.

Example 7.9 ***Inertia Extraction from Capacitor: VSM Approach***

The approach of Figure 7.76 using the system and CC parameters of Example 7.7 is studied here. The inertia constant γ is selected at 25 (to produce results comparable with Example 7.8). The controller gains are $D' = DCV_c = \frac{\mu C V_c}{V_o^2}$ and $M = \frac{\mu}{V_o}$ where $V_c = 500$ is the nominal capacitor voltage and $V_o = 120\sqrt{2}$ is the nominal peak grid voltage. The virtual impedance is designed using (7.87) for $k_1 = 0.1$ and $k_2 = 0.5$ which results in $L_v = 0.725$ mH and $R_v = 0.137\ \Omega$. The gain μ is designed using the root-locus approach and the corresponding equation (7.86) for $\theta = 0$.[66] The value of μ

66 Here, when the inertia extraction from capacitor is performed using the approach of Figure 7.76, the real power corresponds to the frequency and the reactive power to magnitude, thus θ must be set to zero.

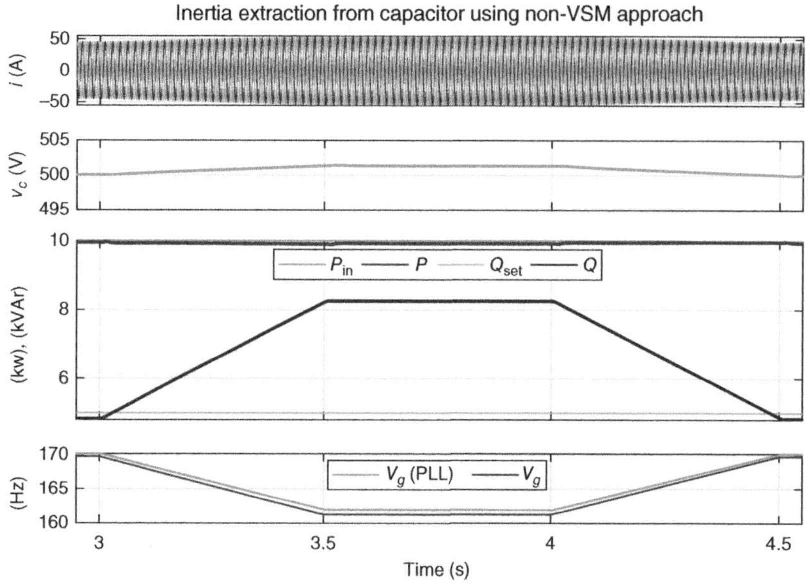

Figure 7.80 Responses of capacitor voltage controller with inertia: non-VSM approach. Grid voltage swing during $3 \leq t < 4.5$ seconds.

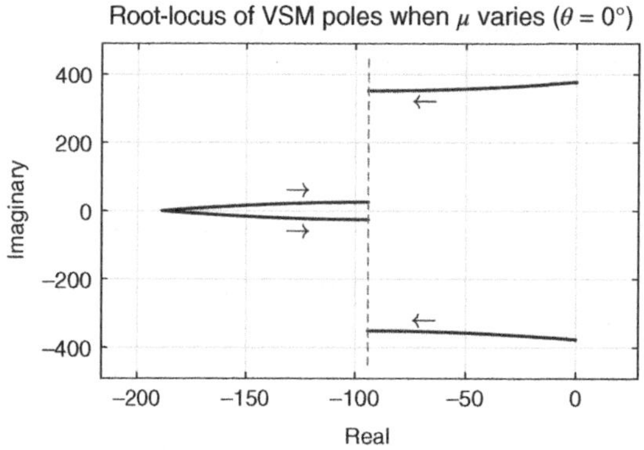

Figure 7.81 Variation of the closed-loop poles, i.e. eigenvalues of the matrix in (7.85), when μ varies from zero to (7.86).

will be 17.2 and the poles of system (as shown in Figure 7.81) are at $-95 \pm j25$ and $-95 \pm j352$. The reactive power reference is $Q_{\text{ref}} = Q_{\text{set}} + K_v(V_o - V_g)$ in which V_g is the grid voltage and the constant K_v is chosen as $K_v = \frac{S}{0.2V_o} = 417$ VAr/V.

Figure 7.82 shows the starting and also real/reactive power tracking performances of the converter. Smooth and accurate responses are observed. At the extreme power jump instance ($t = 0.5$ seconds), there is a transient of about one cycle where the current is also distorted. This is due to the nonlinearities as well as the rather slower design used.

Figure 7.83 shows its responses to a grid frequency swing of 0.5 Hz during the time $1 < t < 2.5$ seconds. Figure 7.83 shows that the converter provides an inertia power response to the frequency changes. The responses are similar to Figure 7.79 and the explanation given there applies here as well.

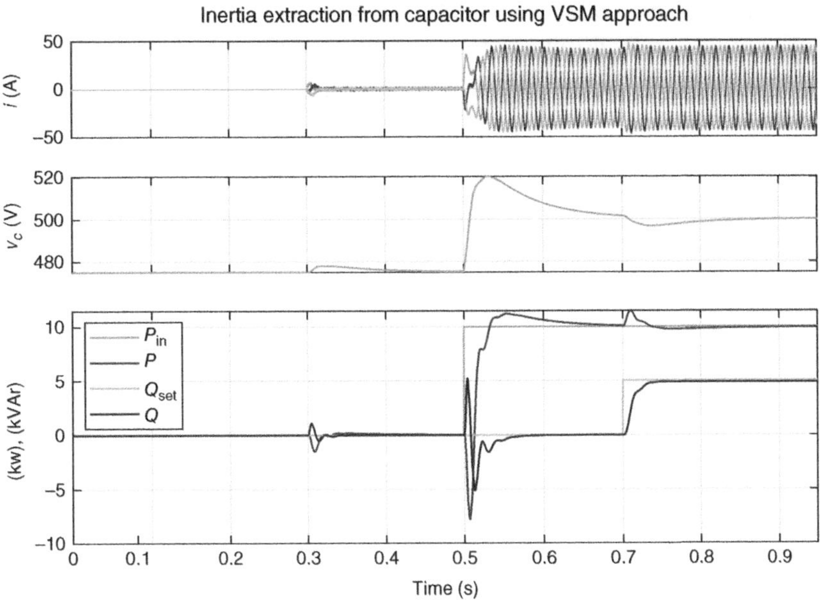

Figure 7.82 Responses of capacitor voltage controller with inertia: VSM approach. Starting synchronization during $0 \leq t < 0.3$ seconds; inverter starts at $t = 0.3$ seconds; real and reactive power set-points at $t = 0.5$ and 0.7 seconds.

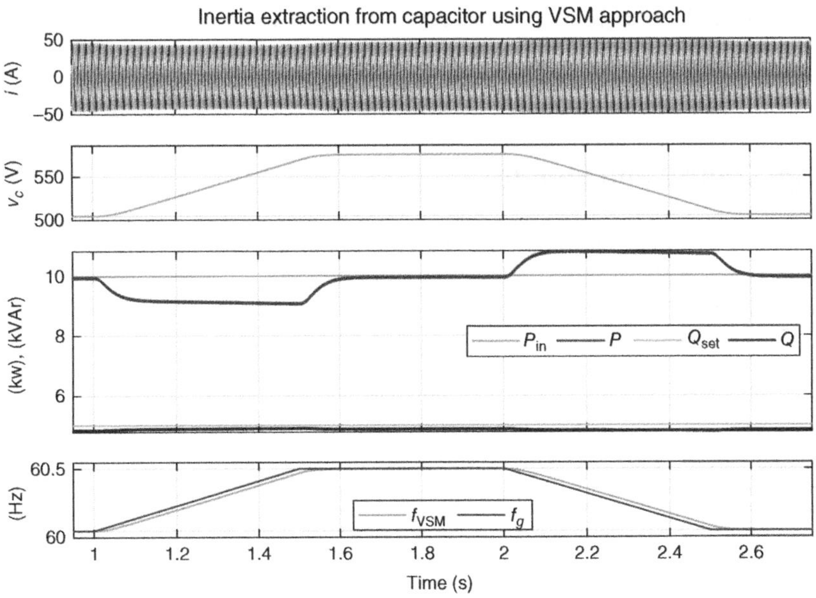

Figure 7.83 Responses of capacitor voltage controller with inertia: VSM approach. Grid frequency swing during $1 \leq t < 2.5$ seconds.

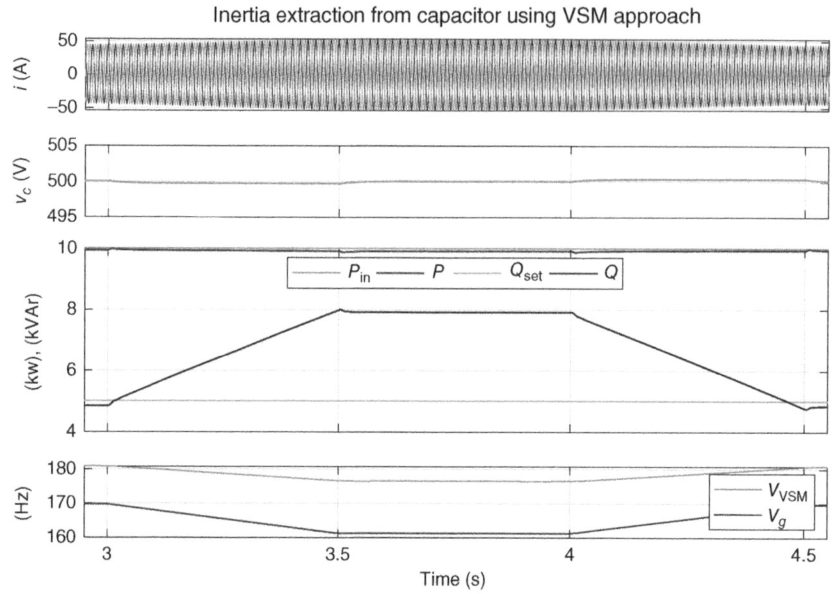

Figure 7.84 Responses of capacitor voltage controller with inertia: VSM approach. Grid voltage swing during $3 \le t < 4.5$ seconds.

Figure 7.84 shows its responses to a grid voltage swing of 5% (8.5 V) during the time $3 < t < 4.5$ seconds. Figure 7.84 shows that the converter provides a volt-VAr (reactive power) support in response to this swing. Again, the responses are similar to Figure 7.80 and the explanation given there applies here as well.

7.7 Load Voltage Control and Support

A three-phase VSC connected to an ac load is shown in Figure 7.85 where L and C constitute the interfacing output filter (per each phase of the three-phase system).[67] The dc side voltage is v_c. The voltage across the load is $v_g(t)$ and the converter current is $i(t)$.[68]

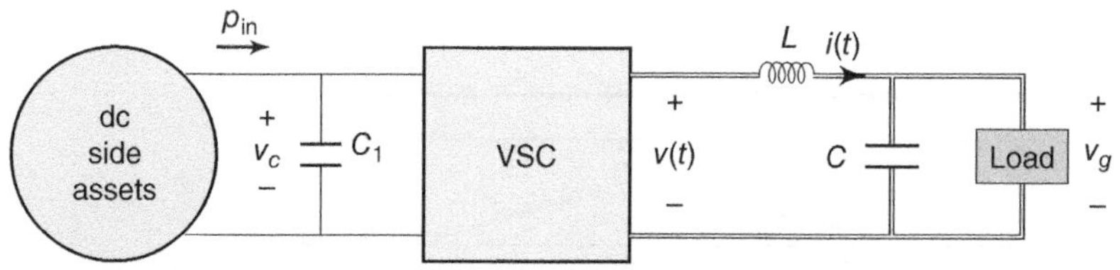

Figure 7.85 Three-phase VSC connected to a three-phase load.

67 The parasitic resistances of the filter components do not impact the control system studies in this section and are not shown for simplicity.
68 We have used the same notation v_g for the ac side voltage that we used for grid-connected mode. This will simplify the notations and also better allows the subsequent discussions and transition to the derivation of grid-forming controllers.

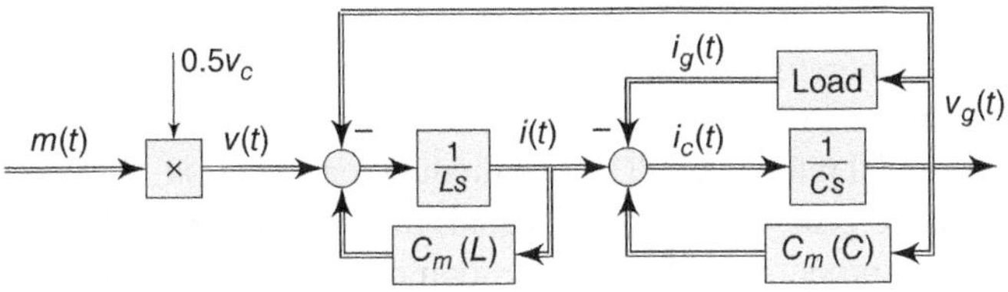

Figure 7.86 Control block diagram of the VSC connected to an ac load.

The control objectives can be stated as follows.

- Control the output voltage v_g: a three-phase (balanced) sinusoid with regulated magnitude and frequency at the desired values V_o and ω_o.
- Maintain this objective robustly despite system uncertainties and changes including those in the filter (L, C), load, and the dc side.
- VSC should respond to load variations in a quick yet smooth manner.
- Prevent converter over-current.

The control block diagram of the system is shown in Figure 7.86. This diagram is valid for both $\alpha\beta$ and dq frames with the note that the coupling matrix C_m is zero in $\alpha\beta$ and is equal to $C_m(x) = \begin{bmatrix} 0 & x\omega \\ -x\omega & 0 \end{bmatrix}$ in dq domain for x being L or C.

State space equations of the system of Figure 7.86 are listed as

$$\begin{aligned} L\tfrac{d}{dt}i(t) &= v(t) - v_g(t) + C_m(L)i(t) \\ C\dot{v}_g(t) &= i(t) - i_g(t) + C_m(C)v_g(t), \end{aligned} \tag{7.102}$$

where $v(t) = 0.5v_c m(t)$ and each variable (including i, v, v_g, and i_g) is a two-dimensional vector, either in $\alpha\beta$ or in dq frame. Equation set (7.102) comprises four first-order differential equations. In the stationary ($\alpha\beta$) domain, the coupling matrix C_m does not exist. Therefore, the α-axis equations are decoupld from the β-axis equations and the system reduces to two SISO system. In the synchronous (rotating) (or dq) domain, however, the matrix C_m causes the equations to become coupled. The first coupling term, i.e. $C_m(L)i(t)$, can be cancelled at the input. The second one, i.e. $C_m(C)v_g(t)$, cannot be easily cancelled. Therefore, one approach is simply to ignore it and design a SISO controller for it. The other approach will be to design a MIMO controller. All these controllers are discussed in this section.

Regarding the load current $i_g(t)$, two approaches may be considered: if the load model is known, it can be included, otherwise, it can be considered as a disturbance which the controller rejects it and resists its changes.[69]

7.7.1 Direct Voltage Control Approach

In the direct voltage control (DVC) approach, the output voltage is controlled directly without an internal CFL. This means that the controller is not explicitly equipped to limit the converter current. The DVC can be done in stationary or in synchronous frame. In the stationary frame, two identical SISO control loops are used because the two axes are decoupled. In the rotating frame, the controller can be a SISO or a MIMO as discussed in this section.

69 Notice that, theoretically speaking, i_g is not a disturbance (with the classical definition) because it is a function of the output voltage v_g. However, since the controller is supposed to robustly stabilize v_g, treating i_g as a disturbance seems acceptable as the simulations confirm.

7.7.1.1 DVC in Stationary Frame

In this frame, the variables are sinusoidal. Therefore, a resonant controller (on the output feedback error) combined with state feedback as shown in Figure 7.87 may be used for each channel α and β to achieve accurate tracking of the sinusoidal reference.[70] The two modulation signals m_α and m_β will then be transformed to *abc* and used to generate the gating signals of the converter switches. The controller in Figure 7.87 is structured properly to allow application of the LQT method. The ac voltage reference is denoted by $v_o(t)$ which may be expressed as $v_{o_\alpha}(t) = V_o\cos(\omega_o t)$ and $v_{o_\beta}(t) = V_o\sin(\omega_o t)$, as shown in Figure 7.88.

State space differential equations of this system may be expressed as

$$\dot{x}_1(t) = x_2(t) \qquad \dot{x}_2(t) = -\omega_o^2 x_1(t) + x_4(t) - v_o(t)$$
$$\dot{x}_3(t) = -\tfrac{1}{L}x_4(t) + \tfrac{1}{L}u(t) \quad \dot{x}_4(t) = \tfrac{1}{C}x_3(t) - \tfrac{1}{C}i_g(t),$$

where $x_4 = v_g$, $x_3 = i$, and x_2 and x_1 are the state variables of the resonant controller. The control input is

$$u(t) = -k_1x_1(t) - k_2x_2(t) - k_3x_3(t) - k_4x_4(t) = -Kx(t)$$

which is in the standard form of a linear full state feedback law. Using the LQT approach, the operator $\frac{d^2}{dt^2} + \omega_o^2$ is applied to arrive at[71]

$$\dot{z}_1(t) = z_2(t) \qquad \dot{z}_2(t) = -\omega_o^2 z_1(t) + z_4(t)$$
$$\dot{z}_3(t) = -\tfrac{1}{L}z_4(t) + \tfrac{1}{L}w(t) \quad \dot{z}_4(t) = \tfrac{1}{C}z_3(t),$$

where

$$z_i(t) = \left(\frac{d^2}{dt^2} + \omega_o^2\right)x_i(t),\ i = 1,2,3,4, \quad w(t) = \left(\frac{d^2}{dt^2} + \omega_o^2\right)u(t).$$

Notice particularly that

$$z_1(t) = e(t),\ z_2(t) = \dot{e}(t),$$

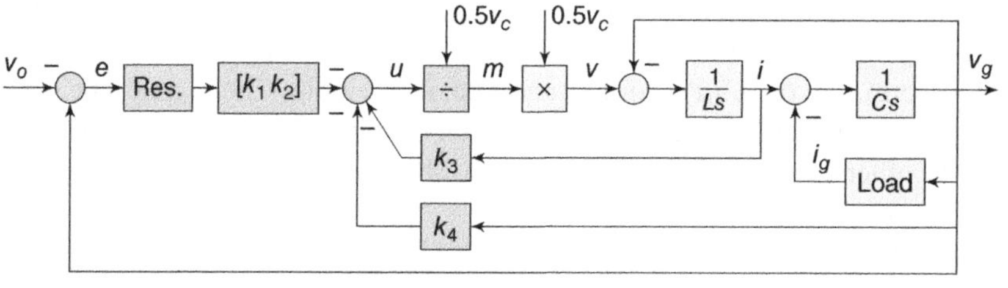

Figure 7.87 Direct voltage control in stationary frame (two identical channels for α and β).

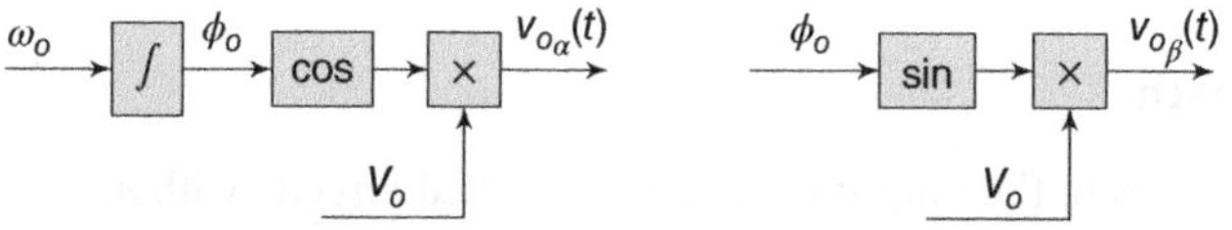

Figure 7.88 Voltage reference generation.

---

70 The division by $0.5v_c$ may be replaced by $0.5V_c$ (its rated value) if this voltage is stable.

71 Notice that, rigorously speaking, this transformation does not null the load current because it is not an external signal. Thus, the presented analysis is approximate as mentioned. Furthermore, the load current is distorted if the load is nonlinear. In this case, the control structure and the applied transformation can be enhanced (if need be) to ensure a clean output voltage. All this analysis and design procedure follows the same steps that were presented for grid-connected inverters with distorted grid and are not detailed here.

where e is the tracking error. In order to minimize the cost function

$$J = \int_0^{\infty} \left[q e(t)^2 + q_2 \dot{e}(t)^2 + q_3 z_3(t)^2 + q_4 z_4(t)^2 + w(t)^2 \right] dt,$$

we choose

$$A = \begin{bmatrix} 0 & 1 & 0 & 0 \\ -\omega_o^2 & 0 & 0 & 1 \\ 0 & 0 & 0 & -\frac{1}{L} \\ 0 & 0 & \frac{1}{C} & 0 \end{bmatrix}, \; B = \begin{bmatrix} 0 \\ 0 \\ \frac{1}{L} \\ 0 \end{bmatrix}, \; Q = \begin{bmatrix} q & 0 & 0 & 0 \\ 0 & q_2 & 0 & 0 \\ 0 & 0 & q_3 & 0 \\ 0 & 0 & 0 & q_4 \end{bmatrix}$$

and solve in Matlab: ≫ `K = lqr(A, B, Q, 1)`.

Example 7.10 ***Direct Voltage Control in Stationary Frame***
An example of a design is shown in Figure 7.89. The system parameters are chosen as $L = 2$ mH, $C = 40\ \mu$F, and $\omega_o = 120\pi$ rad/s. The Q matrix elements are consecutively adjusted as specified in Figure 7.89. At the final point, the eigenvalues are located at about $-1530 \pm j3370$, $-790 \pm j546$. The controller gains are $K = [7.9e5 \;\; 1905 \;\; 9.28 \;\; 0.55]$.[72]

A sample of time-responses of the designed inverter is shown in Figure 7.90. The inverter starts at $t = 0.02$ seconds at no load condition. A parallel RL load (12 kW, 9 kVAr) is switched on at $t = 0.04$ seconds. The inverter succeeds to generate a sinusoidal voltage with 120 rms (line-to-neutral) and 60 Hz frequency across the load terminals. At $t = 0.08$ seconds, the load terminals are short-circuited through a small impedance of 3 Ω. This causes the converter to increase its current to over 100 A to be able to maintain the voltage at the given level of 120 V (rms). Thus, the converter will be subject to an over-current.[73] At $t = 0.12$ seconds, the short-circuit clears and the current comes back to normal.

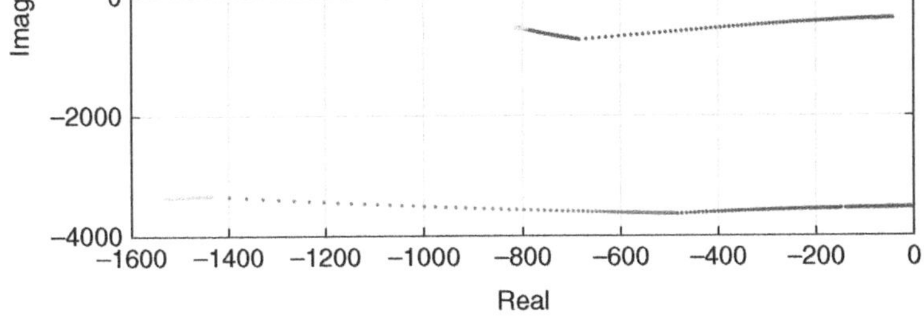

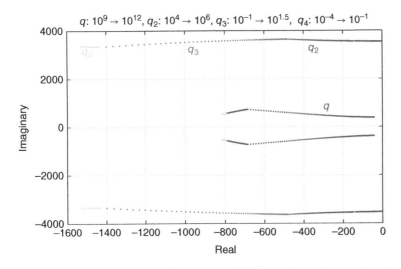

Figure 7.89 Evolution of closed-loop poles of DVC in stationary frame when elements of Q vary.

72 Notice that the LC resonance mode is actively damped by the controller. The pole location $-1530 \pm j3370$ indicates a damping ratio of over 0.4.
73 Assuming a converter current limit of 70 A.

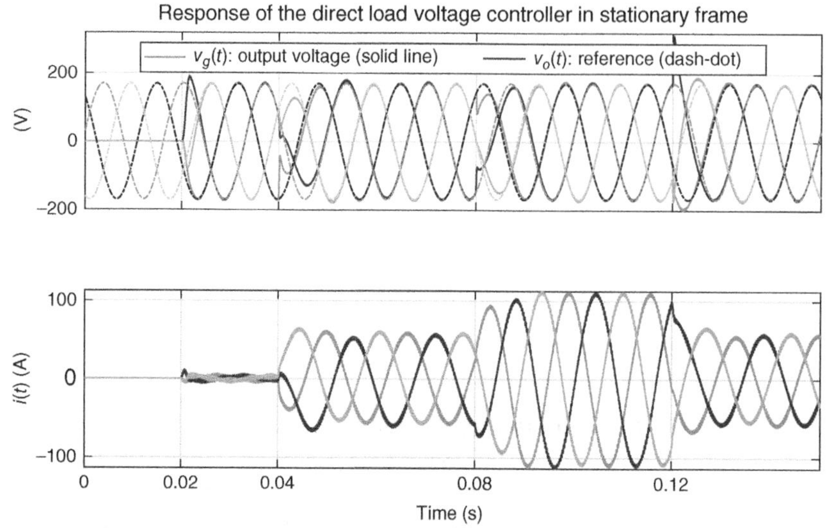

Figure 7.90 Responses of the DVC in stationary frame: inverter starts at $t = 0.02$ seconds (no load), load is switched on at $t = 0.04$ seconds, a nonzero impedance short-circuit occurs at $t = 0.08$ seconds, clears at $t = 0.12$ seconds. Over-current happens during the short-circuit condition.

7.7.1.2 DVC in Rotating Frame

State space equations of the system of Figure 7.86 are listed in (7.102) which, for the rotaing frame, are rewritten as

$$\begin{aligned} L\frac{d}{dt}i_d(t) &= v_d(t) - v_{g_d}(t) + L\omega i_q(t) \\ L\frac{d}{dt}i_q(t) &= v_q(t) - v_{g_q}(t) - L\omega i_d(t) \\ C\dot{v}_{g_d}(t) &= i_d(t) - i_{g_d}(t) + C\omega v_{g_q}(t) \\ C\dot{v}_{g_q}(t) &= i_q(t) - i_{g_q}(t) - C\omega v_{g_d}(t), \end{aligned} \tag{7.103}$$

where $v_d = 0.5v_c m_d$ and $v_q = 0.5v_c m_q$. The obvious choice for the dq transformation angle is $\phi_o = \int \omega_o dt$ shown in Figure 7.88. There are two approaches as far as controlling this system is concerned: SISO approach and MIMO approach.

SISO-DVC Approach in Rotating Frame Here the coupling terms associated with the factor $L\omega$ are compensated at the input but those associated with the factor $C\omega$ are simply ignored. Therefore, the equations are reduced/approximated to

$$\begin{aligned} L\frac{d}{dt}i_d(t) &= v_{d_1}(t) - v_{g_d}(t), \quad v_{d_1} = v_d + L\omega i_q \\ C\dot{v}_{g_d}(t) &= i_d(t) - i_{g_d}(t) \\ L\frac{d}{dt}i_q(t) &= v_{q_1}(t) - v_{g_q}(t), \quad v_{q_1} = v_q - L\omega i_d \\ C\dot{v}_{g_q}(t) &= i_q(t) - i_{g_q}(t), \end{aligned} \tag{7.104}$$

which indicates two decoupled SISO systems. Therefore, they can be controlled (using a simple integrating compensator) as shown in Figure 7.91 by two similar controllers for d and q axes. The only differences between these two controllers is in their reference values (V_o versus 0) and the sign of decoupling term.

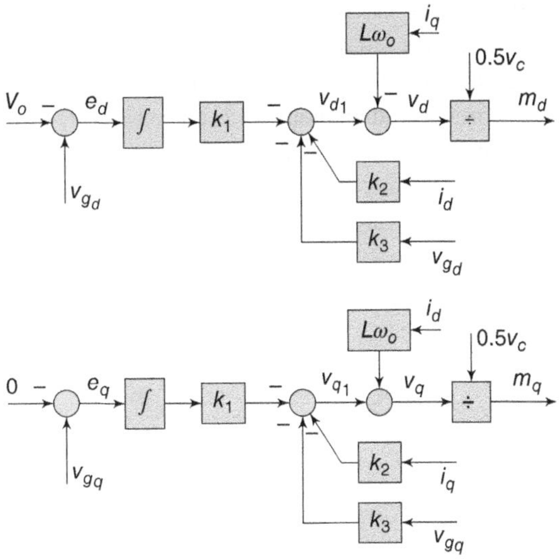

Figure 7.91 Direct voltage control in rotating frame (SISO approach).

Design of k_1 to k_3 may be done using the LQT approach. To this end, the state space differential equations are expressed as[74]

$$\dot{x}_1(t) = x_3(t) - V_o$$
$$\dot{x}_2(t) = -\tfrac{1}{L}x_3(t) + \tfrac{1}{L}v_{d_1}(t)$$
$$\dot{x}_3(t) = \tfrac{1}{C}x_2(t) - \tfrac{1}{C}i_{g_d}(t),$$

where $x_3 = v_{g_d}$, $x_2 = i_d$, and x_1 is the integrator's output in the controller. The control input is

$$v_{d_1}(t) = -k_1x_1(t) - k_2x_2(t) - k_3x_3(t) = -Kx(t)$$

which is in the standard form of a linear full state feedback law. Using the LQT approach, the operator $\frac{d}{dt}$ is applied to arrive at

$$\dot{z}_1(t) = z_3(t)$$
$$\dot{z}_2(t) = -\tfrac{1}{L}z_3(t) + \tfrac{1}{L}w(t)$$
$$\dot{z}_3(t) = \tfrac{1}{C}z_2(t),$$

where

$$z_i(t) = \frac{d}{dt}x_i(t),\ i = 1,2,3,4,\quad w(t) = \frac{d}{dt}u(t).$$

Notice particularly that $z_1(t) = e_d(t)$ is the tracking error on channel d. In order to minimize the cost function

$$J = \int_0^{\infty} \left[qe_d(t)^2 + q_2z_2(t)^2 + q_3z_3(t)^2 + w(t)^2\right] dt,$$

we choose

$$A = \begin{bmatrix} 0 & 0 & 1 \\ 0 & 0 & -\frac{1}{L} \\ 0 & \frac{1}{C} & 0 \end{bmatrix},\quad B = \begin{bmatrix} 0 \\ \frac{1}{L} \\ 0 \end{bmatrix},\quad Q = \begin{bmatrix} q & 0 & 0 \\ 0 & q_2 & 0 \\ 0 & 0 & q_3 \end{bmatrix}$$

and solve in MATLAB: ≫ `K = lqr(A, B, Q, 1)`.

---

74 The equations are written for the d axis. For the q axis, V_o must be replaced with 0. The design process will be the same for both axes.

Example 7.11 ***Direct Voltage Control in Rotating Frame: SISO Approach***

For the same system parameters of Example 7.10, Figure 7.92 show the evolution of closed-loop poles when the Q matrix elements are consecutively adjusted as shown. At the final point, the eigenvalues are located at about $-1534 \pm j3364$, -914 (around the same area as those of Example 7.10). The controller gains are $K = [1000 \quad 7.96 \quad 0.318]$.[75]

A sample of time-responses of the designed inverter is shown in Figure 7.93. The inverter starts at $t = 0.02$ seconds at no load condition. The simulation scenario is the same as defined in Figure 7.90 and the responses of both systems are much similar.

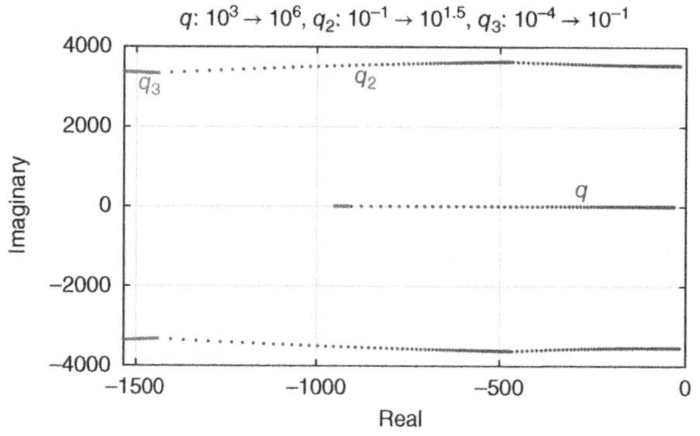

Figure 7.92 Evolution of closed-loop poles of DVC in rotating frame when elements of Q vary: SISO approach.

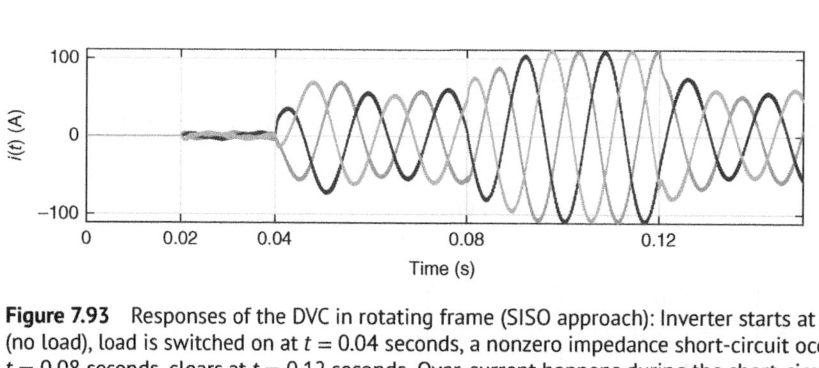

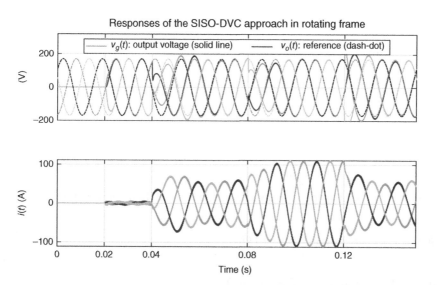

Figure 7.93 Responses of the DVC in rotating frame (SISO approach): Inverter starts at $t = 0.02$ seconds (no load), load is switched on at $t = 0.04$ seconds, a nonzero impedance short-circuit occurs at $t = 0.08$ seconds, clears at $t = 0.12$ seconds. Over-current happens during the short-circuit condition.

75 Notice that the LC resonance mode is actively damped by the controller. The pole location $-1534 \pm j3364$ indicates a damping ratio of over 0.4.

MIMO-DVC Approach in Rotating Frame Here the complete system equations given in (7.103) are used without any decoupling or approximation. A two-input two-output integrating compensator combined with a state-feedback compatible with LQT is used as shown in Figure 7.94. Here, the controller gain K is a matrix of dimensions 2×6.

Design of k may optimally be done using the LQT approach. To this end, the state space differential equations are expressed as

$$
\begin{aligned}
&\dot{x}_1(t) = x_5(t) - V_o && \dot{x}_2(t) = x_6(t) - 0 \\
&\dot{x}_3(t) = \omega x_4(t) - \tfrac{1}{L}x_5(t) + \tfrac{1}{L}v_d(t) && \dot{x}_4(t) = -\omega x_3(t) - \tfrac{1}{L}x_6(t) + \tfrac{1}{L}v_q(t) \\
&\dot{x}_5(t) = \tfrac{1}{C}x_3(t) + \omega x_6(t) - \tfrac{1}{C}i_{g_d}(t) && \dot{x}_6(t) = \tfrac{1}{C}x_4(t) - \omega x_5(t) - \tfrac{1}{C}i_{g_q}(t),
\end{aligned}
$$

where $x_6 = v_{g_q}, x_5 = v_{g_d}, x_4 = i_q, x_3 = i_d$, and x_2 and x_1 are the integrators' outputs on q and d channels, respectively. The control input is $v_{dq}(t) = -Kx(t)$ which is in the standard form of a linear full state feedback law. Using the LQT approach, the operator $\frac{d}{dt}$ is applied to arrive at

$$
\begin{aligned}
&\dot{z}_1(t) = z_5(t) && \dot{z}_2(t) = z_6(t) \\
&\dot{z}_3(t) = \omega z_4(t) - \tfrac{1}{L}z_5(t) + \tfrac{1}{L}w_d(t) && \dot{z}_4(t) = -\omega z_3(t) - \tfrac{1}{L}z_6(t) + \tfrac{1}{L}w_q(t) \\
&\dot{z}_5(t) = \tfrac{1}{C}z_3(t) + \omega z_6(t) && \dot{z}_6(t) = \tfrac{1}{C}z_4(t) - \omega z_5(t),
\end{aligned}
$$

where

$$z_i(t) = \frac{d}{dt}x_i(t),\ i = 1, \ldots, 6, \quad w(t) = \frac{d}{dt}v(t).$$

Notice particularly that $z_1(t) = e_d(t)$ and $z_2(t) = e_q(t)$ are the tracking errors on channels d and q, respectively. In order to minimize the cost function[76]

$$J = \int_0^\infty \left\{q\left[e_d(t)^2 + e_q(t)^2\right] + q_2\left[z_3(t)^2 + z_4(t)^2\right] + q_3\left[z_5(t)^2 + z_6(t)^2\right] + w_d(t)^2 + w_q(t)^2\right\} dt,$$

we choose

$$A = \begin{bmatrix} 0 & 0 & 0 & 0 & 1 & 0 \\ 0 & 0 & 0 & 0 & 0 & 1 \\ 0 & 0 & 0 & \omega & -\frac{1}{L} & 0 \\ 0 & 0 & -\omega & 0 & 0 & -\frac{1}{L} \\ 0 & 0 & \frac{1}{C} & 0 & 0 & \omega \\ 0 & 0 & 0 & \frac{1}{C} & -\omega & 0 \end{bmatrix}, \quad B = \begin{bmatrix} 0 & 0 \\ 0 & 0 \\ \frac{1}{L} & 0 \\ 0 & \frac{1}{L} \\ 0 & 0 \\ 0 & 0 \end{bmatrix}, \quad Q = \begin{bmatrix} qI_2 & 0 & 0 \\ 0 & q_2I_2 & 0 \\ 0 & 0 & q_3I_2 \end{bmatrix},$$

where I_2 is the identity matrix of order 2. Then, we solve in Matlab: ≫ `K = lqr(A, B, Q, I2)`.

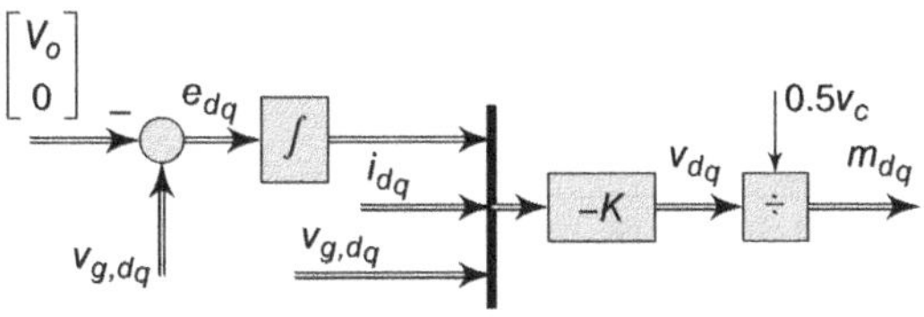

Figure 7.94 Direct voltage control in rotating frame (MIMO approach).

76 In this cost function, we have given equal weight factors to d and q channel variables. Different weights can be chosen for the two channels if need be.

Example 7.12 ***Direct Voltage Control in Rotating Frame: MIMO Approach***
For the same system parameters of Example 7.10, Figure 7.95 shows the evolution of closed-loop poles when the Q matrix elements are adjusted as shown. At the final point, the eigenvalues are located at about $-1534 \pm j3728, -1531 \pm j3004, -919 \pm j29.5$ (around the same area as those of Example 7.11). The controller gains are

$$K = \begin{bmatrix} 993.6 & -112.9 & 7.97 & 0 & 0.319 & -0.028 \\ 112.9 & 993.6 & 0 & 7.97 & 0.028 & 0.319 \end{bmatrix}.$$

A sample of time-responses of the designed inverter is shown in Figure 7.96. The simulation scenario is the same as defined in Figure 7.90 and the responses of systems are much similar.

It is expected that the MIMO controller performs better than SISO approach in more severe working conditions. One example is shown in Figure 7.97 where a LPF with transfer function of

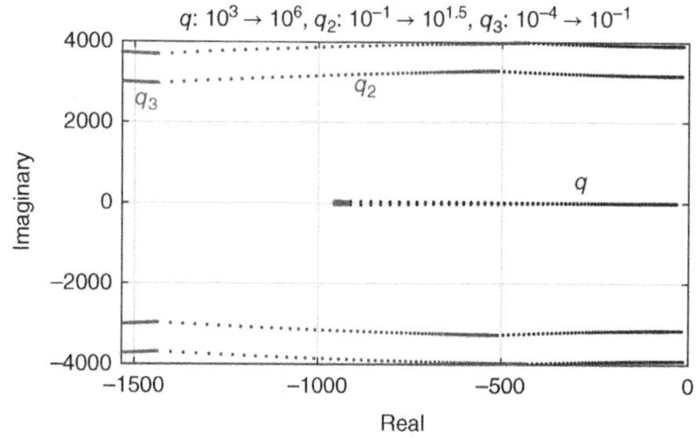

Figure 7.95 Evolution of closed-loop poles of DVC in rotating frame when elements of Q vary: MIMO approach.

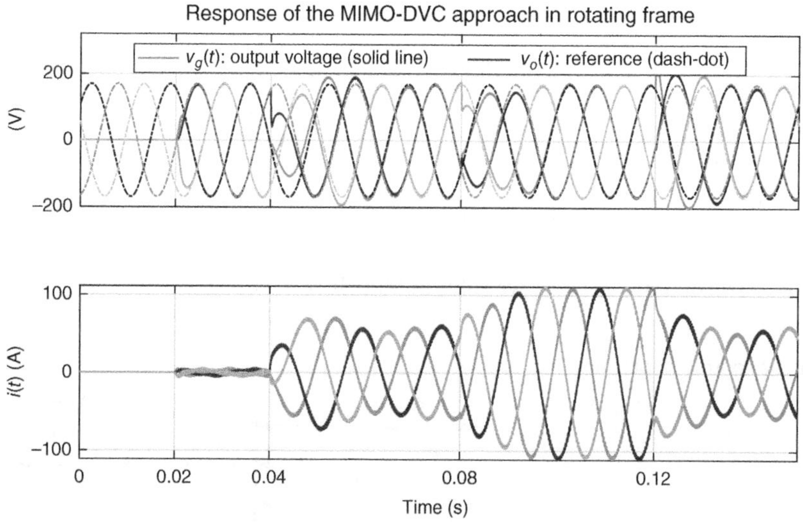

Figure 7.96 Responses of the DVC in rotating frame (MIMO approach): inverter starts at $t = 0.02$ seconds (no load), load is switched on at $t = 0.04$ seconds, a nonzero impedance short-circuit occurs at $t = 0.08$ seconds, clears at $t = 0.12$ seconds.

$\frac{1}{0.000\,75s+1}$ is inserted and operates on the modulation indices m_a, m_b, and m_c before forwarding them to the PWM. This can roughly model a combination of delays and filters that may exist in a practical setting, in addition to the delay caused by a digital controller. Figure 7.97 shows that the MIMO controller can still tolerate this amount of delay. The SISO controller, on the other hand, as shown in Figure 7.98, fails to stabilize the output voltage at no load condition when this delay is present.

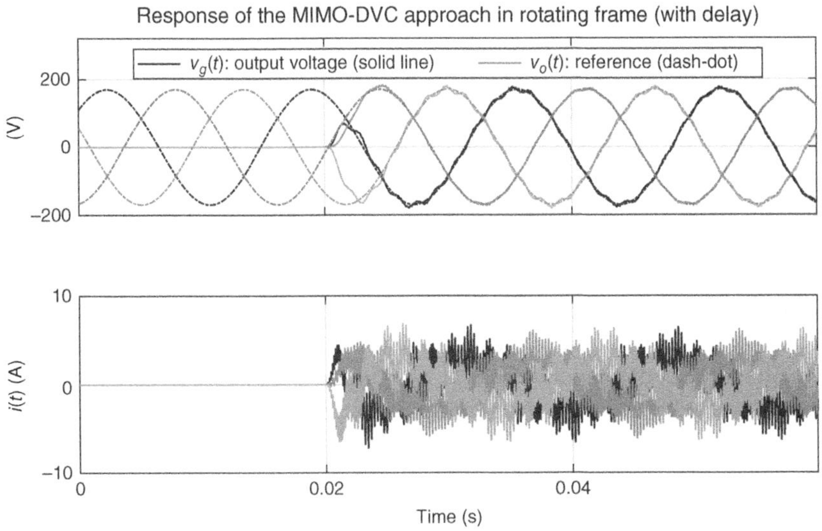

Figure 7.97 Responses of MIMO-DVC in rotating frame when a delay is inserted in the control loop: inverter starts at $t = 0.02$ seconds (no load).

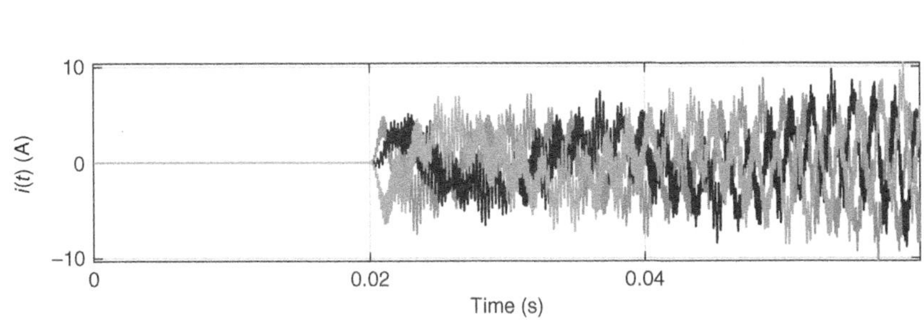

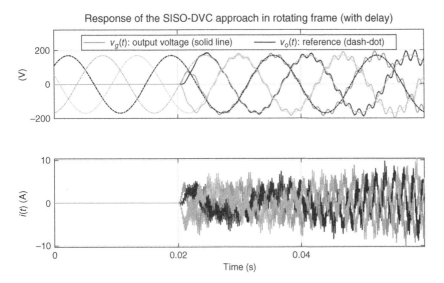

Figure 7.98 Responses of SISO-DVC in rotating frame when a delay is inserted in the control loop: inverter starts at $t = 0.02$ seconds (no load).

7.7.2 Voltage Control with Current Limiting

As illustrated in Section 7.7.1, the DVC has one rather major problem: it cannot limit the converter current during short-term faults or short-circuits happening at the load terminals. In order to achieve converter current limiting, we will need to have a CFL as an internal loop to an external voltage control loop. Moreover, we must allow the load voltage v_g to drop from its reference value v_o during such transient incidents.

To achieve this goal, one approach is to generate the reference current by limiting an intermediary virtual current i_v. This virtual current satisfies[77]

$$v_o - v_g = R_v i_v, \quad i_{\text{ref}} = \text{limit}\ \{i_v\}, \tag{7.105}$$

where R_v is a positive number playing the role of a virtual resistance. During normal operation, i_{ref} and i_v are equal and they are closely followed by i (assuming the CFL is sufficiently fast and accurate). Thus, R_v plays the role of a resistance between the reference voltage v_o and the actual voltage v_g. This means that flow of current will cause v_g to drop below v_o. The virtual resistance R_v must be small enough to keep the voltage within an acceptable range during normal operation.

During transient incidents where v_g tends to drop excessively, i_v also becomes excessively large, as observed from (7.105), while i_{ref} (and hence i) is limited at the maximum value.

The proposed structure is shown in Figure 7.99 where CFL stands for current feedback loop, and CML stands for current magnitude limiting. The CFL may be in stationary frame, discussed in Section 7.3, or in rotating frame, discussed in Section 7.4. The required dq transformations are performed using the reference angle $\phi_o = \int \omega_o dt$ as shown in Figure 7.88.

The CML structure is shown in Figure 7.100 for further clarification. The two synchronizing vectors s_{d_p} and s_{q_p} are calculated using the reference angle ϕ_o as

$$s_{d_p} = \begin{bmatrix} \cos(\phi_o) \\ \sin(\phi_o) \end{bmatrix}, \quad s_{q_p} = \begin{bmatrix} \sin(\phi_o) \\ -\cos(\phi_o) \end{bmatrix}.$$

Example 7.13 ***Voltage Control with Current Limiting***

For the same system studied in the examples of Section 7.7.1, assume that it is desired to limit the converter current under 70 A. Thus, the controller of Figure 7.99 is used. The same CFL gains used in Example 7.1 are used (in stationary domain). The system parameters are chosen as $L = 2$ mH, $C = 40\ \mu$F, $V_o = 120\sqrt{2}$ V, and $\omega_o = 120\pi$ rad/s. A resistance of $R = 5\ \Omega$ is added in series with the filter capacitance for passive damping of resonance mode.[78] Also, the switching frequency is

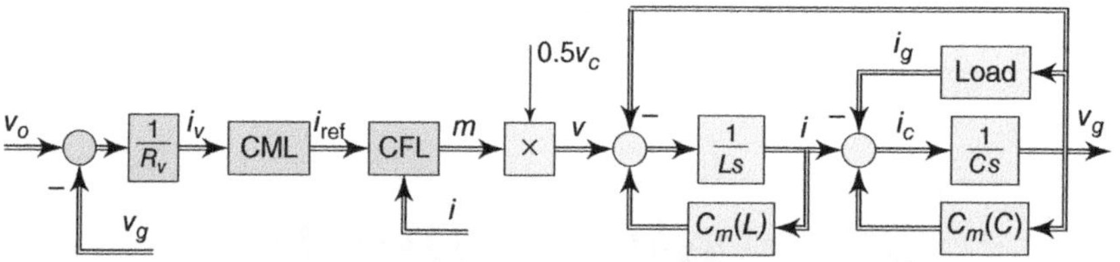

Figure 7.99 Voltage control of VSC connected to ac load with current limiting.

77 In a more general case, an RL impedance may be used which replaces R_v with $L_v s + R_v$.
78 This way, the LC filter characteristics will be $LCs^2 + RCs + 1 = 0$ which indicates a damping ratio of $\zeta = \frac{R}{2}\sqrt{\frac{C}{L}} = 0.35$. This resistance introduces a power loss of below 0.5% at full power of 10 kW. This power loss has two components: at fundamental frequency, and at the switching frequency. Of course, the current controller may also be upgraded, or using aother available approaches, to have active damping if removing this losses is necessary. See Section 7.7.2.1 for an active damping approach.

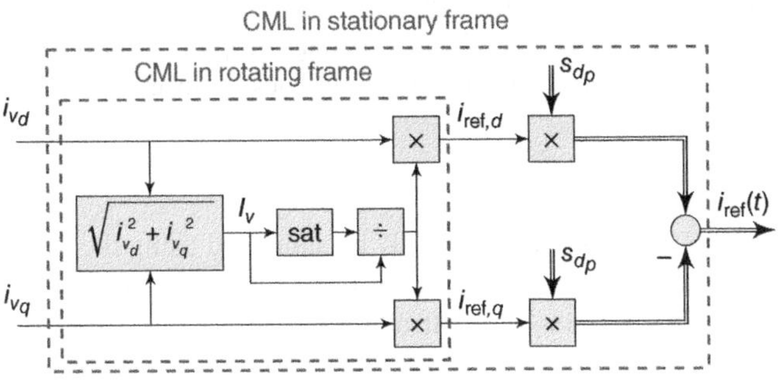

Figure 7.100 Current magnitude limiting (CML).

increased to 8 kHz for better voltage quality.[79] The virtual resistance R_v is selected at 0.5. Assuming that the converter power rating is $10\sqrt{2}$ kVA, its full rated current is about 40 A (rms). This current will cause a voltage drop of $40R_v$, in this case 20 V (rms), across the virtual resistor which is reasonable. The peak rated current is about 56 A, and the limiter is set to 70 A.

A sample of time-responses of the designed inverter is shown in Figure 7.101. The inverter starts at $t = 0.02$ seconds at no-load condition. The inverter generates a sinusoidal voltage with 120 rms and 60 Hz frequency across the output terminals.[80] A series RL load (12 kW, 9 kVAr) is switched on at $t = 0.04$ seconds. The inverter succeeds to generate a sinusoidal voltage with a magnitude about 107 V (rms) across the load terminals.[81] At $t = 0.08$ seconds, the load terminals are short-circuited

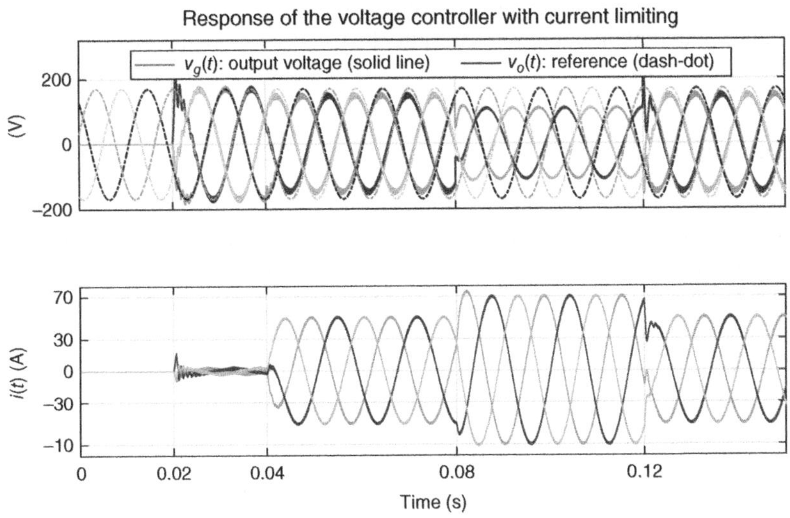

Figure 7.101 Responses of the voltage controller with current limiting: inverter starts at $t = 0.02$ seconds (no load), load is switched on at $t = 0.04$ seconds, a nonzero impedance short-circuit occurs at $t = 0.08$ seconds, clears at $t = 0.12$ seconds. Current is successfully limited at 70 A during the short-circuit condition.

79 The damper resistance added in series with the capacitor causes the switching ripples of the load voltage to go up.
80 Notice that the initial ringing is due to the resonance mode of L and C.
81 The voltage drop is due to $v_o - v_g = R_v i$.

through a small impedance of 3 Ω. This causes the converter to increase its current but is successfully limited to 70 A by the controller. As a result, of course, the load voltage drops to about 110 V (peak). At $t = 0.12$ seconds, the short-circuit clears and the current comes back to normal.

7.7.2.1 Active Damping of Resonance Mode

Active damping of resonance method of *LC* filter can systematically be addressed using the LQT approach as discussed in Section 5.9.2. Here, however, we are presenting a heuristic approach by inspecting how a passive damping can be restructured in the controller to achieve active damping. Figure 7.99 shows the control loop without active damping. The passive damping approach adds a resistor *R* in series with the capacitor, as shown by the green block in Figure 7.102.

In the proposed active damping approach here, the actual resistance is emulated by two control blocks shown in red color in Figure 7.102. The capacitor current i_c is multiplied into the required damping resistance *R* and added to the CFL and also to the voltage feedback to emulate the same effect. The following example shows the simulation results of this method.

Example 7.14 ***Voltage Control with Current Limiting and Active Damping***

For the same system studied in Example 7.13, the active damping approach of Figure 7.102 is implemented in this example. The system parameters are same except that there is no resistance added in series with the filter capacitance. Instead, a value of $R = 5$ is used in the red-colored blocks in Figure 7.102 to emulate a 5 Ω resistor. Also, the switching frequency is lowered back to its original value of 4 kHz.[82]

A sample of time-responses of the designed inverter is shown in Figure 7.103. This is the same simulation scenario as defined in Figure 7.101. The responses are much similar to those of Figure 7.101 and indicate that the emulation process (active damping) is done correctly. The voltage quality is even higher than that of Figure 7.101 despite the fact that the switching frequency is lower.

7.7.3 Deriving Grid-Forming Controllers*

The DVC of Figure 7.87 tightly regulates the output voltage v_g to the reference voltage v_o. The voltage control with current limiting approach shown in Figure 7.99 allows the output voltage v_g to deviate from v_o through the equation $v_g = v_o - (L_v s + R_v)i$. Now, assume a situation that v_g is not exclusively controlled by this converter. Other entities (possibly other converters, loads, and even a

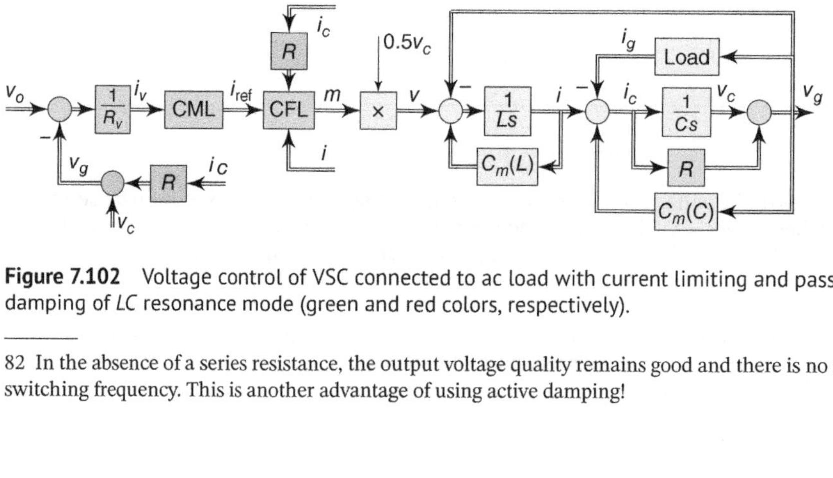

Figure 7.102 Voltage control of VSC connected to ac load with current limiting and passive/active damping of *LC* resonance mode (green and red colors, respectively).

82 In the absence of a series resistance, the output voltage quality remains good and there is no need to increase the switching frequency. This is another advantage of using active damping!

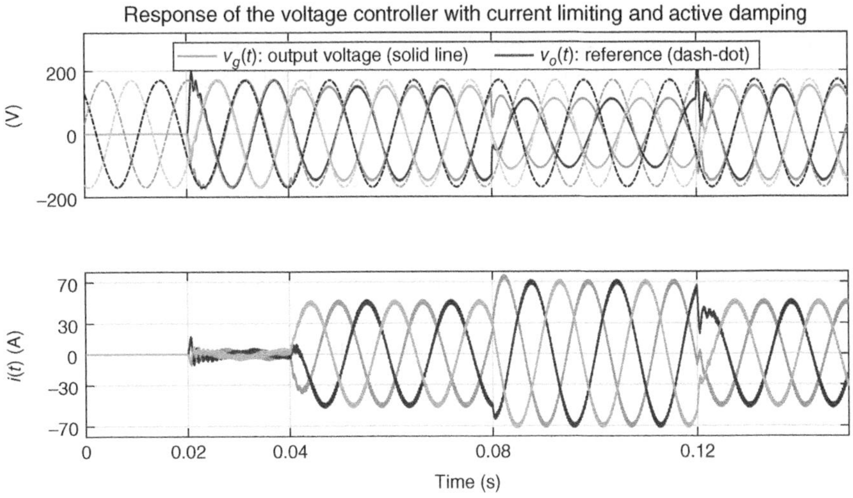

Figure 7.103 Responses of the voltage controller with current limiting and active damping: Inverter starts at $t = 0.02$ seconds (no load), load is switched on at $t = 0.04$ seconds, a nonzero impedance short-circuit occurs at $t = 0.08$ seconds, clears at $t = 0.12$ seconds. Current is successfully limited at 70 A during the short-circuit condition. Resonance is successfully damped.

grid connected at some distance) may be present. In this scenario, the converter should "support" the output voltage rather than "controlling" it. To this end, the reference voltage v_o must be further relaxed to allow a cooperative impact of all entities on v_g. A couple of strategies to determine the output voltage reference v_o is discussed here and shown that this extended voltage control method will converge to the grid forming controllers such as the powers droop approach, or the VSM.

7.7.3.1 Power Droops Strategy

The previous way of generating the voltage reference $v_o(t)$ is shown in Figure 7.88 where its frequency and magnitude are fixed at ω_o and V_o. The common power droop laws manipulate these two quantities according to a linear droop of real and reactive powers expressed by

$$\omega = \omega_o + K_p(P_{\text{set}} - P), \quad V = V_o + K_q(Q_{\text{set}} - Q) \tag{7.106}$$

as shown in Figure 7.104. Here, K_p and K_q are positive, P_{set} and Q_{set} signify some real and reactive power set-points, and P and Q are actual real and reactive powers of the converter.

The reference voltage is then generated according to

$$v_o(t) = \begin{cases} [V \;\; 0]^T, & \text{in rotating frame,} \\ [V\cos(\phi) \;\; V\sin(\phi)]^T, & \text{in stationary frame,} \end{cases} \tag{7.107}$$

where the rotating frame is with the angle ϕ. The droop principles are inspired from the operation of synchronous generators. The virtual frequency ω will be analogous to the rotor speed. It will copy the frequency of the output voltage which may be under the influence of other entities. An

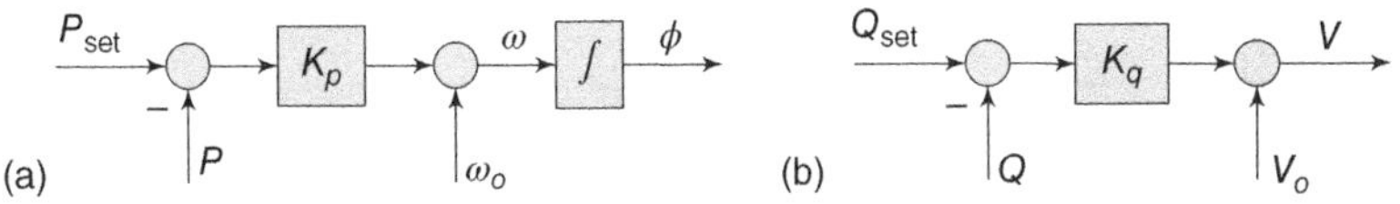

Figure 7.104 Voltage reference generation using the power droops.

increase in this frequency indicates an excess of power in the system, according to the principles of operation of synchronous generators. This will, thus, linearly decrease the converter output power. A similar argument may also be made for the reactive power and its relationship to the voltage magnitude.

As we discussed in Sections 6.3 and 7.5, in an inverter-dominated power system, these relationships need not be held necessarily. For instance, it will be plausible and even advantageous to have a "reverse" droop law between the frequency and reactive power, and also the voltage magnitude and real power. However, if SGs are still present and constitute a sizeable share in the system generation, the conventional droop terms may be combined with the "reverse" ones through a controlling angle as discussed in Sections 6.3 and 7.5.

The real and reactive powers P and Q in Figure 7.104 are calculated according to $P = 1.5Vi_{v_d}$ and $Q = -1.5Vi_{v_q}$ where $i_{v_{dq}}$ are shown in Figure 7.99. It is common to use a simple first-order LPF before applying these powers to the droop laws, as shown in Figure 7.105.

7.7.3.2 Swing Equation Strategy (VSM Approach)

Another approach to generate the reference voltage is based on directly emulating the swing equation of a synchronous machine. This is $P_m - P_e = J\omega\dot{\omega}$ where P_m is the input mechanical power, P_e is the output electrical power, ω is the rotor speed, and J is its moment of inertia. This approach is shown in Figure 7.106. Notice that the division by ω is replaced by ω_o since this frequency is not supposed to change much, anyway. Moreover, inspired from the governor function of the synchronous generator, a droop term $K_f(\omega_o - \omega)$ is added to the power set-point P_{set} to form the total input power (analogous to P_m).

In Figure 7.106, the magnitude dynamics is formed, again inspired from the automatic voltage regulator (AVR) function in a synchronous generator, by adding a droop term $K_v(V_o - V_g)$ to the reactive power set-point. Here, V_g is the measured output voltage magnitude. Figure 7.106 shows the basic structure of what is called the virtual synchronous machine (VSM) and is also called the synchronverter in some literature [10]. In Section 7.7.3.1, we establish the close equivalence between this structure and the filtered-powers droop approach shown in Figure 7.105.

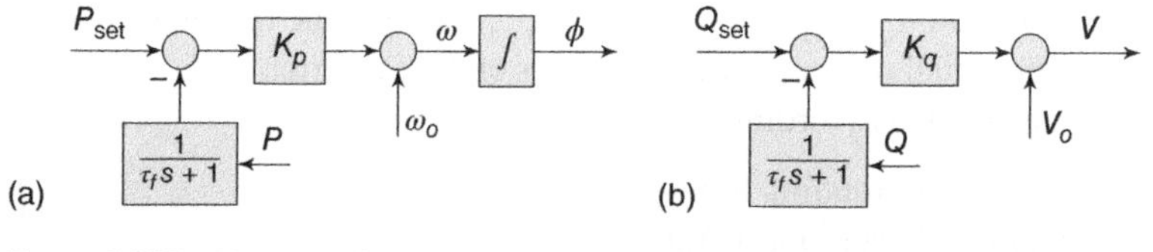

Figure 7.105 Voltage reference generation using the filtered-powers droop approach.

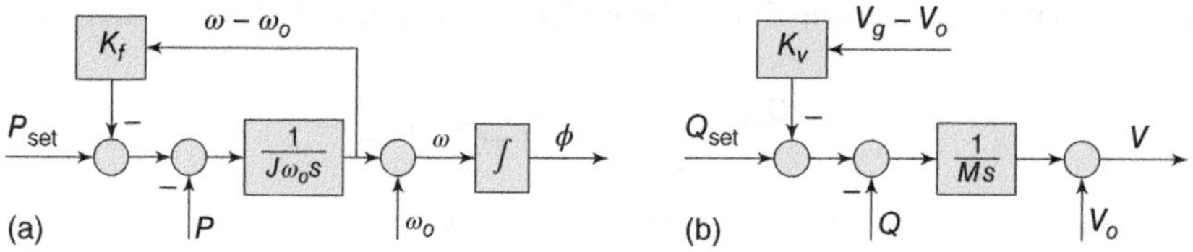

Figure 7.106 Voltage reference generation using the swing equation (VSM).

7.7.3.3 Analogy Between Filtered-Powers Droop and VSM Approaches

In the method based on filtered-powers droops shown in Figure 7.105, the equation describing the dynamics of internal frequency ω is

$$\omega = \omega_o + K_p\left(P_{\text{set}} - \frac{P}{\tau_f s + 1}\right) \Rightarrow \tau_f\dot{\omega} + \omega = \omega_o + K_p(P_{\text{set}} - P). \tag{7.108}$$

Notice that ω_o and P_{set} are constant and their derivatives are zero. For the method based on swing equation shown in Figure 7.106, dynamics of ω are represented by

$$J\omega_o\dot{\omega} = P_{\text{set}} + K_f(\omega_o - \omega) - P \Rightarrow J\omega_o K_f^{-1}\dot{\omega} + \omega = \omega_o + K_f^{-1}(P_{\text{set}} - P). \tag{7.109}$$

It is apparent that the two equations are identical for

$$\boxed{\tau_f = J\omega_o K_f^{-1},\ K_p = K_f^{-1}.} \tag{7.110}$$

In the method based on filtered power droops shown in Figure 7.105, the equation describing the dynamics of internal magnitude V is

$$V = V_o + K_q\left(Q_{\text{set}} - \frac{Q}{\tau_f s + 1}\right) \Rightarrow \tau_f\dot{V} + V = V_o + K_q(Q_{\text{set}} - Q). \tag{7.111}$$

Notice that V_o and Q_{set} are constant and their derivatives are zero. For the method based on swing equation shown in Figure 7.106, dynamics of V are represented by

$$M\dot{V} = Q_{\text{set}} + K_v(V_o - V_g) - Q \Rightarrow MK_v^{-1}\dot{V} + V_g = V_o + K_v^{-1}(Q_{\text{set}} - Q). \tag{7.112}$$

It is apparent that the two equations are "almost" identical for

$$\boxed{\tau_f = MK_v^{-1},\ K_q = K_v^{-1}.} \tag{7.113}$$

The only difference is that V on the left side of (7.111) is replaced by V_g on the left side of (7.112).

It is concluded from this discussion that the filter time-constant τ_f is directly related to the "virtual" moment of inertia. When the filter is excluded, it corresponds to zero-inertia dynamics. In general, presence of some level of inertia is useful to allow more flexible and robust dynamics.

7.7.3.4 Damping Strategies

The VSM discussed above suffers from lack of adequate damping. Particularly, when K_f is not sufficiently large, the responses can become very oscillatory as the moment of inertia increases.[83] Therefore, an additional damping term is required as stated in a number of publications [18–21].

An additional damping term may be introduced as shown in Figure 7.107. As discussed in Section 7.5.3, the way this term is used facilitates the stability analysis and allows the conversion of equations to some LTI equations. We used this fact to develop an approach to design K_d.

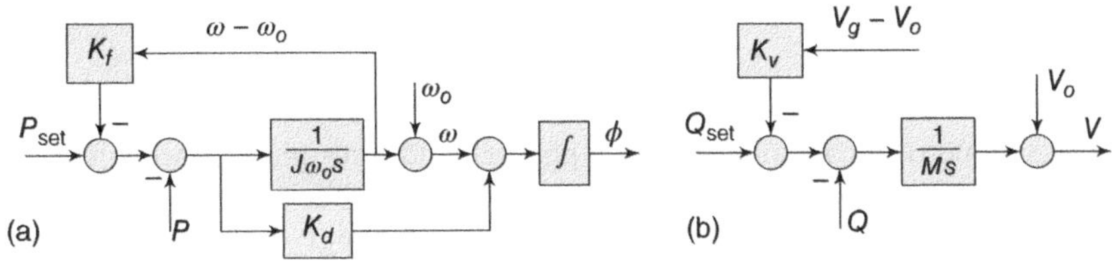

Figure 7.107 Additional damping term K_d.

83 The constant K_f relates to the range of frequency variation to the range power variations of the VSM. Larger K_f means smaller range of frequency variations for a given range of power variations. In large synchronous generators, the range of frequency variations can be as large as 5%.

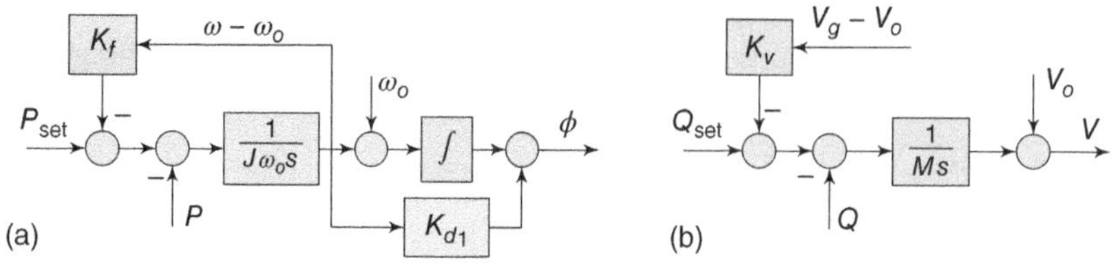

Figure 7.108 Alternative way of including additional damping term K_d.

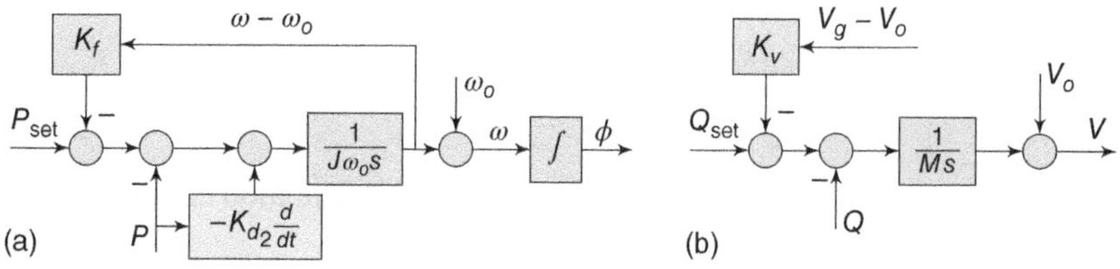

Figure 7.109 Additional damping term K_d operating on derivative of power.

An alternative damping strategy is shown in Figure 7.108. This is equivalent with the one in Figure 7.107 if the coefficient K_{d_1} is properly adjusted to count for the gain $J\omega_o$.[84] We already used this method in the method for extracting inertia from capacitor in Section 7.6.3.

There are several other damping strategies that have been introduced in the literature. In [19], the derivative of real power P is added to ω as shown in Figure 7.109. It is quite easy to understand why this performs a damping: the real power P can be written as $P_{\max} \sin \delta$. Thus, $\dot{P}$ will be proportional to $\dot{\delta}$ which is a measure of $\omega - \omega_g$, the difference between rotor speed and grid frequency.

In [20], the damping term is added in the magnitude loop. The derivative of $v_g \cdot s_q = v_{g_\alpha} \sin(\phi) - v_{g_\beta} \cos(\phi)$ is added to the magnitude V. Again, it is readily observed that $v_g \cdot s_q$ contains the term $\sin(\phi - \phi_g)$ and thus, its derivative will supply the term $\omega - \omega_g$.

7.7.3.5 Discussion

This section showed that how by adding a current-limiting and also relaxing the reference voltage, the load voltage controller for "standalone" operation converged to the grid forming controller (i.e. the VSM controller) that we derived for "grid-connected" operation in Section 7.5. This should of course not be a surprise because a GFM controller must be able to operate the converter in both modes.

7.8 Summary and Conclusion

This chapter discussed the modeling and control aspects of three-phase converters. First, a review of three-phase systems, symmetrical components, space phasors, and three-phase powers is presented. Then, the concept of power control was discussed. A CFL or CC is used as an internal loop to make it possible to limit the converter current. The three-phase ePLL was reviewed as an efficient means of synchronizing the converter with the grid, and also of estimating the grid parameters. The controller was developed in both stationary and rotating frames. The stationary frame is largely similar to the single-phase ac; while the rotating frame is largely similar to dc systems. The

84 See Problem 7.18.

difference is, however, that the synchronous rotating frame leads to two coupled dc systems. Therefore, a multivariable (MIMO) control approach is also introduced and used. In stationary frame, as well, when the PLL model is added, the two control channels become coupled. Enhancing the controller by including the PLL state variables in order to strengthen its performance in weak grid conditions were discussed in both frames.

The power control approach, as presented above, is a GFL or grid-feeding approach. This means that the converter assumes the presence of a grid and follows it, or feeds into it. It cannot operate when the grid is not available. This can be done by GFM converters which, not only they can operate when a grid is available, they can also form (or establish) a grid (or a stable voltage) on their own when the external grid is not available. Thus, we introduced grid-support functions and derived the three-phase GFM controllers, e.g. the VSM approach, using a grid-support perspective. Stability analyses and design of three-phase VSM controller was discussed in details.

Toward the end, the chapter dealt with the problem of load voltage control in a standalone (grid-isolated) application. This can be done again in both stationary and rotating frames. It was shown that, interestingly enough but not surprisingly, the GFM controllers are also emerged from the load voltage controllers when the reference voltage is relaxed and properly calculated; as they emerge from power controllers when grid-support functions are considered.

In all the discussions and proposed controllers, we have made a particular attention to making sure that an internal CFL is present. This seems to be a requirement for the practically important aspect of keeping the converter current limited during short-term faults and similar incidents. Moreover, we have developed a systematic way of designing the CFL control gains using a convenient and efficient optimal control approach, called the linear quadratic tracker (LQT). The LQT approach is further extended for a MIMO control design. The LQT ensures robust performance, and convenient design, without engaging in the concerns about stability during the design process. This approach has been consistently used across the many examples in the chapter and its desired features are demonstrated.

Problems

7.1 [Symmetrical Components] Consider the three-phase sinusoidal function

$$x(t) = \begin{bmatrix} x_a(t) \\ x_b(t) \\ x_c(t) \end{bmatrix} = \begin{bmatrix} X_a \cos(\theta_a) \\ X_b \cos(\theta_b) \\ X_c \cos(\theta_c) \end{bmatrix}, \tag{7.114}$$

where $\dot{\theta}_a = \dot{\theta}_b = \dot{\theta}_c = \omega$. Write $\theta_a = \omega t + \delta_a$, $\theta_b = \omega t + \delta_b$, and $\theta_c = \omega t + \delta_c$. The phasor of this function is defined as

$$\vec{X} = \begin{bmatrix} \vec{X}_a \\ \vec{X}_b \\ \vec{X}_c \end{bmatrix} = \begin{bmatrix} X_a \angle \delta_a \\ X_b \angle \delta_b \\ X_c \angle \delta_c \end{bmatrix}, \tag{7.115}$$

where the angle notation means $\angle \alpha = e^{j\alpha}$. Show that the equation

$$\vec{X} = \vec{X}_p + \vec{X}_n + \vec{X}_z, \tag{7.116}$$

where

$$\vec{X}_p = \begin{bmatrix} X_p\angle\delta_p \\ X_p\angle\left(\delta_p - \frac{2\pi}{3}\right) \\ X_p\angle\left(\delta_p - \frac{4\pi}{3}\right) \end{bmatrix}, \quad \vec{X}_n = \begin{bmatrix} X_n\angle\delta_n \\ X_p\angle\left(\delta_n + \frac{2\pi}{3}\right) \\ X_n\angle\left(\delta_n + \frac{4\pi}{3}\right) \end{bmatrix}, \quad \vec{X}_z = \begin{bmatrix} X_z\angle\delta_z \\ X_z\angle\delta_z \\ X_z\angle\delta_z \end{bmatrix}, \tag{7.117}$$

has a unique solution for $(X_p, \delta_p, X_n, \delta_n, X_z, \delta_z)$. Derive a matrix equation to obtain these six quantities.
(Hint: write

$$\vec{X}_p = \begin{bmatrix} X_p\angle\delta_p \\ X_p\angle\left(\delta_p - \frac{2\pi}{3}\right) \\ X_p\angle\left(\delta_p - \frac{4\pi}{3}\right) \end{bmatrix} = \begin{bmatrix} X_p\angle\delta_p \\ (X_p\angle\delta_p)\left(e^{-j\frac{2\pi}{3}}\right) \\ (X_p\angle\delta_p)\left(e^{-j\frac{4\pi}{3}}\right) \end{bmatrix} = \begin{bmatrix} 1 \\ e^{-j\frac{2\pi}{3}} \\ e^{-j\frac{4\pi}{3}} \end{bmatrix} X_p\angle\delta_p \tag{7.118}$$

and also write similar equations for $\vec{X}_n$ and $\vec{X}_z$. Then, substitute in (7.116) and write it in a matrix equation form.)

7.2 Show that for a positive-sequence or a negative-sequence signal $x(t)$, the sum of all three phase values is zero, i.e. $x_a(t) + x_b(t) + x_c(t) = 0$ for all t.

7.3 [Real and Reactive Powers in a Three-phase System]
(a) Show that a three-phase system with no neutral connection is mathematically equivalent to three single-phase subsystems. From there, prove (7.7) and (7.8).
(b) Assume that the system is balanced. Replace the expression for $p_a(t)$, $p_b(t)$, and $p_c(t)$ (in terms of their constant and double-frequency components) in (7.7) to prove that $p(t)$ is constant.
(c) Assume that the system is balanced. Replace the expression for the current and voltage phasors (of individual phases) in (7.8) to prove obtain (7.9).

7.4 [Real and Reactive Powers in an Unbalanced System] Prove (7.10).

7.5 [Condition to Remove Ripples of Power] Consider (7.10) and assume that the voltage is given. The positive-sequence component of the current is also given. Find conditions on the negative-sequence component of the current, i.e. on I_n and ϕ_n versus V_p, V_n, I_p, and ϕ_p, such that the double-frequency ripples of the power vanish.

7.6 Prove (7.12) and (7.13). Also prove (7.14) and (7.15).

7.7 Prove (7.26).

7.8 Prove that the signal U_d in the ePLL of Figure 7.5 is identical to the LPF output in Figure 7.4.

7.9 Prove (7.29).

7.10 Prove (7.31).
Hint: Write (7.30) in the form of $\dot{x} = Ax + Bu$ and calculate $T(s) = (sI - A)^{-1}B$.

7.11 Prove (7.32).

7.12 Prove (7.37).
Hint: Use the following fact to derive an expression for set($sI - A$). For a block diagonal matrix

$$X = \begin{bmatrix} A_1 & A_2 \\ A_3 & A_4 \end{bmatrix},$$

if $A_3A_4 = A_4A_3$, then $\det(X) = \det(A_1A_4 - A_2A_3)$.

7.13 Show that the magnitude of reactive power pulsations is larger when the real power pulsations are removed, compared to the case where the current is balanced and both powers pulsate.

7.14 Prove Eq. (7.52).

7.15 Prove (7.57).
Hint: notice that $V_g s_d = x_{\text{PLL}} = [x_5 \;\; x_6]^T$ but $V_g s_q = [x_6 \;\; -x_5]^T$.

7.16 Prove (7.82).
Hint: see Problem 6.6.

7.17 Prove that the value of μ that vertically aligns all four eigenvalues of the 4×4 matrix in (7.85) is given by (7.86).
Hint 1: Use the hint given in Problem 7.12.
Hint 2: Write the characteristic equation of the matrix in the form of $(s^2 + 2\alpha s + \beta^2)$ $(s^2 + 2\alpha s + \gamma^2)$ (to have aligned roots) and balance its coefficients with to find an expression for μ and α.

7.18 Derive a relationship between K_d and K_{d_1} in Figures 7.107 and 7.108 such that they produce same level of damping.

7.19 Derive a relationship between K_d and K_{d_2} in Figures 7.107 and 7.109 such that they produce same level of damping (at the linear level).

7.20 Derive a relationship between K_d and K_{d_3} in Figures 7.107 and 7.110 such that they produce same level of damping (at the linear level).

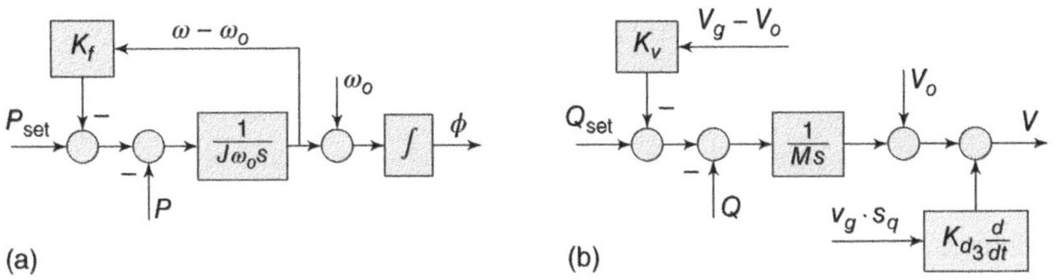

Figure 7.110 Additional damping term K_d added in the magnitude loop.

References

1 Charles L Fortescue. Method of symmetrical co-ordinates applied to the solution of polyphase networks. *Transactions of the American Institute of Electrical Engineers*, 37(2):1027–1140, 1918.

2 IEEE standard for interconnection and interoperability of distributed energy resources with associated electric power systems interfaces. *IEEE Std 1547-2018 (Revision of IEEE Std 1547-2003)*, pages 1–138, 2018.

3 Pedro Rodriguez, Adrian V Timbus, Remus Teodorescu, Marco Liserre, and Frede Blaabjerg. Flexible active power control of distributed power generation systems during grid faults. *IEEE Transactions on Industrial Electronics*, 54(5):2583–2592, 2007.

4 D Energinet. Technical regulation 3.2. 2 for PV power plants with a power output above 11 kW. *Energinet, Fredericia, Denmark, Tech. Rep*, 2015.

5 Sushil Silwal, Masoud Karimi-Ghartemani, Roshan Sharma, and Houshang Karimi. Impact of feed-forward and decoupling terms on stability of grid-connected inverters. In *28th International Symposium on Industrial Electronics (ISIE)*, pages 2641–2646. IEEE, 2019.

6 Masoud Karimi-Ghartemani and Houshang Karimi. A robust multivariable approach for current control of voltage-source converters in synchronous frame. *IEEE Journal of Emerging and Selected Topics in Power Electronics*, 2020.

7 Sushil Silwal, Masoud Karimi-Ghartemani, Houshang Karimi, and Masoud Davari. A multivariable controller in synchronous frame including phase-locked loop for three-phase grid-connected inverters. *IEEE Transactions on Power Electronics*, 9(5): 6174–6183, 2021.

8 Hans-Peter Beck and Ralf Hesse. Virtual synchronous machine. In *International Conference on Electrical Power Quality and Utilisation*, pages 1–6. IEEE, 2007.

9 Fang Gao and M Reza Iravani. A control strategy for a distributed generation unit in grid-connected and autonomous modes of operation. *IEEE Transactions on Power Delivery*, 23(2):850–859, 2008.

10 Qing-Chang Zhong and George Weiss. Synchronverters: inverters that mimic synchronous generators. *IEEE Transactions on Industrial Electronics*, 58(4):1259–1267, 2010.

11 Masoud Karimi-Ghartemani. Universal integrated synchronization and control for single-phase DC/AC converters. *IEEE Transactions on Power Electronics*, 30(3):1544–1557, 2014.

12 Masoud Karimi-Ghartemani, Sayed Ali Khajehoddin, Prasanna Piya, and Mohammad Ebrahimi. Universal controller for three-phase inverters in a microgrid. *IEEE Journal of Emerging and Selected Topics in Power Electronics*, 4(4):1342–1353, 2016.

13 Prasanna Piya, Mohammad Ebrahimi, Masoud Karimi-Ghartemani, and Sayed Ali Khajehoddin. Fault ride-through capability of voltage-controlled inverters. *IEEE Transactions on Industrial Electronics*, 65(10):7933–7943, 2018.

14 Thaer Qunais and Masoud Karimi-Ghartemani. Systematic modeling of a class of microgrids and its application to impact analysis of cross-coupling droop terms. *IEEE Transactions on Energy Conversion*, 34(3):1632–1643, 2019.

15 Masoud Karimi-Ghartemani, Sayed Ali Khajehoddin, Praveen Jain, and Alireza Bakhshai. A systematic approach to DC-bus control design in single-phase grid-connected renewable converters. *IEEE Transactions on Power Electronics*, 28(7):3158–3166, 2013.

16 F Blaabjerg, R Teodorescu, M Liserre, and A V Timbus. Overview of control and grid synchronization for distributed power generation systems. *IEEE Transactions on Industrial Electronics*, 53(5):1398–1409, 2006.

17 Ali Zakerian, Roshan Sharma, and Masoud Karimi-Ghartemani. Improving stability and power sharing of bidirectional power converters by relaxing DC capacitor voltage. *IEEE Journal of Emerging and Special Topics in Power Electronics.*

18 Prasanna Piya and Masoud Karimi-Ghartemani. A stability analysis and efficiency improvement of synchronverter. In *Applied Power Electronics Conference and Exposition (APEC)*, pages 3165–3171. IEEE, 2016.

19 Shuan Dong and Yu Christine Chen. Adjusting synchronverter dynamic response speed via damping correction loop. *IEEE Transactions on Energy Conversion*, 32(2):608–619, 2017.

20 Mohammad Ebrahimi, S Ali Khajehoddin, and Masoud Karimi-Ghartemani. An improved damping method for virtual synchronous machines. *IEEE Transactions on Sustainable Energy*, 10(3):1491–1500, 2019.

21 Shuan Dong and Yu Christine Chen. A method to directly compute synchronverter parameters for desired dynamic response. *IEEE Transactions on Energy Conversion*, 33(2):814–825, 2018.

8

Summary and Conclusion

The power system is undergoing and shall continue to undergo a substantial paradigm shift. The existing paradigm of (i) generating the bulk electric energy in large, central power plants (by burning fossil fuels and from nuclear processes) and (ii) transmitting it over long physical distances using extensive high-voltage transmission infrastructure, is not an environmentally-friendly approach and cannot be sustained forever. The technical solution to this problem is to deploy renewable resources, primarily solar and wind energies, combined with appropriate energy storage devices and mechanisms. Consumer education as well as suitable demand response strategies can also significantly contribute to address this problem.

The existing power system can still serve as a backbone to help realize this solution. To extract the maximum benefit from this solution, it must mainly be implemented in a "distributed" manner, hence the concept of distributed energy resources (DERs), which are dispersed energy systems interfaced with the power system at the distribution (that is low to medium voltage) levels. Large penetration of DERs into the power system will, however, have major impacts on the way the existing power system is maintained. Specifically, the bidirectional power flow at the nodes that used to allow only a unidirectional flow, can complicate the control, protection, and repair/maintenance tasks. Synchronism of the DERs with the ac power system in dynamic conditions (such as grid voltage disturbances and weak grid terminals) and the stability of power system (at the local DER level as well as the level of overall power system) are crucial to achieve.

On account of the major impacts of large penetration of DERs on the operation of power system, these resources must fully and effectively be "integrated." A passive connection of DERs is not going to work. The DERs must actively participate in all major grid functions such as black start, reactive power support, power quality (current and voltage quality), frequency response and inertia, voltage regulation, and overall stability. They must be able to ride through transient faults and provide a support. Fortunately, most modern DERs use a power electronic converter (PEC), that facilitates power extraction, conversion, and control with convenience and flexibility. The PEC is governed by a control program, called the controller, that dictates how it functions and also determines its response to various practical operating conditions.

It is clear from the aforementioned description that "the controller" plays the most significant role in achieving the various objectives intended from adopting DERs. Such objectives must first be translated into corresponding DER responses, and then into control actions that lead to those responses. Structuring an appropriate control system and designing the controller parameters are thus the focal tasks, that are treated systematically in this textbook.

Modeling and Control of Modern Electrical Energy Systems, First Edition. Masoud Karimi-Ghartemani.

The textbook is organized in three parts: power electronics, control theory and tools, and DERs. Parts I and II serve as the fundamental basis for Part III. In Part III, the DERs are subsequently treated according to dc DERs, single-phase ac DERs, and three-phase DERs. The dc DERs also form a basis for three-phase DERs controlled in synchronous rotating frame. In each chapter, the text starts off with providing necessary definitions and backgrounds. The power control approach which is the most common one is then fully addressed. In order to enable converter current-limiting by the controller (during transient grid faults and disturbances, and other system transients), and protect converter switches and/or avoid unnecessary shut-down of the converter caused by such transients, an internal current feedback loop (CFL) is used.

The text systematically treats aspects such as high-order converter filters, e.g. LCL filter, and passive and active damping of their resonance modes, the impact of control and system delays, and compensation of harmonics and dc component. In all such studies, the textbook takes advantage of an extremely capable control theory tool: linear quadratic tracker (LQT). The LQT approach allows systematic design of control system parameters in a much convenient way, while ensuring an optimal and robust performance of the entire control systems. Guidelines and algorithms are presented to facilitate full adoption of this approach.

The power control approach using the medium of converter current is what became known as the grid-following (GFL) control, or correspondingly, the GFL converters. They do not inherently provide grid support functions unless such functions are built into them by properly manipulating the power references; a subject that is also treated in the textbook. However, the textbook proceeds to rigorously study the grid support functions, including voltage and frequency support and inertia response. From there, it builds up and derives a whole family of grid-forming (GFM) controls which are known as virtual synchronous machine (VSM) control approach. While the existing literature derives the VSM by mimicking the principles of a synchronous machine, this textbook takes another starting point: realizing the grid support functions. A novel stability analysis of the VSM is presented and is used to efficiently design the VSM parameters. The internal CFL is maintained in the presented VSM approach to ensure converter current-limiting by the controller.

With reference to the GFM and GFL converters, the textbook also shows how the GFM approach can be derived from GFL approach applied to a standalone application. This provides further insight into the relationships between these two approaches. The textbook also presents an effective presentation of enhanced phase-locked loop (ePLL) tools for single-phase and three-phase applications. Their analysis, design, and adoption for generating desired reference current for converters operating under different specifications are explained.

The textbook is filled with many examples, most of which showing results of computer simulations of the corresponding system conditions. The simulation files of all such examples, including the MATLAB codes to perform the design stage, and the SIMULINK files (version R2018b), are professionally produced by the author and are made available by the publisher. They constitute a valuable set of resources to facilitate in-depth understanding of the text materials.

The book is intended to be a textbook for an independent graduate course, a senior-level undergraduate course, a split-level (shared graduate and undergraduate) course, or a continuing-education course for technicians and engineers working in relevant industries. For each case, an appropriate set of materials may be selected since the book is written with a view to each section and chapter being independent. The sections currently marked by an asterisk may be skipped in the first-reading and left for a more advanced course.

While most of the DER treatment in the textbook may be either directly applied to various converter topologies for different resources, such as solar, wind, and battery resources, or properly tailored to fit a particular application, the presentation in the textbook has attempted to maintain its generality as much as possible. This not only maximizes the breadth of the work, it will also deepen the education intended for a first and fundamental course in understanding the modeling and control aspects of DERs. Building on what is presented in this textbook, a subsequent textbook (or a set of subsequent books) may delve into details of particular applications, such as various isolated and non-isolated dc/dc converters, dedicated PV converters at different power levels, dedicated wind turbine converters, dedicated battery converters, electric vehicle (EV) chargers, etc.

Index

IEEE Press Series on Power and Energy Systems

Series Editor: Ganesh Kumar Venayagamoorthy, Clemson University, Clemson, South Carolina, USA.

The mission of the IEEE Press Series on Power and Energy Systems is to publish leading-edge books that cover a broad spectrum of current and forward-looking technologies in the fast-moving area of power and energy systems including smart grid, renewable energy systems, electric vehicles and related areas. Our target audience includes power and energy systems professionals from academia, industry and government who are interested in enhancing their knowledge and perspectives in their areas of interest.

1. *Electric Power Systems: Design and Analysis, Revised Printing*
   Mohamed E. El-Hawary
2. *Power System Stability*
   Edward W. Kimbark
3. *Analysis of Faulted Power Systems*
   Paul M. Anderson
4. *Inspection of Large Synchronous Machines: Checklists, Failure Identification, and Troubleshooting*
   Isidor Kerszenbaum
5. *Electric Power Applications of Fuzzy Systems*
   Mohamed E. El-Hawary
6. *Power System Protection*
   Paul M. Anderson
7. *Subsynchronous Resonance in Power Systems*
   Paul M. Anderson, B.L. Agrawal, J.E. Van Ness
8. *Understanding Power Quality Problems: Voltage Sags and Interruptions*
   Math H. Bollen
9. *Analysis of Electric Machinery*
   Paul C. Krause, Oleg Wasynczuk, and S.D. Sudhoff
10. *Power System Control and Stability, Revised Printing*
    Paul M. Anderson, A.A. Fouad
11. *Principles of Electric Machines with Power Electronic Applications*, Second Edition
    Mohamed E. El-Hawary
12. *Pulse Width Modulation for Power Converters: Principles and Practice*
    D. Grahame Holmes and Thomas Lipo
13. *Analysis of Electric Machinery and Drive Systems*, Second Edition
    Paul C. Krause, Oleg Wasynczuk, and S.D. Sudhoff

14. *Risk Assessment for Power Systems: Models, Methods, and Applications*
 Wenyuan Li
15. *Optimization Principles: Practical Applications to the Operations of Markets of the Electric Power Industry*
 Narayan S. Rau
16. *Electric Economics: Regulation and Deregulation*
 Geoffrey Rothwell and Tomas Gomez
17. *Electric Power Systems: Analysis and Control*
 Fabio Saccomanno
18. *Electrical Insulation for Rotating Machines: Design, Evaluation, Aging, Testing, and Repair*
 Greg C. Stone, Edward A. Boulter, Ian Culbert, and Hussein Dhirani
19. *Signal Processing of Power Quality Disturbances*
 Math H. J. Bollen and Irene Y. H. Gu
20. *Instantaneous Power Theory and Applications to Power Conditioning*
 Hirofumi Akagi, Edson H. Watanabe, and Mauricio Aredes
21. *Maintaining Mission Critical Systems in a 24/7 Environment*
 Peter M. Curtis
22. *Elements of Tidal-Electric Engineering*
 Robert H. Clark
23. *Handbook of Large Turbo-Generator Operation and Maintenance,*
 Second Edition
 Geoff Klempner and Isidor Kerszenbaum
24. *Introduction to Electrical Power Systems*
 Mohamed E. El-Hawary
25. *Modeling and Control of Fuel Cells: Distributed Generation Applications*
 M. Hashem Nehrir and Caisheng Wang
26. *Power Distribution System Reliability: Practical Methods and Applications*
 Ali A. Chowdhury and Don O. Koval
27. *Introduction to FACTS Controllers: Theory, Modeling, and Applications*
 Kalyan K. Sen and Mey Ling Sen
28. *Economic Market Design and Planning for Electric Power Systems*
 James Momoh and Lamine Mili
29. *Operation and Control of Electric Energy Processing Systems*
 James Momoh and Lamine Mili
30. *Restructured Electric Power Systems: Analysis of Electricity Markets with Equilibrium Models*
 Xiao-Ping Zhang
31. *An Introduction to Wavelet Modulated Inverters*
 S.A. Saleh and M.A. Rahman

32. *Control of Electric Machine Drive Systems*
    Seung-Ki Sul

33. *Probabilistic Transmission System Planning*
    Wenyuan Li

34. *Electricity Power Generation: The Changing Dimensions*
    Digambar M. Tagare

35. *Electric Distribution Systems*
    Abdelhay A. Sallam and Om P. Malik

36. *Practical Lighting Design with LEDs*
    Ron Lenk and Carol Lenk

37. *High Voltage and Electrical Insulation Engineering*
    Ravindra Arora and Wolfgang Mosch

38. *Maintaining Mission Critical Systems in a 24/7 Environment*, Second Edition
    Peter Curtis

39. *Power Conversion and Control of Wind Energy Systems*
    Bin Wu, Yongqiang Lang, Navid Zargari, and Samir Kouro

40. *Integration of Distributed Generation in the Power System*
    Math H. Bollen and Fainan Hassan

41. *Doubly Fed Induction Machine: Modeling and Control for Wind Energy Generation Applications*
    Gonzalo Abad, Jesús López, Miguel Rodrigues, Luis Marroyo, and Grzegorz Iwanski

42. *High Voltage Protection for Telecommunications*
    Steven W. Blume

43. *Smart Grid: Fundamentals of Design and Analysis*
    James Momoh

44. *Electromechanical Motion Devices*, Second Edition
    Paul Krause, Oleg Wasynczuk, and Steven Pekarek

45. *Electrical Energy Conversion and Transport: An Interactive Computer-Based Approach*, Second Edition
    George G. Karady and Keith E. Holbert

46. *ARC Flash Hazard and Analysis and Mitigation*
    J.C. Das

47. *Handbook of Electrical Power System Dynamics: Modeling, Stability, and Control*
    Mircea Eremia and Mohammad Shahidehpour

48. *Analysis of Electric Machinery and Drive Systems, Third Edition*
    Paul C. Krause, Oleg Wasynczuk, S.D. Sudhoff, and Steven D. Pekarek

49. *Extruded Cables for High-Voltage Direct-Current Transmission: Advances in Research and Development*
    Giovanni Mazzanti and Massimo Marzinotto

50. *Power Magnetic Devices: A Multi-Objective Design Approach*
S.D. Sudhoff

51. *Risk Assessment of Power Systems: Models, Methods, and Applications*, Second Edition
Wenyuan Li

52. *Practical Power System Operation*
Ebrahim Vaahedi

53. *The Selection Process of Biomass Materials for the Production of Bio-Fuels and Co-Firing*
Najib Altawell

54. *Electrical Insulation for Rotating Machines: Design, Evaluation, Aging, Testing, and Repair*, Second Edition
Greg C. Stone, Ian Culbert, Edward A. Boulter, and Hussein Dhirani

55. *Principles of Electrical Safety*
Peter E. Sutherland

56. *Advanced Power Electronics Converters: PWM Converters Processing AC Voltages*
Euzeli Cipriano dos Santos Jr. and Edison Roberto Cabral da Silva

57. *Optimization of Power System Operation*, Second Edition
Jizhong Zhu

58. *Power System Harmonics and Passive Filter Designs*
J.C. Das

59. *Digital Control of High-Frequency Switched-Mode Power Converters*
Luca Corradini, Dragan Maksimoviæ, Paolo Mattavelli, and Regan Zane

60. *Industrial Power Distribution*, Second Edition
Ralph E. Fehr, III

61. *HVDC Grids: For Offshore and Supergrid of the Future*
Dirk Van Hertem, Oriol Gomis-Bellmunt, and Jun Liang

62. *Advanced Solutions in Power Systems: HVDC, FACTS, and Artificial Intelligence*
Mircea Eremia, Chen-Ching Liu, and Abdel-Aty Edris

63. *Operation and Maintenance of Large Turbo-Generators*
Geoff Klempner and Isidor Kerszenbaum

64. *Electrical Energy Conversion and Transport: An Interactive Computer-Based Approach*
George G. Karady and Keith E. Holbert

65. *Modeling and High-Performance Control of Electric Machines*
John Chiasson

66. *Rating of Electric Power Cables in Unfavorable Thermal Environment*
George J. Anders

67. *Electric Power System Basics for the Nonelectrical Professional*
Steven W. Blume

68. *Modern Heuristic Optimization Techniques: Theory and Applications to Power Systems*
Kwang Y. Lee and Mohamed A. El-Sharkawi

69. *Real-Time Stability Assessment in Modern Power System Control Centers*
Savu C. Savulescu

70. *Optimization of Power System Operation*
Jizhong Zhu

71. *Insulators for Icing and Polluted Environments*
Masoud Farzaneh and William A. Chisholm

72. *PID and Predictive Control of Electric Devices and Power Converters Using MATLAB®/Simulink®*
Liuping Wang, Shan Chai, Dae Yoo, Lu Gan, and Ki Ng

73. *Power Grid Operation in a Market Environment: Economic Efficiency and Risk Mitigation*
Hong Chen

74. *Electric Power System Basics for Nonelectrical Professional*, Second Edition
Steven W. Blume

75. *Energy Production Systems Engineering*
Thomas Howard Blair

76. *Model Predictive Control of Wind Energy Conversion Systems*
Venkata Yaramasu and Bin Wu

77. *Understanding Symmetrical Components for Power System Modeling*
J.C. Das

78. *High-Power Converters and AC Drives*, Second Edition
Bin Wu and Mehdi Narimani

79. *Current Signature Analysis for Condition Monitoring of Cage Induction Motors: Industrial Application and Case Histories*
William T. Thomson and Ian Culbert

80. *Introduction to Electric Power and Drive Systems*
Paul Krause, Oleg Wasynczuk, Timothy O'Connell, and Maher Hasan

81. *Instantaneous Power Theory and Applications to Power Conditioning*, Second Edition
Hirofumi, Edson Hirokazu Watanabe, and Mauricio Aredes

82. *Practical Lighting Design with LEDs*, Second Edition
Ron Lenk and Carol Lenk

83. *Introduction to AC Machine Design*
Thomas A. Lipo

84. *Advances in Electric Power and Energy Systems: Load and Price Forecasting*
Mohamed E. El-Hawary

85. *Electricity Markets: Theories and Applications*
Jeremy Lin and Jernando H. Magnago

86. *Multiphysics Simulation by Design for Electrical Machines, Power Electronics and Drives*
Marius Rosu, Ping Zhou, Dingsheng Lin, Dan M. Ionel, Mircea Popescu, Frede Blaabjerg, Vandana Rallabandi, and David Staton

87. *Modular Multilevel Converters: Analysis, Control, and Applications*
Sixing Du, Apparao Dekka, Bin Wu, and Navid Zargari

88. *Electrical Railway Transportation Systems*
Morris Brenna, Federica Foiadelli, and Dario Zaninelli

89. *Energy Processing and Smart Grid*
James A. Momoh

90. *Handbook of Large Turbo-Generator Operation and Maintenance*, 3rd Edition
Geoff Klempner and Isidor Kerszenbaum

91. *Advanced Control of Doubly Fed Induction Generator for Wind Power Systems*
Dehong Xu, Dr. Frede Blaabjerg, Wenjie Chen, and Nan Zhu

92. *Electric Distribution Systems*, 2nd Edition
Abdelhay A. Sallam and Om P. Malik

93. *Power Electronics in Renewable Energy Systems and Smart Grid: Technology and Applications*
Bimal K. Bose

94. *Distributed Fiber Optic Sensing and Dynamic Rating of Power Cables*
Sudhakar Cherukupalli and George J Anders

95. *Power System and Control and Stability,* Third Edition
Vijay Vittal, James D. McCalley, Paul M. Anderson, and A. A. Fouad

96. *Electromechanical Motion Devices: Rotating Magnetic Field-Based Analysis and Online Animations,* Third Edition
Paul Krause, Oleg Wasynczuk, Steven D. Pekarek, and Timothy O'Connell

97. *Applications of Modern Heuristic Optimization Methods in Power and Energy Systems*
Kwang Y. Lee and Zita A. Vale

98. *Handbook of Large Hydro Generators: Operation and Maintenance*
Glenn Mottershead, Stefano Bomben, Isidor Kerszenbaum, and Geoff Klempner

99. *Advances in Electric Power and Energy: Static State Estimation*
Mohamed E. El-hawary

100. *Arc Flash Hazard Analysis and Mitigation*, Second Edition
J.C. Das

101. *Maintaining Mission Critical Systems in a 24/7 Environment,* Third Edition
Peter M. Curtis

102. *Real-Time Electromagnetic Transient Simulation of AC-DC Networks*
Venkata Dinavahi and Ning Lin

103. *Probabilistic Power System Expansion Planning with Renewable Energy Resources and Energy Storage Systems*
Jaeseok Choi and Kwang Y. Lee

104. *Power Magnetic Devices: A Multi-Objective Design Approach*, Second Edition
Scott D. Sudhoff

105. *Optimal Coordination of Power Protective Devices with Illustrative Examples*
Ali R. Al-Roomi

106. *Resilient Control Architectures and Power Systems*
Craig Rieger, Ronald Boring, Brian Johnson, and Timothy McJunkin

107. *Alternative Liquid Dielectrics for High Voltage Transformer Insulation Systems: Performance Analysis and Applications*
Edited by U. Mohan Rao, I. Fofana, and R. Sarathi

108. *Introduction to the Analysis of Electromechanical Systems*
Paul C. Krause, OlegWasynczuk, and Timothy O'Connell

109. *Power Flow Control Solutions for a Modern Grid using SMART Power Flow Controllers*
Kalyan K. Sen and Mey Ling Sen

110. *Power System Protection: Fundamentals and Applications*
John Ciufo and Aaron Cooperberg

111. *Soft-Switching Technology for Three-phase Power Electronics Converters*
Dehong Xu, Rui Li, Ning He, Jinyi Deng, and YuyingWu

112. *Power System Protection*, Second Edition
Paul M. Anderson, Charles Henville, Rasheek Rifaat, Brian Johnson, and Sakis Meliopoulos

113. *High Voltage and Electrical Insulation Engineering*, Second Edition
Ravindra Arora and Wolfgang Mosch

114. *Modeling and Control of Modern Electrical Energy Systems*
Masoud Karimi-Ghartemani

Printed and bound by CPI Group (UK) Ltd, Croydon, CR0 4YY
16/04/2025
14658353-0003